Reviews of
Accelerator Science
and Technology

Volume 2

Reviews of Accelerator Science and Technology

Volume 2

Medical Applications of Accelerators

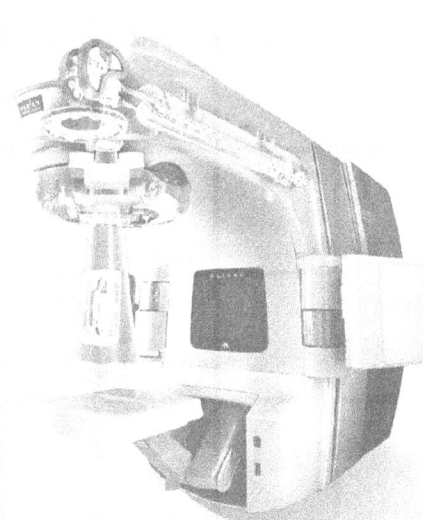

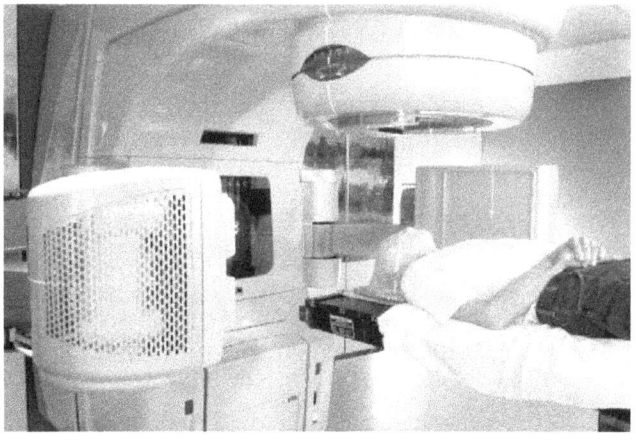

Editors

Alexander W. Chao
SLAC National Accelerator Laboratory, USA

Weiren Chou
Fermi National Accelerator Laboratory, USA

 **World Scientific**

NEW JERSEY · LONDON · SINGAPORE · BEIJING · SHANGHAI · HONG KONG · TAIPEI · CHENNAI

Published by

World Scientific Publishing Co. Pte. Ltd.

5 Toh Tuck Link, Singapore 596224

USA office: 27 Warren Street, Suite 401-402, Hackensack, NJ 07601

UK office: 57 Shelton Street, Covent Garden, London WC2H 9HE

British Library Cataloguing-in-Publication Data

A catalogue record for this book is available from the British Library.

REVIEWS OF ACCELERATOR SCIENCE AND TECHNOLOGY
Volume 2: Medical Applications of Accelerators

ISBN-13 978-981-4299-34-3

Contents

Reviews of Accelerator Science and Technology
Vol. 2 (2009) vii–viii
© World Scientific Publishing Company

Editorial Preface

This is the second volume of the journal *Reviews of Accelerator Science and Technology*. While the previous volume gave an overview of the accelerator field, this volume (and all later ones) is focused on a specific sub-field. The theme of this volume is *Medical Applications of Accelerators*. We chose this theme because of its enormous importance to human health and its deep impact on our society.

Ever since the discovery of x-rays more than a century ago, people began to contemplate using them for medical purposes. The invention of particle accelerators in the early 20th century created a whole new world for producing energetic x-rays, electrons, protons, neutrons and other particle beams. Immediately these beams found applications in medicine. There are two important yet distinct medical applications. One is that accelerators produce radioisotopes for various medical tests in nuclear medicine and positron emission tomography (PET) used to diagnose millions of patients each year. The other is that accelerators produce particle beams for radiation therapy for the treatment of cancer. The particle beams can be x-rays (generated by high-energy electrons), protons, neutrons or heavy ions such as carbon. Today there are more than 5,000 accelerators routinely used in hospitals all over the world for nuclear medicine and cancer therapy.

The great potential of accelerator applications in medicine can hardly be exaggerated. Take proton therapy as an example. Because the Bragg peak of protons in the human body is much sharper than for x-rays, treatment can be localized, tumor targeting accuracy improved, and the irradiation of sensitive neighboring tissues and side effects reduced. However, despite these indisputable advantages the number of proton therapy patients is much smaller than the number treated using x-rays. Each year several million cancer patients receive x-ray treatment. In contrast, the number of patients treated by protons, neutrons and ions over the past several decades is less than 50,000. The reason is simple — an x-ray therapy machine can readily fit in a hospital room, whereas a proton (or ion) therapy facility occupies an entire building. If future proton therapy machines can be made as small and cheap as x-ray machines so that every hospital could afford one, this would revolutionize oncology!

This volume contains fourteen articles, all written by distinguished scholars. The first three articles are overviews by physicians (Suit *et al.*, Ruth, and Slater *et al.*). They review the status of radiation therapy, radioisotopes in nuclear medicine and hospital-based facilities, respectively. The following six articles describe in detail various types of accelerators used in medicine: electron linacs for x-ray therapy (Whittum), accelerator systems for heavy particle radiotherapy (Tsujii *et al.*), high frequency linacs for hadrontherapy (Amaldi *et al.*), medical cyclotrons (Friesel *et al.*), synchrotrons for hadrontherapy (Pullia), and beam delivery systems (Schippers). They are followed by three articles discussing future medical accelerators: laser acceleration of ions (Tajima *et al.*), FFAG (Trbojevic) and the dielectric wall accelerator (Caporaso *et al.*). There is also an important article on the Superconducting Super Collider (Wojcicki), which is a continuation of the first part published in Volume 1.

In each volume we dedicate one article to a prominent figure of the accelerator community. In Vol. 1, it was Pief Panofsky. In this volume, it is Robert Wilson. Wilson is selected because of his 1946 seminal paper "Radiological use of fast protons" that began the whole field of proton therapy. This important event in accelerator history is described in Slater's article. But Wilson is chosen for much deeper reasons. From his unique close association with Wilson over many years, Ned Goldwasser in his article vividly portrays Wilson as a charismatic and imaginative leader. His article is not a chronological biography of Wilson, nor is it intended to be, but a beautiful piece that presents an illuminating sketch of a great man. We feel it fits perfectly in this volume.

Alexander W. Chao
SLAC National Accelerator Laboratory, USA
achao@slac.stanford.edu

Weiren Chou
Fermi National Accelerator Laboratory, USA
chou@fnal.gov

Editors

Reviews of Accelerator Science and Technology
Vol. 2 (2009) 1–15
© World Scientific Publishing Company

Physical and Biological Basis of Proton and of Carbon Ion Radiation Therapy and Clinical Outcome Data

Herman Suit*, Thomas F. Delaney† and Alexei Trofimov‡

Department of Radiation Oncology,
Massachusetts General Hospital and Harvard Medical School,
Boston, MA, USA
*hsuit@partners.org
†tdelaney@partners.org
‡atrofimov@partners.org

There is a clear basis in physics for the clinical use of proton and carbon beams in radiation therapy, namely, the finite range of the particle beam. The range is dependent on the beam initial energy, density and atomic composition of tissues along the beam path. Beams can be designed that penetrate to the required depth and deliver a uniform biologically effective dose across the depth of interest. The yield is a superior dose distribution relative to photon beams. There is a *potential* clinical advantage from the high linear energy transfer (LET) characteristics of carbon beams. This is based on a lower oxygen enhancement ratio (OER) and a flatter age response function. However, due to uncertainties relating OER with relative biological effectiveness (RBE), there is no clinical evidence to date that carbon ion beams have an advantage over proton beams. We strongly support performance Phase III clinical trials of protons *vs* carbon ion beams designed to feature a single variable, LET. Dose fractionation would be identical in both arms and dose distribution would be similar for the sites to be tested. For sites for which the carbon beam has a demonstrated important advantage in comparative treatment planning due to the narrower penumbra would not be selected for the clinical trials.

Keywords: Protons; carbon ions; beams; radiation therapy.

1. Introduction

The intent in implementation of a new radiation treatment method is to increase the probability of eradication of the irradiated tumor with no increase or preferably a lesser risk of treatment-related morbidity. This goal has been realized in multiple steps throughout the history of radiation oncology by technical advances that have provided progressively improving distributions of the dose. In parallel, major clinical gains have been made by combining radiation with other treatment modalities, viz. surgery, chemicals and biological/genetic agents.

At present, high interest is directed to the possibility of important gains in curative radiation therapy by moving from x-ray to proton and carbon ion beams. This is evident from the 61,112 patients treated by protons and the 5342 patients by carbon ion beams as of March 2009 (M. Jermann,[a] personal communication, 2009). These treatments have been delivered at 26 proton and 3 carbon ion therapy centers. Further, many proton and carbon ion centers are being planned or under construction. Here, we assess the relative merits of these two particle beams for radiation therapy by considering some of the relevant physics, radiation biology and clinical outcome data. This is not based on data from clinical trials, as there are none.

There have been earlier reviews of particle beam radiation therapy, such as Refs. 8, 17, 26, 50, 63 and 65.

---

[a]M. Jermann is the Secretary of the Particle Therapy Co-operative Oncology Group. See also his website, PTCOG.web.psi.ch.

2. Physics

2.1. *General considerations*

The physical basis for the enthusiasm for ^{1}H and ^{12}C therapy is the finite range of protons and carbon ions. Thus, the radiation dose is near zero for all tissues deep to the target for each beam path of ^{1}H beams, but not quite so low for ^{12}C beams because of their short and low dose fragmentation tails. The finite range is determined by the initial energy of the ^{1}H or ^{12}C ions and the densities and stopping powers along the beam path. The dose increases slowly with the depth until approximately the terminal $\sim 10\,$mm of the range and then rises very steeply to produce a Bragg peak. The dose decrease deep to the Bragg peak is precipitous. These points are illustrated by plots of depth–dose curves for pristine ^{1}H and ^{12}C beams in Fig. 1.

Use of these beams makes feasible the planning and delivery of biologically effective dose (BED) distributions which are manifestly superior to those by the highest technology x-ray therapy for all but a small fraction of tumors. This means that for a defined dose and dose distribution to the target, there is a lesser dose to uninvolved normal tissues.

For proton and carbon ion beams, the Bragg peak decreases in height with beam energy, i.e. the depth of penetration, as shown in Fig. 2 for pristine ^{12}C ion beams of 135 MeV, 270 MeV and 330 MeV, from Ref. 54. This is due to energy straggling and

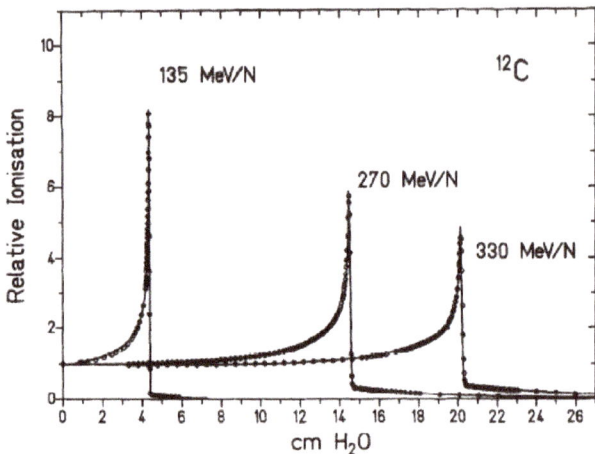

Fig. 2.   A plot of dose vs. depth of the Bragg peaks of pristine ^{12}C ion beams of 135 MeV, 270 MeV and 330 MeV [54].

nuclear interactions; additionally for carbon beams there is fragmentation of the ^{12}C ions, and additionally the peaks broaden with the depth of the peak.

A clinical beam is designed to provide a nearly uniform BED over the range in depth that covers the target from its most proximal to its most distal aspect plus a small margin for errors in positioning the target on the beam and in estimation of the beam range in the patient. This is achieved by the layering of beams of graded energies along the depth of interest so as to generate a flat dose — say, $\pm 5\%$ — in the patient for most anatomic situations, over that length of the beam. This flat region is designated the spread-out Bragg peak (SOBP). This layering of proton beams of selected energies of Bragg peaks was described by Koehler and Preston [30] and is shown in Fig. 3(a). To appreciate the large and clear advantage of a particle beam in radiation therapy, examine the depth–dose curves for a clinical proton beam (with its SOBP) and a high energy x-ray beam, in Fig. 3(b). The x-ray dose is markedly higher distal to the target, due to the fact that it decreases exponentially in the beam path and exits the body. This contrasts with the near-zero dose from a proton beam deep to the target. Additionally, the dose proximal to the target is lower for the ^{1}H beam except for the initial few mm. An important point is that this advantage obtains for each beam path. A similar set of curves could be drawn for ^{12}C ion beams. This difference between the depth–dose curves of particle and x-ray beams is neither subtle nor trivial.

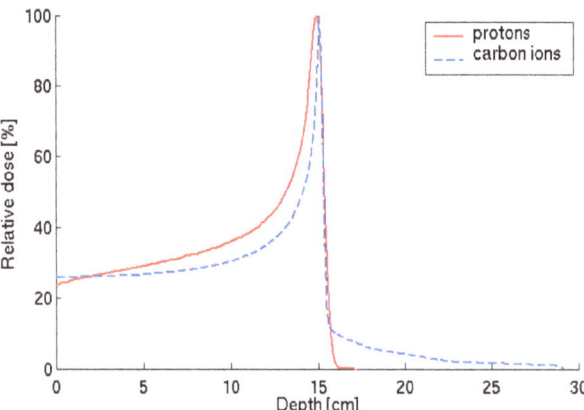

Fig. 1.   Depth–dose curves for pristine ^{1}H and ^{12}C beams, demonstrating the very steep Bragg peak and extremely sharp fall in the dose immediately beyond the Bragg peak. On the ^{12}C curve, there is a fragmentation tail. (*Prepared by Trofimov from MGH data and data from GSI, Ref. 31.*)

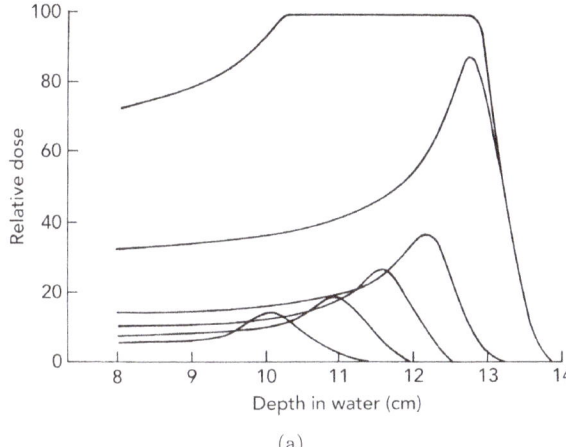

(a)

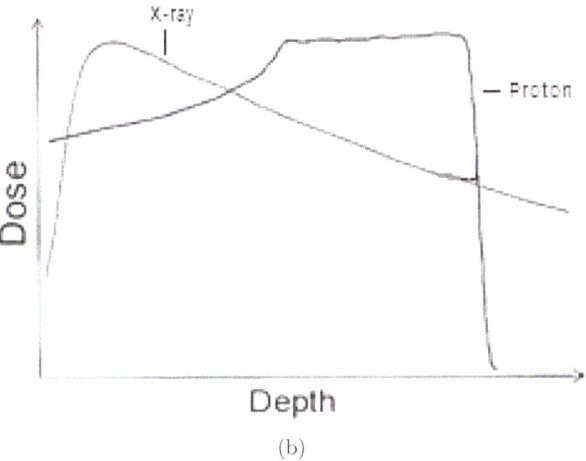

(b)

Fig. 3. (a) The SOBP for a ^{1}H beam is produced by layering proton beams of the required energy distribution to achieve a flat (±5%) dose over the depth that includes all of the target plus a thin margin [30]. (b) Depth–dose curves for a representative clinical proton and a high energy x-ray beam.

The biological effectiveness of ^{1}H and x-ray beams is quite similar. Namely, the median RBE determined at clinical dose levels on *in vivo* experimental systems has been estimated to be 1.10. This is based on an analysis of all published data by Paganetti *et al.* [42]. The International Commission on Radiological Units (ICRU) has recommended that the RBE for clinical proton therapy be 1.10 and be used as a generic RBE. That is to say, this one RBE value should be employed for all dose levels, tissues, end points and other parameters. Clinically relevant is the fact that this RBE of 1.10 is less than or equal to the RBE of 1.10–1.15 of 250 kVp x-radiations.

This difference in the dose distribution between x-ray and ^{1}H beams is the one and only rationale for ^{1}H therapy, viz. superior dose distributions of low LET radiations. This dose distribution rationale obtains for ^{12}C as for ^{1}H therapy. Further, ^{12}C beams are high-LET and consequently have a high RBE relative to ^{1}H beams. High-LET–high-RBE radiation provides the prospect of a greater effect on tumors than on normal tissues, *vide infra*.

Further, for high technology x-ray therapy, e.g. intensity-modulated x-ray therapy (IMXT), there are more beams than for particle therapy for nearly all treatment plans. Figure 4 shows an example of the larger number of fields employed for IMXT than for IMPT (intensity-modulated proton therapy). An important aspect of ^{1}H and ^{12}C treatment is the more stringent requirement for accurate positioning due to the steep decline of the dose at the end of the beam's range. That is to say, a small error could result in an almost-zero dose to a small segment of the target and thus a lower tumor control probability (TCP).

The ^{12}C beam fragmentation tail is produced by atomic fragments secondary to ^{12}C ion collisions with atoms in the beam path. These fragments are of moderate-to-low energy and are projected

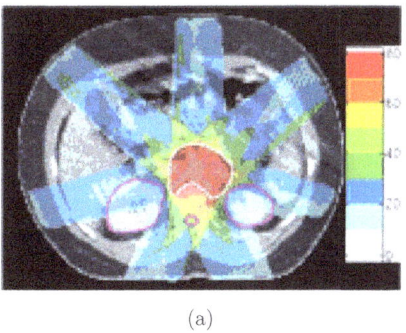

(a)

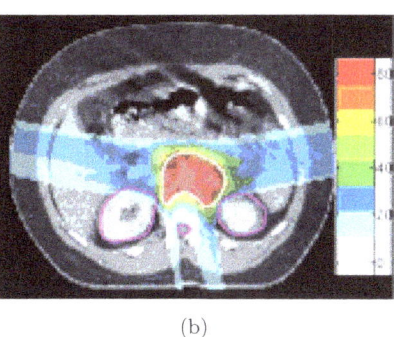

(b)

Fig. 4. A treatment plan for irradiation of a prevertebral mass by IMXT and IMPT.

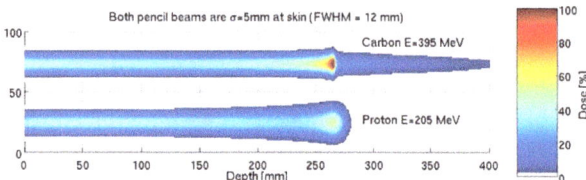

Fig. 5. Ionization along a ^{1}H and a ^{12}C beam and extension of the dose by nuclear fragments generated upstream from the Bragg peak to form the ^{12}C fragmentation tail, but no tail on the ^{1}H beam.

principally in the forward direction. They are predominately protons plus a lower fluence of helium-to-boron nuclei [20]. The physical dose and BED in the tail are low. The fragmentation tail is the one aspect of the BED distribution by ^{12}C beams that is less favorable than that by ^{1}H beams. These fragmentation tails are not judged to constitute an important issue for ^{12}C ion therapy.

Particle beams are generated by cyclotrons, synchrotrons or synchrocyclotrons. The clinically required distribution of energies is achieved by two general techniques. Presently, the most commonly employed ^{1}H beam is designated passive energy modulation (PEM), viz. a broad beam of the required energy transits a rotating range modulator of graded thickness designed to yield the distribution of particle energies that penetrates to the planned depth. Additionally, to generate the energy distribution to provide a uniform dose over the 3D distal contour of the irregularly shaped target, the beam next transits a 3D compensator that defines the particle range over each small segment of the target. For a near-uniform dose distribution transversely across the target, scatterers are employed. To define the lateral borders of the field and reduce the penumbra, i.e. sharpen the beam edge, specially made collimators are prepared for each field in each patient treated by a PEM beam.

There is rapidly increasing interest in the second beam category, viz. pencil beam scanning (PBS). This technique scans a narrow pristine beam over the field of interest with changing particle fluence and energy in order to provide the desired dose distribution throughout the target. Lomax *et al.* [35] wrote of the gains in dose distributions expected from PBS that he and colleagues at the Paul Scherrer Institute have demonstrated. Their results encourage conversion from PEM to PBS for some treatment rooms

and for new accelerators to come with PBS capability. Note that the dose distributions over the distal surface of the target are virtually equivalent for the PEM and PBS techniques. The dose conformation around the proximal margin is readily obtained by PBS but not by PEM.

2.2. *Need for gantries*

A central factor in striving for the very best dose distribution in the patient is the use of gantry systems. The patient is in most instances in the supine position, the most comfortable. Further, the patient is not required to move, nor is the tabletop for other than non-coplanar fields. Immobilization procedures are more effective if applied to comfortable patients. Gantry systems are standard features of current x-ray linear electron accelerators. Although gantries for particle beam therapy are large, complex and costly, they are important for achieving the best feasible dose distribution. A gantry system should provide the same flexibility of planning and delivery of the radiation dose as that of conventional x-ray therapy in terms of beam number, beam direction, intensity modulation and 4D IGRT. Thus, to employ particle beams with only a fixed horizontal and/or vertical beam(s) loses some of their dose distribution advantage. Several aspects of the complexities of planning ^{12}C ion therapy have been reviewed by Kramer *et al.* [31, 32].

2.3. *Penumbra*

The penumbra width of ^{1}H beams varies quite widely with the particular design of the beam line, the use of collimators, and the distance between the beam defining mechanism and the patient surface and the depth in tissue being considered. Collimated PEM ^{1}H beams with 4 cm compensators and air gaps of 1, 11 and 21 cm have penumbras [80%–20%] at 10 cm depths of 6, 9.3 and 13 mm. Penumbras at 30 cm are 12, 14 and 17 mm [48]. For comparison, the penumbra for a 6 MV x-ray beam increases from 6.8 to 10.1 mm at depths of 10 and 30 cm. The penumbras are wider for higher energy x-ray beams; for example, the penumbra of an 18 MV x-ray beam at 10 cm is 7.9 mm. With few exceptions, the ^{12}C penumbra is more narrow than for ^{1}H beams, e.g. 1.5–3 mm covering a wide range in depth. This varies only slightly with the various factors mentioned for

[1]H beams. For further consideration, see Ref. 28. The BED lateral to the penumbra for the PEM proton beam is largely due to low energy neutrons and is similar to or slightly lower than for high energy x-ray beams. The lateral scattered doses for proton beams are low energy neutrons; we have used the neutron BEDs provided by the authors. Also, the lateral dose from carbon beams is less than that from x-ray beams. The papers concerning this lateral dose include Refs. 37, 57, 68, 69 and 71.

2.4. *Heterodensities in the beam path*

A problem in achieving the planned dose distribution for particle beams is the impact of high and low density structures in the beam path, e.g. bone or air cavities. The positions of the margins of structures of differing density in the beam path can be defined in 3D or 4D. Additionally, the density throughout the structure can be determined. With this information, the required beam energy in each voxel can be computed to avoid a shortfall in the range due to, say, bone or an overshoot due to, say, a sinus cavity. To allow for uncertainty in the position of the structure of concern, there will be a narrow rim of the overshoot near the projected margin of the structure of concern. Goitein *et al.* [18] and Urie *et al.* [64] have defined this problem in treatment planning and have developed methods to correct for the perturbations in the dose secondary to objects with density differing from that of soft tissues.

3. Radiation-Biological Considerations

3.1. *Slopes of dose–response curves*

The consequence of a superior dose distribution is an increased tolerance of radiation by the patient, and hence a higher dose can be delivered to the target. This results in a higher TCP with only modest changes in the dose to nontarget tissues due to steep dose gradients between target and normal tissues. Were the TCP judged acceptable but the risk of radiation complications high, the target dose could be unchanged (no alteration in the TCP), but a clinically important decrement in the dose given to the nontarget tissues and accordingly provide a lower NTCP (normal tissue complication probability).

An appropriate question is: What is the likely gain in the TCP from a specific increment of the radiation dose? The answer is that the gain will be a function of the slope of the dose–response curve and the position on that curve of the TCP for the reference treatment. There are, of course, no dose–response curves from studies on human patients. The most secure strategy for considering this question is to examine results from determinations of dose–response curves in laboratory animals. The experiments to consider are those based on spontaneous tumors from inbred animals growing as very early generation transplants in syngenic hosts that are of a narrow age range. The tumor(s) would be transplanted at one time from tumor cells prepared from a small number of tumors of early generation transplants. The resulting tumors would be assigned by a random number process to doses expected to yield TCPs in the range 0.1–0.9. This wide range in planned TCPs increases the accuracy of estimating the slope of the TCP vs. dose curve. The tumors at irradiation should be of a quite narrow range in volume and the radiation distributed uniformly throughout the tumor. Further, the animals need to be monitored for local regrowth for a time that is long relative to the time for regrowth at the highest dose level employed. In the laboratory setting there would be no or only rare subjects "lost to followup." The result is that the dose–response curve would be based on tumors that are virtual clones of a single tumor.

Results from such experiments have been reported and they indicate a steep dose–response curve, viz. γ_{50} of 4.[b] Dose–response curves are S shaped when plotted on a linear–linear grid with the steepest portion in the midrange of ~ 0.2–0.75, as shown in Fig. 6 [58]. The curve is nearly a straight line if the TCP is plotted on a logit–log dose grid. Thus, were $\gamma = 4$ and the dose increased by 5%, the TCP would be raised by ~ 20 percentage points. For example, were the TCP of the reference treatment 35% and the dose increased by 5% by a new high technology treatment, the TCP would be raised to $\sim 55\%$. The maximum steepness of the curve is at a TCP of 37%. It is much more shallow in the TCP ranges of 1–10% and 85–95%.

---

[b]The γ factor represents the percentage point increase in the response probability for a 1% increase in the dose. γ_{50} is the γ factor at the 50% response point on the dose–response curve.

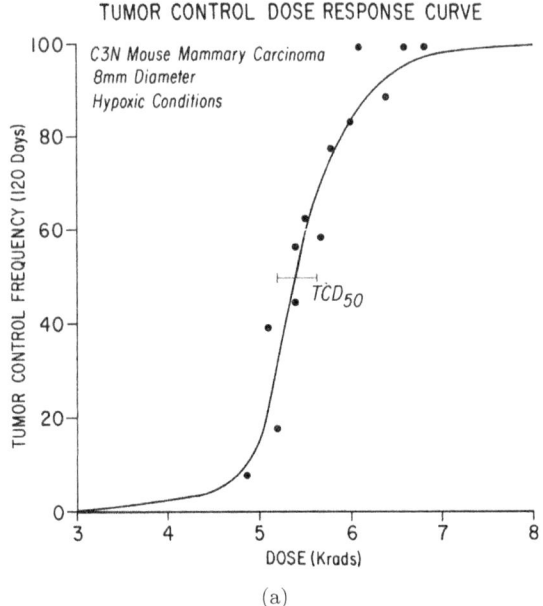

(a)

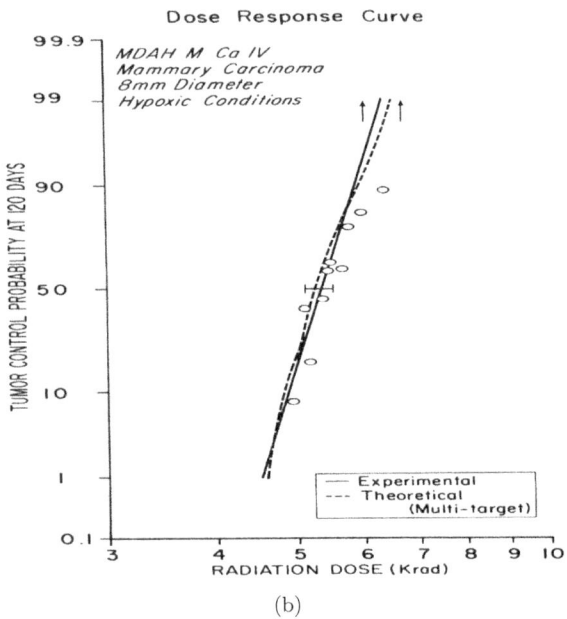

(b)

Fig. 6. Dose–response curves for early generation syngenic transplants of a C_3H mouse mammary carcinoma irradiated at an 8 mm diameter and under conditions of local tissue hypoxia. The curve in Fig. 6(a) is a linear–linear plot, and in Fig. 6(b) a logit–log dose plot [58].

The slope of the dose–response curve is based on tumors in a heterogenous population that have some level of heterogeneity in tumor characteristics and treatment delivered and that results in a flattening of the curve, depending on the degree of heterogeneity. The impact of population heterogeneity is

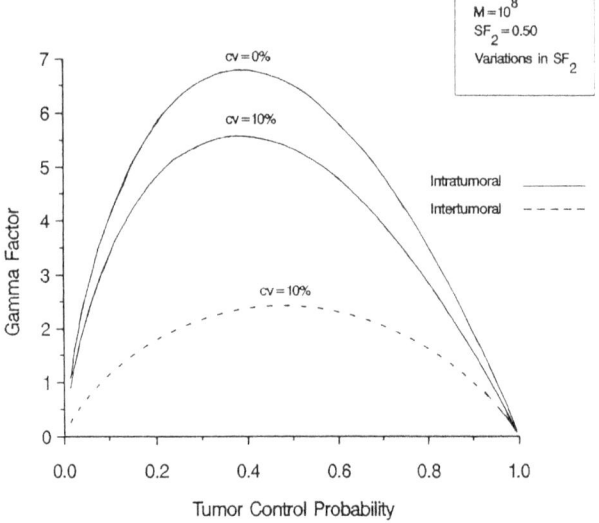

Fig. 7. Curves illustrating the variation of the γ factor for the TCP of a tumor of 10^8 clonogens, and $SF_{0.5}$ is 2 Gy. The three curves are for with zero variation in $SF_{0.5}$, with an intratumoral or an intertumoral 10% coefficient of variation [59].

illustrated in Fig. 7, showing the γ factor vs. tumor control probability for a model tumor system of 10^8 clonogens characterized with a zero or with an intratumoral or an intertumoral coefficient of variation of 10% in SF_2 [59].

The obvious implication is that the examination for a change in the TCP for a specified percentage increase in the dose would be most likely to be successful were the TCP of the reference treatment near the midpoint of the dose–response curve and with the minimum feasible heterogeneity in the tumor/subjects.

3.2. *LET and RBE*

Clinical ^{12}C ion beams are high-LET, viz. up to 120–200 keV/μm extremely close to the end of the range. This adds significant complexity to treatment planning due to the dependence of cell/tissue response on an array of parameters on LET. These are RBE, dose, α/β, the specific cell or tissue, pO_2 and cell age in the cell replication cycle. This complexity is enhanced by the variation of LET over the entire beam path, viz. the entrance to the end of the fragmentation tail, as discussed above.

Clinical practice has essentially employed a generic RBE, as there has not been an adequate database that would make estimation of an integrated RBE feasible for the complex of different tissues in the treatment volume. Fortunately, the generic RBEs employed have proven to be clinically effective, i.e. the frequency of severe radiation injuries has been judged to be acceptable.

RBE increases slowly with LET over most of the range and then rather sharply over the terminal 10 mm of the range, peaking at 100–200 keV/μm.

There is not a uniform RBE–LET relationship, as demonstrated by published RBE vs. ^{12}C LET for difference cell lines. Furasawa *et al.* [16] determined $SF_{0.1}$ of V79, HSG and T1 cell lines for single dose irradiation under aerobic conditions by ~ 40 and 100 keV/μm ^{12}C ion and 200 kVp x-rays under aerobic conditions. The $RBE_{0.1}$ values for the three cell lines at 40 keV/μm were 2.36, 1.80 and 1.72, respectively. At 80 keV/μm, the $RBE_{0.1}$ values were 3.10, 2.61 and 3.18. Weyrather *et al.* [66] also reported large RBE differences between V79 and CHO cell lines over the LET range 40–80–200 keV/μm. Thus LET, as defined by keV/μm for a specific beam, does not provide a certain guide to the RBE of a different cell line irradiated by the same beam.

The design of a clinical ^{12}C beam requires the adjustment of the physical dose to balance the impact of RBE increasing over the depth to be in the SOBP. This is illustrated in Fig. 8 for a 135 MeV ^{12}C ion beam whose range is 4 cm and planned for a 3 cm SOBP, from Ref. 27. The RBE is that for V79 cells and the end point is $SF_{0.1}$. This is the process employed for the design of most clinical ^{12}C beams.

3.3. *RBE and dose*

For high LET beams, there is a clinically very important inverse relationship between RBE and dose. The essential point is that over the clinical dose/fraction range of 1–10 Gy, the RBE increases as the dose is decreased. Figure 9 is a plot of RBE for injury to the normal kidney of female pigs for 42 MeV neutrons and 250 kVp x-rays [46]. The radiation was given as a single dose and in 6, 12 and 30 fractions in 39 days. The RBE moved up from 1.2 for a single dose to 4.6 Gy as the x-ray dose/fractions decreased from 7.9 to 1.4 Gy. This is a big effect and merits careful consideration in treatment planning.

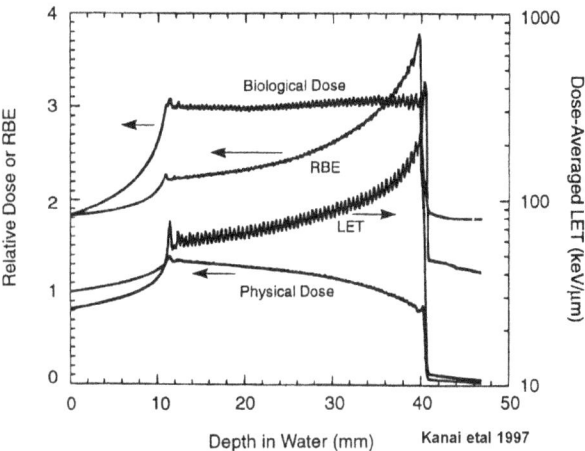

Fig. 8. The basic factors in the design of a clinical ^{12}C ion beam are shown here; namely, the increasing LET and RBE are balanced by a declining physical dose across the SOBP [27].

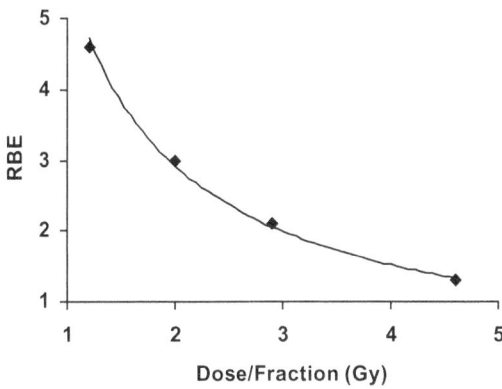

Fig. 9. RBE 42 MeV neutrons vs. x-ray dose per fraction for injury of the pig kidney given in 1–30 fractions [46].

3.4. *OER*

There has been serious and sustained interest in high LET radiation for some five decades, principally due to three facts: (1) many epithelial and mesenchymal tumors have hypoxic foci; (2) OER^c is ~ 2.5–3 for x-radiation; and (3) OER is lower for high LET radiations. Thus, the RBE would be higher for cells that are hypoxic than for similar cells that are normally oxygenated. Accordingly, the effect of high LET irradiation of tumors containing hypoxic cells would be greater than that of low LET irradiation. The prediction would accordingly be an increased TCP with

---

^{c}OER is the ratio of the dose to inactivate a cell under hypoxic conditions to that under aerobic conditions.

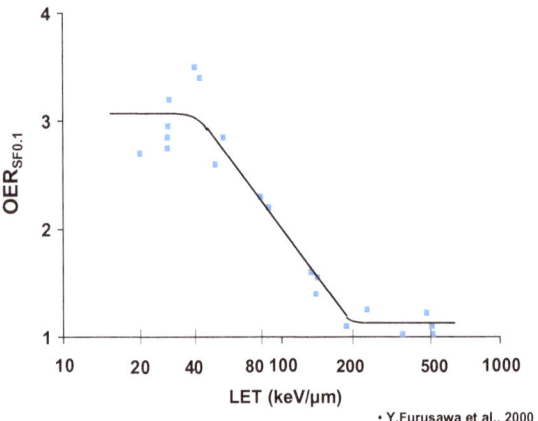

OER vs. LET of HSG Human Cell *in vitro* for ¹²C ion Beams

• Y.Furusawa et al., 2000

Fig. 10.   OER vs. LET for the HSG cell line [16].

no increased NTCP for high LET irradiation of such tumors.

OER varies widely over a rather narrow range of LET, as shown in Fig. 10. This is a plot of OER vs. LET for a human salivary gland tumor cell line and is drawn from Ref. 16. The OER is relatively constant with rising LET up to about $\sim 40 \, \text{keV}/\mu\text{m}$. Then, as LET increases from ~ 50 to 200, OER falls to ~ 1.1. For ¹²C ion therapy requiring SOBPs of 6–12 cm, OER at the mid-SOBP is likely to be $\geq \sim 2.2$. Thus, for most of the SOBP, OER would be expected to be 2–2.5, a relatively modest but certainly not negligible change in the desired direction.

An early finding of cellular radiation biology was a variation of radiation sensitivity with position in the cell replication cycle, i.e. the age response function. The magnitude of this variation is cell-line-dependent. The late S phase is generally the most resistant and the M phase the most sensitive phase.

There is a flattening of the age response function with increasing LET, and at very high LET it is almost flat. (See Ref. 45.) This is a factor for tissues with a high proportion of cells actively dividing, viz. acute responding normal tissues and some tumors. In contrast, this would be a small factor in the response of late responding tissues.

4. Clinical Outcome Data

We consider the local control and treatment–related morbidity results from reports on seven groups of tumors: chordoma and chondrosarcoma of the skull base and upper cervical spine, uveal melanoma, head and neck carcinoma (squamous cell and adenocystic), non-small-cell lung carcinoma (NSCLC), hepatocellular carcinoma (HCC) and prostate carcinoma. For several of the series, results are given for an approximate median of doses applied. Regarding the stage of disease, for some reports the statement is that the patients had locally advanced tumors. The column headed "Serious radiation injury" gives the data for $\geq$ GII or $\geq$ GIII complications, or for some series the injuries are listed as serious.

Note that ¹²C radiation has been deployed as hypofractionated therapy and ¹H predominately as "standard fractionation," viz. doses of 1.8–2 Gy (RBE).

4.1. *Chordoma*

A large number of patients have been treated by ¹H and ¹²C beams for chordoma of the skull base and upper cervical and spine, especially considering their rarity. The data in Table 1 give 5-year ¹²C local control rates of 91% and 100% by 60 Gy (RBE) for doses/fractions of 3–3.8 Gy (RBE), compared with

Table 1.   Chordoma of the skull base and upper cervical spine.

Beam	Number of patients	Dose Gy (RBE)		Local control at years	Late injury at years	Reference
		Total	Per Fx			
¹H	115	69	1.8	59% at 5	Not given	[60]
	42	74	1.8–2	81% at 5	High grade 6% at 5	[1]
¹²C	19	61	3–3.8	91% at 5	None	[21]
	10	(48–58)		60% at 5		[21]
	12	> 61–790	3.0–3.5	100% at 5	2 patients	[52]
	84	≤ 60		63% at 5		
x	37	67	1.8	50% at 5	Serious in 1 (3%)	[7]

54% and 81% for ^{1}H doses of 69 Gy and 74 Gy (RBE) at conventional doses per fraction. Importantly, the TCP values are higher and apparently the NTCP a bit lower for ^{12}C than for ^{1}H-treated patients.

Results are available from ^{1}H and ^{12}C irradiation of sacral chordoma. Imai *et al.* [25] treated 30 patients for chordoma of the sacrum by ^{12}C ion beams to 70.4 Gy (RBE) in 16 fractions and no surgery. They have had 95% local control of tumor in the defined PTV. That is to say, there has been one in-field failure and that was of a 1470 ml lesion. There were, however, 6 failures in relatively nearby tissues, e.g. gluteus, coccyx, spinal muscle, buttock femoral muscle (2 patients) and lumbar spine. These results suggest that some of these might well have represented wider subclinical extensions of disease than appreciated at the start of treatment. Park *et al.* [43] have reported on 21 patients treated by ^{1}H + x-radiation and surgical resection for sacral chordoma. The local control results were 12/14 and 1/7 for primary and locally recurrent (after prior surgery). In the ^{12}C series 2 patients required skin grafts but no other serious persisting treatment-related morbidity. Six patients had temporary neurological symptoms. In 12 patients with a primary tumor of the ^{1}H + surgery series there were 4 colostomies and 2 with incontinence; whether these adverse events were principally the result of the radiation or surgery cannot be definitely stated.

4.2. *Chondrosarcoma*

Results from ^{1}H treatment were markedly better for chondrosarcoma than for chordoma of the skull base, viz. 5-year local control rates, 94–98% vs. 54–81%. ^{1}H therapy has a modest numerical gain relative to ^{12}C therapy, Table 2.

Table 2.   Chondrosarcoma of the skull base and upper cervical spine.

Beam	Number of patients	Dose Gy (RBE)		Local control at years	Reference
		Total	Per Fx		
^{1}H ± x*	200	72	1.8–2	98% at 5	[47]
	22	68	1.8–2	94% at 5	[1]
^{12}C	54	60	3	90% at 4	[49]

*Surgery was gross total removal in 5% of patients.

4.3. *Uveal melanoma*

A group of tumors for which several radiation methods have provided quite high local control rates is uveal melanoma as documented in Table 3. The TCP values have been > 95% and the NTCP (enucleation) rate 6–7% for treatment with ^{1}H and ^{12}C [9, 11, 12, 19]. These treatments have been quite-high-dose, viz. 60 Gy (RBE) in 4 fractions or 70 Gy (RBE) in 5 fractions. The dose of 70 Gy (RBE) in 5 fractions would be estimated to be the BED equivalent of 280 Gy at 2 Gy/fraction for $\alpha/\beta = 2$ and 140 Gy for $\alpha/\beta = 10$. For 60 Gy (RBE) in 4 fractions, the BED equivalents would be 255 Gy and 125 Gy. The yield for SRT at 2.8 years in a series of 158 patients was 98%, but with the enucleation rate of 13% [10]. With really large numbers of patients, and very high local control and eye preservation rates at 10 and 15, ^{1}H therapy appears, at present, to be the most effective radiation treatment for uveal melanoma.

4.4. *Head and neck*

Here, the results of irradiation of squamous cell carcinoma (nonskin) appear to give a definite advantage to ^{1}H therapy over ^{12}C therapy, Table 4. This is based on the reported 5-year local control rates of 56% for ^{12}C vs. $\sim$93% for ^{1}H therapy [39, 56]. Note that these results from ^{12}C therapy are based on small patient numbers. Even allowing for the small patient numbers, these results do not support the expectation that ^{12}C radiation is especially effective against tumors, with head/neck carcinomas even though characterized as having a high probability of hypoxic foci.

The reported results of particle beam irradiation of locally advanced adenocystic carcinomas indicate a higher local control rate for ^{1}H + x than for ^{12}C therapy, viz. 93% vs. 78–79% [39, 44, 51]. However, these ^{1}H results have been associated with a quite high incidence of serious radiation injury, i.e. developing in 16 of the 23 patients. This has been discussed with Dr. A. Chan, an MGH (personal communication, 2009) head/neck radiation oncologist, who states that in patients treated after 2002 the treatment has been 1.8 Gy (RBE) once daily (no BID), substantially smaller CTVs, dose levels of $\sim$70 Gy (RBE) (protons only) and combined with chemotherapy. The frequency of radiation injury is now much lower. Of note is the report by Schulz-Ertner of local

Table 3. Uveal melanoma.

Beam	Number of patients	Dose Gy (RBE)		Local control at years	Enucleation	Reference
		Total	Per Fx			
^{1}H	2069	70	14	95% at 15	7%	[19]
	2435	60	15	95% at 10	∼7%	[11, 12]
	1406	60	15	96% at 5	8%	[9]
^{12}C	57	70 (60–85)	∼14	97% at 3	5%	[62]
x	132	60	12	98% at 2.8	13%	[10]
	25	70	14			

Table 4. Head and neck.

Beam	Number of patients	Dose Gy (RBE)		Stage	Local control at years	Injury Glll	Reference
		Total	Per Fx				
Squamous cell carcinoma							
^{1}H ± x	29	76	∼1.7	T1–T2 13 T3 10 T4 6	12/13 9/10 6/6	3 patients	[56]
^{1}C	15	57.6–64	3.6–4	Advanced	56 % at 5	None	[39]
Adenocystic carcinoma							
^{1}H + x	23	76	1.8 1.6 BID	Gross tumor at RT in 20	93% at 5	16 patients*	[44]
^{12}C	90	58–64	3.6–4	Advanced	79%	None	[39]
^{12}C + x	29	72	1.8 x 3 × 6 ^{12}C	Advanced	78% at 4	1 patient	[51]
x	34	66	1.8		25% at 4	2 patients	

*1 GIV retinopathy, 3 required surgery, 10 CNS (7 seizure, 3 short term memory), 2 other.

control rates of 78% by 72 Gy (RBE) vs. 25% by 66 Gy [51], i.e. dose is important. These data are not from a random assignment of patients to one of the two dose levels. Recognition must be given to the lack of knowledge in any detail of the extent of tumors in these diverse series and, hence, interpretation of the data can only be tentative.

4.5. Non-small-cell lung carcinoma

A substantial number of patients have been treated by ^{1}H or ^{12}C beams for their early stage NSCLC, Table 5. Studies on both beams employed hypofractionation, i.e. 10 or 12 fractions [4, 22, 38]. Higher success rates were obtained by ^{12}C therapy for stages T1 and T2, viz. 95% at 5 years for ^{12}C ions using 8 fractions to 72 Gy (RBE). The ^{1}H results were 87% and 49% for T1 and T2 tumors at lower BED dose levels. Related morbidity in particle beam therapy

has been low, at least up to this time. The extremely aggressive x-ray treatment by Fakiris et al. of 60–66 Gy in 3 fractions obtained local control of 88% at 3 years. Note that the BED for 60 Gy in 3 fractions would be 330 Gy at 2 Gy/fraction for $\alpha/\beta = 2$ and 150 Gy for $\alpha/\beta = 10$ [13]. The serious injury rate was 10% for peripherally sited lesions but 27% for central lesions. Notably, five of their patients had a Grade V radiation injury Forquer et al. [14]. have reported a nonnegligible frequency of brachial plexopathy following hypofractionated x-irradiation of the apical region of the thorax.

4.6. Hepatocellular carcinoma

Primary liver cancer is one of the most frequent carcinomas in humans and is a disease for which the role of radiation has traditionally been limited to palliation. However, results from ^{1}H and ^{12}C therapy have

Table 5. Non-small-cell lung carcinoma.

Beam	Number of patients	Dose Gy (RBE)		Stage	Local control at years	Serious or ≥ GIll	Reference
		Total	Per Fx				
^{1}H	68	51–60	5.1–6	T1 + T2 T1 T2	74% at 3 87% 49%	None	[4]
	21	55* or 66	5.5 or 6.6	T1 A + B	95% at 2	None	[22]
^{12}C	51	72	8	1	95% at 5	1 (skin)	[38]
x	70	60 66	20 22	T1 + T2	88% at 3	≥ GIII Peripheral 10% Central 27%	[13]

*Dose was specified in Gy in their paper. These have been converted to Gy (RBE) by multiplying by 1.1.

Table 6. Hepatocellular carcinoma.

Beam	Number of patients	Dose Gy (RBE)		Stage	Local control at years	Serious or Injury ≥ GIll	Reference
		Total	Per Fx				
^{1}H	51	66	6.6	Child–Pugh A (80%) B (20%)	88% at 5	8 of 41 Child–Pugh A → B None to C	[15]
	162	72 (median)	∼4.5	60% stage II–IIIB	87% at 5	5 with ≥ GII	[5]
^{12}C	24	∼64.5	4.2	Not given	81% at 5	Child–Pugh ↑	[29]
	86	∼62–50	∼5.2–12.6		87% at 5	≥ 2 points in 18%	
	47	52.8	13.2		96% at 5	No serious skin, GI injuries	

to be rated as impressive. For two ^{1}H therapy studies at Tsukuba, the 5-year local control results have been 87–88% [5, 15]. High grade treatment–related morbidity has been of low frequency. The more recent ^{12}C dose escalations study by Kato *et al.* [29] from Chiba has yielded progressive gains in the TCP as the dose/fraction was increased from ∼ 49.5 to 79.5 in 15 fractions [3.6 Gy (RBE)/fraction] to 52.8 Gy (RBE) in 4 fractions [13.2 Gy (RBE)]. Namely, the TCP rose from 81% to 96% (47 patients) at this very high dose.

4.7. *Prostate carcinoma*

Among US men, this is a quite common tumor and the proportion diagnosed in an early stage is high. The local control and bNED rates for early stages of tumors by high dose radiation are high. This is achieved at a low frequency of serious radiation-related morbidity. For ^{1}H + x radiation treatment

of stage T1c–T2c prostate cancer treated by 74 Gy at 2 Gy (RBE)/fraction, the 5-year bNED result was 74% [55]. The critical role of doses is evident by the 5-year bNED rates of 61% and 80% following 70.2 vs 79.2 Gy (BRE) to stages 1b–2b prostate cancers in a Phase III clinical trial [73]. A higher success rate has been reported by Tsujii *et al.* [62] for ^{12}C ion treatment, viz. 5-year bNED rate of 92% and with no serious complications for 295 patients with stage T1 and T2 disease given 66 Gy (RBE) at 3.3 Gy (RBE)/fraction. Note that for IMXT treatment to 81 Gy by Zelefsky *et al.* [72], the 8-year bNED rate for 425 and 101 patients classed as ASTRO risk groups, low and intermediate stage, were 85% and 76%, respectively, Table 7.

The first study of ^{1}H beam treatment of prostate patients was done by Shipley *et al.* [53] for T3–T4 tumors, through a phase III trial with treatment by x-ray alone to 67.2 Gy or 50.4 Gy + a 25.2 Gy (RBE) ^{1}H boost; the total dose was 75.6 Gy (RBE). Local

Table 7.   Prostate carcinoma.

Beam	Number of patients	Dose Gy (RBE)		Stage	Local control at years	≥ GIII injury	Reference
		Total	Per Fx				
x + ^{1}H			1.8x → 50.4	T3–T4			[54]
vs.	93	75.6	2.1 ^{1}H		77% at 8	R^a 9% vs. 2%	
x + x	96	67.2	2.1 x		60% at 8	U 4% vs. 2%	
x + ^{1}H	1255	74	2	T1c–T2c	bNED 74% at 5	GIII 2%	[56]
x + ^{1}H	197	50.4 x 28.8 ^{1}H	1.8	T1b–T2b	bNED 80% at 5	1% 1%	[74]
	96	50.4 x 19.8 ^{1}H	1.8		bNED 61% at 5	2%	
^{12}C					bNED	None	[63]
	295	66	3.3	T1–T2	92% at 5		
	162			T3	81% at 5		
x	561	81	1.8	Risk (ASTRO)	bNED at 8 85%	3%	[73]
	425			Low	76%		
	101			Medium	72%		
	35			High			

^{a}R is late rectal bleeding. U is late urethral stricture.

control at 8 years was reported as 77% and 60% for the high and low dose groups.

5. Discussion

Proton and carbon ion beams provide clearly superior dose distributions for most sites relative to that achievable by the highest technology x-ray beam therapy, due to the physical fact that the particle beams have finite ranges. This results in the advantage of near-zero doses deep to the target for each beam path. This distal dose can be contoured to match closely the deep margins of the target plus the PTV. Further, for PBS dose delivery the high dose zone can be contoured to the proximal margins. For gantry-equipped treatment rooms, these particle beams have the same flexibility in dose delivery as modern x-ray machines, viz. beam number, direction, intensity modulation and 4D IGRT. ^{12}C beams have one physical advantage over ^{1}H beam, and that is a narrower penumbra, except for penetration depths of only a few cm. The narrower penumbra is judged to be clinically significant in radical dose therapy where important late responding normal tissues are in close proximity to the target. The late responding tissues tend to be low α/b and hence at increased risk of radiation injury by high LET beams, e.g. ^{12}C. Thus, the slimmer penumbra of ^{12}C beams, i.e. 1.5–3 mm compared with the much wider penumbras of ^{1}H beams, is rated as an advantage. This is so as the tissues included in the penumbra would be normal tissues, and in many instances late responding tissues.

The attraction of high LET beams for radiation therapy has been focused on a reduced OER. The OER is reduced from ∼3 to 2–2.5 at the mid-SOBP, i.e. a modest but nonzero reduction. The data from a good number of clinical studies of the efficacy of fast neutron (high LET) and x-ray therapy for head/neck carcinomas have not yielded outcome data favoring neutron therapy. These data are in part from clinical trials with comparable depth–dose characteristics for the neutron and x-ray beams. The general finding has been excessive rates of serious injury to normal tissues with no important gain in TCP. The relevant studies include those by Cohen [6], Hussey et al. [23, 24] and Maor et al. [36]. Head/neck squamous cell carcinomas were heavily represented in several of these studies. Multiple studies on untreated squamous cell carcinoma of the head/neck have demonstrated hypoxic regions. These studies include those by Becker et al. [2]; Brizel et al. [3] and Nordsmark et al. [41]. There is one apparently quite bright outcome for neutron therapy, and that is for treatment of salivary gland carcinomas. The local controls result is higher for neutrons than for x-rays in a prospective trial [33]. However, that result is perhaps not related to the reduced OER of neutrons. Wijffels *et al.* [67] examined in considerable detail a series of salivary gland carcinomas for evidence of hypoxic cells and found none. The overall experience in neutron therapy has been a setback for high LET radiation therapy. This shift in attitude is reflected by the fact that in the US there were eight centers and five have now closed. The status of neutron therapy has been reviewed by Laramore *et al.* [34], who judged that there was ample basis for continued study of neutron therapy for salivary gland tumors. They also suggested that further study be directed to sarcomas of soft tissue and bone, as well as selected carcinomas of the lung.

Consider the substantial difference in local control of squamous cell carcinomas by ^{1}H and ^{12}C ion therapy given in Table 4, viz. $\sim 90\%$ for ^{1}H and 56% for ^{12}C. Despite the small numbers, these results do constitute a disappointment.

The results obtained by ^{12}C ion therapy for chordoma of the skull base and sacrum are definitely impressive. The high local control rate associated with no serious complication rate (at least till 2008) appears to constitute a clinical gain. The ^{12}C dose was evidently highly successful in the inactivation of chordoma in the designated PTV. The frequency of local regrowths in the adjacent tissues raises concern regarding our ability to estimate the extent of subclinical disease.

Also, the success of stage 1 NSCLC appears positive. However, the results from ^{1}H irradiation are, at present, at least as good as that from ^{12}C therapy for chondrosarcoma of the skull base, uveal melanoma, squamous cell carcinoma of the head/neck, hepatocellular carcinoma, and also prostate cancer. The very high success rates in the irradiation of two additional tumors could also be classed as noteworthy. One is a small series of 10 patients treated by ^{12}C ions for their primary renal cell carcinoma [40], and obtained a 5-year local control rate of 100%. The second is the 5-year local control of 84% in a series of 72 patients treated by Yanagi *et al.* [70] for mucosal melanoma to ~ 58 Gy (RBE) [~ 3.6 Gy (RBE)/fraction] by ^{12}C ions.

In summary, the clinical data presented constitute a manifest need for phase III clinical trials of ^{1}H vs. ^{12}C therapy. High LET is the one characteristic that clearly distinguishes ^{12}C from ^{1}H beams. The trial should be a prospective randomized study designed to determine if high LET radiation provides a clinical advantage for one or more types of cancer. To achieve this, the design should have identical dose fractionation for the ^{1}H and ^{12}C arms. In addition, there should be fully comparable technology for defining the margins of the target and the alignment of the target on the beam. Further, any combined modality therapy must be the same for the two arms. The protocol should include plans for long term followup examinations. This is critical, as the question is: is there a gain in the TCP for a specified NTCP?

Special effort should be directed toward selecting some tumors for the trials whose TCPs by protons therapy are in the range of 0.2–0.75, i.e. the steepest portion of the dose–response curve. This provides a higher probability of detecting a difference in the TCP of 10% points. In accord with this concern, efforts should be made to have a highly homogeneous group of patients and of tumors.

The rationale for high LET radiation therapy is the expectation of greater efficacy against the gross tumor and no expected clinical benefit for irradiation of grossly normal tissues. Based on this, our opinion that the most effective strategy would be to employ low LET ^{1}H beams for the dose to the CTV (GTV + normal tissues judged to be infiltrated by a subclinical tumor) and then high LET ^{12}C for the boost dose to the GTV. As the object of the trial would be to examine for a positive impact of the highest feasible LET and RBE by ^{12}C therapy, the boost dose should be administered at a low dose/fraction, e.g. ~ 2–2.5 Gy (RBE). As noted earlier, the RBE increases steeply as the dose/fraction is reduced to 2 and 1 Gy (RBE). This high LET boost dose could be given as a concomitant boost, a BID schedule. As the point is to examine for gain by high LET and RBE, the trial should use ^{12}C in the highest feasible RBE mode.

Efforts to assess the cost of particle beam therapy need to consider it as one component of the total societal cost of managing cancer patients. They include screening, diagnosis, patient evaluation, combined modality therapies (chemical, biological and genetic strategies) and the cost of failure, in terms of either recurrence of the tumor or treatment-related morbidity. These are in addition to the actual cost of administering particle beam treatment.

Acknowledgments

The authors are extremely appreciative of the really important contributions to the effort made in preparing this manuscript by B. Clasie, G. Chen, L. Gerweck, S. Goldberg, A. Niemierko and H. Paganetti. In addition, we are pleased to acknowledge the partial support of this work by the US National Cancer Institute through grant No. PO1CA021239.

Conflict-of-interest statement: Dr. Delaney has received honoraria from IBA Proton Therapy for speaking at the Industry-Sponsored Symposia on Proton Beam Radiation Therapy.

References

[1] C. Ares, E. B. Hug, A. J. Lomax *et al.*, *Int. J. Radiat. Oncol. Biol. Phys.*, Apr. 20, 2009 (Epub ahead of print).

[2] A. Becker, G. Hänsgen, M. Bloching *et al.*, *Int. J. Radiat. Oncol. Biol. Phys.* **42**(1), 35 (1998).

[3] D. M. Brizel, G. S. Sibley, L. R. Prosnitz *et al.*, *Int. J. Radiat. Oncol. Biol. Phys.* **38**(2), 285 (1997).

[4] D. A. Bush, J. D. Slater, B. Shin *et al.*, *Chest* **126**(4), 1198 (2004).

[5] T. Chiba, K. Tokuuye, Y. Matsuzaki *et al.*, *Clin. Cancer Res.* **11**, 3799 (2005).

[6] L. Cohen, *Int. J. Radiat. Oncol. Biol. Phys.* **8**, 2173 (1982).

[7] J. Debus, D. Schulz-Eernter, L. Schad *et al.*, *Int. J. Radiat. Oncol. Biol. Phys.* **47**, 591 (2000).

[8] T. F. Delaney and H. M. Kooy, *Proton and Particle Radiation Therapy* (Lippincott Williams and Wilkins, Philadelphia, 2008).

[9] R. Dendale, L. L. L. Rouic, G. Noel *et al.*, *Int. J. Radiat. Oncol. Biol. Phys.* **65**(3), 780 (2006).

[10] K. Dieckmann, D. Georg, M. Zehetmayer *et al.*, *Strahlenther. Onkol.* **183**, 11 (2007).

[11] E. Egger, A. Schalenbourg, L. Zografos *et al.*, *Int. J. Radiat. Oncol. Biol. Phys.* **51**(1), 138 (2001).

[12] E. Egger, L. Zografos, A. Schalenbourg *et al.*, *Int. J. Rad. Oncol. Biol. Phys.* **55**(4), 867 (2003).

[13] A. J. Fakiris, R. C. McGarry, C. T. Yiannoutsos *et al.*, *Int. J. Radiat. Oncol. Biol. Phys.* (2009) Feb. 27.

[14] J. A. Forquer, A. J. Fakiris, R. D. Timmerman *et al.*, *Int. J. Radiat. Biol.* **77**(6), 713 (2001); *Radiother. Oncol.* (2009) May 17 (Epub ahead of print).

[15] N. Fukumitsu, S. Sugahara, H. Nakayama *et al.*, *Int. J. Radiat. Oncol. Biol. Phys.* **74**(3), 831 (2009).

[16] Y. Furusawa, K. Fukutsu, H. Itsukaichi *et al.*, *Radiat. Res.* **154** 485 (2000).

[17] M. Goitein, *Radiother. Oncol.* (2009) Jul. 4 (Epub ahead of print).

[18] M. Goitein, *Int. J. Radiat. Oncol. Biol. Phys.* **4**, 499 (1978).

[19] E. Gragoudas, L. I. Wenjun, M. Goitein *et al.*, *Arch. Opthalmol.* **120**, 1665 (2002).

[20] E. Haettner, H. Iwase, D. Schardt *et al.*, *Radiat. Prot. Dosimetry* **122**(1–4), 485 (2006).

[21] A. Hasegawa, J. Mizoe, R. Takagi *et al.*, Carbon ion radiotherapy for skull base and paracervical tumors, in *Proc. NIRS MD Anderson Symposium on Clinical Issues for Particle Therapy* (2008), pp. 84–89.

[22] M. Hata, K. Tokuuye, K. Kagel *et al.*, *Int. J. Radiat. Oncol. Biol. Phys.* **68**(3), 786 (2007).

[23] D. H. Hussey, J. H. Jardine, G. O. Raulston *et al.*, *Int. J. Radiat. Oncol. Biol. Phys.* **3**, 2083 (1982).

[24] D. H. Hussey, G. H. Fletcher and J. B. Caderao, *Cancer* **34**, 65 (1974).

[25] R. Imai, T. Kamada, H. Tsuji *et al.*, *Clin. Cancer Res.* **10**, 5741 (2004).

[26] B. Jones, *Br. J. Radiol.* **79**, 24 (2006).

[27] T. Kanai, Y. Furusawa, K. Fukutsu *et al.*, *Radiat. Res.* **147**, 78 (1997).

[28] T. Kanai, M. Endo, S. Minohara *et al.*, *Int. J. Radiat. Oncol. Biol. Phys.* **44**(1), 201 (1999).

[29] K. Kato, H. Tsujii, T. Miyamoto *et al.*, *Int. J. Radiat. Oncol. Biol. Phys.* **59**(5), 1468 (2004).

[30] A. M. Koehler and W. M. Preston, *Radiology* **104**(1), 191 (1972).

[31] M. Kramer, O. Jakel, T. Haberer *et al.*, *Radiother. Oncol.* **73**(Suppl. 2), 80 (2004).

[32] M. Krämer, O. Jäkel and T. Harberer, *Phys. Med. Biol.* 3299 (2000).

[33] G. E. Laramore, J. M. Krall, T. W. Griffin *et al.*, *Int. J. Radiat. Oncol. Biol. Phys.* **27** 235 (1993).

[34] G. E. Laramore and T. W. Griffin, *Int. J. Radiat. Oncol. Biol. Phys.* **32**(3), 879 (1995); **35**, 599 (1995).

[35] A. J. Lomax, E. Perdoni, H. Rutz and G. Goitein, *Z. Med. Phys.* **14**, 147 (2004).

[36] M. H. Maor, D. Errington, R. J. Caplan *et al.*, *Int. J. Radiat. Oncol. Biol. Phys.* **32**(3), 599 (1995).

[37] G. Mesoloras, G. A. Sandison, R. D. Stewart *et al.*, *Med. Phys.* **33**, 2479 (2006).

[38] T. Miyamoto, M. Baba, N. Yamamoto *et al.*, *Int. J. Radiat. Oncol. Biol. Phys.* **67**(3), 750 (2007).

[39] J. Mizoe, A. Hasegawa, R. Takagi *et al.*, Carbon ion radiotherapy for head and neck tumors, in *Proc. NIRSMD Anderson Symposium on Clinical Issues for Particle Therapy* (Mar. 21–22, 2008), pp. 9–15.

[40] T. Nomiya, H. Tsuji, N. Hirasawa *et al.*, *Int. J. Radiat. Oncol. Biol. Phys.* **72**(3), 828 (2008).

[41] M. Nordsmark, S. M. Bentzen, V. Rudat *et al.*, *Radiother. Oncol.* **77**(1), 18 (2005).

[42] H. Paganetti, A. Niemierko, M. Ancukiewicz *et al.*, *Int. J. Radiat. Oncol. Biol. Phys.* **53**(2), 407 (2002).

[43] L. Park, T. F. Delaney, N. J. Liebsch *et al.*, *Int. J. Radiat. Oncol. Biol. Phys.* **65**(5), 1514 (2006).

[44] P. Pommier, N. J. Liebsch, D. G. Deschler *et al.*, *Arch. Otolaryngol. Head Neck Surg.* **132**, 1242 (2006).

[45] M. R. Raju, *Heavy Particle Radiotherapy* (Academic, New York, 1980).

[46] M. E. Robbins, D. W. Barnes, D. Campling *et al.*, *Br. J. Radiol.* **64**(765), 823 (1991).

[47] A. Rosenberg, G. P. Nielsen, S. B. Keel *et al.*, *Am. J. Surg. Path.* **23**(11), 1370 (1999).

[48] S. Safai, T. Bortfeld and M. Engelsman, *Phys. Med. Biol.* **53**(6), 1729 (2008).

[49] D. Schulz-Ertner, A. Nikoghosyan, H. Holger *et al.*, *Int. J. Radiat. Oncol. Biol. Phys.* **67**(1), 171 (2007).

[50] D. Schulz-Ertner and H. Tsujii, *J. Clin. Oncol.* **28**, 8 (2007).

[51] D. Schulz-Ertner, A. Nikoghosyan, B. Didinger *et al.*, *Cancer* **104**(2), 338 (2005).

[52] D. Schulz-Ertner, C. P. Karger, A. Feuerhake *et al.*, *Int. J. Radiat. Oncol. Biol. Phys.* **68**(2), 449–457 (2007).

[53] W. U. Shipley, L. J. Verhey, J. E. Munzenrider *et al.*, *Int. J. Radiat. Oncol. Biol. Phys.* **31**(1), 3 (1995).

[54] L. Sihver, C. H. Tsao, R. Silberber *et al.*, *Adv. Space. Res.* **17**(2), 105 (1996).

[55] J. D. Slater, C. J. Rossi, L. T. Yonemoto *et al.*, *Int. J. Radiat. Oncol. Biol. Phys.* **59**(2), 348 (2004).

[56] J. D. Slater, L. T. Yonemoto, D. W. Mantik *et al.*, *Int. J. Radiat. Oncol. Biol. Phys.* **62**(2), 494 (2005).

[57] M. Stovall, C. R. Blackwell, J. Cundiff *et al.*, *Med. Phys.* **22**, 63 (1995).

[58] H. D. Suit, *Radiation Biology: A Basis for Radiotherapy — Textbook of Radiotherapy*, G. H. Fletcher, 2nd ed. (Lea and Febiger, Philadelphia, l973), pp. 75–l2l.

[59] H. D. Suit, S. Skates, A. Taghian *et al.*, *Radiother. Oncol.* **25**, 251 (1992).

[60] A. Terehara, A. Niemierko, M. Goitein *et al.*, *Int. J. Radiat. Oncol. Biol. Phys.* **45**(2), 351 (1999).

[61] H. Tsuji, H. Ishikawa, T. Yanagi *et al.*, *Int. J. Radiat. Oncol. Biol. Phys.* **67**(3), 857 (2007).

[62] H. Tsuji, H. Kato, T. Yanagi *et al.*, Carbon ion radiotherapy for prostate cancer, in *Proc. NIRS-MD Anderson Symposium on Clinical Issues for Particle Therapy* (Mar. 21–22, 2008), pp. 62–71.

[63] I. Turesson, K. A. Johansson and S. Mattsson, *Acta Oncologica* **42**, 107 (2003).

[64] M. Urie, M. Goitein and M. Wagner, *Phys. Med. Biol.* **29**(5), 553 (1984).

[65] U. Weber and G. Kraft, *Cancer J.* **15**(4), 325 (2009).

[66] W. K. Weyrather, S. Ritter, M. Scholz and G. Kraft, *Int. J. Radiat. Biol.* **75**(11), 1357 (1999).

[67] K. Wijffels, I. J. Hoogsteen, J. Lok *et al.*, *Int. J. Radiat. Oncol. Biol. Phys.* **73**(5), 1319 (2009).

[68] A. Wroe, B. Clasie, H. Kooy *et al.*, *Int. J. Radiat. Oncol. Biol. Phys.* **71**, 306 (2009).

[69] A. Wroe, A. Rosenfeld and R. Schulte, *Med. Phys.* (2007) **34**, 3449. Dosage error in article: (2008) **35**, 3398.

[70] T. Yanagi, J. E. Mizoe, A. Hasegawa *et al.*, *Int. J. Radiat. Oncol. Biol. Phys.* (2008) 1.

[71] S. Yonai, N. Matsufuji, T. Kanai *et al.*, *Med. Phys.* **35**, 4782 (2008).

[72] M. J. Zelefsky, H. Chan, M. Hunt *et al.*, *J. Urol.* **176**, 1415 (2006).

[73] A. L. Zietmann, M. L. DeSilvio, J. D. Slater *et al.*, *JAMA* **294**(10), 1233 (2005).

Herman Suit is a member of the faculty of the MGH and of Harvard Medical School. He has been active in the start of clinical proton radiation therapy in 1973 and that interest has continued unabated.

Thomas F. Delaney is the Director of the Francis H. Burr Proton Therapy Center of the MGH. He is a specialist in tumors of the connective tissues and bones. He is a member of the faculty of the MGH and of Harvard Medical School.

Alexi Trofimov is a clinical physicist in the proton program with special interest in complexities of rt planning and biomathematical modeling. He is a member of the faculty of the MGH and of Harvard Medical School.

Reviews of Accelerator Science and Technology
Vol. 2 (2009) 17–33
© World Scientific Publishing Company

The Production of Radionuclides for Radiotracers in Nuclear Medicine

Thomas J. Ruth

TRIUMF and BC Cancer Agency, Vancouver, BC, Canada
truth@triumf.ca

Medical applications represent the vast majority of the uses for radiotracers. This review addresses how accelerators are employed for the production of high purity radionuclides that are used in basic biomedical research, as well as for clinical medicine both for diagnosing disease and for treatment.

Keywords: Radioisotopes; irradiation; nuclear medicine; specific activity; accelerators; cyclotron; imaging; therapy.

1. Introduction

The building of the cyclotron in the 1930s made possible the routine production of radionuclides that would find use in a variety of applications, including medicine, industry, agriculture, and basic physical and biological research. With the high power of the charged particles (energy and flux or beam current) available in the cyclotron, it is possible to produce abundant quantities of a wide variety of radionuclides.

Immediately after World War II, almost all radionuclides and radioisotopes in use were made in a reactor. The production of radionuclides in cyclotrons for medical applications revived in the 1950s, due in large part to the discovery that ^{201}Tl could be used as an analog of potassium ions and thus is an ideal tracer for detecting myocardial perfusion. Thallous chloride labeled with ^{201}Tl remains the standard for measuring cardiac blood flow despite the availability of ^{99m}Tc myocardial perfusion agents. The preparation of ^{18}FDG in the mid-1970s and its use for studying glucose metabolism was a major breakthrough, leading to the development of the now-widely-used imaging modality called positron emission tomography (PET). ^{18}F-FDG when used along with the PET camera yields excellent quality images of the brain (for studying both normal function and functional abnormalities), the heart (for studying viability function), and tumors (for detecting metastasis). A large number of other ^{18}F- and ^{11}C-labeled radiopharmaceuticals were developed subsequently, and the quest for newer and more effective ones continues today.

In addition to the use of PET and single proton emission tomography (SPECT) radionuclides for diagnostic imaging studies, cyclotron-produced radionuclides are finding extensive therapeutic applications. An example is the use of dedicated cyclotrons with large beam currents for the production of ^{103}Pd for brachytherapy applications. Another example of cyclotron-produced radionuclides being used for treatment is the production of alpha-particle-emitting isotopes, notably ^{211}At and ^{213}Bi, for targeted therapy of cancer.

The applications of cyclotron-produced isotopes have been expanding at a much faster pace in the last 15 years, as seen by the large number of new machines being installed for isotope production. Some cyclotrons are dedicated to the production of a single isotope such as ^{18}F or, as indicated above, ^{103}Pd. The International Atomic Energy Agency (IAEA) has published a directory on cyclotrons used for isotope production in 1998, and a revised version of it was published in 2006 (IAEA-DCR/CD) that documents the cyclotrons available in the member states.

This article will describe the fundamental operation of the cyclotron as it is used for producing radionuclides for medical purposes. Some background information on the imaging technologies is provided for completeness.

2. Radioisotope/Radionuclide Production

Radionuclide production is indeed true alchemy that is converting the atoms of one element into those of another. This conversion involves altering the number of protons and/or neutrons in the nucleus (target). If a neutron is added without the emission of particles, then the resulting nuclide will have the same chemical properties as the target nuclide and thus is an isotope of that element. If, however, the target nucleus is bombarded by a charged particle, such as a proton, the resulting nucleus will usually be that of a different element. The exact type of nuclear reactions a target undergoes depends on a number of parameters, including the type and energy of the bombarding particle. A more complete description of the process of radionuclide production is given below.

The binding energy of nucleons in the nucleus is, on average, of the order of 8 MeV. Therefore, if the incoming projectile has more than this amount of energy, the resulting reaction will cause other particles to be ejected from the target nucleus. By carefully selecting the target nucleus, the bombarding particle and its energy, it is possible to produce a specific radionuclide.

2.1. *Specific activity* [1, 2]

Specific activity (SA) is a measure of the number of radioactive atoms or molecules as compared to the total number of those atoms or molecules present in the sample. It is usually expressed in terms of radiation units per mass unit. The traditional units have been Ci/mole (or Ci/g) or a fraction thereof (now expressed as GBq/mole). If the only atoms present in the sample are those of the radionuclide, then the sample is referred to as carrier-free. For example, a compound labeled with ^{211}At will be carrier-free since there are no stable isotopes of astatine.

However, in most cases there are small quantities of nonradioactive atoms which serve as a carrier or molecules that have a similar chemical behavior and can act as a *pseudocarrier*. The SA of an isotope or radiopharmaceutical is important in determining the chemical/biological effect which the substance may have on the system under investigation.

The number of radioactive atoms in a sample can be calculated from the relationship of radioactivity to quantity expressed as

$$\frac{dN}{dt} = -\lambda N, \qquad (1)$$

where dN/dt is the disintegration rate in seconds and λ is the decay constant in reciprocal seconds [$\lambda = \ln(2)/t_{1/2}$, where $t_{1/2}$ is the half-life].

As an example of SA, assume that glucose has been labeled with 10 mCi of ^{11}C with a half-life of 20.3 min. The carrier-free SA can be calculated from the number of atoms contained in the 10 mCi. The number of atoms will be

$$^{11}C = N = \frac{dN/dt}{\lambda} = \frac{(10\,\text{mCi})(3.7 \times 10^7\,\text{dps/mCi})}{\dfrac{\ln(2)}{(20.3\,\text{min})(60\,\text{s/min})}}$$

$$= 6.5 \times 10^{11}\,\text{atoms}$$

Using Avogadro's number, the number of moles is then 1.08×10^{-12} and SA $= 9.3 \times 10^9$ Ci/mol or 9.3×10^3 Ci/μmol.

If the radionuclide had been ^{14}C with its 5715-year half-life and following the same process but using the decay constant ($\lambda = 3.82 \times 10^{-12}\,\text{s}^{-1}$) for ^{14}C, then the SA would be 62 Ci/mol. Therefore, it is easy to see that the short-lived radioisotopes potentially have a much higher SA. If, however, the radiolabeled glucose had been prepared in a plant, the naturally occurring glucose would have lowered the SA due to the nonradioactive glucose molecules.

3. Accelerators

The transformation of one element into another was first demonstrated by Ernest Rutherford in 1919 [3], when he directed the α particles emanating from a sample of polonium onto nitrogen gas and detected the protons being emitted (having produced O-17). The future of accelerator production of radioisotopes reached a turning point with the construction of the cyclotron by Ernest Lawrence in 1931 [4, 5]. With the cyclotron, it became possible to produce radioactive isotopes of a wide variety for the first time. Researchers from all over the world came to Berkeley to use the artificially produced radiotracers such as radioactive sodium and iodine in the late 1930s. Cyclotron-produced radionuclides for biomedical research were used in the late 1930s for some clinical research and for basic research in biochemistry. In 1936 the University of California officially established the Radiation Laboratory as an independent entity within the Physics Department. The reorganized laboratory was dedicated to nuclear science rather than, as in its first incarnation, to accelerator physics. A center for nuclear medicine already existed at the University of California Hospital in San Francisco, where J. G. Hamilton and

Robert Stone were using radioactive sodium clinically in 1937. The use of these artificially produced radiotracers continued with Hamilton and Stone. In 1938 S. Hertz, A. Roberts and R. D. Evans used radioactive iodine in the study of thyroid physiology, followed in 1939 by J. H. Lawrence, K. G. Scott and L. W. Tuttle studying leukemia with radioactive phosphorus. By 1940 Hamilton and M. H. Soley were performing studies on iodine metabolism by the thyroid gland *in situ* by using radioiodine in normal subjects and in patients with various types of goiters [2].

They were joined by Lawrence's brother John, who had been interested in the biological effects of neutrons during a visit to Berkeley in the summer of 1935. Funding for the machine promised to Lawrence in 1936 was raised on the grounds of its utility in medicine [6, 7].

Even with this background, in the early years, cyclotrons were mainly used in physics research. Radionuclides for medical applications were a sidelight. The first cyclotron dedicated to medical applications was installed at Washington University, St. Louis in 1941, where radioactive isotopes of phosphorus, iron, arsenic and sulfur were produced. During World War II, a cyclotron at Cambridge, Massachusetts also provided a steady supply of radionuclides for medical purposes. In the mid-1950s a group at Hammersmith Hospital in the United Kingdom put into operation a cyclotron wholly dedicated to radionuclide production. The major change occurred in the early and mid-1960s, when the work in hot atom chemistry (such as the *in situ* chemistry of nucleogenic atoms occurring in a target being bombarded) laid the foundation for the synthesis of organic compounds labeled with positron emitters. A 1966 article by Ter-Pogossian and Wagner focused on the use of carbon-11 [8]. As the field of nuclear medicine has progressed, the number of available types of particle accelerators with varying characteristics dedicated to radionuclide production for nuclear medicine has also expanded. The major classes of accelerators are the positive and negative ion cyclotrons. More recent innovations include superconducting magnet cyclotrons, small low energy linacs, tandem cascade accelerators and helium-particle-only linacs. These types of accelerators have not gained wide acceptance.

Wolf and colleagues [9–11] have reviewed the application of cyclotrons for the production of

Table 1. Classification of accelerators.

Classification	Characteristics	Proton energy	Comments
Level I	Single particle, p or d (some dual particle)	10 MeV	
Level II	Single or multiple particle, p, d	20 MeV	^{3}He, ^{4}He — not usually available
Level III	Single or multiple particle, p, d	50 MeV	^{3}He, ^{4}He — may be available
Level IV	Usually p only	70–500 MeV	

radionuclides and suggested that the accelerators can be classified into four levels reflecting the particle type and energy of these particles. These are listed in Table 1 (adapted from Refs. 10 and 11).

The principle advantage of accelerator-produced radioisotopes is the high SAs that can be obtained through the (p, xn) and (p, α) and other reactions involving charged particles that result in the product being a different element than the target. Another significant advantage is that a smaller amount of radioactive waste is generated from charged particle reactions.

Cyclotrons designed for producing medical radioisotopes were initially capable of accelerating protons, deuterons, ^{3}He^{+2} and α particles (the nucleus of ^{4}He). However, the principal radioisotopes currently used in medical applications can all be produced by protons. The simplicity of design for proton-only cyclotrons has resulted in cyclotrons which are capable of generating two or more simultaneous beams of varying energies and intensities. The modern cyclotron is completely controlled by a computer and can be in continuous operation for many days with minimal attention.

3.1. Development of the linac

The concept of the linac arose because there is a practical limit to the energy which can be supplied to a particle between two single electrodes. No matter how well the electrodes are insulated, there will be a discharge of the potential to ground. To overcome this limitation, R. Wideroe devised a system in Germany in the 1930s which would allow the particles to be accelerated in many small steps adding up to a much greater potential than one giant push [1].

3.1.1. *Principles of operation*

The principle of acceleration used in all accelerators is the fact that a charged particle has its energy changed when it is acted on by an electric field. In the linac, this change in energy is applied by an alternating potential which must be applied in exactly the proper sequence to keep accelerating the particle. In practice, this is achieved with the use of hollow electrodes called drift tubes, which allow the particle to drift at constant velocity within the tube and then be accelerated between the tubes. The particle is accelerated into the tube by an electric field which is opposite in sign to the charge on the particle. As the particle passes through the hollow tube, the phase of the electric field is changed and, at the exit of the tube, the particle is accelerated with a push from the field, which now has the same sign as the particle.

One fact which helps to maintain the timing of the acceleration is what is referred to as phase stability. The potential at each stage of the accelerator can be set so that the maximum potential is applied just after the particles have passed a point. If the particle arrives too early, the potential applied will be slightly less than optimal and the particle will traverse the next section more slowly and will be in phase for the next accelerating potential. This allows for some margin of error in the timing. If the particle is moving too slowly and arrives a little late, the phase of the accelerator can give it a little extra push and again the particle will move to be in phase at the next section of the accelerator.

This effect will result in the particles emerging from the end of the linac to be in bunches. The magnitude of this effect can sometimes be a problem in the design of targetry for the linac, since the instantaneous power delivered to a target can be quite high.

3.1.2. *Radio frequency acceleration*

The radio frequency (RF) power for the current designs of medical linacs is supplied at high frequency (200–500 MHz) which allows the overall length of the linac to be much shorter. The power for the RF system is usually supplied with a bank of power tubes contained within a coaxial cavity. The use of very high frequencies allows the use of shorter sections for the drift tubes, and also allows the linac to be shorter. This is an extremely important factor, since

the particles must traverse the length of one drift tube in one cycle of the RF. This implies that the length of the drift tube must be

$$l = \beta\lambda, \qquad (2)$$

where l is the length of the tube, β is the fraction of the speed of light for the particle and λ is the free space wavelength of the RF.

In order to build a linac with reasonable energies and of reasonable size, a high frequency is essential. Power supplies which can provide high power at high frequency have only recently become available.

3.1.3. *Current linacs*

In the late 1980s the United States Department of Defense supported research and development of new accelerators based on the "Star Wars" technology. There were three funded projects, all of which were of a linear design [12]. The aim was to make use of the technology that could produce a very high density of particle beams of low energy. These new accelerators were to compensate for the low production cross-sections at low energy (< 10 MeV) with increased beam current (100–1000 μA).

While the accelerator technology had advanced to achieve these beam currents, the target technology had not been tried under these severe conditions. Science Applications International Corporation, San Diego, California built an 8 MeV ^{3}He^{++} RFQ accelerator. Its unique features included simplicity in design and operation with a low neutron field from the accelerator [no inherent neutrons from the accelerating particle or the nuclear reactions to be utilized — (^{3}He, ^{4}He) and (^{3}He, p)]. The machine had particle energy of 10 MeV. AccSys Technology Incorporated, Pleasanton, California proposed a linac, also powered by RFQ, but accelerating protons. A variety of energies could be achieved by varying the length of the accelerator (adding on accelerating cavities). Science Research Laboratory Inc., Somerville, Massachusetts proposed a 3–4 MeV tandem cascade accelerator (TCA) that would accelerate deuterons for ^{15}O and ^{13}N production and protons for ^{18}F production.

The TCA is an electrostatic accelerator that starts with negative ions that pass through a charge stripper to convert to positive ions, which doubles the energy for the same potential difference. At the same time Ion Beam Applications,

Louvain-la-Neuve, Belgium built a 3 MeV deuteron⁺ cyclotron dedicated to the production of ¹⁵O. Several of these small cyclotrons have been situated in Europe. Of the "Star Wars" machines, only the TCA was built, installed and operated on a routine basis to produce radioisotopes for PET.

3.2. *Development of the cyclotron*

Cyclotrons are the most commonly used devices for the acceleration of particles to energies sufficient for bringing about the required nuclear reactions. It was the remarkable idea of Lawrence to bend the path of the particles in a linear accelerator into a circle and therefore use the same electrode system over and over again to accelerate the particles. This idea is the basis of all modern cyclotrons and has made the cyclotron the most widely used type of particle accelerator. The first model was built in 1930, with proof of particle acceleration being provided by M. S. Livingston in 1931.

Unfortunately, the literature on cyclotrons for medical purposes is somewhat sparse. The book by Livingood published in 1961 [13] and a more recent review by Scharf [14] are general texts on cyclotrons and other particle accelerators. Detailed information on advances in cyclotrons and other accelerators is available as a series of symposium papers [15]. Cyclotrons for biomedical radionuclide production have been reviewed by Wolf and Jones in 1983 [11].

3.2.1. *Principles of cyclotron operation*

The principle of the cyclotron is based on the application of small accelerating voltages repeatedly. See Fig. 1 for a schematic of the cyclotron's principal components. Hollow cavities called dees (because of their shape) serve as the electrodes for the acceleration. An RF oscillator is connected to the

dees such that the electrical potential on the dees is alternatively positive and negative with respect to each other. By placing the dees between the poles of a strong magnet so that the magnetic field is perpendicular to the plane of the charged particle undergoing acceleration will move in a circular path. As the particle gains energy it moves in a spiral outward from the center. With the source of negative ions at a point in the center of the cyclotron the positive dee will accelerate the ions toward that dee with a magnetic field, forcing them to move in a curved path. Once inside the cavity the particles no longer experience an electric force. Continuing in the circular path, the particles will exit the dee and enter the gap between the dees where the second dee has changed its potential to be an attracting force, accelerating the particles to that dee.

The dees reverse their potential when the particles are inside them, so that at each crossing of the gap the particles receive an increase in energy of the order of 20–50 keV. Lawrence discovered the equations defining this principle of operation in 1929 and built the first cyclotron in 1931.

$$Bev = \frac{mv^2}{r}, \qquad (3)$$

where

$$r = \frac{mv}{Be}.$$

Since angular velocity,

$$\omega = \frac{v}{r}$$
$$\omega = \frac{Be}{m}, \qquad (4)$$

where m is the mass of the ion, e is its charge and v its velocity with B equaling the magnetic field, and r is the radius of the ion's orbit. Thus the orbit of the particle is directly proportional to the particle momentum, and the particle orbit frequency is constant and independent of energy. This principle breaks down under relativistic effects where the mass is not constant.

While the basic components of modern cyclotrons are essentially the same as the original designs (RF cavities, vacuum tank, magnet, ion source, extraction system), there have been some innovations in the last few decades that have had a major impact on the design of the modern cyclotron. The two most significant changes have occurred in getting

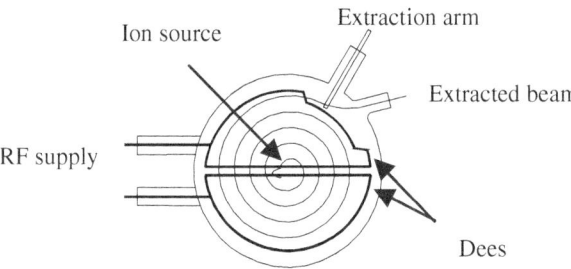

Fig. 1. Schematic of a cyclotron, showing the key components. See text for discussion.

the ions into the cyclotron (ion source) and out of the cyclotron (extraction system).

Nearly all modern cyclotrons now use a negative ion source. Ions are generated by passing the source gas through an electric field that generates negative and positive ions (for example, in the case of H_2, the resulting ions will be H^+ or protons and H^- ions, a proton with two electrons). The advantage of negative ions resides in the ability to easily have a variable energy cyclotron, to have nearly 100% extraction (see below), and to extract multiple beams, simultaneously. The design of the ion source has also changed, in that the ion source can reside inside of the cyclotron, where the ions are generated at the center of the cyclotron (center region) or from outside of the cyclotron (external ion source) and subsequently injected into the center region for acceleration. There are obviously advantages and disadvantages to each approach. With an external ion source the vacuum can be operated at very low pressures with very little beam loss due to stripping of the negative ion by the residual gas. However, the vacuum system must be of a very clean nature to maintain this high vacuum. With an external ion source, maintenance can be performed without opening the cyclotron or breaking the vacuum. In addition, the center region is not disturbed as in the case of the internal ion source that is part of the center region.

The simplicity of the design for proton-only cyclotrons resulted in cyclotrons which accelerate H^- ions capable of two or more simultaneous beams of varying energies and intensities. The modern cyclotron is completely controlled by a computer and is capable of running for many days with minimal attention. The major drawback of these proton cyclotrons lies in the fact that in some cases an enriched target material must be used for sufficient products to be generated.

3.2.2. *Energies and particles*

The energy of the accelerator needed again depends on the demands of the program. However, this increase in the number of radionuclides which may be produced comes with a price both in the equipment expenses and in the infrastructure. In addition, the number of side channel reactions rises as the energy increases and unwanted radionuclides can be produced. This is especially true at energies greater than 30 MeV.

The production of the traditional radioisotopes used in nuclear medicine (^{201}Tl, ^{67}Ga, ^{123}I and ^{111}In) has been via proton reactions for more than 25 years. However, many of the most useful radionuclides can be produced with proton energies below 30–40 MeV. Higher energy cyclotrons (greater than 40 MeV) are usually installed only at large laboratories, government facilities or large commercial facilities where radionuclides are produced for sale. The particular choice of particle(s) and energy will depend on the envisioned program. The selection of a cyclotron will also depend on which radionuclides are needed to prepare the radiopharmaceuticals used in the clinical and research programs, and whether these radioisotopes will be distributed to other locations.

3.3. *Choice of an accelerator*

The question we are left with is: Which type of accelerator is best for my situation? There are some practical considerations when one is making that decision. The main factors are the initial cost, the reliability, the radiological hazards, the operating costs, the installation costs and the available support from the manufacturer. Some of these aspects are outlined in the next sections.

3.3.1. *Comparison between cyclotrons and other accelerators* [1]

There are several aspects to consider when one is choosing an accelerator. Some of the characteristics that may be considered are given in Table 2.

The first consideration with any accelerator is whether or not it is capable of producing sufficient quantities of radionuclides for particular needs of the facility. Regardless of what type of accelerator is installed, it must be kept in mind that the accelerator delivers protons, deuterons or, less commonly, helium-3 and helium-4 ions. Cross-sections of nuclear reactions for production of most radioisotopes are well characterized, and the practical yields as a function of the particle energy are also well known.

The selection of an accelerator is determined in practice by the energy of the particle beam required for the desired nuclear reaction. For example, for a facility wishing to produce only the conventional PET isotopes (^{11}C, ^{13}N, ^{15}O and ^{18}F), the level

Table 2. Comparisons of accelerators for radionuclide production.

Accelerator type	Advantages	Disadvantages
Positive ion cyclotron	Proven record Versatility Ease of maintenance	High cost High activation
Negative ion cyclotron	Extraction efficiency Low activation Beam uniformity	High cost High vacuum requirements Maintenance of stripper foil
Superconducting cyclotron	Compact size Low power	Liquid helium Maintenance
Linac	Stable operation Low power	Targetry Size
Tandem cascade	Low power Low cost	Targetry
Helium-3 linac	Low power Stable operation	Low specific activity Targetry

I cyclotron, which delivers a 10 MeV proton beam, may be sufficient.

On the other hand, a center wishing to manufacture the SPECT isotopes (Tl, ^{67}Ga, ^{123}I, ^{111}In, etc.) must consider a cyclotron capable of delivering a higher energy particle beam (level II or level III).

Of all the various accelerators, cyclotrons are by far the most extensively used for the production of PET and SPECT radioisotopes. It is therefore to be expected that there is a great deal more information concerning the application and reliability of the cyclotron in comparison with the other types of accelerators. In general, cyclotrons have proven to be reliable accelerators and to provide optimum conditions for consistent isotope production. The linacs and Van de Graaff accelerators have also been used for production of isotopes, but not commonly.

Although linacs are also stable machines, they have been used in centers where there are accelerator physicists to address any problems and do not have a history of radioisotope production facilities. As these accelerators are placed, however, a better idea can be obtained as to how reliable they will be.

The tandem cascade accelerator and the He-3 linac RFQ are accelerators which were built and tested, but proved to have very limited application in radionuclide production.

4. Medical Applications

Diagnostic nuclear medicine makes use of the fact that certain radioisotopes emit gamma rays with sufficient energy that the gamma rays can be detected outside of the body.

If these radioisotopes are attached to biologically active molecules, the resulting compounds are called radiopharmaceuticals. They can either localize in certain body tissues or follow a particular biochemical pathway. The following discussion will concentrate on the uses of radioactive substances for the diagnosis or therapeutic treatment of human pathology.

4.1. *Historical background* [2]

Nuclear medicine has its origins in the pioneering work of the Hungarian physician George de Hevesy, who, in 1924, used radioactive isotopes of lead as tracers in bone studies. Shortly thereafter, R. H. Stevens made intravenous injections of radium chloride to study malignant lymphomas.

As indicated above, the first medical cyclotron was installed in 1941 at Washington University, St. Louis, where radioactive isotopes of phosphorus, iron, arsenic and sulfur were produced. With the development of the fission process during the Second World War, most radioisotopes of medical interest began to be produced in nuclear reactors. After the War the wide use of radioactive materials in medicine established the new field of what was then called atomic medicine, which later became known as nuclear medicine. Radioactive carbon, tritium, iodine, iron, and chromium began to be used more widely in the study of disease processes.

Ben Cassen, in 1951, developed the concept of the rectilinear scanner, which opened the way to obtaining in a short amount of time the distribution of radioactivity in a subject. This was followed by production of the first gamma camera by Hal Anger in 1958. The original design was modified in the late 1950s to what is now known as the anger scintillation camera, thus heralding the modern era of gamma cameras, whose principles are still in use today.

Powell Richards developed the ^{99}Mo/^{99m}Tc generator system at Brookhaven National Laboratory in 1957. Technetium-99m produced via this generator system has become the most widely used radionuclide in nuclear medicine today, accounting for as many as 85% of all diagnostic procedures.

The modern era of nuclear medicine has become known as molecular medicine, since it is now possible to translate advances in molecular biology and biochemistry into an understanding of human physiology, and from there into clinical treatment, and the diagnosis of pathology and anatomical abnormalities. The advent of clinical PET for cancer diagnosis makes use of sophisticated tracers to unravel cancer biology.

4.2. Radionuclides for imaging

Nuclear medicine imaging differs from other types of radiological imaging, in that the radiotracers used in nuclear medicine map out the function of an organ system or metabolic pathway and, thus, imaging the concentration of these agents in the body can reveal the integrity of these systems or pathways. This is the basis for the unique information that a nuclear medicine scan (described in Table 3) provides with various scanning procedures for the various organ/functional systems of the body.

Table 4 provides the various low energy production routes along with the half-lives of the radioisotopes. Technetium-99m is included, since this isotope alone accounts for nearly 85% of all nuclear medicine imaging studies. There have been a number of proposals suggesting that ^{99m}Tc could be produced at an accelerator. The economics of producing ^{99m}Tc at an accelerator can never compete with the extremely low costs of producing it at a reactor. While there is concern about the ability to build new reactors and thus jeopardize the availability of this important isotope, and the recent leakage problems with the primary reactors producing Mo-99, there is growing concern about alternatives. Consequently, the prospect for producing Mo-99 or Tc-99m directly via charged particle reactions or from photons needs to be revisited.

Iodine-123 has been of interest for nearly three decades. Its unique chemistry that makes it possible to attach this isotope to a wide variety of molecules and the γ-ray energy (159 keV), which is well matched to SPECT cameras. The ability to produce this isotope in high purity from enriched ^{124}Xe targets made it possible to ship ^{123}I over long distances and still have high SA ^{123}I available for labeling. However, the production costs are still very high

Table 3.   Typical radioisotopes and their uses for imaging.

Radioisotope	Half-life	Uses
Technetium-99m	6 h derived from ^{99}Mo parent 66 h	Used to image the skeleton and heart muscle, in particular; but also for the brain, thyroid, lungs (perfusion and ventilation), liver, spleen, kidneys (structure and filtration rate), gall bladder, bone marrow, salivary and lachrymal glands, heart blood pool, infection and numerous specialist medical studies.
Cobalt-57	272 d	Used as a marker to estimate organ size and for *in vitro* diagnostic kits.
Gallium-67	78 h	Used for tumor imaging and localization of inflammatory lesions (infections).
Indium-111	67 h	Used for specialist diagnostic studies, e.g. brain, infection, and colon transit studies.
Iodine-123	13 h	Increasingly used for diagnosis of thyroid function, it is a gamma emitter without the beta radiation of ^{131}I.
Krypton-81m	13 s from ^{81}Rb 4.6 h	^{81m}Kr gas can yield functional images of pulmonary ventilation, e.g. in asthmatic patients, and for the early diagnosis of lung diseases and function.
Rubidium-82	65 h	Convenient PET agent for myocardial perfusion imaging.
Strontium-92	25 d	Used as the "parent" in a generator to produce ^{82}Rb.
Thallium-201	73 h	Used for diagnosis of coronary artery disease and other heart conditions, such as heart muscle death and for location of low-grade lymphomas.
Carbon-11 Nitrogen-13 Oxygen-15 Fluorine-18	20.4 m 9.97 m 2 m 110 m	These are positron emitters used in PET for studying brain physiology and pathology, particularly for localizing the epileptic focus, and in dementia, psychiatry and neuropharmacology studies. They also have a significant role in cardiology. ^{18}F in FDG has become very important in the detection of cancers and the monitoring of progress in their treatment, using PET.

Table 4.   Routes to production for imaging radionuclides.

Radionuclide	$t^{1/2}$	Reaction	Energy (MeV)
^{99m}Tc	6.0 h	^{100}Mo$(p, 2n)$	30
^{123}I	13.1 h	^{124}Xe$(p, 2n)^{123}$Cs	27
		^{124}Xe$(p, pn)^{123}$Xe	
		^{124}Xe$(p, 2pn)^{123}$I	
		^{123}Te$(p, n)^{123}$I	15
		^{124}Te$(p, 2n)^{123}$I	25
^{201}Tl	73.1 h	^{203}Tl$(p, 3n)^{201}$Pb $\rightarrow ^{201}$Tl	29
^{11}C	20.3 m	^{14}N(p, α)	11–19
		^{11}B(p, n)	10
^{18}F	110 m	^{18}O(p, n)	15
		^{20}Ne(d,α)	14
		^{nat}Ne(p, X)	40
^{64}Cu	12.7 h	^{64}Ni(p, n)	15
		^{68}Zn$(p, \alpha n)$	30
		^{nat}Zn$(d, \alpha x n)$	19
		^{nat}Zn$(d, 2pxn)$	19
^{124}I	4.14 d	^{124}Te(p, n)	13
		^{125}Te$(p, 2n)$	25

in comparison with other radioisotopes, which will limit its use for the foreseeable future. While ^{123}I can be produced for local use via the ^{123}Te(p, n) or ^{124}Te$(p, 2n)$ reactions, the coproduction of ^{124,125}I limits the product's shelf-life.

Thallium-201 was developed at Brookhaven National Laboratory, where it was shown that ^{201}Tl could be used as a tracer for detecting myocardial perfusion. Thallous chloride labeled with ^{201}Tl has been used extensively for more than 30 years and remains the gold standard for measuring blood flow. Over this period there have been numerous reports of its demise due to the availability of ^{99m}Tc-labeled alternatives, yet the growth in demand for this isotope is still upward, especially during shortages of Tc-99m.

The remaining isotopes listed are used in PET imaging. Carbon-11 is extremely attractive because, in principle, one can replace an existing carbon atom in the molecule of interest with the radioactive isotope without altering the biochemistry of the molecule. However, because of the short half-life, its availability will be limited to those sites possessing an accelerator or which are in the vicinity of one.

The demand for ^{18}F exceeds its availability. To overcome this shortage, a number of central distribution centers have been placed in large metropolitan areas in North America, Europe and Asia. Although

several nuclear reactions are possible, the (p, n) reaction is the choice for producing large quantities of ^{18}F. If the availability of ^{18}F continues to grow, ^{18}F-labeled compounds may begin to compete with SPECT agents such as ^{123}I.

The two other isotopes (^{64}Cu and ^{124}I) are candidates for both PET imaging and possible use in therapy (see below). The interest in these two is primarily related to the relatively long half-lives. Such properties would enable studies to be performed where the kinetics are slow and exceed the ability to image with ^{18}F. The disadvantages include the low production rate (^{124}I) and the need for expensive enriched target material (^{64}Ni, ^{124}Te). Results from Washington University in St. Louis have shown that even with the high energy β^+ particles associated with ^{124}I decay and other photons in coincidence with the β^+ decay, they can still be imaged at high resolution (^{64}Cu) [16, 17].

While there is a wide range of radionuclides that are used in imaging, a relatively small number make up the vast majority of all studies in SPECT and PET imaging.

Table 4 lists the most widely used radionuclides for imaging, along with a couple of potentially useful radionuclides.

PET imaging has been in use for several decades for human brain and whole body imaging, first only as a research tool, now gaining acceptance as a diagnostic imaging modality in selected applications such as oncology and, very recently, as an aid in the diagnosis of Alzheimer's disease. All of these advances have been made possible through the improvement in resolution and sensitivity of the scanners but, more importantly, through the development of more specific tracers.

4.3. *Radionuclides for therapy*

The idea of a radioisotope used in therapy is based on the desire to link a radionuclide which has a high linear energy transfer (LET) associated with its decay products, such as Auger electrons, β particles or α particles, to a biologically active molecule that can be directed to a tumor site. Since the β^--emitting radionuclides are neutron-rich, they have, in general, been produced in reactors. Table 5 lists some of the radionuclides that have been proposed as possible radiotoxic agents and their routes of production via charged particle reactions.

Table 5. Charged particle production routes and decay modes for selected therapy isotopes.

Radionuclide	$t^{1/2}$	Decay mode	Reaction	Energy (MeV)
^{77}Br	2.4 d	Auger electrons	^{75}As$(a, 2n)$	27
			^{77}Se(p, n)	13
			^{78}Se$(p, 2n)$	24
			^{79,81}Br$(p, xn)^{77}$Kr	45
			^{nat}Mo$(p,$ spall.$)$	> 200
^{103}Pd	17.5 d	Auger electrons	^{103}Rh(p, n)	19
			^{nat}Ag(p, xn)	> 70
^{186}Re	90.6 h	β^-	^{186}W(p, n)	18
			^{186}W$(d, 2n)$	20
			^{197}Au$(p,$ spall.$)$	> 200
			^{nat}Au$(p,$ spall.$)$	> 200
			^{nat}Ir$(p,$ spall.$)$	> 200
^{211}At	7.2h	α	^{209}Bi$(a, 2n)$	28
			^{209}Bi$(7$Li$, 5n)^{211}$Rn	60
			^{232}Th$(p,$ spall.$)^{211}$Rn	> 200

Table 6. Examples of generator systems available today.

Generator	Parent $t_{1/2}$	Daughter $t_{1/2}$	Uses
^{99}Mo/^{99m}Tc	66 h	6 h	Tc-99m is the most widely used radionuclide in nuclear medicine, single photon emitter.
^{68}Ge/^{68}Ga	270 d	68 m	In equilibrium as a long-lived positron source; Ga metal chemistry
^{82}Sr/^{82}Rb	25.5 d	76.4 s	Cardiac blood flow
^{81}Rb/^{81m}Kr	4.58 h	13 s	Lung ventilation studies
^{225}Ac/^{213}Bi	10.0 d	45.6 m	Radionuclide therapy (α particles)
^{62}Zn/^{62}Cu	9.26 h	9.7 m	Blood flow, hypoxia

The attractive feature of ^{77}Br is its chemical versatility in addition to its half-life. Production rates are relatively low and purity may be an issue since ^{76}Br is often coproduced. The demand for ^{103}Pd, which is used in treating prostate cancer, is continuing to grow. A large number of low energy (19 MeV) cyclotrons are dedicated solely to the production of this isotope.

Rhenium-186 is attractive for a number of reasons. Besides the decay characteristics, rhenium is in the same chemical family as technetium, and thus much of the extensive chemistry developed for technetium can be applied to rhenium.

The production rates from all of the reactions for this radioisotope listed in Table 6 are very low. Thus, the only practical route to this potentially important isotope is via neutron capture in a reactor. This route results in a very low SA product, which severely limits its utility.

And, finally, the α-emitting isotopes have been of interest for use in therapy because of the high LET associated with the α decay.

Astatine is of interest because it possesses many properties of halogens and each decay of ^{211}At has an α particle associated with it. Because of its short half-life, multiple production sites would be required for practical applications. Thus, the interest in producing its parent radionuclide (^{211}Rn, $t_{1/2} = 14.6$ h) has been suggested as a way of producing and shipping ^{211}At to remote sites.

4.4. Radioisotope production rates and yield considerations

The rate of radionuclide production is dependent on a number of factors, including the magnitude of the reaction cross-section as a function of energy, the incident particle energy, the thickness of the target in nuclei per cm^2 which will determine the exit particle energy, and the flux (related to the beam current) of incoming particles. The rate of production is given by

$$-\frac{dn}{dt} = R = nI(1 - e^{-\lambda t}) \int_{E_s}^{E_0} \frac{\sigma(E)}{dE/dx} dE, \quad (5)$$

where:

- R is the number of nuclei formed per second;
- n is the target thickness in nuclei per cm^2;
- I is the incident particle flux per second and is related to the beam current;
- λ is the decay constant and is equal to $\ln 2/t_{1/2}$;
- t is the irradiation time in seconds;
- σ is the reaction cross-section, or the probability of interaction, expressed in cm^2, and is a function of energy;
- E is the energy of the incident particles;
- x is the distance traveled by the particle in the target;
- $\int_{E_s}^{E_0}$ is the integral from the initial to the final energy of the incident particle along its path.

It is of historical interest to note that the unit for the cross-section is the barn, which is equivalent to 10^{-24} cm^2. The word barn comes from the fact that the probability for a neutron to interact with a target is proportional to the area of the nucleus, which, compared to the size of the neutron, appeared *as big as a barn*.

For routine production of radioisotopes, the practical yield can be quite different from the saturation yield of a radionuclide usually found in the literature. The rate of production is, of course, affected by the fact that the resulting nuclide is radioactive and thus undergoes radioactive decay. For short-lived nuclides the competing reaction rates, production and decay will achieve equilibrium at sufficiently long bombardment times since the rate of decay is proportional to the number of radionuclei present. The point where equilibrium is reached is called saturation. This means that there is no benefit to longer irradiations, as the production rate equals the rate of decay, and therefore no additional product will be formed. At shorter irradiation times the fraction of the product that is yielded is related to the saturation factor given by $(1 - e^{-\lambda t})$, where λ is the decay constant of the decaying nuclide and t is the bombardment time. It is evident that an irradiation equivalent to one half-life would result in a saturation factor of 50%. For practical reasons, an irradiation rarely exceeds three half-lives (90% saturation) except for the shortest-lived radionuclides.

For long-lived species, the quantity produced is usually expressed in terms of the integrated dose or total beam flux (μA-h). For example, with a long-lived radionuclide such as ^{82}Sr ($t_{1/2} = 25$ d) the amount produced will be essentially the same whether it is produced from 100μA in 1 h or 50μA in 2 h (both represent 100μA-h of the beam).

The chemical form of the radionuclide is also of major interest in considering the attributes of any machine. The target material must withstand the intensely ionizing particle beam, and must also be able to withstand the intense radiation field accompanying the particle bombardment. The question of the chemical form of the radioisotope coming from the target, which target material to use and the specific bombardment conditions have been a matter of research since the early 1950s. Target conditions can be manipulated to some extent to provide the desired precursors for syntheses of labeled compounds directly from the target.

Another aspect relating to chemistry is the choice between proton-only machines and two-particle (proton, deuteron) machines. If a proton-only machine is chosen, enriched isotopes are needed to produce some of the four PET radionuclides. If a dual particle machine is chosen, the production of ^{15}O and ^{13}N can be done with natural abundance target materials. Other radioisotopes may require enriched isotope targets with either protons or deuterons.

Cyclotrons employed for producing medical radionuclides were initially designed for physics experiments and used only part-time for medical applications. These cyclotrons were capable of accelerating protons, deuterons, ^{3}He^{+2} and α particles (the nucleus of ^{4}He). As can be seen from Table 4, however, the PET radionuclides are produced from either proton or deuteron reactions. In the early 1980s, small compact proton-only cyclotrons became available and cyclotrons specifically designed for producing PET radionuclides were installed in many hospitals.

One of the major drawbacks to the widespread availability of PET is the high capital cost associated with the cyclotrons and scanners. However, the success of the small low energy cyclotron encouraged research into the design of even lower energy accelerators, i.e. linear accelerators and cyclotrons of a-few-MeV extracted energy. To date, there are very few of these machines in routine use.

4.5. *Generators*

Finally, the other source of radionuclides used in medicine is the generator. A radioactive generator takes advantage of the cases where one longer-lived (parent) radionuclide decays, usually by β^- emission, to a shorter-lived (daughter) radionuclide. The chemical differences in the two elements are exploited to separate the daughter product from the parent. The parent radionuclide is produced by one of the methods described above and then attached to an inert substance, from which the desired product can be eluted or washed off the support. The product can be used directly, as in the case of ^{82}Rb$^+$ from the Sr/Rb generator, or after undergoing a chemical reaction, as in the case of ^{99m}Tc from the Mo/Tc generator (see Table 6).

The equilibrium equations that reflect the relative radioactivity of parent and daughter are given by the general equation

$$A_d(t) = A_p(0)\left[\frac{(\lambda_d)(e^{-\lambda_p t} - e^{-\lambda_d t})}{\lambda_d - \lambda_p}\right] + A_d(0)e^{-\lambda_d t},$$

$$(6)$$

where A is the radioactivity of the daughter (d) and parent (p) respectively. A_d is equal to the product of the decay constant, λ_d, and the number of radioactive nuclei, N, present ($\lambda_d N$). The first term accounts for the growth of the daughter as a function of the decay of the parent as well as the disappearance of the daughter due to its own decay. The last term accounts for the presence of daughter nuclei at zero time.

All generator systems used routinely in nuclear medicine form an equilibrium between parent and daughter radionuclei. In the case of the ^{99}Mo/^{99m}Tc generator, the parent (^{99}Mo) decays at a rate sufficiently similar to that of the daughter (^{99m}Tc). With a half-life of 66 h for ^{99}Mo vs. 6 h for ^{99m}Tc, there is an appreciable decay of the parent before the daughter reaches the steady state. This steady state condition is referred to as transient equilibrium. With transient equilibrium, the daughter radioactivity grows in and surpasses that of the parent before equilibrium is reached. The ratio of the daughter radioactivity to that of the parent is given by the equation

$$\frac{A_d}{A_p} = \frac{T_p}{T_p - T_d},$$

$$(7)$$

where T is the half-life for each species, respectively.

For the situation where the parent has a half-life much longer than the daughter, e.g. ^{68}Ge/^{68}Ga and ^{82}Sr/^{82}Rb, the change in the amount of the parent during the time for the steady state to be reached will be negligible; the steady state condition is referred to as secular equilibrium. The quantity of daughter activity at any time is then expressed by the equation

$$A_d(t) = A_d(0)(1 - e^{-\lambda_d t}).$$

$$(8)$$

Thus, in secular equilibrium, when $e^{-\lambda t} \approx 0$, the daughter and parent radioactivity are approximately equal.

From Table 6, it is easy to see that generators have a wide variety of uses and half-lives of both parent and daughter nuclides.

Obviously, from an end user perspective, the long-lived parent makes it possible to have a single generator in use for an extended period of time.

The utility of the generator is actually based primarily on the daughter's half-life and the chemistry required to provide the radionuclide in a useful species. The simplest systems make use of the daughter nuclide directly; ^{82}Rb$^+$ and ^{81m}Kr are used directly as a K$^+$ ion analog and as an inert gas ventilation tracer, respectively.

The parent radionuclides are or can be all produced in accelerators. Ac-225 is extracted from the decay chain of Th-232 but efforts are underway to produce this radionuclide by irradiation of a radium target.

Of course, the most widely used generator system is the ^{99}Mo/^{99m}Tc pair, where over 80% of all nuclear medicine procedures performed worldwide use Tc-99m as the imaging radionuclide. There are numerous Tc-99m kits for producing tracers to examine the brain, kidney, heart, bone, liver, lung, red blood cells and TcO$_4^-$ for the thyroid. Based on the successful use of Tc-99m in radiopharmaceuticals for diagnostic purposes, similar tracers are being developed based on the ^{68}Ge/^{68}Ga generator system, where the parent Ge-68 is produced in a cyclotron.

5. Imaging

While a detailed discussion on imaging is beyond the scope of this review, it is worth pointing out a few of the specifics in order to place the choice of radionuclide into context.

5.1. *Planar imaging*

By far the most common imaging device in nuclear medicine is the planar camera or the Anger camera. The basic components of the camera include a thin crystal of NaI scintillator coupled to a cluster of photomultiplier tubes (PMTs), an X, Y positioning circuit and a readout device that may be an oscilloscope or a photographic film. The NaI scintillator design minimizes multiple interactions with the incident γ-rays so that the position of interaction can be determined with great accuracy. Typical scintillation cameras have detectors 25–45 cm in diameter and 0.64–1.27 cm in thickness. The thickness is determined to match the photopeak for low energy photons emanating from Tc-99m.

The X, Y positioning circuit relies on the light output from the many (19–91) PMTs mounted on the back of the scintillator. The PMT located nearest to the γ-ray interaction will receive the maximum amount of light, and the other PMTs will receive light in proportion to the solid angle subtended by the tube at the point of interaction. The positioning circuit sums the output of the PMTs and produces X and Y pulses proportional to X and Y coordinates of the γ-ray interaction.

In between the radioactive source and the detector is a collimator constructed of dense metal such as lead or tungsten. The collimator has one or more holes drilled through it to allow the passage of γ-rays. Since γ-rays cannot be bent or focused, the collimator's function is to absorb those γ-rays that do not pass through the openings.

In its simplest form the collimator has a single pin hole and acts like a camera lens. Other collimators are constructed with the holes converging, diverging or parallel to the imaginary line connecting the object to be imaged and the camera face. The converging collimator has the effect of magnifying the imaged object, while the diverging collimator minifies the object. The parallel collimator is used for high resolution. Regardless of which collimator is used they all absorb a large fraction of the photons emitted by the radiotracer in the patient.

5.2. *Single photon emission computed tomography*

Single photon emission computed tomography (SPECT) acquires views of the emitted photons from many different angles and reprojects these views to reconstruct the image or distribution of radioactivity in the object or patient. In SPECT, the radiopharmaceuticals used contain radionuclides such as Tc-99m that emit single photons (ones that are not in timed coincidence with one another). Directional information is achieved by collimating the photons incident on the detector of the Anger camera. The collimator thus reduces the sensitivity of the camera because all of the photons not parallel to the holes in the collimator are prevented from reaching the detector surface.

Since the reconstructed image contains the three-dimensional information on the distribution of radioactivity, SPECT also has the potential for quantification. The factors in this capability are similar to those in PET, e.g. system sensitivity, dead time, spatial resolution, sampling interval, reconstruction filters and the size of the object being imaged. Also, as

in PET, the photons emerging from the subject are attenuated by the amount of matter between their origin and the detector, and, of course, they have a definite probability of being scattered along their path.

Because of the inherently lower energies (100–150 keV) of the photons emitted by radionuclides used in SPECT, the effect of attenuation can be quite dramatic, with reductions as great as a factor of 5 or more. Thus, with single photon emitters, it is difficult to determine whether data reflect a weak source near the surface or a stronger source located at a greater depth. In addition, the amount of scatter is strongly dependent on the energy of the photons, with the photons from ^{201}Tl having a scatter fraction of as much as 40–50%, depending on the depth of the source. It is for these reasons that attenuation and scatter are the most significant and difficult nonlinear effects to correct for.

Whereas scintillation camera images show the distribution of the radiopharmaceutical in defined regions in the planar view, they suffer from the superimposition of organs and background contributions to the areas of interest. It is because of these shortcomings in planar imaging that SPECT has a major role to play in diagnostic imaging regardless of whether SPECT can achieve the difficult task of providing quantitative information. The ability to view the distribution in three dimensions greatly affects the interpretation of the images.

5.3. *Positron emission tomography*

PET imaging makes use of the self-collimating nature of positron decay, as two nearly collinear photons are utilized to define the location of an annihilation event. PET cameras are typically made up of a ring of detectors that are in timed coincidence (resolving time of a few nanoseconds), allowing a line of response to define the cord along which the positron was annihilated (the location of the emission is not known because of the short distance the positron travels before annihilation). By mathematically back-projecting the lines of response, a density map can be generated that reflects the distribution of the positron emitter.

There are several physical limitations inherent in PET technology. Firstly, as the emitted positron has kinetic energy, varying from a few hundred keV to several MeV, depending upon the radionuclide, it will travel a few millimeters to centimeters before annihilating with an atomic electron. As such, the site of annihilation is not the site of emission, thus resulting in a limitation when one is defining the origin of the decay. Another limitation is the fact that the positron–electron pair are not at rest when the annihilation occurs, and thus by conservation of momentum the two photons are not exactly collinear. Although the lack of colinearity becomes increasingly important with greater detector separation, this effect is ignored, for the most part, in existing tomographs because the detector ring diameter is less than a meter, at which distance the deviation from 180° is a fraction of a millimeter.

One of the major strengths that PET has over SPECT is the ability to measure, directly, the attenuation effect of the object being viewed. This is the result of requiring that both photons be detected. Thus, if one photon of the pair is not observed then there is no line of response. Along the path to the detectors, one or both photons (511 keV each, the rest mass of the electron) can undergo absorption by the photoelectric effect or Compton scattering when interacting with surrounding material. Thus, in order to be detected as an event, both photons must be detected in temporal coincidence. By using an external source of the positron emitter, the attenuating (absorbing) extent of the object to be measured can be determined. However, that advantage has been eliminated, since all commercial PET (and many SPECT) cameras are now built with a CT scanner (x-ray tomography) so that a merged image of structure and function can be obtained. In addition, as the CT image is a measure of electron density, it is used to calculate the necessary coefficients for attenuation correction. However, the calculated attenuation coefficients are difficult to perform in the thorax. Nevertheless, the use of the CT image is standard for attenuation corrections now, although its primary function is to provide a detailed view of the section of the body under investigation. Figure illustrates the power of this approach.

Once the attenuation of the object is measured and the radiotracer is injected, the temporal and spatial distribution of the tracer may be determined. However, to make a quantitative estimate of the distribution, other corrections are required. First of all, for true quantitative extraction of information the detector system must be normalized to account for the nonuniform response of the detector system. This is achieved by placing a cylindrical flood phantom of known tracer concentration in the field of view and measuring the responses of all detector pairs.

Other corrections needed are to account for scattered photons, which for modern systems can be anywhere from 30 to 50% of the events. The amount of scatter can be reduced by selecting a narrow energy window of acceptance so as to eliminate large angle scatter (large angle scatter results in lower energy of the scattered photon). This will, however, reduce the efficiency. The remaining scatter profile is removed by analytical techniques, discussion of which is beyond this article.

Finally, there are random coincidences that must be subtracted. Because of the finite timing window for defining the coincidences, there is the possibility of unrelated events arriving within the timing window. The amount of random events is related to the size of the timing window and the number of events in any one detector. Random events can be reduced by using fast detectors and electronics which enable a short timing window to be employed. Randoms are usually estimated by monitoring the single event rate and subtracting globally from the image.

6. Functional Imaging

Functional imaging using PET started as a research tool in neuroscience in the late 1970s and still remains a major research tool for the current day neurosciences. However, its major impact recently has been in the diagnosis of cancer. While simple tracer molecules such as water, carbon monoxide and carbon dioxide had been used for many years, the first complex molecule to be used extensively was the glucose analog, ^{18}F-fluorodeoxyglucose (FDG), developed at Brookhaven National Laboratory in collaboration with researchers at the National Institutes of Health in the US and the University of Pennsylvania around 1975. Since the human brain uses glucose as its primary energy source, the availability of the tracer led to groundbreaking work for studying the human brain in health and disease. This effort was driven by the success at using C-14-labeled deoxyglucose developed at the NIH by Louis Sokolov in the 1960s. Since C-14 is not imageable *in vivo*, the effort went into developing a labeled analog that could be shipped from a cyclotron facility (BNL in this case) to the PET camera (the University of Pennsylvania) [2].

Today, many more tracers are used to investigate the various neuronal systems probing both the presynaptic and postsynaptic pathways. Several hundred tracers have been prepared and tested for the utility in investigating various enzymatic and receptor systems, while only a handful are routinely used. There are tracers specifically designed to monitor cell proliferation, the hypoxic nature of cells, and cell apoptosis.

Because diagnostic imaging is driven by a digital approach (present/absent, yes/no), the desire to have uncluttered images resulting from PET is strong. Nevertheless, the true power of PET is its ability to track the distribution of a tracer over time, and extract detailed kinetic data as in a physical chemistry experiment where rate constants are determined. So the conflict between using the technology for clinical diagnosis and using PET as an *in vivo* biochemistry tool will not be easily resolved, nor should it be.

With the advances in the technology enabling increasingly better resolution, it has become possible to build PET scanners capable of imaging small animals. The pharmaceutical industry has recognized the power of using such small animal PET scanners as a screening tool for their preclinical research. PET can be used as a surrogate to monitor changes in metabolism or receptor occupation or by labeling the drug directly and determining the distribution and time course of the compound, *in vivo*. One of the strengths of PET in this regard is that animals can be used many times so that they can serve as their own controls and changes due to interventions monitored. Such an approach increases the statistical power of the study.

Pharmaceutical companies also recognize that human PET scanning can be used as surrogates for monitoring the therapeutic efficacy of drugs in phases II and III drug trials. By performing base line scans and scans at intervals following intervention, the PET data can often reveal biochemical changes much sooner than the clinical signs — thus shortening the assessment time. Most often, surrogate markers are used to monitor a particular functional change.

As the physical limit of detection are approached, the remaining avenue is to increase signal-to-noise by utilizing tracers that are uniquely suited to imaging the function in question, and otherwise clears rapidly from surrounding tissue. To this end, the development of more specific tracers is believed to be the most critical component of PET.

7. Radiotracer and Chemistry Development

Tracer development is an extremely important component of PET and SPECT imaging. The scanner measures only radioactive decays and cannot by itself identify a biological process of interest. This is accomplished by careful radiotracer design and development to make it as specific as possible for the relevant biological sites and processes, while minimizing its binding to other tissue types [18, 19]. As the imaging instrumentation becomes more powerful, there is an increasing demand for new tracers as more sites and processes become potentially observable *in vivo*. In addition to undergoing *in vitro* validation, however, the new tracers must undergo a rigorous validation of their behavior *in vivo* and, where necessary, new imaging protocols and analysis methods must be developed. Presently there are a number of small molecules that have been used in human scanning for years.

7.1. *Radiopharmaceuticals*

The term "radiopharmaceutical" is derived from the fact that a radionuclide has been attached to a biologically active compound as opposed to using radiotracers that are elements, or their analogs, found in the body and suitable for human use. Radiopharmaceuticals differ in one major aspect from regular pharmaceuticals: they are given in such small concentrations that they do not elicit any pharmacological response. Because of this there have been a number of attempts to change the name used to describe these substances, such as "radiotracers." Present day radiopharmaceuticals are used for diagnostic purposes in about 95% of the cases, and the remainder are used in therapy. However, the use of radiopharmaceuticals in therapy is seen as the next major area of growth in the use of radionuclides.

In order for a radiotracer (radiopharmaceutical) to be used in humans safely, it must meet the quality standards, which include chemical and radiochemical purity, that it be sterile and free from pyrogenic material.

The ideal diagnostic radiopharmaceutical for imaging should:

(1) Be readily available at a low cost.
(2) Be a pure gamma emitter, i.e. no particle emission such as α and β. These particles contribute a radiation dose to the patient while not providing any diagnostic information (see the section on dosimetry). This is, of course, not followed with PET.
(3) Have a short effective half-life, so that it is eliminated from the body as quickly as possible.
(4) Have a high target-to-nontarget ratio, so that the resulting image has a high contrast, i.e. the background does not blur the image.
(5) Possess the proper metabolic activity, in that it follows or is trapped in the metabolic process of interest.

The ability to measure regional biochemical function requires a careful design process with these principles in mind. However, in reality it is not possible to meet all of these criteria. For example, all decay processes involve the emission of particles, as in the case of the pure γ-emitters which have Auger electrons emitted during some fraction of the decays. Thus, it is necessary to address the following steps [20] in the development of a biochemical probe:

(1) Develop a radiotracer that binds preferentially to a specific site;
(2) Determine the sensitivity of the radiotracer to a change in biochemistry;
(3) Find a biochemical change as a function of a specific disease that matches that sensitivity.

A large number of radiotracers have been synthesized to probe metabolic turnover such as oxygen consumption, glucose utilization and amino acid synthesis. Enzymatic activity, neurotransmission, receptor density and occupancy have all been measured via appropriately designed radiotracers. It should be pointed out that the development of radiotracers for PET fundamentally violates rule No. 2 for the ideal tracer because PET radionuclides, by nature, emit β^+ particles. However, the resulting coincident γ-rays from the β^+ annihilation form the basis for the technique.

In addition to consideration of the above principles, the synthetic chemist must plan how to insert the radionuclide into the molecule at a point in the synthetic process where there is minimal handling, yet late enough in the synthesis to minimize loss due to chemical yield and radioactive decay. For these reasons the preparation of radiopharmaceuticals requires planning and techniques not encountered by traditional synthetic chemistry.

The development and use of PET tracers can be viewed as covering two major areas: (1) tracers that can be used as surrogate markers for biological processes and (2) tracers that are specific to a particular process, whether it is intended to measure enzyme activity or receptor concentration or the expression protein synthesis. A major hindrance to tracer development is the complex nature of the synthesis process itself. While major strides have been made to simplify the synthesis steps, there are still areas in need of improvement, such as miniaturization of the synthesis instrumentation. Miniaturization provides the opportunity to use small amounts of starting materials and radioactivity that would make the purification simpler and easier. Simple solid phase columns could be used instead of cumbersome high performance liquid chromatography. In addition, if the miniaturization can be realized it is conceivable that multiple compounds could be prepared in parallel for testing with a single supply of radionuclides. This can be viewed as the radiochemist's attempt at screening compounds.

8. Future Directions

In order for PET to develop, the availability of F-18 and other positron-emitting radionuclides must be secured. Several commercial networks have been established throughout North America and Europe and parts of Asia. They have concentrated on supplying F-18, although some suppliers have made I-124 and Cu-64 available in small quantities.

The need for C-11 will probably remain of interest primarily to the research community and drug developers. The short half-life of C-11 makes it almost impossible to transport it over long distances; thus, if C-11 is to have an impact on clinical care, a small inexpensive accelerator will be required, as will modules for the rapid synthesis of C-11-labeled agents. As indicated above, such machines are being explored.

With superconducting technology reaching maturation, a number of areas in accelerator development will become possible, including building very small cyclotrons and using electron linear accelerators for radionuclide production.

The challenges of building new reactors for radionuclide production will force alternative routes to be explored using accelerators, especially with respect to Mo-99 and Tc-99m. Recent papers indicate that local production and distribution of Tc-99m may be possible on a limited basis [21, 22]. The availability of powerful e-linacs may indeed be a solution through the photofission of U-238 if the ability to handle megawatt power in the converter and targets can be addressed. The interesting aspect of the photofission approach is that the distribution of radionuclides is almost identical to the thermal neutron fission of U-235. The challenge is that the yields are several orders of magnitude lower for photons than for neutrons.

References

[1] D. J. Shyler and T. J Ruth, Resource manual on cyclotron and production of radioisotopes. IAEA Tech. Doc. (2006).

[2] T. J. Ruth, *Rep. Prog. Phys.* **71** (2008).

[3] E. Rutherford, *The London, Edinburgh and Dublin Philosophical Magazine and Journal of Science* **37**, 581 (1919).

[4] E. O. Lawrence and N. E. Edlefsen, *Science* **72**, 376 (1930).

[5] E. O. Lawrence and M. S. Livingston, *Phys. Rev.* **38**, 834 (1931).

[6] J. H. Lawrence, *Radiology* **35**, 51 (1940).

[7] J. L. Heilbron, R. W. Seidel and B. R. Wheaton (1981) *Lawrence and His Laboratory: A Historian's View of the Lawrence Years* (LBL newsmagazine 1981 publication, Web publication, 1996, W. E. Johnston, M. Wooldridge, J. Kahn, B. Frane, M. Thompson and R. Kolb).

[8] M. M. Ter-Pogossian and H. N. Wagner Jr., *Nucleonics* **24**, 50 (1966).

[9] A. P. Wolf (1977), Cyclotrons for biomedical radioisotope production, in *Medical Radionuclide Imaging*, Vol. I (IAEE Symposium, Los Angles, California; Oct. 25–29, 1976) (Vienna, 1977), pp. 343–353.

[10] A. P. Wolf, *Ann. Neurol.* **15**(Suppl.), S19 (1984).

[11] A. P. Wolf and W. B. Jones, *Radiochim. Acta.* **34**, 1 (1983).

[12] T. J. Ruth, B. D. Pate, R. Robertson and J. K. Porter, *Nucl. Med. Biol.* **16**, 323 (1989).

[13] J. J. Livingood, *Principles of Cyclic Particle Accelerators* (D. Van Norstrand, New York, 1961).

[14] W. Scharf, *Biomedical Particle Accelerators* (AIP Press, New York, 1994).

[15] J. L. Duggan and I. L. Morgan, *Nucl. Instrum. Methods B* **56**, 57 (1990). See also prior years 1988, 1986, 1984.

[16] D. W. McCarthy, R. F. Shefer, R. E. Klinkowstein, L. A. Bass, W. H. Margenau, C. S. Cutler, C. J. Anderson and M. J. Welch, *J. Nucl. Med. Biol.* **24**, 35 (1997).

[17] M. R. Lewis, M. Wang, D. B. Axworthy, L. J. Theodore, R. W. Mallet, A. R. Fritzberg, M. J. Welch and C. J. Anderson, *J. Nucl. Med.* **44**, 1284 (2003).

[18] K. Kawamura, P. H. Elsinga, T. Kobayashi, S. Ishii, W. F. Wang, K. Matsuno, W. Vaalburg and K. Ishiwata, *Nucl. Med. Biol.* **30**, 273 (2003).

[19] S. M. Okarvi, *Eur. J. Nucl. Med.* **28**, 929 (2001).

[20] W. C. Eckelman, *Int. J. Rad. Appl. Instrum. B* **18**, iii (1991).

[21] S. Takács, Z. Szücs, F. Tárkányi, A. Hermanne and M. Sonck, *Radioanal. Nucl. Chem.* **257**(1), 195 (2003).

[22] S. Takács, F. Tárkányi, M. Sonck and A. Hermanne, *Nucl. Instrum. Methods Phys. Res. B* **198**, 183 (2002).

Thomas J. Ruth is a Senior Research Scientist at TRIUMF and Senior Scientist at the British Columbia Cancer Research Centre. He is a leader in the production and application of radioisotopes for research in the physical and biological sciences. He has served on a multitude of committees, including the Institute of Medicine's Committee on Medical Isotopes and on the NAS's Committee on the State of the Science in Nuclear Medicine and the panel for the Production of Medical isotopes without Highly Enriched Uranium. He serves as an expert on radioisotope production for the IAEA. Most recently he served on the Subcommittee of the NSAC on Isotopes for the Nuclear Physics Program of the DOE. He has published more than 250 peer reviewed papers and book chapters. Dr. Ruth received his Ph.D. in nuclear spectroscopy from Clark University.

Reviews of Accelerator Science and Technology
Vol. 2 (2009) 35–62
© World Scientific Publishing Company

Proton Radiation Therapy in the Hospital Environment: Conception, Development, and Operation of the Initial Hospital-Based Facility

James M. Slater

Department of Radiation Medicine and
Radiation Research Laboratories,
Loma Linda University Medical Center,
Loma Linda, CA, USA
jmslater@dominion.llumc.edu

Jerry D. Slater

Department of Radiation Medicine,
Loma Linda University Medical Center, Loma Linda, CA, USA
jslater@optivus.com

Andrew J. Wroe

Department of Radiation Medicine and
James M. Slater, M.D. Proton Treatment and Research Center,
Loma Linda University Medical Center, Loma Linda, CA, USA
awroe@dominion.llumc.edu

The world's first hospital-based proton treatment center opened at Loma Linda University Medical Center in 1990, following two decades of development. Patients' needs were the driving force behind its conception, development, and execution; the primary needs were delivery of effective conformal doses of ionizing radiation and avoidance of normal tissue to the maximum extent possible. The facility includes a proton synchrotron and delivery system developed in collaboration with physicists and engineers at Fermi National Accelerator Laboratory and from other high-energy-physics laboratories worldwide. The system, operated and maintained by Loma Linda personnel, was designed to be safe, reliable, flexible in utilization, efficient in use, and upgradeable to meet demands of changing patient needs and advances in technology. Since the facility opened, nearly 14,000 adults and children have been treated for a wide range of cancers and other diseases. Ongoing research is expanding the applications of proton therapy, while reducing costs.

Keywords: Proton; synchrotron; accelerator; hospital-based; treatment center; radiotherapy.

1. Introduction

Nearly two decades have passed since the Loma Linda University Medical Center (LLUMC) proton treatment facility — the first hospital-based proton center in the world — began clinical operations. Nearly 40 years have passed since Loma Linda University (LLU) investigators began their work toward realizing a heavy-charged-particle delivery system for a hospital environment. Then, in 1970, treatments with protons and other heavy charged particles had been offered to patients for about 15 years, but always in facilities based in physics laboratories and conceded to be investigational in nature. As the first decade of the 21st century comes to a close, five hospital-based proton treatment facilities are operational in the United States, others are under

construction, and several other proton and heavy-ion treatment centers exist on other continents [1]. After several years of skepticism, proton radiation is now becoming regarded as a routine standard treatment option for several cancers and benign tumors, and is being investigated as a therapeutic option for some benign conditions, such as epilepsy.

Scientists were the primary early enthusiasts, most probably because of their clear understanding of the characteristics of photon beams as compared to those of protons and other heavy charged particles, in terms of inherent controllability. Physicians have long understood that normal-tissue injury has been the cause of radiation sickness and complications from photon (x-ray or gamma-ray) therapy, but few understood the potential benefit of highly controllable light ions.

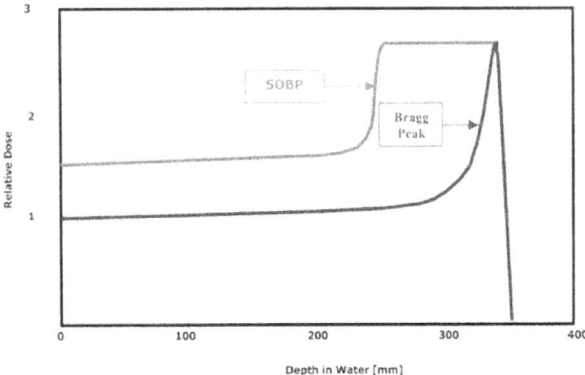

Fig. 1. Idealized representation of the pattern of energy deposition from a single 250 MeV proton beam showing the Bragg peak and spread-out Bragg peak (SOBP). Relatively little energy is deposited until the beam nears its end of range; energy deposition then rises sharply, peaks around 340 mm, and falls off rapidly. In clinical practice the SOBP is used to encompass the tumor, but overall energy deposition is the same: a relatively small dose is deposited until the beam reaches the target, the maximum energy is deposited in the target, and no radiation is deposited beyond the target.

The fundamental reasons for the medical utility of proton radiation therapy were first described by Robert R. Wilson in 1946. Wilson's analysis built on the work of William H. Bragg, who first identified the curve (Fig. 1) that bears his name, and described what came to be called the Bragg peak [2].

In Wilson's words:

The proton proceeds through the tissue in very nearly a straight line, and the tissue is ionized at the expense of the energy of the proton until the proton is stopped. The dosage is proportional to the ionization per centimeter of path, or specific ionization, and this varies almost inversely with the energy of the proton. Thus the specific ionization or dose is many times less where the proton enters the tissue ... than it is in the last centimeter of the path where the ion is brought to rest.... (T)he biological effects near the end of the range will be considerably enhanced due to greater specific ionization, the degree of enhancement depending critically upon the type of cell irradiated.... (M)achines now under construction should have little difficulty in producing (appropriate) currents.... It will be simple to collimate proton beams to less than 1.0 mm diameter or to expand them to cover any area uniformly [3].

The first clinical application of a proton beam occurred at Lawrence Berkeley Laboratory (LBL), at the University of California at Berkeley, in 1954. Investigations led by John H. Lawrence and Cornelius A. Tobias continued during the 1950s, beginning with protons and extending to heavier ions [4, 5]. Tobias, a nuclear physicist concerned with applying physics to biology and medicine, was instrumental in the development of heavy-charged-particle therapy. He helped to spread the research interest in proton beams by traveling to universities and laboratories around the world to share the discoveries made at Berkeley. The second application of a physics research accelerator for proton therapy occurred in Sweden in 1957 [6]. In the United States, physicians at Massachusetts General Hospital (MGH), working with physicists at Harvard Cyclotron Laboratory (HCL), began employing the HCL proton beam for neurological radiosurgery in 1961; pituitary adenomas were first treated at MGH-HCL in 1963 [7], followed by other malignant tumors in 1973, when Herman D. Suit expanded the program and began to investigate large-field proton radiation therapy. Proton-beam treatments began at Dubna, USSR, in 1967; subsequently, facilities started operating at Moscow (ITEP), USSR, in 1969 and St. Petersburg, USSR, in 1975 [8].

It was in this milieu that, in 1970, the first author (JMS) began working toward development of a hospital-based proton treatment center, to be designed and constructed in such a manner that patients' total needs would be carefully planned for. This concept required that the overall accelerator, beam transport system, patient delivery system, and control system adhere to a new set of design requirements, including the capability of precisely focusing and spreading the Bragg peak in any targeted, irregularly shaped treatment volume.

The present paper relates that developmental process and its subsequent applications in the clinic, with reference to fundamental objectives guiding our efforts from the beginning. Among these were to develop a treatment system featuring: (1) patient and personnel safety during treatment delivery; (2) high reliability, to meet the everyday demands of multiple patient treatments in a hospital environment; (3) high efficiency, to accommodate a high patient throughput, not only to meet demand but also to develop a sufficient body of data for research purposes; (4) upgradeable components and systems, to avoid obsolescence by adapting to changes in technology and patients' needs; and (5) technology capable of accommodating a comprehensive research program of basic, translational, and clinical studies aimed at revised, improved, and sometimes novel clinical practices.

2. Preparation Phase

The first steps toward development of the LLUMC proton treatment center began even before 1970, when JMS, then a radiation oncology resident, became dissatisfied with many techniques and subsequent outcomes of photon-beam treatments. There was a clear need for improvements: side effects of radiation were common, and they often were of such severity that treatment courses had to be interrupted or the total doses necessary for eradicating cancer could not be given. There were four fundamental reasons for such unacceptable outcomes: (1) inadequate imaging for target delineation; (2) computer-assisted treatment planning did not then exist and, therefore, precision in treatment planning was not possible; (3) patient positioning techniques were not subject to daily reproducibility; and (4) the absorption characteristics of photon beams did not permit the physician to exercise sufficient control to conform the ionization to the disease and its pattern of local spread. These deficiencies caused excessive normal-tissue injury in virtually all patients treated, hence leading to the common and unacceptable treatment side effects and complications of conventional radiation therapy. All four issues had to be addressed if optimal treatment using ionizing radiation was to be developed.

In 1970, after completing residency training, JMS returned to LLU and undertook a program to correct these deficiencies. He began by consulting with National Cancer Institute (NCI) personnel with regard to developing a heavy-charged-particle facility in the hospital environment. NCI recommended that he consult with Duncan Pruitt, a physicist and specialist in developing radiotherapy facilities. Pruitt agreed that a feasibility study should be undertaken to determine whether a heavy-charged-particle treatment facility might be constructed on the hospital campus at Loma Linda. JMS realized that physics laboratories were experimenting with heavy charged particles and, after visiting most of the sites doing so, he decided that the plausibility of designing a patient-dedicated facility at that time was premature because certain requirements for such a facility did not exist. The deficiencies identified

were: (1) imaging modalities were inadequate for to visualizing the true extent of tumors and, therefore, accurate target volumes could not be identified; (2) because computer-assisted treatment planning systems did not exist, the physician could not visualize the radiation distribution on the patient's anatomy and thereby develop an optimal treatment plan; and (3) existing computing technology was inadequate for controlling a complex heavy-charged-particle treatment facility.

In 1970, as noted, proton therapy was being delivered at several high-energy-physics facilities around the world. It was not clear then, however, that the particle of choice for a hospital-based facility would be the proton; at the time, investigations of pi-meson, helium-ion, and heavier-ion therapy were underway at some research centers, and there was some enthusiasm for neutrons (albeit we never seriously considered the latter particle owing to the lack of a charge as a means of controlling the beam). Regardless of the heavy charged particle that might be selected, however, the various deficiencies noted above militated against development of any of them; further development of these missing requirements would be needed.

JMS had begun working on the problem of therapy planning in the late 1960s, during his residency. At the time, he consulted with Ivan Neilsen, a physicist at LLU. When JMS was recruited by LLUMC to develop a radiation-oncology section, Neilsen and he, assisted by others, including a second physicist, William Chu, continued this effort, and in 1971 began using a computer-assisted treatment-planning system based on digital data taken from

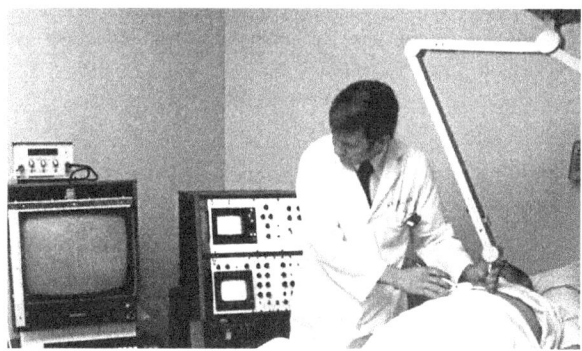

Fig. 2. James M. Slater demonstrating use of the initial, ultrasound-based, computer-assisted treatment planning system, developed at LLU and first employed clinically in 1971. This photograph was taken in 1972–1973.

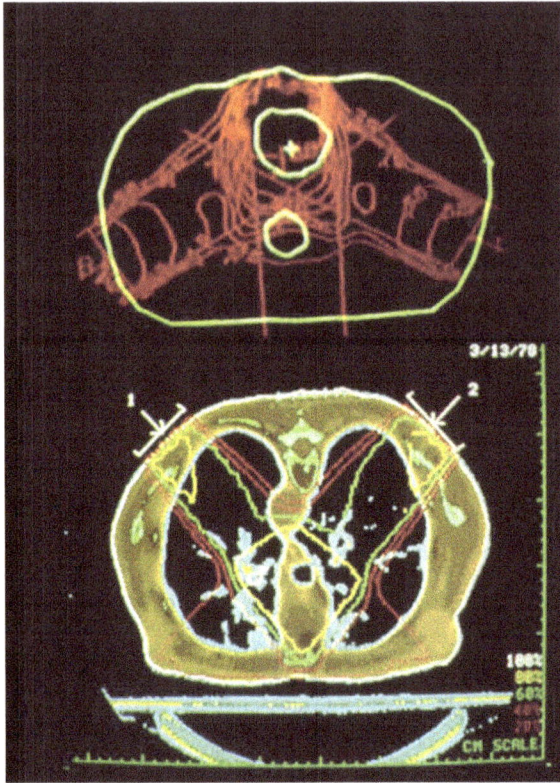

Fig. 3. Treatment-planning image from the first (ultrasound) LLU planning system, for a patient treated in 1973 (top), and a planning image from the second LLU system, which employed CT scans, for a patient treated in 1978 (below). The second system used color to aid in assessing planning options. In addition to better reproduction of the patient's anatomy, the CT-based system allowed assessment of density variations as the x-ray beams passed through tissue.

patients' ultrasound scans (Fig. 2). The system enabled more-precise planning of photon-beam treatments [9] but remained inadequate due to lack of detail and density data from the ultrasound-based images.

When computed tomography (CT) became available in the mid-1970s, we purchased one of the first General Electric CT scanners and interfaced it with the computer-assisted treatment-planning system (Fig. 3) [10]. This hard-wired interface was done in spite of opposition from General Electric, which warned that it would not honor its warranty in the event of malfunction. However, when the interface proved to be successful, the company used the newly developed LLU treatment-planning system to advertise its CT scanners. (It should be noted here that this entire episode owed its success to the financial support of the chairman of the department

of radiation sciences at the time, Melvin Judkins.) The pioneering work was recognized in awards given to the LLU investigative team at multiple international meetings; in subsequent years CT-based radiation-therapy planning was adopted by radiation oncologists worldwide [11, 12], and many accelerator manufacturers developed computer-assisted treatment-planning systems.

The importance of precision treatment planning cannot be overstated. These achievements led to a sea change in the practice of radiation oncology. Therapy planning is, in essence, a prescription for delivering ionizing energy to tissue volumes where it is needed and avoiding normal tissues as much as beam controllability allows. The exactitude with which these twin requirements can be met is a function of both the dose-distribution characteristics of the treatment beam and the precision with which target volumes are identified. The former cannot be exploited fully without the latter, but the latter increases the precision of all forms of radiotherapy — photon and heavy-charged-particle alike. Advances that have been made over the years in photon radiation therapy, such as intensity-modulated radiation therapy (IMRT), derived from the increased precision with which target volumes could be delineated. Although IMRT concentrates the high dose in the target volume, it does so at the expense of a greater volume of normal tissues receiving unwanted radiation (i.e. an increased normal-tissue volume integral dose) than with conventional photon-beam therapy techniques. Nonetheless, IMRT's ability to concentrate the high-dose volume would not be possible without modern computer-assisted treatment planning.

Throughout the 1970s and early 1980s, as precision therapy planning was becoming used more and more commonly, we monitored heavy-charged-particle therapy as it was being practiced in research settings with beams of protons, pi-mesons (pions), helium ions, and heavier ions. Early in the investigation pions seemed an attractive option [13], and JMS was for a time a part-time faculty member at Los Alamos National Laboratory (LANL), where he assisted in patient treatments with pion beams. He also visited LBL, where helium ions and heavy ions were employed, and, as a member of the laboratory's Bevatron-Bevalac Users' Association, referred some patients for treatment of tumors for which a high-LET[a] beam might be advantageous. As the studies at Los Alamos and Berkeley progressed, it became clear that pions and heavy ions had desirable properties but were nonetheless not suitable for routine radiation therapy, mainly owing to excessive damage to normal cells amongst which tumor cells are invading, which poses a major problem for most patients beyond the earliest stages of tumor growth. It is well understood that photon and proton beams, both of which have relatively sparse ionization tracks (low LET), allow for significant repair of radiation injury to normal cells. High-LET radiation may have a role to play, particularly in cases where the tumor forms a large solid mass, thus becoming somewhat radioresistant due to the poorly oxygenated environment that occurs at such times, in turn requiring higher doses if low-LET particles are used.

Economic considerations also became apparent: it would be significantly more expensive to develop a pion or heavier-charged-particle facility than a proton center. Finally, both Los Alamos and Berkeley accrued too few patients to demonstrate a clear clinical advantage with either modality. The proton was selected as the particle of choice for a hospital-based facility owing to a combination of capabilities: a low entrance dose; a Bragg peak that could be modulated to encompass a broad range of target volumes; a sharp dose fall-off that afforded protection to tissues distal to the beam; and low-LET behavior in tissue, similar to that of photon beams. These properties would allow the physician to deliver needed doses to any target volume while sparing much of the nearby and admixed normal tissue. Since it is generally accepted that any form of ionizing radiation will, if given in sufficient doses, kill any cell, a superior dose distribution was deemed of paramount utility for routine radiotherapy, in addition to the factors of lower cost, less technical risk (particularly in isocentric beam delivery), and biologic behavior similar to that of well-understood photon beams.

By the mid-1980s, development of the prerequisites for hospital-based proton therapy had proceeded to the point of feasibility. Computer-assisted, CT-based radiotherapy planning systems were becoming common, and other modalities, such as magnetic resonance imaging (MRI), and later, positron emission tomography (PET) imaging, promised further precision in delineating extent of spread and identifying target volumes. Further, computing capabilities had increased exponentially; the technology was capable of being used to develop complex control systems, not only for the accelerator and beam transport systems but also for distributing the complex beams required for patients in a treatment room. These cumulative developments heightened JMS's interest in proton and heavier-ion therapy. He therefore hired his first physician associate, John O. Archambeau, to help in the work toward developing a hadron treatment facility at Loma Linda. Archambeau, like JMS, had a long-standing interest in proton therapy, and had published papers on the topic during the previous decade [14, 15].

In late 1984, JMS and Herman Suit conferred about their mutual interest in proton therapy and agreed that a meeting or symposium on the subject would be timely and would help to focus the emerging interest in a hospital-based facility. They approached key physicists at Fermilab, including the deputy director, Philip Livdahl, about the possibility of organizing a meeting; subsequently, the Fermilab administration issued invitations to physicists and high-energy-physics laboratories throughout the world.

In January 1985, 93 individuals from several high-energy-physics laboratories all over the world met to discuss the design requirements for a hospital-based proton treatment facility. The symposium generated such enthusiasm that a second meeting was held at Fermilab the subsequent August, followed by the formation of a professional interest group, the Proton (now Particle) Therapy Co-operative Group (PTCOG), which began meeting semiannually at various institutions around the world and continues to do so today. During this early period JMS brought several LLU faculty leaders, administrators, and members of the Board of Trustees to Fermilab to complete their educational process and become better prepared for presenting a review to the full Board of Trustees, a presentation that was rapidly approaching reality. These individuals were overwhelmed by the complexity and beauty of the

---

[a]LET: linear energy transfer. Bragg curves look essentially the same for protons and heavier ions, but scale approximately as the square of the charge of the ion. This translates to higher ionization density for heavier ions (and hence greater biological damage) in the vicinity of the particle tracks.

Fig. 4. Leon Lederman, director of Fermilab at the time the synchrotron was built. Lederman's support was vital to the project.

laboratory, and the lecture presented by the laboratory's director, Leon Lederman (Fig. 4). The visits alleviated many of the concerns that some LLU personnel had about the project.

PTCOG was and is a remarkable example of collegiality. Members met to share knowledge and information, and to encourage the development of hospital-based, treatment-dedicated, heavy-charged-particle systems. Personal agendas, if any, were largely subordinated to the general interest and greater good. For example, when the group was formed, members agreed that "proton" should be part of the name because proton systems were most likely to be realizable at the time, even though some members preferred treatment systems based on other particles. Three major committees were formed within PTCOG, concentrating on accelerator design, facility design, and clinical trials.

Perhaps the essential attitude of PTCOG was expressed by Richard Wilson, then chair of physics at Harvard, who remarked to JMS, "If we want to see these treatments prosper, it doesn't matter that much to us whether it prospers at MGH down the street or at Loma Linda on the other side of the country." After the project was announced and PTCOG members used it as a focus of discussion, Stanley Schriber, of LANL, said, "It's the first time I've ever seen in my career where everybody pulled together and helped one project." Although PTCOG did not play a direct role in the development of the LLUMC proton treatment facility, it performed a valuable consulting role and added a cachet of scientific approval to the project, inasmuch as its

membership comprised some of the eminent physicists in the world.

3. Developmental Phase

The initial attempts to begin a hospital-based project at LLUMC consisted of inquiries to major medical manufacturers. Representatives of approximately 35 industrial and engineering firms, such as General Electric and Siemens, were invited to visit LLU for a presentation on the proposed project. The firms were asked, in essence, whether they would partner in developing a pioneering accelerator and facility dedicated to proton treatments for patients. All declined; later, JMS learned from a senior engineer at Siemens that the project would be too great a change of direction for any of the companies to undertake. Subsequently, rather than drop the project, JMS began a systematic series of visits to the major high-energy-physics laboratories in the United States, assessing the potential of each as a possible collaborator in the project. The decision was made to approach Fermilab.

JMS participated in the 1985 meetings at Fermilab and was actively involved in PTCOG, chairing one of its three committees. During these meetings he became better acquainted with several Fermilab personnel, including Lederman, Livdahl, and several other physicists, such as Lee Teng, Fred Mills, and Miguel Awschalom. All had been colleagues of Fermilab's founding director, Robert R. Wilson, and shared Wilson's vision for proton therapy. JMS hoped that an active collaboration could be established with Fermilab, inasmuch as the latter was among the world's premier designers and fabricators of synchrotrons and related components.

The accelerator would be a first-time venture, designed for operation in a hospital, where an accelerator should run virtually full-time with few interruptions and had to be safe, reliable, easily maintained, as simple to use as possible, capable of being upgraded as technology changed, and yet, because it would have to be controlled for the variable everyday demands of patient treatments rather than physics research, would have different design requirements than an accelerator system designed for high-energy-physics work. Such a venture required superb expertise, the kind that Fermilab personnel possessed.

Fig. 5. Left to right: Lee Teng, James M. Slater, and Philip Livdahl, at Fermilab during accelerator construction, 1988. Teng was the principal designer of the LLUMC synchrotron, and Livdahl, then deputy director of the laboratory, was an early and enthusiastic supporter whose help in realizing the project was invaluable.

In conversations between JMS and Livdahl, the latter reported that a collaboration could be accomplished under the Department of Energy's "work for others" policy, which could be invoked if the outcome was technology transfer for the general good and if expertise to perform the task did not exist in the private sector. Given the fact that several major manufacturers had declined to participate, the latter question was answered. The expectation that proton therapy would accomplish cancer control while reducing side effects addressed the former. Furthermore, Livdahl indicated that several individuals at Fermilab were favorably disposed toward a project to build a hospital machine; these included himself, Teng, Mills, and Awschalom, among others, all of whom were veteran physicists and engineers. Lederman supported the venture, and later termed it "the quintessential example of technology spinoff." Further approvals came from the Department of Energy (DOE) and the Universities Research Association, which operated Fermilab for DOE. LLU and Fermilab signed an agreement for a conceptual design study in January 1986, followed by an agreement for an engineering design study about a year later. JMS announced the agreement at the first PTCOG meeting in 1986. José Alonso, a physicist at LBL who was a steadfast supporter of the project, applauded the decision. Alonso became a very valuable asset for discussing multiple design issues and questions regarding the

delivery system. His personal conversations reduced many concerns throughout the project development as well.

The pros and cons of various accelerator types, mainly cyclotrons and synchrotrons, had been discussed at the Fermilab meetings. A hospital-based proton accelerator must be compact, reliable, require little maintenance, and operate at low cost. The beam intensity must be sufficient to keep each patient's treatment time short. Beam energy must be rapidly variable to permit total flexibility in beam delivery. The latter was a salient reason for choosing a synchrotron rather than a cyclotron or linear accelerator; beam energy could be varied easily and precisely, and the radioactive buildup in the accelerator region would be minimized. Changing energy was the key to treating a variety of tumors of various depths and sizes within the body. The need for a synchrotron was an added reason why we desired an association with Fermilab, given the extensive history Fermilab physicists and engineers had with that type of accelerator. Even so, the task facing the project would not simply be a matter of downsizing a research synchrotron of the sort that Fermilab personnel had so much experience with. The project would be a pioneering one: at the time, no synchrotron had ever been designed and built for treating patients, and none had ever been so small as was needed in the hospital environment. Further, a clinical accelerator for medical purposes had different design requirements than a physics-research machine.

Specifically, these design requirements included: (1) *safety* — it had to be safe for the patient and for radiation-therapy personnel to operate; (2) *machine longevity* — it had to be easily maintained and upgradeable; (3) the *control system* had to be complex and interfaced with a treatment-planning system to fully control each patient's setup and treatment, yet nonetheless had to be simple to operate; (4) *reliability* — it had to be able to deliver treatments to at least 200 patients per day and also support multiple research projects.

Several Fermilab physicists and engineers worked on the design and construction of the proton accelerator. The fundamental design, however, was Lee Teng's (Fig. 6).

The machine is a 250 MeV weak-focusing, eight-magnet, four-sector synchrotron. It had to fit within a relatively small space, and Teng's design satisfied

Fig. 6. The synchrotron at LLUMC. In the drawing (top), the main magnets are shown in light gray and some components are identified: (A) ion source, (B) RFQ, (C) injection septum, (D) radiofrequency accelerating cavity, (E) Lambertson split extraction magnet, (F) 20° bending magnet to treatment rooms. The parts A, B (black letter), C, D, and E are also indicated in the photographs.

the minimum space requirements at LLUMC. A compact radiofrequency quadrupole (RFQ) accepts the protons from the injector and accelerates the low-energy proton beam to 2 MeV in a 1.6 m distance, simultaneously focusing and accelerating the beam. The ion source ionizes hydrogen gas, using a 35 KeV voltage for creating a flow of protons to the RFQ. The synchrotron provides a pulsed, 300 m beam to the designated treatment room every 2.2 s.

In the synchrotron ring, the protons gain energy as they pass through an RF cavity with each turn until they reach the desired energy, ranging from 70 to 250 MeV, depending on the desired depth of penetration in the patient; at that point, the magnetic dipole field and RF frequency are held constant until the protons are extracted and transported to the treatment rooms. After extraction, the magnetic dipole fields return to injection level for the next batch of 2 MeV protons.

Fermilab's responsibility in the project was to design and fabricate the synchrotron and the transport system for carrying the beam to four treatment rooms. Three of the rooms would have gantries to rotate the beam through 360°; the other would have two fixed horizontal beam lines. Beam was also to be transported to a fifth room, containing three beam lines for research purposes.

To assist in transferring the accelerator and its associated technology from Fermilab to the medical center, LLUMC acquired an industrial partner, Science Applications International, Inc. (SAIC), selected by Fermilab; SAIC's mission, under the leadership of its project manager, Ben Pritchard, was to assist Fermilab personnel with the setup and startup of the accelerator. Secondly, they were responsible for disassembling the accelerator at Fermilab and, with the Loma Linda engineering team, reassembling it at LLUMC.

An area of concern during development was the control system. This concern had been anticipated by us and by some at Fermilab; in fact, one of the pre-eminent physicists at the laboratory, Helen Edwards, in a discussion with JMS, expressed her concern about the control system meeting Loma Linda's needs. We understood from the beginning that the design requirements for a medical accelerator's control system, like the accelerator itself, were different than the requirements for controlling a physics-research accelerator: the need to vary the energy and the beam intensity, and the need to change both rapidly and to make decisions as dictated by the computer-assisted planning system, were at the heart of this difference. Additional differences were anticipated as advances in treatment techniques were made. Edwards's point was not entirely unexpected, therefore.

JMS took proactive steps to deal with the issue of developing a control system for medical use. Three engineers, Jon Slater, David Lysena, and James Nusbaum, were sent to Fermilab by LLU to help with the design effort. They were to learn about the accelerator control system used at the laboratory and to begin designing a control system for the treatment rooms. The ultimate objective was for these engineers to devise an overall integrated control system, one based on the needs of a medical facility, and to be the on-site caretakers of the entire system when it was in clinical operation. Differences of opinion arose, owing primarily to fundamentally different conceptions of what the control system must do, as seen by the physicists and engineers familiar with the high-energy-physics environment at Fermilab and the computer scientists and engineers familiar with the medical milieu. The engineers representing LLU understood that a hospital-based treatment facility needed a control system that medical personnel could operate, and that could function effectively and routinely in a clinical setting. The control system had to be simple to operate and understandable to medical personnel or to operators who would be working with medical personnel every day. Interfaces had to enable the operator (ideally, a radiation-therapy technologist) to control not only the machine but also the treatment situation, in which several rooms would contain patients awaiting treatment, and the accelerator, therefore, had to be ready to respond to each treatment room as expeditiously as possible.

Following their work at Fermilab, the engineers worked with physicists and engineers at LBL, studying the control systems used there for helium and heavy-ion therapy. From this experience and the experience that Jon Slater had accumulated in working with radiation physicists at LLU and the University of Utah, the engineers formulated the system ultimately employed.

Timely development was emphasized throughout the process; timing in planning, fabrication, and installation was as important to the project's success as the successful operation of its component parts, because of the high monthly construction costs. Prebuilding arrangements with the purchasing department, suppliers, labor unions, and all primary and secondary contractors avoided costly delays and disputes, and resulted in the ability to take early partial occupancy for installing certain heavy subsystems, such as the massive gantries (Fig. 7).

Fig. 7. James M. Slater at LLUMC during construction, 1989. Behind him is the steel framework of one of the three isocentric gantries. When completed, each gantry was three stories tall and weighed more than 90 tons.

The LLUMC proton synchrotron was fabricated at Fermilab. It was not constructed in isolation, however. From the outset, it was clear that the entire project — building, accelerator, delivery systems, and patient-care areas — needed to be designed as an integrated whole. A structure was required to both house the treatment-delivery systems and manage patient flow efficiently. Close collaboration between the architectural and equipment design teams was imperative; the shielding concrete and steel walls could not be adjusted, nor could there be any significant changes in equipment design to accommodate building errors. Accordingly, the various engineering design teams accomplished their tasks in a collegial manner; regular intercommunication among teams and team members was maintained. Eight teams were formed, each having tasks relevant to the overall engineering design; LLUMC, Fermilab, and the architectural firm, NBBJ, of Seattle, Washington, provided staff for each team. Weekly meetings were supplemented with monthly meetings with hospital and university administrators, and periodic meetings were held with outside reviewers; these meetings helped to keep the teams communicating and aware of progress and problems.

The building was designed to treat patients safely and efficiently. It was also designed to be erected within budget; to provide facilities for patients, practitioners, and for scientist education and research; and to be modifiable when new technology appeared. Design requirements stemmed from clinical treatment, research, and educational needs, plus marketing and financial projections. These resulted in a facility with, as noted, five therapy beam lines in four rooms, three containing gantries for rotating the beam around the patient, as well as a separate research room with three beam lines. Clinics, treatment rooms, patient-preparation areas, waiting rooms, animal-research facilities, and offices were sited to promote optimal efficiency.

Aside from the proton synchrotron itself, among the salient features of the facility are its three, three-story-tall, rotating gantries, the first ever used to transport a proton beam (Fig. 8). Large though they are, the gantries are nonetheless an exercise in saving space for clinical use: they utilize a "corkscrew" design devised by Andreas Koehler, of HCL, in which the beam path is folded. The design, patented by Harvard and for which LLUMC paid a royalty, was executed by SAIC, notably in the person of Rudy Prechter, to whom goes much of the credit for actualizing Koehler's design. A local machine shop in Riverside, California, namely Martinez and Turek, did a superb job of constructing the gantries. SAIC also assisted in transferring all

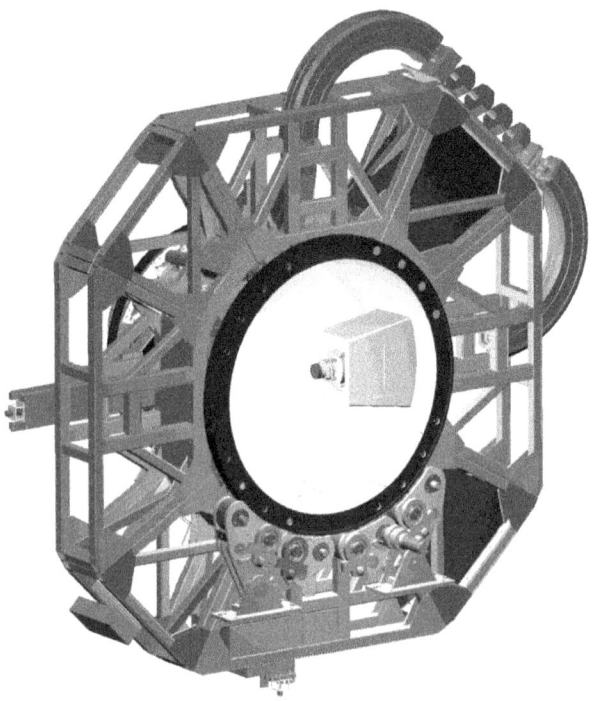

Fig. 8. The compact isocentric gantry at LLUMC provides 360° access to the patient.

accelerator components from Fermilab, in the summer and autumn of 1989, for installation in the new proton treatment center, then under construction.

Critical to the success of proton therapy was a system for delivering the beam to the patient in the treatment room. A cadre of beam-shaping devices, collectively called the nozzle, was developed by LLU engineers with valuable guidance from physicists and engineers at LBL. The nozzle conforms the proton beam to the patient's target volume; it works in tandem with the prescribed beam energy, set according to the depth of penetration desired. Among the beam-shaping devices, an aperture is created for each treatment portal and is shaped to conform the tumor dose in the dimension lateral to the beam direction. A compensator bolus controls penetration of the beam and, in essence, conforms the dose in the distal direction. A modulator wheel spreads the Bragg peak to accommodate the extent of the tumor along the beam direction, i.e. the wheel helps conform the dose throughout the tumor. Sets of modulator wheels are prefabricated, and the appropriate wheel for a given portal is selected by the treatment-planning system. Apertures and boluses are custom-designed for each portal of every patient; the designs for these devices, again, are generated by the treatment-planning system. With the aid of these devices, we can, in principle, deliver a three-dimensional conformal dose to a tumor with each proton portal. Clinical constraints, such as the location and extent of the tumor and the surrounding critical structures, often require more than one portal to deliver a conformal dose to the tumor and avoid an unnecessary dose to the surrounding structures, but, in general, proton therapy requires fewer portals compared to photons.

The entire facility was ready for clinical operation in 1990. George Coutrakon, one of the team of physicists working on the accelerator at Fermilab, agreed to come to Loma Linda with the accelerator and assume the position of chief accelerator physicist. He devoted his career to the proton synchrotron for nearly 20 years.

By 1990, the engineering team had grown to approximately 30 engineers, as they prepared to maintain and carry out upgrades to the accelerator and proton center. LLU administration desired that the team form an independent company to reduce their expenses. Accordingly, the firm known today as Optivus Proton Therapy, Inc., of San Bernardino, California, was formed.

3.1. *Accelerator performance and maintenance*

The synchrotron has been in almost continual operation since 1990. The commissioning phase, before clinical operations began, was extended by some problems with the RFQ, and some early clinical operations were delayed for the same reason. Since then, however, the LLUMC facility has accumulated a remarkable record of reliability. Engineers at Optivus have upgraded the accelerator and facility regularly since installation, and the present therapy system is as state-of-the-art now as it was almost two decades ago. Its "uptime" has exceeded 98% throughout the years, 24 h per day, 5–7 days per week. Patients are treated from 5:30 a.m. to 10:30 p.m.; during the overnight hours, all five rooms, including the research room, are available for research studies. (At this writing, one of the gantry treatment rooms is shut down for installation of a robotic patient positioning and alignment system, described later in this account. Patient throughput has not been compromised, however, owing to extraordinary efforts by staff to maintain extended operations in the remaining rooms. That shutdown is the first such event in the history of the facility.)

The accelerator is located in a radiation-controlled room, called the vault, to which access is limited and none is permitted when the beam is on. Most power supplies for the accelerator are placed in a non-radiation-controlled area and can be accessed for service while the beam is being accelerated. Very few items within the radiation-controlled vault, however, require routine maintenance. The item needing the most attention is the cryogenic vacuum system. This system, which includes components for the ion source, RFQ, and ring, is usually regenerated once per week, but two weeks can elapse without regeneration if scheduling is severe. This procedure is performed near locations at the synchrotron where measured dose rates are very low; it typically takes less than 15 min to connect the external pumps and initiate the cold finger warmup and pumping, and another 15 min to disconnect the external pumps.

Another common maintenance item is replacement of the cooling water filters for the cryopumps. These filters, also located within the accelerator vault, are typically changed once per month. The deionized cooling water system for the synchrotron is located on the floor above the accelerator, outside the restricted radiation area.

A third maintenance task within the accelerator vault is replacement of the cooling water hoses. This is not a frequent task, however: the cooling hoses connected to the split Lambertson magnet were replaced only three times during the first 17 years of the facility's operation; the hoses for an octopole near one of the synchrotron's long straight sections have been replaced only twice.

During the first eight years of the facility's operation, an additional maintenance item was a monthly cleaning of a ceramic cable passthrough to the extraction wire septum. In 1998, however, engineers conducted a redesign in which components were sealed in an SF6 environment. The cleaning requirement was eliminated.

Only infrequently have engineers and maintenance personnel found it necessary to conduct corrective maintenance procedures within the radiation-controlled area where the synchrotron sits. Work on the extraction wire septum, for example, was performed only twice in the first 17 years of the facility's life. The monitor ionization chamber placed at the entrance to the beam dump was serviced five times during that period, and the Faraday cup was removed and reinstalled three times. A fan for the ring radiofrequency accelerating cavity was replaced three times. In the first year of operation (1990–1991), the injection septum was serviced once, but it has not required attention since. The item within the radiation-controlled vault requiring the most attention has been the vacuum system, but work on that system is performed in areas away from locations where a higher exposure rate occurs after beam shutdown. Exposure rates have been determined to be low anywhere in the facility.

A paper published in 2009 indicates that the design of the LLU synchrotron and the layout of the LLUMC facility have resulted in minimal production of residual radiation [16]. This was demonstrated by measurements around the accelerator immediately after shutdown (Table 1) and by reviewing personnel dosimetry records of maintenance staff. The low exposure rate provides the ability to treat patients and perform research 24 h per day, 5–7 days per week.

Table 1. Average dose equivalent rates, in Sv/h, at various locations in or near the LLUMC synchrotron, as measured with ion chamber (IC) and Geiger–Müller (GM) survey meters, between 3 and 19 min after shutdown.[*]

Measurement site	IC avg.	±1 s.d.	GM avg.	±1 s.d
Background	1.9×10^{-7}	1.9×10^{-7}	9.7×10^{-8}	9.7×10^{-8}
Ion source	9.7×10^{-7}	1.9×10^{-7}	5.8×10^{-7}	9.7×10^{-8}
Exit of RFQ	1.8×10^{-6}	2.9×10^{-7}	1.3×10^{-6}	2.9×10^{-7}
Center of ring	2.2×10^{-6}	2.9×10^{-7}	1.7×10^{-6}	9.7×10^{-8}
Injection septum	1.6×10^{-6}	5.8×10^{-7}	7.8×10^{-7}	1.9×10^{-7}
Long straight, #2; pinger plate	4.1×10^{-6}	5.8×10^{-7}	4.1×10^{-6}	2.9×10^{-7}
Long straight, #2; octopole	6.5×10^{-6}	1.8×10^{-6}	5.6×10^{-6}	3.9×10^{-7}
Extraction wire septum	9.1×10^{-6}	2.7×10^{-6}	1.3×10^{-5}	3.8×10^{-6}
RF accelerating cavity	5.0×10^{-6}	9.7×10^{-7}	4.3×10^{-6}	9.7×10^{-8}
Long straight, #3; sextupole	5.1×10^{-6}	3.9×10^{-7}	4.7×10^{-6}	7.8×10^{-7}
Short straight, #3; quadrupole	2.7×10^{-6}	2.9×10^{-7}	2.2×10^{-6}	3.9×10^{-7}
Lambertson; upstream outside	1.5×10^{-5}	1.9×10^{-6}	1.7×10^{-5}	5.4×10^{-6}
Lambertson; downstream	9.6×10^{-6}	4.3×10^{-6}	1.3×10^{-5}	2.7×10^{-6}
Lambertson; center inside ring	7.6×10^{-6}	2.4×10^{-6}	7.5×10^{-6}	1.9×10^{-6}
20° bending magnet	4.2×10^{-6}	1.1×10^{-6}	4.7×10^{-6}	7.8×10^{-7}
Beam dump	2.1×10^{-6}	6.8×10^{-7}	1.8×10^{-6}	6.8×10^{-7}

[*]Adapted from Ref. 16; used with permission.

At several locations within the beam-transport switchyard (not the radiation-controlled synchrotron vault), multiwire ionization chambers used for tuning the beam have had to be repaired or replaced, owing to the finite lifetime of radiation-exposed wires held under tension. On the few occasions where access to the controlled area was required during scheduled patient-treatment times, access was immediate, with no time lost to clinical usage while maintenance personnel waited for decay of radioactivated components.

Within the accelerator radiation-controlled area, in point of fact, more time has been spent performing upgrades than on preventive or corrective maintenance. Upgraded or new items installed within the accelerator radiation-controlled area included the ion source, sextupoles, vacuum system, and uninterruptible power supplies. The low residual radiation levels in the controlled accelerator vault have allowed these upgrades to be performed with little exposure to installation and test personnel.

Maintenance personnel that work on the synchrotron also work on the gantries, nozzles, and patient positioners in the treatment rooms, and equipment located in the switchyard where beam is transported to the various treatment rooms. Therefore, even though dose-equivalent readings for these personnel are extremely low, the fraction of the dose equivalent due to residual activation of the accelerator is even smaller. These findings indicate that the prime requirement for the accelerator — safety — is being met.

4. Clinical Operations

We treated our first patient in October 1990. In designing treatments, we built on the experience generated over several decades at the various high-energy-physics research facilities around the world. During those years, proton therapy had developed a reputation for excellence in treating certain kinds of tumors and diseases, usually difficult-to-treat conditions such as ocular melanomas and lesions intimate to vital structures. When planning the proton facility, however, we intended that protons would be used to treat virtually any localized or regional solid tumor in any anatomic site. The ability to design and deliver highly conformal treatments to intended target volumes, we reasoned, should be exploited whenever possible; the benefits of delivering effective therapeutic doses to intended volumes while minimizing side effects by avoiding normal tissues would apply to almost all cases that were ordinarily treated with conventional photon irradiation.

At present, the facility treats approximately 150 patients per day. The total number of patients treated with proton beams at LLUMC as of July 1, 2009, approaches 14,000, the largest total from any institution. Most patients come from the southwestern region of the United States, mainly

California, but patients have appeared from throughout the United States and from many other countries in the world. This large clinical experience has been made possible in part by the reliability that was designed into the treatment system; by the efficiency that was designed into the facility, in terms of placement of clinics, simulation rooms, dressing rooms, physician offices, control rooms, and the like; and by the capacity designed into the facility, permitting us to develop a sufficient body of patient data to validate outcomes. In general, the effectiveness of medical procedures is determined and validated by results observed and analyzed in large populations. We foresaw, therefore, that several beam lines and treatment rooms would be needed, not only to serve the populous southern California region and patients from elsewhere, but also to develop meaningful clinical data.

The high volume of patients treated at LLUMC is not merely a function of the several treatment rooms and the design of the treatment areas (Fig. 9). Several years ago, to satisfy the growing demand for proton treatment at Loma Linda, a decision was made to operate in two shifts. This is not routine practice in clinical radiation oncology, but hospital administration were willing to enable the second shift in an effort to accommodate the greater demand.

The large patient throughput is a function of the efficiency of the facility, which is the result not only of the initial facility design but also of the upgrades that have occurred throughout the life of the center.

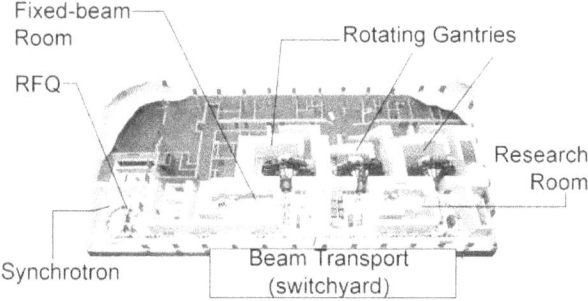

Fig. 9. Cutaway model of the proton treatment floor at LLUMC. The fixed-beam room has two beam lines, for eye and for head-and-neck treatments. The research room has three beam lines. The unlabeled rooms at the top of the model include dressing rooms for patients, control rooms for the gantries and the fixed-beam room, and rooms for physicians and technologists to evaluate plans and consult on individual cases.

Technological advances have been exploited whenever possible, and upgrades in the facility have occurred virtually since the day the center opened. For example, the power supplies for the accelerator have been upgraded several times in the past two decades; new gantry devices have been installed, notably hardware modifying the manner in which beam exits the gantry; the beam control system has been upgraded; the treatment planning system has had two major upgrades, including the current system, marketed under the name Odyssey[TM] by Per-Medics, Inc. (San Bernardino, California). One of the major advances facilitating patient throughput and accuracy was a computer-assisted digital alignment system, which improved daily operational efficiency by about one-third. More recently, a completely new accelerator control system was implemented; it will enable further upgrades in the near future, including full beam scanning. Our continual efforts to maintain optimal operating efficiency have permitted high patient throughput while retaining highly individualized treatment of each patient.

Efforts to exploit the therapeutic potential of protons began when the facility opened in 1990. The initial goals were to (1) build on the long experience with photon radiation and previous experience with protons at laboratory facilities by investigating additional clinical sites that might benefit from protons; (2) develop protocols to treat anatomic sites and evaluate outcomes; and (3) improve or develop technology that would allow additional anatomic sites to be investigated. Clinical studies had two objectives. The first was to employ protons to reduce treatment-related morbidity for patients having conditions for which curative treatment options exist but have been associated with significant morbidity. Investigators pursued this outcome irrespective of whether proton therapy increased control rates. The second objective was to use protons to improve control rates for tumors that were not well controlled by other modalities. Studies proceeded in small, progressive steps. Initial dose-escalation studies for prostate cancer, for example, explored whether total doses only 10% higher than had been used in conventional photon regimens could be administered without affecting treatment-related morbidity. If the desired outcomes occurred, our investigators pursued further escalation studies. Proton studies on other anatomic sites sought to determine whether lower morbidity rates could be obtained given the same total dose as had been delivered with photons; dose-escalation studies were conducted if the data showed that morbidity had in fact been reduced.

In both forms of investigations, we started from the presumption that ionizing radiation from any source will destroy a targeted tissue volume if the total dose is high enough. The salient consideration was whether such doses could be delivered to patients without causing unacceptable permanent damage to untargeted tissues. That is where the main advantage from protons was expected: physicians could control the physical dose distribution so as to permit such outcomes; the slightly higher proton radiobiologic effect (RBE; it was 1.1 relative to gamma radiation from cobalt sources) was not a significant factor in employing them. Accordingly, we presumed further that the proton RBE was essentially the same as that of photons, and that, consequently, irradiated normal cells would have the same repair capacity as cells exposed to photon radiation.

In the early years of the facility, relatively few tumors were treated with protons. More anatomic sites were added as experience accumulated and as technological advances occurred in treatment delivery and control systems. Today we use protons to treat approximately 50 tumor sites and other diseases, in most anatomic regions of the body.

4.1. The central nervous system and the base of the skull (in adults): stereotactic radiosurgery

Proton radiosurgery is used at Loma Linda to treat brain metastases and arteriovenous malformations (AVMs). In comparison with photon beams from a linear accelerator or a gamma knife, protons produce less normal-tissue radiation, particularly at larger volumes and for peripheral lesions; offer better coverage for irregularly shaped volumes; and yield more uniformity of the dose within the target volume [17–19]. One radiosurgical application of protons has been in the treatment of large AVMs, for which one report suggests that fractionated proton radiosurgery was effective in managing such lesions [20]. At present, investigators from our departments of radiation medicine, neurosurgery, and neuroradiology are collaborating with investigators from the departments of neurosurgery and neuroradiology at Stanford University to evaluate a program for treating large (> 3 cm) AVMs with surgery, embolization, and hypofractionated protons. The program has been pursued since 1994; doses of 20–25 GyE (gray equivalent) in 1–5 fractions are given, using stereotactic protons; the dose is based on the volume of the target.

4.2. Fractionated proton therapy for tumors of the central nervous system

Fractionated protons are used for most lesions occurring in the central nervous system (CNS) and the base of the skull, including chordomas, chondrosarcomas, acoustic neuromas, meningiomas, and pituitary adenomas. Protocols have built on prior experience at other institutions. Our experience with these lesions began in 1992. From March of that year through January 1998, LLUMC radiation oncologists treated 58 chordoma and chondrosarcoma patients with protons, with doses ranging from 64.8 GyE to 79.2 GyE; 44 patients were treated for primary disease, and 12 for recurrent disease. Analysis revealed five-year local control and survival rates of 59% and 79%, respectively, for chordomas, and local control and survival rates of 75% and 100%, respectively, for chondrosarcomas. The control rate was related to tumor size, as had been reported by other investigators: for tumors ≤ 25 ml, the local control rate was 100%; for tumors > 25 ml, it was 55% [21, 22]. A collaboration was undertaken with colleagues from MGH and LBL on a Phase III dose-escalation trial of proton radiation for chordomas and chondrosarcomas. Patients were randomized into one of three treatment arms, to receive total radiation doses of 66, 72, or 79 GyE, stratified according to histology, site, boost volume, and sex. Accrual was completed in 1999; followup and data analysis are ongoing.

Acoustic neuromas are tumors for which complete surgical removal or radiosurgery yields control rates greater than 90%. However, this outcome is associated with some toxicity: facial nerve dysfunction has been noted to occur in 5–38% of patients treated; postsurgical deaths have been reported in up to 6%; and useful hearing retention has been observed in 30–65% of patients having tumors < 1.5 cm in the greatest dimension [23–25]. We use protons to treat patients having recurrent or unresectable acoustic neuromas, or for patients who refuse surgery. There is no size limit. Doses of 1.8 GyE per fraction are delivered, to total doses of 50.4 or 59.4 GyE, depending on pretreatment analysis of hearing. Of the first 30 patients who had completed treatment through February 2000, 18 had been followed for periods ranging from 1 to 7 years after treatment (median 3.5 years). All lesions were controlled, and no patient had permanent cranial nerve injury. Patients continue to be followed, to evaluate long-term outcomes in hearing [26]. Protons also are used to treat patients with meningiomas who present with recurrent or unresectable disease, or who refuse surgery. The treatment regimen consists of protons at 1.8 GyE per fraction; the total dose delivered ranges from 54.0 to 63.0 GyE in 6 weeks.

Analysis of 47 patients treated with protons for pituitary adenoma revealed that 42 of them underwent prior surgical resection and 5 were treated with primary radiation. Approximately 50% of the tumors were functional. The median dose was 54 GyE. Tumor stabilization occurred in all 41 patients available for followup imaging; 10 of these had no residual tumor and three showed greater than 50% reduction in tumor size. Seventeen patients with functional adenomas had normalized or decreased hormone levels; progression occurred in three patients. Six patients died; two of the deaths were attributed to functional progression. Complications included temporal lobe necrosis in one patient, new significant visual deficits in three patients, and incident hypopituitarism in 11 individuals. Investigators concluded that fractionated conformal proton radiation achieved effective radiologic, endocrinologic, and symptomatic control. Significant morbidity was uncommon, with the exception of postirradiation hypopituitarism, which was attributed in part to concomitant risk factors for hypopituitarism present in the patient population [27].

We employ proton radiation as one of the primary treatment options for patients with benign CNS tumors, who traditionally have been viewed as candidates for primary surgery. Sometimes combined treatment is used; sometimes protons alone are used. Although stereotactic treatment approaches, featuring one or a few treatments, are often used in treating these lesions, the precision of the proton beam and modern methods of positioning patients make fractionated proton therapy as precise, in effect, as stereotactic methods [28].

4.3. *Diseases of the eye and tumors of the head and neck*

Proton irradiation has an established role in treating patients with ocular melanoma: this modality is an alternative to enucleation. Control rates exceeding 95% are common for small tumors, and rates of eye retention typically reach 90%. We evaluated the efficacy and safety of proton radiotherapy for medium-size and large choroidal melanomas. A retrospective review revealed that the 5-year local control rate was 91% and the 5-year disease-specific survival rate was 76%. Eye preservation was achieved in 75.3% of patients, with useful ($> 20/200$) visual acuity obtaining in 49%. The patient's initial visual acuity, the proximity of the tumor to the optic disk, and the total dose received by the optic disk and fovea were all significant prognostic factors for maintaining useful visual acuity following treatment. The diameter of the tumor at its base was related significantly to survival but did not impair local tumor control or visual acuity. Data suggested, therefore, that protons were effective and safe for medium size and large melanomas, and can preserve the eye and its function in a reasonable percentage of patients [29].

We employ proton radiation to treat locally advanced oropharyngeal cancer. In one report, patients were treated under a Phase I/II study employing protons as a boost treatment for squamous-cell carcinomas of the oropharynx. We assessed accelerated fractionation with a concomitant boost using photon and proton radiation to improve local control and reduce complications. Twenty-nine patients received accelerated photon (50.4 GyE to the clinical target volume) and proton radiation (25.5 GyE to the gross tumor volume), yielding a total dose of 75.9 GyE in 45 fractions administered in 5.5 weeks, to the primary disease, involved lymph nodes, and potential areas of subclinical spread (three patients were administered a prescribed total dose of 74.4 GyE). The five-year actuarial control rate for local disease was 88%, and for neck node disease, 96%, yielding a locoregional control rate of 84% at five years. Four patients developed distant metastases. The actuarial two-year disease-free survival rate was 81%; the rate was 65% at five years. Investigators concluded that protons, used as a concomitant boost with photons, effectively delivered an accelerated time-dose schedule to the cancer with a more tolerable schedule to surrounding normal tissues; preliminary results revealed increased locoregional control without increased toxicity as compared to other radiation techniques delivering lower doses [30]. Further studies are being done to evaluate the optimal time-dose schedule.

Protons are also used for nasopharyngeal cancers, including retreatment following photon radiation. In one report, 16 patients with nasopharyngeal carcinoma initially treated with 50.0–88.2 Gy photons were retreated with protons to additional doses of 59.4–70.2 GyE. Twenty-four-month actuarial overall and local–regional progression-free survival rates were both 50%. Actuarial overall survival rates for patients with "optimal" dose-volume histogram coverage versus "suboptimal" coverage were 83% and 17%, respectively ($P = 0.006$). Doses to critical structures were low; no CNS side effects supervened. We concluded that adequate tumor coverage is the most important variable influencing local–regional control and survival. This could be achieved with protons, while avoiding CNS complications [31].

4.4. Lung, breast, and liver cancer

These disparate sites are here considered together because they illustrate the use of hypofractionation, the delivery of a total dose in fewer fractions, and thus via larger fractions than are given conventionally. The superior dose distribution of the proton beam, and attendant sparing of normal tissues, makes this possible. We are examining this treatment as a possible approach to reducing treatment time and costs, provided that control rates are maintained and side effects do not increase, as is generally the case when photon hypofractionation is employed. Results obtained thus far, while still preliminary, are promising.

Approximately 170,000 new cases of lung cancer are diagnosed every year in the United States. About 20% of patients have clinical Stage I disease at diagnosis, and in approximately 15% of those cases the patients are medically inoperable, even though many of them have tumors that are technically resectable. Although conventional photon irradiation can control early-stage inoperable lung cancer, it often results in injury to functional lung tissue. We conducted a prospective Phase II clinical trial to determine the efficacy and toxicity of high-dose hypofractionated proton radiotherapy for patients

with clinical Stage I lung cancer, all of whom were medically inoperable or refused surgery. Preliminary reports on the use of protons for such cases were encouraging [32, 33], and a later report indicated that excess pulmonary toxicity did not occur when higher-than-conventional doses of radiation at a higher-than-conventional dose per fraction were delivered via conformal radiation techniques with protons [34].

A more recent review described 68 patients in the trial. All were treated with multibeam proton-beam radiation to a target that included the gross tumor volume as seen on a CT scan, with an additional margin to allow for respiratory motion. The delivered treatment was 51 GyE in 10 fractions over 2 weeks to the first 22 patients; the subsequent 46 patients received 60 GyE in 10 fractions over 2 weeks. No symptomatic radiation pneumonitis or late esophageal or cardiac toxicities were seen; the 3-year local control and disease-specific survival rates were 74% and 72%, respectively. There was significant improvement in local tumor control in T1 versus T2 tumors (87% versus 49%), with a trend toward improved survival. Patients with higher performance status, female patients, and patients having smaller tumors had significantly higher survival rates. We concluded that high-dose hypofractionated proton-beam radiotherapy can be administered safely, with minimal toxicity, to such patients, and that local tumor control appears to be improved when compared to historical results of patients treated with conventional radiotherapy, with a good expectation of disease-specific survival three years following treatment [35].

For breast cancer (Fig. 10), we do not use protons to treat the whole breast, as is commonly done in current photon-beam regimens following lumpectomy or partial breast resection. Rather, protons are delivered to a more circumscribed volume around the postoperative site. The rationale for this approach is the small difference in remote breast recurrence following lumpectomy alone as contrasted with lumpectomy and whole-breast irradiation in a subset of women with early breast cancer [36–39]. We conducted a Phase II clinical trial of 50 patients, who received a total dose of 40 GyE in 10 fractions of 4 GyE each; treatment typically was given with 3 or 4 beams, with multiple fields treated each day. The entire dose was confined within the treated breast.

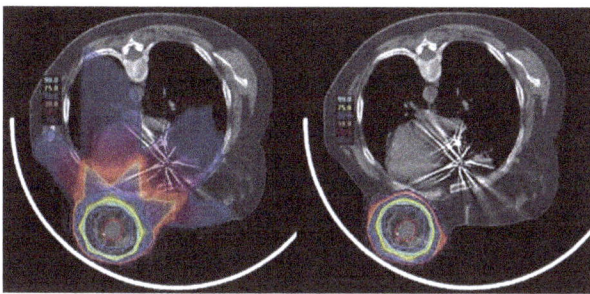

Fig. 10. Comparison of photon (left) and proton (right) treatment plans for postlumpectomy breast cancer. The lumpectomy site is indicated by a red circle within the breast. Color washes indicate that the proton plan allows for complete sparing of both lungs, the heart, and the contralateral breast.

No treatment interruptions were necessary during the hypofractionated course, because no side effects supervened, nor did serious treatment-related sequelae occur.

Favorable initial results warranted a second Phase II trial, now underway. The current trial is identical to the first except that patients with *in-situ* disease or involvement of up to three axillary nodes are permitted entry. Our investigators devised a unique immobilization procedure; details of the procedure and patients' response thereto have been compiled and will be submitted for publication. Early treatment-related toxicity in both trials has been minimal; data on late toxicity, local control, and survival await completion of the second study and followup of patients in both groups.

Primary liver cancers (Fig. 11) are associated with a high mortality rate, partly because many patients are not able to undergo surgery owing to concomitant cirrhosis. Nonconformal photon

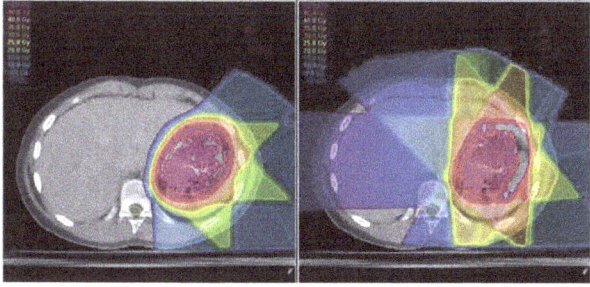

Fig. 11. Comparison of proton (left) and IMRT plans for a tumor in the liver. Both plans conform the high-dose region to the target (red wash), but in the proton plan the volume of body receiving > 2 Gy of radiation is 1038 cm^3; in the IMRT plan the corresponding volume is 2564 cm^3.

radiation often cannot be employed, because the liver outside the target volume often cannot tolerate the high doses required. We undertook a Phase II clinical trial to determine the efficacy and toxicity of proton therapy for patients with locally unresectable hepatocellular carcinoma. Eligible patients included those having T1–T3 hepatocellular carcinomas; selected T4 patients also were accepted. Cirrhotic patients were eligible if they had a Child–Pugh score of 10 or less. Patients with lymph node or distant metastases were not eligible. The target volume encompassed the liver tumor with an additional 1–2 cm margin; the total dose was 63 GyE, administered in 15 fractions. In a published report, two-year actuarial data showed a 75% local tumor control rate and an overall survival rate of 55%. Of patients with elevated pretreatment alpha-fetoprotein (AFP), 85% were found to have declining AFP levels, from a pretreatment mean of 1405 to 35 at six months after treatment. Six patients underwent liver transplantation several months after radiotherapy was completed; two of them demonstrated no evidence of residual carcinoma within the explanted liver. Posttreatment toxicity was minimal and included a small but significant decline in albumin levels and increased total bilirubin. Three patients experienced bleeding when the bowel was immediately adjacent to the treated tumor [40].

4.5. *Cancer of the prostate*

Our extensive experience with proton therapy for prostate cancer illustrates a conservative approach to optimizing use of the proton beam. Because the proton treatment facility was designed to treat up to 200 patients per day, a large clinical experience could be accumulated sooner than had the facility not been able to accommodate a large daily patient load. Many of the advances in proton treatment of prostate cancer at LLUMC have ultimately been made possible by the extensive data that have been collected as treatment strategies for prostate cancer proceeded from a beginning not very different from standard photon protocols of the time. Initial studies used total doses 10% greater than was typical at the time, and preliminary results were encouraging [41]: proton radiation enabled delivery of effective doses to the desired prostatic clinical target volumes while limiting radiation exposure of nearby tissues, thus yielding minimal effects in most patients treated.

A later report, reviewing results in a larger number of patients, treated to 74–75 GyE and followed for periods of up to 12 years, confirmed the earlier observations in terms of biochemical relapse and toxicity. The overall 10-year biochemical disease-free survival rate in this series of 1255 patients was 73%, and was 90% in patients with initial PSA levels of 4.0 or less. The 10-year biochemical disease-free survival rate was 87% in patients with posttreatment prostate-specific antigen (PSA) nadirs of 0.50 or less. Rates dropped with rises in initial and nadir PSA values. Conformal proton radiation therapy at these initial dose levels yielded disease-free survival rates comparable with those of other forms of local therapy, and was associated with minimal morbidity. These results laid the groundwork for dose-escalation trials [42]. A related report addressed the common perception that radiotherapy is preferred for "older" prostate-cancer patients and surgery should be indicated for "younger" men. Again, the large body of data made it possible to analyze biochemical disease-free survival results from more than 1000 patients treated solely with conformal proton radiotherapy to determine whether a difference in outcome supervened for patients younger than 60 years of age versus those older. No statistically significant difference obtained; rather, analysis confirmed the well-known statistically significant predictors of outcome: pretreatment PSA level, clinical stage at diagnosis, and Gleason score. We concluded that age should not be used in and of itself to recommend one type of treatment over another for men with prostate cancer [43].

As experience in treating patients with prostate cancer accumulated, we increasingly reached the conclusion that the precise dose distribution of the proton beam would enable higher doses to be delivered, to increase the probability of controlling the disease, while retaining a low rate of radiation-related side effects. LLUMC and MGH investigators collaborated on a protocol to evaluate the hypothesis that increasing the radiation dose delivered to men with clinically early-stage prostate cancer improves disease outcome. Patients were randomized to receive external-beam radiation, via a combination of conformal photon and proton beams, to a total dose of either 70.2 gray (Gy) (conventional dose) or 79.2 Gy (high dose). The primary outcome measure was PSA level five years after treatment. Sixty-one percent of patients receiving treatment on the conventional dose arm were free from biochemical failure at five years, as opposed to 80% of those receiving high-dose treatment. The difference was significant, and the advantage obtained for both low-risk and higher-risk subgroups. No significant difference was seen in overall survival rates. The two groups had similarly low rates of acute urinary or rectal morbidity, and of severe late morbidity (RTOG Grade 3 or greater).

Study participants concluded that men with clinically localized prostate cancer had a lower risk of biochemical failure if they received high-dose conformal radiation, and that this advantage obtained without an associated increase in RTOG Grade 3 acute or late urinary or rectal morbidity [44].

Dose-response curves for early prostate cancer are still relatively steep, i.e. a 12% increase in the dose yields an 18% increase in disease-free survival at five years for low-risk patients, and a 34% increase in disease-free survival at five years for those at higher risk. Findings such as these suggest that the utility of protons for prostate cancer, and most probably for other diseases, has not yet reached its ultimate application.

At present, most patients treated for prostate cancer receive total doses of 80 GyE or more. At this writing, patients are being enrolled in a clinical trial of hypofractionated proton therapy, which aims to deliver the total dose in about four weeks rather than eight. Our experience accumulated with hypofractionated regimens for other disease sites, noted above, underlies the new trial, as well as the repeated demonstration from dose-escalation studies that high total doses can be delivered without increasing side effects, thanks to the superb ability to spare normal tissues with the proton beam. The same superior proton dose distribution that enables dose-escalation studies also makes hypofractionation studies possible.

In an attempt to quantify as well as illustrate differences between proton and photon irradiation of prostate cancer, in terms of dose distribution and volume integral dose to nontargeted tissues, we randomly selected patients from the cohort that had previously been treated at LLUMC. The aim was to complete comparative proton and IMRT planning of prostate patients, allowing comparative analysis of the mean dose to various volumes including the prostate, bladder, rectum, and external volume. Patients were planned using the Odyssey 4.5

therapy planning system, developed by PerMedics, for IMRT treatment using 6 MV photons. The tumor dose parameters were 81 Gy at 1.8 Gy per fraction for 45 fractions. All patients were treated with protons, while the IMRT plan was constructed to provide the same mean dose to the prostate as delivered using protons as the treatment medium. The beam configuration for proton treatments consisted of two opposing lateral beams, while IMRT treatments consisted of five regularly spaced beams beginning at a gantry angle of 0° (Fig. 12). Comparative results are shown in Table 2.

4.6. *Pediatric neoplasms*

Tumors in children are a variegated mixture of neoplasms that occur in growing tissues. This fact has always presented a special problem for radiation treatment of pediatric tumors: normal-tissue damage can lead to a progressive series of side effects that persist throughout the patient's lifetime. We have explored proton radiation for a variety of pediatric treatment problems, in hopes of exploiting its physical dose distribution in such ways as to spare growing tissues as much as possible. In treating children, avoiding even moderate amounts of irradiation to normal tissues is paramount; our

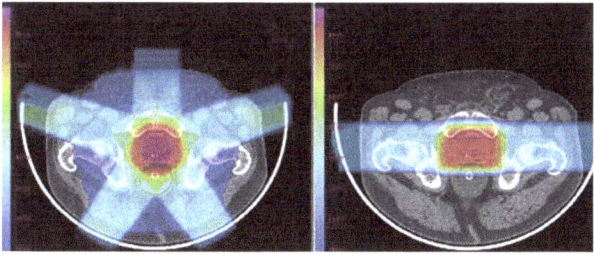

Fig. 12. Comparison in the transverse plane of five-field IMRT (left) and two-field proton plans for prostate cancer.

Table 2. Mean dose received in various anatomic sites by a group of 10 prostate-cancer patients selected at random for comparison planning with IMRT photons and protons. All patients were treated at LLUMC with protons; comparison is illustrative only.

Mean dose (Gy)	IMRT	Proton
Prostate	81.12	80.94
Rectum	46.34	19.42
Bladder	17.10	9.38
External	8.69	4.49

investigators have proceeded on the assumption that conformal 3D planned proton irradiation can contribute to this goal. It is reasonable to expect that a reduced dose and a smaller irradiated volume will reduce radiation effects, but full expression of late effects may occur in children 10 or more years after treatment [45].

Protons have been used to limit treatment-related morbidity in children with tumors in or near the developing brain. In an early study, analysis indicated that instances of early treatment-related morbidity associated with proton therapy were infrequent, albeit tumor progression remained a problem, particularly for histologies such as high-grade glioma [46]. In a study of patients having progressive or recurrent low-grade astrocytoma, proton radiation treatment was generally well tolerated and all children who achieved local control maintained their performance status [47]. This outcome also prevailed in a study of children treated by protons for optic-pathway glioma, a neoplasm for which adequate therapy offers excellent long-term survival rates, making it especially important to avoid treatment-related functional long-term sequelae. A comparison of proton, three-dimensional photon, and lateral photon treatment plans revealed that the proton plans offered a high degree of conformity to target volumes, with steep dose gradients, leading to substantial normal-tissue sparing. Notably, we observed that even in small tumors, conformity of 3D photon irradiation was achieved only at the expense of a larger volume of normal tissues receiving moderate-to-low radiation doses, i.e. the integral volume dose was higher for the photon plans than for the proton plans [48]. In another comparison study, we found similar differences between proton and photon plans for pediatric posterior-fossa tumors, in terms of sparing of auditory structures [49].

In a report of patients between the ages of 3 and 4 years, having Stage M2 or M3 medulloblastoma treated with protons to the craniospinal axis and posterior fossa, a substantially reduced dose to the cochlea and vertebral bodies was noted, as was virtual elimination of the exit dose through the thorax, abdomen, and pelvis. Radiation-related sequelae were minimal, and investigators felt that the technique employed may be especially advantageous in children having a history of myelosuppression (Fig. 13) [50].

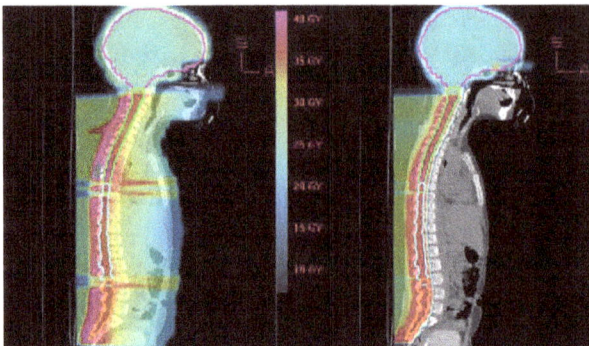

Fig. 13. Comparison of dose distribution using x-rays (left) and protons (right) in the irradiation of the craniospinal axis and posterior fossa for the treatment of meduloblastoma in a three-year-old child. Note the substantial reduction in the dose to the vertebral bodies, and the virtual elimination of the exit dose through the thorax, abdomen, and pelvis.

Similar findings obtained for children with primary skull-base mesenchymal tumors; proton treatment for children with aggressively recurring tumors after major skull-base surgery offered a reasonable prospect of tumor control and survival [51]. Protons are also being used to assist in the management of pediatric craniopharyngioma. A preliminary report indicated that in this instance too, few acute or long-term side effects were observed [52].

Although most of the pediatric neoplasms we have treated thus far have been located in the CNS or the base of the skull, our physicians have used protons in other sites, such as locoregionally advanced, postoperative neuroblastoma, wherein protons have allowed for reducing the dose to uninvolved kidneys, liver, intestine, and spinal cord [53]. Patients in all these studies continue to be followed to assess long-term outcomes.

4.7. *Perspective*

Our fundamental concern in delivering proton radiation is sparing of normal tissues. The greater the extent to which the radiation dose to normal tissues can be reduced or eliminated, the lesser is the likelihood that radiation treatment may have to be compromised because of unacceptable side effects. The ability to reduce or eliminate the radiation dose to normal tissues, thanks to the reduced lateral scatter and sharp dose fall-off of the proton beam, not only allows delivery of the total needed dose but also affords opportunities to deliver that dose in fewer fractions without increasing side effects. The latter

has been borne out in several comparative dosimetry studies and clinical trials at our institution.

The importance of reducing the volume integral dose to normal tissues can be appreciated by studying the work of Philip Rubin. In investigations spanning more than 40 years, Rubin and several collaborators identified the clinicopathologic courses of radiation injury in organs and tissues throughout the body, and also identified tolerance doses for those organs. Tolerances were identified in ranges of total doses in which severe or life-threatening complications were likely to occur within five years of therapeutic radiation; i.e. severe sequelae would likely occur in 5% of patients treated at the lower end of the range $(TD_{5/5})$ and in 50% of patients treated to the dose at the top of the range $(TD_{50/5})$ [54]. Although organs and tissues were separated into categories in accordance with their importance for survival [55], no "safe" dose $(TD_{0/5})$ was identified for any organ; rather, in a classic series of graphs, Rubin and Casarett demonstrated that sublethal doses of radiation initiate a course that tends to clinical manifestations of radiation injury, some of which can progress further to lethality [56].

In later studies, Rubin and colleagues demonstrated early and persistent elevation of cytokine production following pulmonary irradiation. The temporal relationship between elevation of specific cytokines and histological and biochemical evidence of fibrosis illustrated the continuum of response which, the authors speculated, underlies pulmonary radiation reactions and supports the concept of a perpetual cascade of cytokines, produced immediately after radiation treatment and persisting until pathologic and clinical late effects are expressed [57].

5. Research Activities

Research was always intended be an integral component of the proton treatment facility. The inherent characteristics of the proton beam, its precision and conformability, provide a tool for patient care and research, including physical, *in vivo*, and *in vitro* biological studies. Therefore, a carefully developed program of basic, translational, and clinical investigations was deemed necessary.

Three separate proton beam lines, dedicated to research, were part of the original design of the facility; the room housing that beam line has been used for physics, engineering, and radiobiological investigations since the facility opened. In addition, planning for a separate research building, consisting of laboratories and equipment used by scientists, commenced soon after the proton treatment facility opened in 1990. That building opened in 1995; its facilities are used by internal investigators and by visiting researchers, many of whom conduct studies of interest to the National Aeronautics and Space Administration (NASA).

The connection between LLU and NASA began in the early 1990s, when JMS met Walter Schimmerling, Project Scientist for NASA's Space Radiation Health Programs. It was clear that the two parties had similar interest in basic studies of the effect of proton radiation on humans; the LLU interest grew out of proton treatment of patients, while NASA's, as expressed by the then director, Daniel Goldin, arose from its concerns about the effects of long-duration space missions on the health of astronauts. Protons constitute most of the particles encountered in space, owing to solar flares from the Sun and other stars throughout the universe; NASA needed access to a facility that could provide a source of protons for its basic studies, including simulation of solar flares. A series of discussions led in December 1994 to the signing of a Memorandum of Agreement between the two parties, in which LLU agreed to provide beam time to NASA investigators for their research projects.

5.1. *Research strategies*

Basic and preclinical research occurs in five major areas. These are: (1) *physics, chemistry, and molecular-radiobiology studies*, designed to promote better understanding of ionization effects at the molecular, cellular, and organelle levels when proton beams pass through tissue; (2) *modifying results* of proton irradiation by various methods of radiation sensitization and protection; (3) *engineering developments* to exploit technological advances that increase the precision and reproducibility of beam delivery; (4) investigating the therapeutic use of proton radiation for a variety of *nonmalignant* disease processes; and (5) *space-science research*, which yields information of interest for space travel and will be useful for clinical studies. All these investigations have the end goal of stimulating clinical trials and, eventually, making proton radiotherapy more effective and more broadly used.

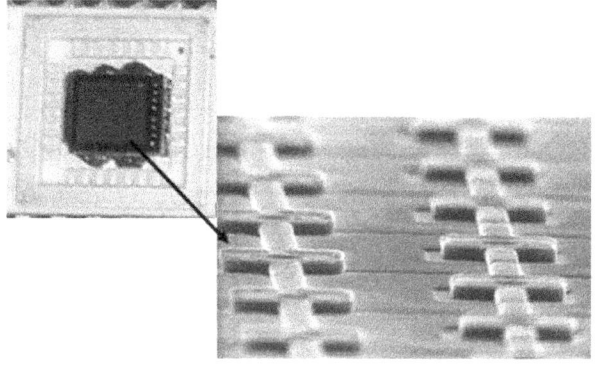

1/3 Diameter of a Human Hair

Fig. 14. Example of a silicon-on-insulator (SOI) microdosimeter — a two-dimensional array of detectors working together to collect radiation events and provide information on the radiation field. The SOI device essentially looks like a computer chip; each holds 150–5000 detector elements. The small size of these detectors enables one to place them within an anthromorphic (man-shaped) radiation-therapy phantom, thus providing information on not only the radiation field at a point in a room or at the surface of the patient, but also within the patient.

5.1.1. *Basic physics*

In the broadest sense, these studies analyze the effects of the proton beam in tissue. These effects can be observed at the microdosimetric and nanodosimetric levels. Investigating phenomena occurring in these tiny volumes helps our scientists to better understand fundamental processes that lead to cell destruction, and also to better understand what is happening in tissues.

In collaboration with Professor Anatoly Rosenfeld at the Centre for Medical Radiation Physics at the University of Wollongong, Australia, we are studying a new approach to microdosimetry using silicon-on-insulator (SOI) technology (Fig. 14) [58]. According to research done thus far, the SOI microdosimeter used at LLU measures the radiation dose as accurately as older, more established dosimeters do, and does so while permitting measurements at distances that are very close to the target volumes. More importantly, these measurements indicate that the out-of-field doses, including scattered neutron radiation, drop off very rapidly from the edge of the field. This useful information, which might not have been available without research into the use of SOI technology, indicates a high degree of conformability, and indicates further that neutron scattering occurs at extremely low levels [59].

In nanodosimetric studies, our investigators, in collaboration with colleagues from the Weizmann Institute of Science, Israel, developed a wall-less ion-counting nanodosimeter, conceived for precise ionization-cluster measurements in an accelerator environment. This nanodosimeter provides an accurate means of counting single radiation-induced ions, in dilute-gas models of condensed matter [60]. In further studies, investigators have equipped the device with a silicon microstrip telescope that tracks the primary particles, allowing one to correlate nanodosimetric data with the particle position relative to the device's sensitive volume. Monte Carlo track structure simulations indicated good agreement between "tracking nanodosimetry" data acquired with the new system and simulated data, thus supporting further development and study of the device [61].

5.1.2. *Modifying results of proton irradiation*

Antioxidant radioprotectants are potentially attractive in managing cancers treated optimally with high total doses of radiation. Although the cure rate for many tumors would be increased by escalating the radiation dose, that outcome has to be balanced against increasing the risk of normal-tissue injury. Local control of prostate cancer, for example, improves significantly when the total radiation dose exceeds 72 GyE, but the risk of severe complications also increases. One way of reducing this untoward outcome is to use a proton beam, which allows for higher total doses, up to 80 GyE or more, and spares much of the normal tissue, but even so, some normal tissues are exposed to radiation. The goal of this research is to protect these normal cells that cannot be shielded from the radiation beam.

Efforts are ongoing to develop new radioprotective drugs. The superoxide dismutases (SODs), constitute one class of such agents. Metal-containing SOD mimetics have emerged as being especially promising. We are studying some of these agents [62, 63].

5.1.3. *Engineering advances*

Modern treatment-planning systems exploit the latest advances in software for optimal utility in the clinical setting. In the mid-1990s, our physicians and physics staff worked with PerMedics to develop an advanced planning system, called Opitrad™, and used it for more than a decade.

When still further treatment-planning capabilities became desired, such as the ability to fuse image data from a variety of sources, including CT, MRI, and PET scanning, we worked again with PerMedics to develop an improved system, Odyssey™. The medical-physics staff evaluated the system's host of manual and semiautomatic tools, including not only image fusion but the ability to display dose-volume histograms (DVHs) for different structures on the same graph or display DVHs for the same structure from alternate plans for easy evaluation. This comprehensive external-beam treatment-planning system can be used with all radiotherapeutic modalities. This new system was commissioned over a period of several months and was deployed in late 2007.

Another aspect of modern technology that has been investigated at Loma Linda is robotics (Fig. 15). Specifically, the department of radiation medicine has collaborated for several years with Optivus to develop a precision patient alignment system (PPAS), which is based on an industrial robot very like those used in plants where automobiles are assembled. The PPAS enables the physician to deliver the required conformal dose precision and to repeat that precision, treatment after treatment. The system also permits shorter overall total treatment times owing to reduced per-treatment patient setup times, which is the most time-consuming part of radiation-therapy delivery. It does so via several

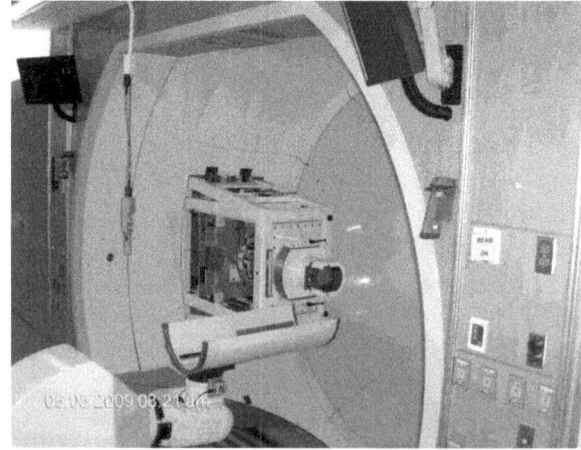

Fig. 15.   Robotic precision patient alignment system (PPAS) during installation in Gantry Room 1 at the LLUMC proton facility, May 2009. It is expected to be in clinical use by September 2009.

components that permit an almost infinite range of positions. Highly sophisticated software controls the positioning of all components of the system according to each patient's unique treatment prescription. The movements of the turntable and the robotic arm, and ultimately the beam delivery, are controlled by software that places the patient support unit in exactly the right position at each treatment session. The system ensures rapid, highly precise, repeatable patient positioning, enhancing the ability of radiation oncologists at LLUMC to deliver proton-radiation treatments to even more patients, without sacrificing quality.

An active beam-delivery system, more commonly called the scanning beam, is nearing completion of a long developmental process at LLU and Optivus. The development process has been essentially on hold while work has proceeded on installation of the PPAS (above), but installation is expected in 2010. A prototype system is operating in the research room at present. The scanning beam will permit radiation oncologists to treat larger volumes and will operate dynamically and without collimation devices inserted in the beam's path: it will provide the ability to change field sizes and shapes solely through three-dimensional electronic control of the beam. The system will thus enable delivery of intensity-modulated proton therapy (IMPT) if desired, and will permit radiation oncologists to treat larger and more irregular volumes, with greater precision, than can generally be accomplished with passive-scattering techniques.

5.1.4. *Protons for non-malignant diseases*

At LLU, investigators are evaluating the potential of proton radiation for treating nontumor processes, particularly those with underlying focal abnormalities. One of these is medically refractory epilepsy. Most of this effort has been focused on developing a stereotactic radiosurgical procedure for curing intractable mesial temporal-lobe epilepsy (TLE).

The clinical course of mesial TLE often begins in infancy. There is usually a long, silent (or drug-responsive) interval, but spontaneous recurrent seizures re-emerge in late childhood or early adolescence. Episodes last a few minutes, but can lead to secondary convulsions. After the seizure, patients may be disoriented; many will have problems speaking or understanding language and will not remember any of the seizure episode except the initial aura. These seizures are often refractory to antiepileptic drugs and can reoccur spontaneously, hundreds of times a month.

Approximately 50% of newly diagnosed patients have partial epilepsy, most commonly of temporal-lobe origin. The temporal lobe is the most epileptogenic region of the brain. Only half of new-onset partial seizures are controlled effectively by the first antiepileptic drug. About 40% of patients continue to have seizures in spite of trials of antiepileptic drugs, and the chance for seizure freedom in these patients is low. Uncontrolled seizures take their toll on normal brain function: patients with refractory TLE typically have deficits in memory function, and these get worse with time. Surgical resection of the amygdala and anterior hippocampus is curative for most patients, but even so, less than half of good surgical candidates opt for surgery. Upward of 100,000 patients in the United States would benefit from an alternative, but remain untreated.

Preclinical studies on rodents have shown that converging proton beams, targeting the hippocampus, amygdala, and parahippocampal gyrus, selectively ablate these mesial epileptogenic regions. The same properties that make protons so appealing for treating cancer are equally important in targeting the mesial temporal-lobe structures, which lie adjacent to several functionally critical and radiosensitive brain regions. Preclinical efforts have led to clinical treatment plans for stereotactic radiosurgical ablation of the amygdala and hippocampus that will be used to treat patients with mesial TLE, and research into means of correcting for imaging uncertainties is being done [64]. Clinical trials of stereotactic amygdala–hippocampal radiosurgery are a logical next step in this ongoing process.

5.1.5. *Space-science investigations*

Protons constitute more than 90% of cosmic irradiation, and the proton accelerator at LLU has been a useful tool for simulating the environment of space. Our proton-beam research room includes an apparatus for simulating solar-particle events (solar flares), and experiments are done with test animals in this environment (Fig. 16). Mice are placed in cages mounted on a rack, in identical conditions and so situated that each receives the same dose as all the

Fig. 16.   Apparatus designed to hold test animals during simulated solar-particle events (simulating solar flares) or during low-dose, low-dose-rate studies.

others. They are exposed to simulated solar-particle-event protons over a period of 36 h to a total dose of 2 Gy; during experiments, proton energies have ranged from 25 to 215 MeV. The same setup has been used to deliver low-dose, low-dose-rate proton radiation (simulating the background radiation astronauts are inevitably exposed to during deep-space missions); the total dose has been as low as 0.01 Gy, delivered at a rate of 0.1 centigray per hour.

In a different set of experiments, we also have analyzed various outcomes of space flight in laboratory animals that have been carried into space on NASA missions. Animals from these missions have been transported to Loma Linda for thorough evaluation of the effects of space flight on most organs.

LLU investigators collaborate with scientists of Brookhaven National Laboratory to examine the effects of all components of solar radiation, from protons to heavy ions. The effects of simulated or real space flight on immune function, antioxidant gene expression, bone and bone marrow, liver cells, and the hematopoietic system, among others, have been analyzed and reported [65–72]. From NASA's perspective, these investigations provide data that will help prepare for the potential effects astronauts may experience on long-duration space missions, such as a trip to Mars. The same basic data on radiation effects

may be relevant to future basic studies of proton or other radiation treatment of patients.

5.2. Future directions

Until 1990, proton treatments were available only in physics research laboratories. As mentioned earlier in this account, enormous technological advances were needed to allow a proton treatment facility to be placed in a hospital environment. Today, hospital-based centers are increasing in the United States. The existence of more large clinical facilities, and the large numbers of patients they will serve, will permit multi-institutional cooperative clinical trials that lead to valuable clinical data about the applications of proton therapy. This amplifies a process begun at LLU: as noted, one of the reasons that our facility was designed to have several treatment rooms and to be located on the hospital campus was to offer a full spectrum of evaluation and treatment services necessary for excellent cancer-patient treatment. This is important to patients and also for developing meaningful clinical data on results.

Further advances in technology, and radiobiological investigations, will exploit proton radiation treatment even more in the future. One such example, noted above and now close to clinical utilization, is the use of robotic patient positioning devices and real-time monitoring of patient and tumor during treatment. An active beam delivery system, also under development, is another. Such developments will increase precision and repeatability, resulting in a even greater potential for minimizing normal-tissue damage.

We also are exploring protons for therapy planning, replacing the photon-based CT systems commonly employed today with a system that uses protons. Because conformal proton radiotherapy requires accurate prediction of the Bragg peak position, proton imaging may be more suitable than conventional x-rays for this task. Our work thus far has shown that a reasonable density resolution for imaging can be achieved with a relatively small dose of protons, one that is comparable to or even lower than that of CT with x-rays [73].

The potential of protons is only beginning to be developed. Technological advances, such as those alluded to above, and collaborative trials will reveal many more applications for protons as the years go on. From our perspective, the LLU experience is but the prolog to even more progress.

6. Summary

Bringing proton radiation therapy into a hospital environment and using advanced technology to do so, has been motivated and driven solely by patients' needs. Hospital-based proton therapy was developed to bring about several benefits to patients. These include:

- *Reducing normal-tissue injury.* This is the objective upon which everything else rests. Sparing normal cells from exposure to radiation is the key to reducing the side effects that sometimes cause intolerable distress, leading to discontinuation of treatment and occasionally to late complications. Reducing exposure of normal cells is also the key to dose-escalation and hypofractionation studies.
- Reducing normal-tissue injury allows *reducing the cost of therapy* because of hypofractionation. The number of fractions is one of the primary factors in reimbursement of delivering radiation therapy. Aside from this cost-saving potential, proton therapy reduces costs by reducing the incidence of side effects, which require costly treatment.
- Reducing normal-tissue injury provides a *basis for using radiation as an adjunct to chemotherapy.* Bone marrow, lung, heart, and other vital organs and tissues can be avoided when using protons for any anatomic site, thereby allowing full-course chemotherapy to be delivered.

The common thread running through all three points is control: the control the physician has over therapeutic tools. X-rays have saved millions of lives worldwide since being discovered by Roentgen in 1895. However, inability to control the beam in three dimensions led to, in many cases, considerable patient and physician distress during and following treatment. It became very clear to JMS, early in his training, that excessive normal-tissue damage was caused primarily by two factors: the lack of advanced treatment-planning technology and the photon's lack of three-dimensional controllability. Heavy charged particles such as protons provide the physician with a major advantage: they proceed to the target and cause ionization in a highly predictable pattern. The ionization process can be subjected to relatively simple computer-assisted modeling, thereby further enhancing the value of heavy charged particles in clinical therapy.

Protons became available for experimental studies only 15 years prior to our initial studies. As noted in this article, the impetus for designing the hospital-based center at LLU, the advances that occurred during the developmental phase, and all improvements that have occurred since, have been driven by patients' needs. The fundamental needs, from which all others arise, can be reduced to two: all patients must receive effective treatment of tumors, and there is a paramount need to spare normal cells to the maximum extent possible.

Reducing injury to normal cells was the key to many other advances in radiation therapy. A controllable and conformable therapy beam was one advance needed to realize that goal. To exploit such a beam to the utmost, however, the physician first had to employ therapy planning that defines the true extent of tumors as accurately as possible.

Computer-assisted planning systems were the first steps in addressing the need to accurately delineate the extent of the local tumor and its regional spread. The initial, ultrasound-based system developed at LLU and employed clinically in 1971 seems crude now, but it was a notable advance over radiotherapy planning as it was practiced till that time. The value of this work became manifest as the 1980s progressed and CT-based, computer-assisted treatment-planning systems became widespread. Further, the use of such systems prompted the development of conformal therapy systems such as IMRT, which reduce the volume of tissues in the high-dose region. Precision therapy planning, therefore, improved photon-beam therapy; subsequently, major manufacturers offered treatment-delivery systems mated to their precision planning systems.

Improvements in photon-beam therapy arising from the use of precision treatment-planning systems, important though they were, were not sufficient. The fact remained that the controllability of photons is, ultimately, inadequate. The need for a more controllable particle, yet one that approximated the radiobiologic effect of photons, guided our efforts. Heavy charged particles offered the potential for the needed three-dimensional control, and thus were regarded as better choices than photons for good medical practice. The proton became our particle of choice because it combined the required controllability with low-LET characteristics, similar to x-rays.

As clinical research proceeded, we pursued hypofractionation studies. The sites for which clinical trials are ongoing at this writing include the prostate, breast, lung, and liver. Historically, hypofractionation has been little used in photon radiotherapy. Early therapeutic uses of x-rays, occurring soon after Roentgen's discovery in 1895, often led to unacceptable damage to normal tissues. Coutard [73], building on earlier work by Regaud, established the value of fractionating the treatment dose, thus providing to normal cells opportunities to recover from radiation injury. This practice, effective because cancer cells had inferior ability to repair radiation injury, became standard in radiation oncology. As noted elsewhere in this article (see Figs. 10–13), protons can be used in a manner that reduces normal tissue injury, owing to reduced scattering to normal tissues and the sharp dose fall-off at the end of the range. We documented the proton advantage in normal-tissue sparing in several sites for which we compared proton and photon treatment plans, including IMRT photons; our observations echoed those reported by others [75–78]. The difference in the volume integral dose to normal tissues suggested to us that we could begin hypofractionating treatments in attempts to reduce overall treatment time and costs. Thus far, our preliminary published and unpublished results have demonstrated improved outcomes in control of disease when hypofractionation is used clinically; the reason is probably the increased ionization density in the targeted tumor volume with hypofractionation. Greater damage to tumor cells is occurring therein, even as untargeted tissues continue to be spared.

In the future, we expect to concentrate on advancing technology for delivery of protons, including: (1) fully computer-assisted treatment rooms, featuring automated patient alignment systems and a scanning beam; (2) continued investigation of alternate fractionation schedules, including hypofractionation; (3) adjunctive radioprotection of crucial anatomic sites to enhance hypofractionated treatments and protect crucial sensitive tissues intimate to radiation fields; and (4) studies to enhance tumor radiosensitivity. In addition, new areas of treatment are being investigated, including the use of protons for benign diseases and neurologic disorders.

The past is indeed prologue. As has been true from the beginning, routine clinical practice at our institution will reflect and apply the findings of research on the optimal use of the proton beam.

References

[1] H. Eickoff and U. Linz, Medical applications of accelerators, in *Reviews of Accelerator Science and Technology*, eds. A. W. Chao and W. Chou (World Scientific, Singapore, 2008), Vol. 1, pp. 143–161.

[2] W. H. Bragg, *Studies in Radioactivity* (Macmillan, London, 1912), pp. 29–38.

[3] R. R. Wilson, *Radiology* **47**, 487 (1946).

[4] J. H. Lawrence, *Cancer* **10**, 795 (1957).

[5] C. A. Tobias *et al.*, *Peaceful Uses At. Energy.* **10**, 95 (1956).

[6] B. Larsson *et al.*, *Acta. Chir. Scand.* **125**, 1 (1963).

[7] R. N. Kjellberg *et al.*, *New Engl. J. Med.* **309**, 689 (1968).

[8] N. K. Ambrosimov *et al.*, *Proc. Acad. Sci. USSR* **5**, 84 (1985).

[9] J. M. Slater *et al.*, *Cancer* **34**, 96 (1974).

[10] I. R. Neilsen *et al.*, An interactive system for radiation treatment-planning utilizing computed tomography data as input to the planning process, in *Medinfo 80: Proc. Second World Conference on Medical Informatics* (North-Holland, Amsterdam, 1980), p. 1072 (abstract).

[11] M. Goitein and M. Abrams, *Int. J. Radiat. Oncol. Biol. Phys.* **9**, 777 (1983).

[12] M. Goitein *et al.*, *Int. J. Radiat. Oncol. Biol. Phys.* **9**, 789 (1983).

[13] P. H. Fowler *et al.*, *Nature* **189**, 524 (1961).

[14] J. O. Archambeau *et al.*, *Radiology* **110**, 445 (1974).

[15] J. O. Archambeau *et al.*, *Acta Radiol. Ther. Phys. Biol.* **13**, 393 (1974).

[16] M. F. Moyers and D. A. Lesyna, *Radiat. Meas.* **44**, 176 (2009).

[17] M. H. Phillips *et al.*, *Int. J. Radiat. Oncol. Biol. Phys.* **18**, 211 (1990).

[18] V. Smith *et al.*, *Int. J. Radiat. Oncol. Biol. Phys.* **40**, 507 (1998).

[19] L. J. Verhey *et al.*, *Int. J. Radiat. Oncol. Biol. Phys.* **40**, 497 (1998).

[20] H. Silander *et al.*, *Acta Neurol. Scand.* **109**, 85 (2004).

[21] E. B. Hug and J. D. Slater, *Neurosurg. Clin. N. Amer.* **11**, 627 (2000).

[22] E. B. Hug *et al.*, *J. Neurosurg.* **91**, 432 (1999).

[23] M. Samii and C. Matthies, *Neurosurgery* **40**, 11 (1997).

[24] J. C. Flickinger *et al.*, *J. Neurosurg.* **94**, 141 (2001).

[25] R. L. Foote *et al.*, *Int. J. Radiat. Oncol. Biol. Phys.* **32**, 1153 (1995).

[26] D. A. Bush *et al.*, *Neurosurgery* **50**, 270 (2002).

[27] B. B. Ronson *et al.*, *Int. J. Radiat. Oncol. Biol. Phys.* **64**, 425 (2006).

[28] W. D. Johnson *et al.*, *Neurosurg. Focus.* **24**, E2 (2008) [review].

[29] M. Fuss *et al.*, *Int. J. Radiat. Oncol. Biol. Phys.* **49**, 1053 (2001).

[30] J. D. Slater *et al.*, *Int. J. Radiat. Oncol. Biol. Phys.* **62**, 494 (2005).

[31] R. Lin *et al.*, *Radiology* **213**, 489 (1999).

[32] D. A. Bush *et al.*, *Am. J. Roentgenol.* **172**, 735 (1999).

[33] D. A. Bush *et al.*, *Chest* **116**, 1313 (1999).

[34] R. B. Bonnet *et al.*, *Chest* **120**, 1803 (2001).

[35] D. A. Bush *et al.*, *Chest* **126**, 1198 (2004).

[36] B. Fisher and S. Anderson, *World J. Surg.* **18**, 63 (1994).

[37] T. Whelan *et al.*, *Int. J. Radiat. Oncol. Biol. Phys.* **30**, 11 (1994).

[38] G. Liljegren *et al.*, *J. Clin. Oncol.* **17**, 2326 (1999).

[39] U. Veronesi *et al.*, *N. Engl. J. Med.* **347**, 1227 (2002).

[40] D. A. Bush *et al.*, *Gastroenterology* **127** (5 Suppl. 1), S189 (2004).

[41] J. D. Slater *et al.*, *Urology* **53**, 978 (1999).

[42] J. D. Slater *et al.*, *Int. J. Radiat. Oncol. Biol. Phys.* **59**, 348 (2004).

[43] C. J. Rossi, Jr. *et al.*, *Urology* **64**, 729 (2004).

[44] A. L. Zietman *et al.*, *JAMA* **294**, 1233 (2005); erratum in *JAMA* **299**, 899 (2008).

[45] E. B. Hug and J. D. Slater, *Strahlenther. Onkol.* **175** Suppl 2, 89 (1999).

[46] B. McAllister *et al.*, *Int. J. Radiat. Oncol. Biol. Phys.* **39**, 455 (1997).

[47] E. B. Hug *et al.*, *Strahlenther. Onkol.* **178**, 10 (2002).

[48] M. Fuss *et al.*, *Int. J. Radiat. Oncol. Biol. Phys.* **45**, 1117 (1999).

[49] R. Lin *et al.*, *Int. J. Radiat. Oncol. Biol. Phys.* **48**, 1219 (2000).

[50] G. E. Yuh *et al.*, *Cancer J.* **10**, 386 (2004).

[51] E. B. Hug *et al.*, *Int. J. Radiat. Oncol. Biol. Phys.* **52**, 1017 (2002).

[52] Q. T. Luu *et al.*, *Cancer J.* **12**, 155 (2006).

[53] E. B. Hug *et al.*, *Med. Pediatr. Oncol.* **37**, 36 (2001).

[54] P. Rubin, The law and order of radiation sensitivity, absolute vs. relative, in *Clinical Radiation Pathology*, ed. J. M. Vaeth (University Park Press, Baltimore, 1988), Vol. 22.

[55] P. Rubin, R. A. Cooper and T. L. Phillips (eds.), *Radiation Biology and Radiation Pathology Syllabus* (American College of Radiology Publications, Chicago, 1978).

[56] P. Rubin and G. W. Casarett, *Clinical Radiation Pathology* (W. B. Saunders, Philadelphia, 1968), Vols. 1 and 2.

[57] P. Rubin *et al.*, *Int. J. Radiat. Oncol. Biol. Phys.* **33**, 99 (1995).

[58] P. D. Bradley *et al.*, *NIM B* **184**, 135 (2001).

[59] A. Wroe *et al.*, *Med. Phys.* **34**, 3449 (2007).

[60] G. Garty *et al.*, *Radiat. Prot. Dosimetry* **99**, 325 (2002).

[61] V. Bashkirov *et al.*, *Radiat. Prot. Dosimetry.* **122**, 415 (2006).

[62] D. S. Gridley *et al.*, *Anticancer Res.* **27**(5A), 3101 (2007).

[63] A. Y. Makinde *et al.*, *Anticancer Res.* **29**, 107 (2009).

[64] D. Kittle *et al.*, *Med. Phys.* **35**, 5708 (2008).

[65] S. A. Lloyd *et al.*, *Adv. Space Res.* **42**, 1889 (2008).

[66] M. T. Ortega *et al.*, *J. Appl. Physiol.* **106**, 548 (2009).

[67] D. S. Gridley *et al.*, *J. Appl. Physiol.* **106**, 194 (2009).

[68] J. S. Willey *et al.*, *Radiat. Res.* **170**, 201 (2008).

[69] D. S. Gridley *et al.*, *Int. J. Radiat. Biol.* **84**, 549 (2008).

[70] E. R. Bandstra *et al.*, *Radiat. Res.* **169**, 607 (2008).

[71] D. S. Gridley *et al.*, *In Vivo* **22**, 159 (2008).

[72] D. S. Gridley *et al.*, *Radiat. Res.* **169**, 280 (2008).

[73] R. W. Schulte *et al.*, *Med. Phys.* **32**, 1035 (2005).

[74] H. Coutard, *Lancet* **2**, 1 (1934).

[75] J. Y. Chang *et al.*, *Int. J. Radiat. Oncol. Biol. Phys.* **65**, 1087 (2006).

[76] A. J. Lomax *et al.*, *Int. J. Radiat. Oncol. Biol. Phys.* **55**, 785 (2003).

[77] K. R. Kozak *et al.*, *Int. J. Radiat. Oncol. Biol. Phys.* **65**, 1572 (2006).

[78] D. C. Weber *et al.*, *Radiat. Oncol.* **1**, 22 (2006).

James M. Slater is Professor and Vice Chairman of the Department of Radiation Medicine, Loma Linda University Medical Center (LLUMC), and Director of the Loma Linda University Radiation Research Laboratories. He spearheaded development of digital-image-based, computer-assisted radiotherapy planning and the world's first hospital-based proton treatment center, which opened in 1990 following more than two decades of research. Dr. Slater developed the radiation oncology clinical program and research laboratories at Loma Linda, retiring as chairman in 2001. He continues research to further enhance the clinical utility of proton therapy. In recognition of his achievements, in 2007 the LLUMC proton treatment center was named in his honor.

Jerry D. Slater is Professor and Chairman of the Department of Radiation Medicine, Loma Linda University Medical Center. He is a member of several professional societies, including the American College of Radiology, the American Radium Society, the American Society of Clinical Oncology, the American Society for Therapeutic Radiology and Oncology, and the Particle Therapy Co-Operative Group. His prime pursuit is optimizing radiation therapy, including proton-beam irradiation; he has designed treatment protocols and clinical research investigations since hospital-based proton therapy was pioneered at Loma Linda in 1990. Among his current research pursuits are image-guided and intensity-modulated proton therapy; hypofractionation studies; and proton radiosurgery.

Andrew J. Wroe is a medical physicist and translational researcher at the James M. Slater, MD Proton Treatment and Research Center, Loma Linda University Medical Center (LLUMC). He is also a faculty member of Loma Linda University School of Medicine, where he teaches radiation physics to radiation therapy students and radiation oncology residents. He is an Honorary Fellow at the University of Wollongong, Australia. His research focuses on microdosimetry and nanodosimetry, in which he obtained his PhD and was awarded a Fulbright Scholarship in 2005 and, in 2007, the Australian Institute for Nuclear Science and Engineering Gold Medal for Excellence in Research.

Reviews of Accelerator Science and Technology
Vol. 2 (2009) 63–92
© World Scientific Publishing Company

Microwave Electron Linacs for Oncology

David H. Whittum

Oncology Systems, Varian Medical Systems
911 Hansen Way C077, Palo Alto, CA 94304-1028, USA
david.whittum@varian.com

The history and technology of medical linacs are reviewed, focusing on machine requirements for radiotherapy. Configurations used in modern machines are described and operational aspects of a gantry-style linac system are illustrated with reference to the state of the art. Aspects of structure design, modeling and testing are discussed.

Keywords: Radiotherapy; dosimetry; computed tomography; respiratory gating; image guided; microwave; biperiodic; Brillouin; klystron; magnetron.

1. Introduction

It was William Hansen who realized that an empty copper box could make huge voltages possible in a small space [1, 2]. An enclosed cavity the size of a baseball can resonate at microwave frequencies. If a high conductivity material such as copper is used, a large quality factor $Q \approx 10^4$ results, and resonant enhancement of the applied electric field by a factor $Q^{1/2} \approx 10^2$ is possible. When the cavity is evacuated, baked and sealed, it can sustain microwave electric fields over $20\,\mathrm{MV/m}$ without arcing. Microwave electron accelerators depend on the resonant excitation of such high electric fields in copper vacuum tubes [3], and because of them can produce beams suitable for cancer therapy with an assembly small enough to rotate around the patient.

In the years following the work of Hansen on the accelerator, and the Varian brothers on the microwave power source [4], large machines were built for high energy physics research [5, 6]. The same researchers at Stanford and others explored application of microwave linear accelerators to radiation therapy. The history of this seminal period in the 1950s has been recounted by one of the original participants, Edward Ginzton, with his colleague Craig Nunan, *circa* 1984 [7]. In the two decades since, while accelerator technology has developed at a dramatic pace for high energy physics, applications in medicine have been slow to change. The improvements in medical linac systems, which have

been dramatic, have for the most part not involved more than evolutionary change in the heart of the machine — the accelerator.

An excellent introduction to medical electron accelerator systems and components has been provided by Karzmark, Nunan and Tanabe, *circa* 1993 [8]. A current historical perspective has been given by Thwaites and Tuohy [9]. In this work we review the concepts and history of medical linacs and the requirements for radiotherapy, with the goal of equipping the reader with a perspective suited to exploration of accelerator technology in radiotherapy, and an appreciation of the physical limitations that have shaped the present-day machines.

1.1. *Why is an accelerator structure needed?*

A brief review of some accelerator basics is in order; longer reviews are available [10, 11]. From Maxwell's equations one can predict that there are three methods of acceleration. Electrostatic acceleration corresponds to the limit where the electric field may be represented as the gradient of an electrostatic potential and Maxwell's equations reduce to Poisson's equation. A common electrostatic accelerator is an electron gun.

The second method of acceleration engages Faraday's law. External currents drive a magnetic field shaped by magnetic materials, and the resulting time-varying magnetic flux induces an electric field.

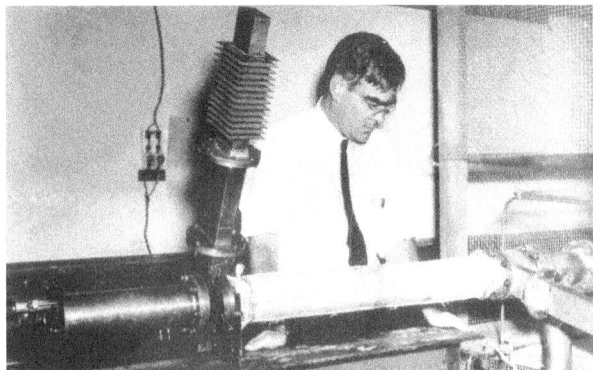

Fig. 1. William Hansen conceived of the high-Q resonant cavity and, based on this "rhumbatron" concept, went on to build the Mark I accelerator shown here.

This mechanism underlies the betatron and induction linacs.

The third method of acceleration engages Maxwell's displacement current and involves a fully electromagnetic excitation, the simplest example of which is an electromagnetic wave in free space. A key point, however, is that free space alone does not support acceleration at first order in the applied field, and material boundaries are required. To see this, consider that we require an electric field, $\mathbf{E}$, to produce any change in particle energy, ε, and energy transfer is governed by

$$\frac{d\varepsilon}{dt} = q\mathbf{E} \cdot \mathbf{V},$$

where $\mathbf{V}$ is the particle velocity and q is the charge. Let us suppose that the particle is already relativistic, so that $V \approx c$ is constant. Denote the direction of particle motion as $\hat{s}$, and parametrize the motion by the length traversed, s, with $ds = V\,dt$. Then the particle coordinates may be expressed as $\mathbf{r} = s\hat{s}$, $t = t_0 + s/V$, where t_0 is the time at which the particle reaches $s = 0$. Energy gain takes the form

$$\frac{d\varepsilon}{ds} = qE_s(\mathbf{r}, t) = qE_s\left(s\,\hat{s}, t_0 + \frac{s}{V}\right),$$

where E_s is the electric field component parallel to the particle motion. Fields in free space may be expressed as a superposition of plane waves, and at linear order we may consider one such component,

$$\mathbf{E} = \Re\tilde{E}\exp(j\omega t - j\mathbf{k} \cdot \mathbf{r}),$$

$$\mathbf{H} = \Re\tilde{H}\exp(j\omega t - j\mathbf{k} \cdot \mathbf{r}),$$

corresponding to a wave with angular frequency ω and wave vector $\mathbf{k}$, varying in time t at position $\mathbf{r}$. Gauss's law requires that the polarization

of the electric field be transverse to the direction of propagation, $\tilde{\mathbf{E}} \cdot \mathbf{k} = 0$, implying two independent polarizations. Faraday's law requires that the magnetic field polarization be transverse to both the direction of propagation and the electric field, $Z_0\tilde{H} = \tilde{\mathbf{E}} \times \hat{\mathbf{k}}$. The quantity $Z_0 = \sqrt{\mu_0/\varepsilon_0} = 376.7\,\Omega$ is the wave impedance of free space. With this we can compute the energy imparted to the particle in passing through any finite region $s_0 < s < s_1$,

$$\Delta\varepsilon = \Re q\tilde{E}_s \int_{s_0}^{s_1} ds \, \exp s\left(j\omega t_0 + j\left[\frac{\omega}{V} - k_s\right]s\right)$$

$$= \Re q\tilde{V}e^{j\omega t_0}$$

and we introduce the *accelerating voltage* phasor $\tilde{V}$. In free space, we extend the limits to infinity to obtain

$$\tilde{V} = 2\pi\tilde{E}_s\delta\left(\frac{\omega}{V} - k_s\right),$$

with δ the Dirac delta function. The dispersion relation $\mathbf{k}^2 = \omega^2/c^2$, and transverse polarization, imply that for $E_s \neq 0$, $\omega/k_s > c$, so that $\omega/k_s > V$, and therefore $\tilde{V} = 0$.

Thus, there is no net acceleration in free space at first order in the applied fields. Material boundaries must figure in some essential way in shaping the fields needed for acceleration. This is why an accelerator structure is needed.

Further, it is evident that acceleration should take place either in a terminated region of space or over an extended region, but with a wave whose phase velocity $\omega/k_s = c$. These two possibilities correspond to *standing wave* and *traveling wave* accelerators. In the case of a region of length L, one may show that $\tilde{V} = \tilde{E}_sLT$, up to an overall phase factor, with the "transit angle" factor

$$T = \frac{\sin\left(\frac{1}{2}\theta\right)}{\frac{1}{2}\theta},$$

expressed in terms of the transit angle $\theta = \omega L/V$. This result quantifies the effect of the changing of the phase from accelerating to decelerating as the electron traverses a cavity. The ideal cavity length is no more than $L \approx \beta\lambda/2$, with $\beta = V/c$.

Our discussion does not exclude the more novel acceleration mechanisms discussed in recent years. Dielectric loading of a conducting waveguide can lower the phase velocity to achieve synchronism, and is the basis for a traveling wave accelerator. Inject-ing an electron beam into a plasma can expel plasma electrons radially outward, and the resulting radial current induces an axial electric field that can accel-erate a follow-on beam (plasma wakefield acceler-ator). A laser beam exerts radiation pressure on charged particles and so can accelerate them at sec-ond order in the applied field, in free space. Where an intense laser beam is injected into a plasma, the radial radiation pressure expels plasma electrons from the beam volume, and the corresponding radial currents induce an axial electric field (laser wakefield accelerator). Meanwhile, to appreciate the require-ments for accelerators in oncology, it is helpful to look at what has been done.

1.2. *How does an accelerator structure work?*

A closed cavity can support electromagnetic modes, each with a characteristic frequency determined by the boundary conditions and the number and loca-tion of the nodes of the standing wave. Excited in the steady state, the oscillation of one such mode will exhibit a cycling of energy between the electric field and the magnetic field, just as in an LC cir-cuit. Among the lowest frequency modes will be one with an electric field extending across the cavity, and only one antinode in the electric field, as sketched in Fig. 2. Wall losses effect a parallel resistance in the LC circuit.

To this RLC circuit we add beam tubes as shown in Fig. 3, and to appreciate how beam current enters the circuit model we first consider a charged particle entering an empty cavity. On leaving the cavity, some electromagnetic energy U may remain in the cavity, and for a small-enough charge we expect that $U = k_l q^2$, for some constant k_l, referred to as the loss factor. The "loss" here has nothing to do with wall losses or dissipation; it refers to the loss of energy of a charged particle transiting a discontinuity on a beamline, in this case a cavity.

After the passage of the particle, the various modes of the cavity are ringing and each mode can be described with a loss factor. Let us ignore all the modes except for the one designed for acceleration. For energy loss to have occurred the charge must have been acted on by an electric field — evidently a beam-induced field — and one may describe the result in terms of a voltage amplitude V in the cavity,

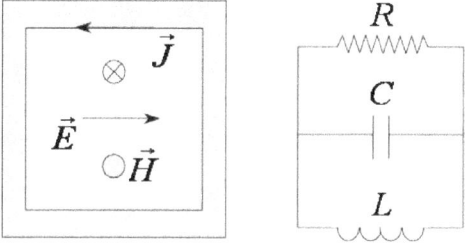

Fig. 2(a). The closed pillbox accelerator. Easy to under-stand. Hard to couple to.

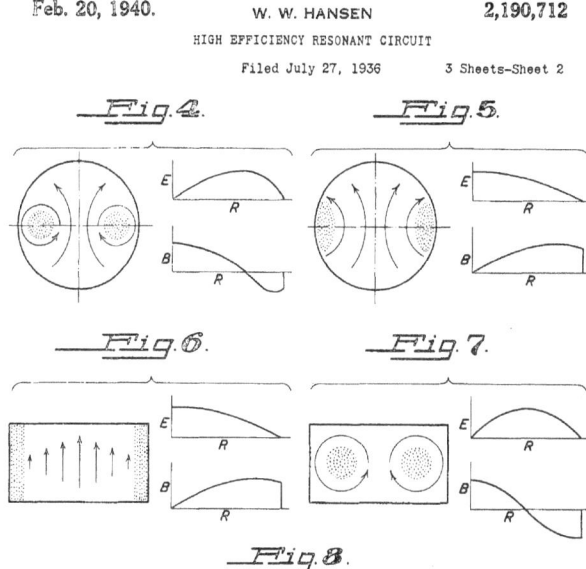

Fig. 2(b). Like Fig. 2(a) but with more verve.

ringing at the natural resonance frequency of the accelerating mode, after the particle has left. This amplitude must satisfy $U = qV/2$, so that $V = 2k_l q$.

From this we can infer a relationship between energy stored in a cavity and the corresponding accelerating voltage in the cavity:

$$U = \frac{V^2}{4k_l}.$$

An accelerating mode designed with a high-loss fac-tor is an excellent accelerating mode, since it requires

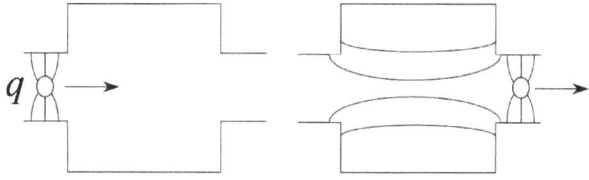

Fig. 3. We open up beam tubes in the closed pillbox of Fig. 2 and consider the passage of a charge q through the cavity.

very little energy to produce a large voltage. However, the same mode will exhibit a large induced voltage ("beam-loading") due to a particle beam.

Extending this analysis of energy transfer to a cavity already excited with a "no-load" voltage $V_{\rm NL}$, the energy change for a charge phased on-crest is

$$\Delta\varepsilon = \frac{V_{\rm NL}^2}{4k_l} - \frac{(V_{\rm NL} - V_b)^2}{4k_l} = qV_{\rm NL} - k_l q^2,$$

with the beam-induced or "beam-loading" voltage $V_b = 2k_l q$. Energy is quadratic in the fields, and if there were no beam-induced voltage, there would be no acceleration at first order in the applied field.

Related to the loss factor is a more convenient quantity referred to as "R over Q":

$$\left[\frac{R}{Q}\right] = \frac{V^2}{\omega U} = \frac{4k_l}{\omega}.$$

The notation, while common for historical reasons, is awkward didactically, because the quantity has nothing to do with Q or wall losses. It can be thought of as a special metric evaluating the suitability of an electromagnetic mode pattern for acceleration. For a particular mode α it takes the explicit form

$$\left[\frac{R}{Q}\right]_\alpha = \frac{\left|\int ds\, \tilde{E}_{s\alpha}(\mathbf{r}_\perp, s)e^{j\omega_\alpha s/V}\right|^2}{\omega_\alpha \int dV\, \frac{1}{2}\varepsilon_0 |\tilde{E}_\alpha|^2}.$$

The integral is evaluated along the design trajectory for the accelerator, parametrized by path length s and transverse coordinate $\mathbf{r}_\perp$. The eigenmode angular frequency is ω_α, and the particle velocity V is assumed constant. The quantity ε_0 is the permittivity of free space, most easily remembered as $\varepsilon_0 = 1/cZ_0$, with $c = 1/\sqrt{\varepsilon_0\mu_0}$ the speed of light in vacuum, 2.9979×10^8 m/s.

To drive the accelerator and establish the no-load voltage $V_{\rm NL}$, the cavity is connected through a small hole to a waveguide network powered by an external source such as a klystron or magnetron. The system may operate at a pulse repetition frequency (PRF) up to 1 kHz, with each pulse a few microseconds long. The start of each microwave pulse results in a transient, as illustrated in Fig. 4. The microwave signal propagating down the waveguide is initially mostly reflected, and a small amount of energy enters the cavity. Over time, as energy builds up in the cavity, the cavity radiates back into the waveguide. The net reverse signal on the waveguide is a superposition

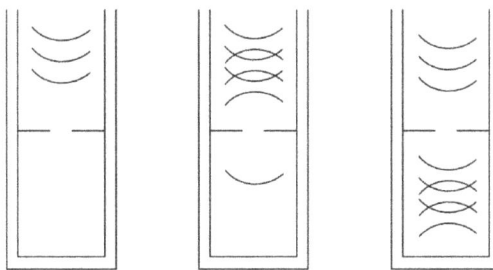

Fig. 4. Illustration of transient filling of a cavity. The beam port is not shown, but would go into the page. In the steady state the reverse voltage on the connecting waveguide consists of two terms: a prompt reflection from the coupling iris, and the radiated signal diffracting through the coupling iris from the filled resonator.

of the prompt reflection from the iris and the signal diffracting through the coupling iris. The dimensions of the coupling iris control the interference of these two terms and can be chosen to provide no net reverse signal in steady state.

To appreciate the relationship between forward power and voltage, consider the limit of negligible beam current where power flowing into the cavity is balanced by power flowing into the copper: the dissipated power $P_{\rm dis}$. In this case one has $P_{\rm dis} = \omega U/Q_w$, with Q_w arising from the lossy copper walls. Using the definition of $[R/Q]$ one may write this in terms of voltage as

$$P_{\rm dis} = \frac{V^2}{Q_w[R/Q]} = \frac{V^2}{R_s},$$

or $V = \sqrt{R_s P_{\rm dis}}$, with R_s the shunt impedance. Evidently, accelerator structure performance is limited by the peak power available and, in addition to the problem of designing the structure, it is critical to have a high peak power source, really a power converter drawing low-frequency power from the wall plug, and transforming it into high-frequency electromagnetic power.

There are limits on shunt impedance, because a cavity needs to be about one-half of a free space wavelength $\lambda/2$ in the transverse dimension to support an oscillation, and should not be longer than about $\lambda/2$ due to the transit time effect. Because cavity dimensions scale in this way with wavelength, it works out that the maximum practical $[R/Q] \approx \frac{2}{3}Z_0$. And this figure is reached only for small beam tube radius a; for $a/\lambda \approx 0.1$, $[R/Q] \approx 170\,\Omega$, depending on other parameters.

The quality factor depends on the cavity shape and is optimum for large radii of curvature where the magnetic field resides. The optimum scales as $Q_w \propto \lambda/\delta \propto \lambda^{1/2}$. The linear density of cavities scales as λ^{-1}, and so shunt impedance per unit length scales as $r \propto Q_w[R/Q]\lambda^{-1}$ or $r \propto \lambda^{-1/2}$. The ultimate voltage achievable in a fixed length L scales as $V \propto \sqrt{rLP} \propto \lambda^{-1/4}P^{1/2}$.

Because of this scaling, the favored approach to high voltage historically has been to aim for the highest frequency where adequate power sources exist, limited by manufacturability, performance, and cost.

Manufacturing tolerances meanwhile have their own scaling, deriving from the tolerance on frequency for a resonant circuit $\delta f/f \approx 1/Q_w$. This translates into dimensional tolerances because, as in Fig. 5, the cavity dimension D sets the frequency according to $f \approx c/2D$. Thus the dimensional tolerance δD is set by $\delta D/D \approx 1/Q_w$ or $\delta D \approx \lambda/2Q_w \propto \lambda^{1/2}$. At S-band (3 GHz), structures tune at 1 MHz per mil in diameter, where a mil is 0.001" or 0.0254 mm. Structures are typically machined with a tolerance in the range of a thousandth, and then hand-tuned cell by cell, to a precision of about 50 kHz. Modern machine shops with computer-numerically-controlled (CNC) lathes and careful control of the temperature of the work can machine S-band parts that do not require tuning. To insure that the finished structure does not require tuning, precise control of the braze process is also required.

At X-band (8–12 GHz), structures tune at 10 MHz per mil, and tuning is generally thought to be required due to the limitations of commercially viable machining and braze processes.

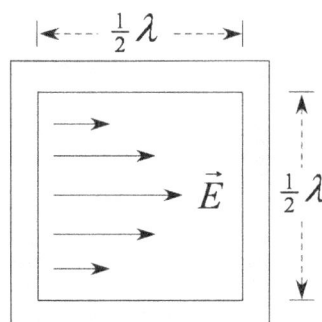

Fig. 5. Cavity dimensions scale with wavelength. Fractional tolerances scale as $1/Q_w$ and machining tolerances become difficult at high frequency.

1.3. *Circuit-equivalent model for a standing wave accelerator*

A single cavity coupled to a waveguide and beam can be described as a driven harmonic oscillator:

$$\left(\frac{d^2}{dt^2} + \omega_0^2\right) V_c$$
$$= -\frac{\omega_0}{Q_w}\frac{dV_c}{dt} + \frac{\omega_0}{Q_e}\frac{d}{dt}\left(V_F - V_R\right) - 2k_l\frac{dI_b}{dt},$$

with I_b the rf modulated beam current. The cavity voltage V_c responds to an incident voltage V_F defined at the plane of the detuned short, as depicted in Fig. 6 [12], where a reverse voltage V_R will also be present, satisfying a condition deriving from continuity of the transverse electric field in the waveguide:

$$V_c = V_F + V_R.$$

This plane is uniquely defined modulo 1/2 guide wavelength, as the reference plane which acts like a short circuit when the rf drive is tuned off-resonance (detuned). On resonance, and in the limit of low loss, the same plane acts like an open circuit.

The term Q_e characterizes the coupling of the cavity to the waveguide. Combining terms one has

$$\left(\frac{d^2}{dt^2} + \frac{\omega_0}{Q_L}\frac{d}{dt} + \omega_0^2\right) V_c = 2\frac{\omega_0}{Q_e}\frac{dV_F}{dt} - 2k_l\frac{dI_b}{dt},$$

where the loaded Q is given by

$$\frac{1}{Q_L} = \frac{1}{Q_w} + \frac{1}{Q_e}.$$

Common circuit equivalents for this system are seen in Fig. 7.

In the steady state, with drive on resonance, this result can be cast in the form of the "load line"

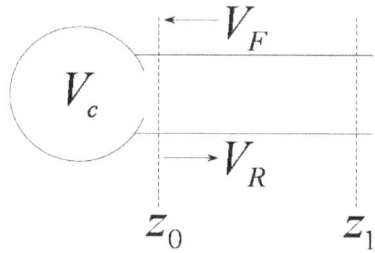

Fig. 6. Schematic for a cavity coupled to a waveguide, depicting a forward voltage V_F, reverse voltage V_R and cavity voltage V_c.

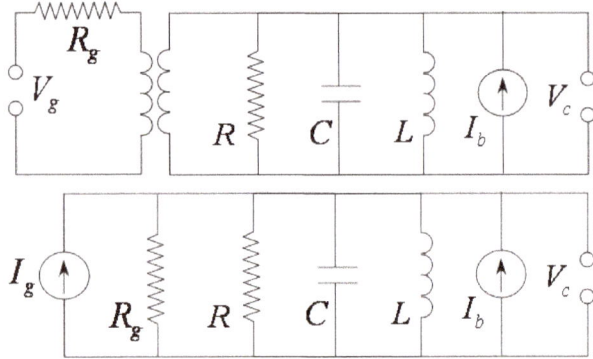

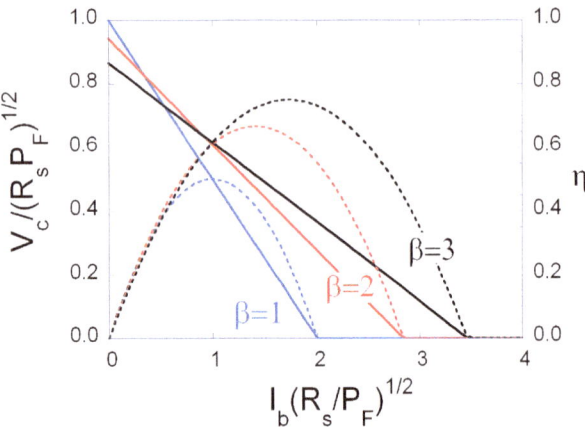

Fig. 7. Two common circuit-equivalent models for a cavity coupled to a beam and a waveguide.

Fig. 8. Variation of loaded cavity voltage V_c, and efficiency η, with beam current I_b, for several values of the coupling parameter β.

equation in terms of toroid current I_{bo}

$$V_c = \frac{2\beta^{1/2}}{1+\beta}\sqrt{RP} - \frac{R}{1+\beta}I_b = V_{\rm NL} - mI_{bo}.$$

The latter form applies to both standing wave and traveling wave linacs. This scaling is illustrated in Fig. 8 for various values of the coupling parameter β. The quantity η is the electronic efficiency.

1.4. *Cold test*

Measurements on accelerator structures for manufacturing process control are made without a beam, on a workbench, typically using an automatic vector network analyzer. This measurement equipment permits a sweep of the drive angular frequency ω and easy display of steady state excitation. At one

frequency, in steady state terms take the forms

$$V_F = \Re\tilde{V}_F e^{j\omega t},$$
$$V_R = \Re\tilde{V}_R e^{j\omega t},$$
$$V_C = \Re\tilde{V}_C e^{j\omega t}.$$

And our equation takes the form

$$\left(j\frac{\omega\omega_0}{Q_L} + \omega_0^2 - \omega^2\right)\tilde{V}_C = 2j\frac{\omega\omega_0}{Q_e}\tilde{V}_F.$$

The difference in drive frequency from the resonant frequency of the cavity is described by the *tuning angle* ψ,

$$\tan\psi = Q_L\left(\frac{\omega_0}{\omega} - \frac{\omega}{\omega_0}\right),$$

in terms of which

$$\tilde{V}_C = \frac{2\beta}{1+\beta}e^{j\psi}\cos\psi\,\tilde{V}_F,$$

and the coupling parameter $\beta = Q_w/Q_e$. The tuning angle is the angle which the cavity voltage phasor $\tilde{V}_C$ makes with the forward drive phasor $\tilde{V}_F$. On resonance, the two are in phase,

$$\tilde{V}_C = \frac{2\beta}{1+\beta}\tilde{V}_F,$$

and a reflected signal is propagating back up the waveguide toward the source:

$$\tilde{V}_R = \tilde{V}_C - \tilde{V}_F = \frac{\beta-1}{\beta+1}\tilde{V}_F.$$

If $\beta > 1$ the external Q is lower than the wall Q, and the cavity is said to be *overcoupled*. In this case the reverse signal is in phase with the forward drive. If $\beta < 1$ the cavity is said to be *undercoupled*, and in this case the reverse signal is 180° out of phase with the drive. If $\beta = 1$ the cavity is said to be *critically coupled*, and there is no reverse signal in steady state. The steady state depicted in Fig. 4 shows critical coupling, with no net reverse signal after the transient.

With this result for cavity voltage, we may compute the reflection coefficient using the continuity condition, $\tilde{V}_R = \tilde{V}_C - \tilde{V}_F$:

$$S_{11} = \frac{\tilde{V}_R}{\tilde{V}_F} = \frac{2\beta}{1+\beta}e^{j\psi}\cos\psi - 1.$$

The notation S_{11} is conventional in microwave measurement, and a display option on modern network analyzers. Measurement of the (complex) S matrix can permit one to determine the resonant frequency

Fig. 9. Cold test tune stand in the days of the Clinac-4. Temperature-controlled water is provided to the structure to stabilize the frequency. Before the advent of the automatic vector network analyzer, a VSWR meter was employed to check the frequency of each side cavity.

ω_0, the coupling factor β, and the loaded Q. Thus one can extract as well the external Q.

This result implies that the "trajectory" defined by $\tilde{S}_{11}(\omega)$ in the complex S plane is a circle centered at $s_c = -1/(1 + \beta)$, with radius $s_r = \beta/(1 + \beta)$ crossing the real axis at

$$S_R(\omega_0) = \frac{\beta - 1}{\beta + 1},$$

enclosing the origin, if overcoupled. The voltage standing wave ratio (VSWR) may be calculated from the reflection coefficient on resonance $\rho = |\tilde{S}_{11}(\omega_0)|$ as VSWR $= (1 + \rho)/(1 - \rho)$, which, after some algebra, yields

$$\text{VSWR} = \begin{cases} \beta; & \beta > 1, \\ \dfrac{1}{\beta}; & \beta < 1. \end{cases}$$

Observation of the Q circle in the Smith chart format permits one to determine whether the resonance is overcoupled (encloses the origin) or undercoupled; this permits one to determine β from the VSWR.

With a figure for β in hand, one may proceed to determine Q_L as follows. Locate the drive frequencies, $\omega_\pm$, where $S_R = \pm S_I$, i.e. where S_R and S_I are equal in magnitude in the upper $(-)$ (frequency below resonance) and lower $(+)$ (frequency above resonance) half-plane. The condition $S_R = \pm S_I$

takes the form

$$\psi_\pm = \pm \frac{1}{2} \left\{ \cos^{-1} \left(\frac{1}{\beta\sqrt{2}} \right) - \frac{\pi}{4} \right\}.$$

After a great deal of algebra one can show that $\tan \psi_+ = F(\beta)$, where

$$F(\beta) = \frac{2\beta - 1 - \sqrt{2\beta^2 - 1}}{\sqrt{2\beta^2 - 1} - 1}.$$

Combining this with the observation $\psi_+ = -\psi_-$, one can show that

$$Q_L = \frac{\omega_0}{\omega_+ - \omega_-} F(\beta).$$

A simpler approach to measuring Q_L derives from a measurement of the 3 dB bandwidth in transmission. However, this method requires a second port.

1.5. *Multicell accelerator structures*

It is possible to overcome the limitations on the shunt impedance of a single cavity by coupling cavities together. The simplest such scheme employs direct coupling of adjacent accelerating cavities. Two forms of coupling are common: on-axis coupling, via the beam tube; and off-axis coupling, via slots in the end walls. The first form, shown below, involves coupling in an area where the electric field is large. The second method involves coupling where the magnetic field is large and is referred to as magnetic coupling. A third means of coupling involves the use of additional resonators, and we will discuss that in Sec. 3.

To appreciate what is possible with two cavities coupled through the center hole, consider that the result is still a cavity, and thus has electromagnetic modes, and in the lossless limit these modes simply oscillate. With the analogy of two coupled pendula in mind as in Fig. 10, we can expect to find a 0 mode and a π mode. The phase $\theta = 0$ or π refers to the spatial phase shift from one cavity to the next. We can describe the system in terms of these two independent modes of oscillation,

$$\left(\frac{d^2}{dt^2} + \Omega_0^2 \right) \xi_0 = 0,$$

$$\left(\frac{d^2}{dt^2} + \Omega_\pi^2 \right) \xi_\pi = 0,$$

corresponding to some mode angular frequencies Ω_0, Ω_π. One is free in how one defines the amplitudes ξ_0, ξ_π as long as they have something to do with either the electric or magnetic field amplitudes when

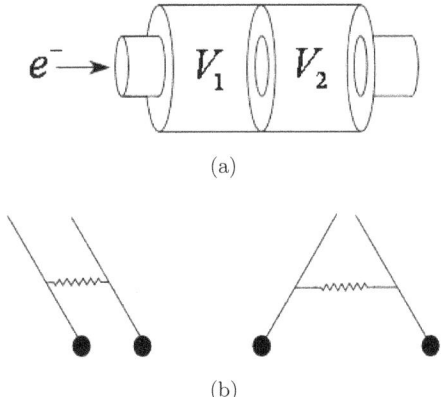

(a)

(b)

Fig. 10. (a) Concept for a two-cavity accelerator structure. (b) The two modes of two coupled resonators.

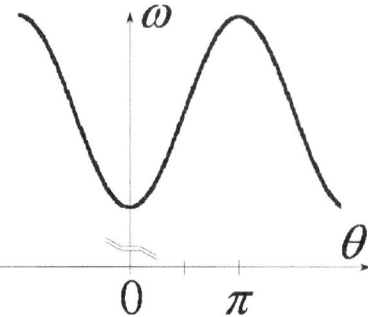

Fig. 11. Brillouin diagram for a forward wave ($\kappa > 0$) periodic structure.

the respective mode is excited. For the purposes here we will refer them to the on-axis electric field amplitude, and since we are interested in acceleration we choose to define them in terms of the single-cavity voltages, as in the limit of small cell-to-cell coupling these quantities would still be well defined. In this limit one has from the symmetry of the modes

$$\xi_0 = V_1 + V_2, \quad \xi_\pi = V_1 - V_2.$$

Substituting these equations into those for the mode amplitudes, and alternately adding and subtracting the results, one finds that

$$\left(\frac{d^2}{dt^2} + \omega_0^2\right) V_1 = \frac{1}{2}\kappa\omega_0^2 V_2,$$

$$\left(\frac{d^2}{dt^2} + \omega_0^2\right) V_1 = \frac{1}{2}\kappa\omega_0^2 V_2,$$

where

$$\omega_0^2 = \frac{1}{2}\left(\Omega_0^2 + \Omega_\pi^2\right),$$

$$\kappa = \frac{\Omega_\pi^2 - \Omega_0^2}{\omega_0^2}.$$

This result points the way to modeling of multiple coupled cells. For a periodic structure one infers that

$$\left(\frac{d^2}{dt^2} + \omega_0^2\right) V_n = \frac{1}{2}\kappa\omega_0^2 (V_{n-1} + V_{n+1}),$$

for the excitation of the nth cell voltage V_n. It is instructive to consider an infinite structure of this form, driven at some cell (call it $n = 0$), in the steady state, at angular frequency ω. Solutions for cell voltages $n > 0$ should take the form $V_n = \Re \tilde{V}_n e^{j\omega t}$, and since the structure is lossless, $|\tilde{V}_n| = |\tilde{V}_{n+1}|$. Thus

adjoining cell voltage phasors differ by a phase factor. Since the structure is periodic, this factor must be the same for each cell, downstream of the drive point. One concludes that $\tilde{V}_n = \tilde{V}_0 e^{-jn\theta}$, and plugging this into the coupled cavity equation one finds that

$$\omega^2 = \omega_0^2(1 - \kappa\cos\theta).$$

This dispersion relation is depicted in Fig. 11 for $\kappa > 0$, corresponding to electric coupling. The slope of the curve is proportional to the group velocity, $V_g = L d\omega/d\theta \propto \kappa\sin\theta$, at which energy propagates at that frequency in this disk-loaded transmission line. The cell period is L. When group velocity and phase velocity $V_\varphi = L\omega/\theta$ have opposite signs, the structure is referred to as a backward wave structure. This is the case for magnetic coupling, where $\kappa < 0$.

This "nearest neighbor" coupling model can be extended to include wall losses and waveguide coupling,

$$\left(\frac{d^2}{dt^2} + \omega_0^2\right) V_n$$

$$= \frac{1}{2}\kappa\omega_0^2(V_{n-1} + V_{n+1}) - \frac{\omega_0}{Q_w}\frac{dV_n}{dt}$$

$$- \frac{\omega_0}{Q_e}\frac{dV_n}{dt}\delta_{n,M} + \frac{2\omega_0}{Q_e}\frac{dV_F}{dt}\delta_{n,M},$$

where cell #M is the coupling cavity.

In a finite structure of N cells, the end cell conditions result in a discrete number of modes, $\alpha = 1, \ldots, N$, populating the Brillouin curve of Fig. 11. One can show that introduction of a port in one cell contributes a correction in the form of an external Q for each mode, $Q_{e\alpha}$, inversely proportional to the squared mode amplitude in that cell. The external features of the N-cell cavity, as determined from measurement of S_{11}, take the form of a sum over the

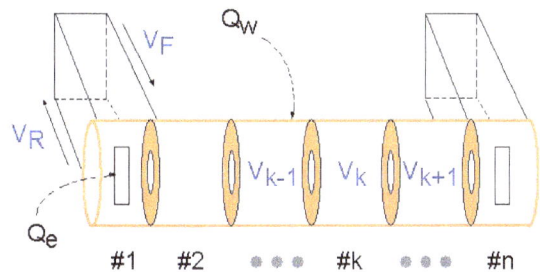

Fig. 12. Illustration of a traveling wave accelerator.

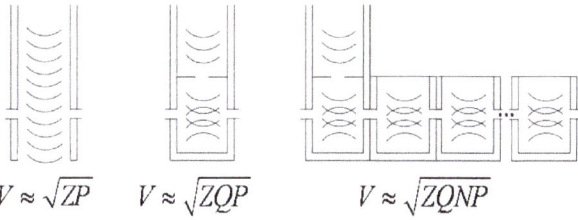

Fig. 13. Illustration of the advantages of a cavity, and coupled cavities, in producing a high accelerating voltage V with limited peak power, P, thanks to a high quality factor Q, and a number of cavities $N > 1$.

modes α of the multicell structure

$$\tilde{S}_{11}(\omega) = \frac{\tilde{V}_R}{\tilde{V}_F} = \sum_{\alpha=1}^{N} \frac{2j\omega\omega_0/Q_{e\alpha}}{(\omega_\alpha^2 + j\omega\omega_\alpha/Q_{L\alpha} - \omega^2)} - 1.$$

Provided that the modes are well separated in frequency, the description of the multicell structure, when operated in one particular mode, can be reduced to the form we started with above, for a single cavity. The shunt impedance, however, will be higher than that for a single cavity, by a factor that can approach N, the number of cells.

The traveling wave accelerator differs from the standing wave structure in that it will have a waveguide feed at one end, at the gun end for electrically coupled (center-hole-coupled) cells, and at the target end for magnetically coupled cells. In addition, it will have an output waveguide at the opposite end, with a high-power load on it. The coupler cell external Q is much lower, to provide a match into and out of what is in effect a broadband multipole filter. What is narrowband in a traveling wave structure is the range of frequencies over which the cell voltages add in phase for the design synchronous particle. Such a particle satisfies the condition that the transit angle matches the geometric phase advance in the design mode of operation, at the design frequency, $\omega L/V = \theta$.

The load line for a travelling wave structure can be cast in a form similar to that for a standing wave structure. Rather than a coupling factor β, one needs to know the attenuation parameter τ, defined such that power to the load is given by $P_{\text{out}} = P_{\text{in}}e^{-2\tau}$. For both constant impedance (CZ) and constant gradient (CG) structures, the fill time for the structure is given by $T_f = 2Q_w\tau/\omega$. No-load voltage is given by

$$V_{\text{NL}} = (R_s P_{\text{in}})^{1/2}(1 - e^{-\tau})\left(\frac{2}{\tau}\right)^{1/2} \quad \text{(CZ)}$$

for constant impedance, and for constant gradient

$$V_{\text{NL}} = (R_s P_{\text{in}})^{1/2}(1 - e^{-2\tau})^{1/2} \quad \text{(CG)}.$$

The loading coefficient is given by

$$m = R_s\left(1 - \frac{1 - e^{-\tau}}{\tau}\right) \quad \text{(CZ)},$$

$$m = R_s\left(\frac{1}{2} - \frac{\tau}{e^{2\tau} - 1}\right) \quad \text{(CG)}.$$

In Fig. 13, we summarize the logic behind the multi-cell accelerator structure. In free space, a *high* peak power is required to produce a high electric field. In structures a high field can be obtained with *lower* peak power than in free space. This is because, with the help of material boundaries, we may *store* energy by resonant excitation. How much energy we may store, or how long we may store it, depends on dissipation in the medium. Thus we require not only a material enclosure, but one with low loss. Low-loss dielectric structures have been studied successfully, but have not yet made their way out of the laboratory. As for conductors, the lowest-loss conductor at room temperature is silver, but within 5% in conductivity, and much lower in price is oxygen-free electronic grade copper, and this has been the material of choice for microwave linacs up to the present era.

The foregoing is the standard lore on microwave accelerator design. In compact commercial linacs some of these simplifying assumptions do not apply. Electrons do not travel at c, and they do not travel at a constant speed $V < c$ either. Consequently opinions vary on how best to design commercial accelerator structures.

2. Overview of Oncology Linacs

In this section we provide an overview of the landscape in which oncology linac development takes place. A previous article in this series has provided a broad overview [13].

Cancer is the second-largest cause of mortality, with over one million new cancer cases annually, on the order of 200,000 each of prostate, lung, breast and colorectal cancers. Radio therapy is applied in half of all cases and has cumulatively treated millions of patients. American men have a 50% lifetime risk of developing cancer, and women about 30%. There are half-a-million deaths each year. The goal of radiotherapy is to reduce a fatal disease to a manageable illness. Ionizing radiation holds the promise of this capability, as it is able to penetrate matter and to effect change at the molecular level with consequences for the chemistry and cell biology underlying the physiologic and homeostatic systems. The challenge in applying radiation is to spare healthy tissue and maximize the dose to cancerous tissue. This challenge has been taken up by medical physicists around the world, and because of improved treatments in validated group trials, great progress is being made. For example, *circa* 1975 55% of children with cancer lived five years after diagnosis. Today the figure is 70% [14].

2.1. *Ionizing radiation*

In general, ionizing radiation refers to any particle flux capable of ionizing the constituents of the medium. For electron linacs the two treatment modalities employed are x-rays and electrons. An additional whole-body dose results from x-ray leakage and, above 9 MV, neutron production. To analyze and monitor the dose, a number of units are employed. The most basic unit of exposure is the roentgen, the flux of photons sufficient to produce 1 esu/cm^3, in dry air at standard-temperature pressure (278 K, 1 atm) or 2.6×10^{-4} C/kg. The dose is a measure of energy absorbed per unit mass with units of rad (radiation-absorbed dose), or Gray, 1 Gy = 1 J/kG or 100 rad. In water, for energies of interest, the exposure to the dose conversion factor is 0.97, and so for the purposes here the distinction is not critical.

Damage to biological mass depends on more than the dose; it also depends on the kind of particle. Particles, such as photons, and electrons, which distribute their energy more uniformly, do less damage than particles such as neutrons. One accounts for this with the radiation-weighting factor, a measure of relative biological effectiveness (RBE). This factor is unity for photons and electrons, and can be as large as 20 for neutrons in the 100 keV–2 MeV range. Multiplying the dose by the radiation-weighting

factor, one obtains the *equivalent dose*. Units for the equivalent dose are the rem (roentgen equivalent man), or 1 Sievert (Sv) = 1 J/kG ≡ 100 rem. An equivalent dose computed in this way is referred to as the whole-body dose. An effective dose is obtained from an equivalent dose by computing a weighted sum over exposed areas. Generally speaking, those areas of the body with the highest tissue weighting factors are those where cells are dividing most rapidly, such as gonads, bone marrow, colon, lung and stomach.

The direct and indirect effects of ionizing radiation on living organisms remain an active area of study. The direct effects within a cell start with ionization resulting in peroxides and other free radicals in the cell cytoplasm, organelles and nucleus. These effects lead to chemical consequences in the environment of the altered molecules and atoms. The resulting biological consequences include altered enzyme activity and modified or inhibited replication. At the extracelluar level, the consequences include cell degeneration and cell death. This results in cell loss, as more differentiated cells are not replaced. At a higher level, the consequences are tissue atrophy, damage to blood vessels, starvation and hypoxia of parenchymal tissue and with it loss of organ function, and systemic disorders.

In general, cell response is dependent on the phase of the cell cycle, with mitosis being the most sensitive period. Since cancer cells tend to divide rapidly, they also tend to be *more sensitive* than normal tissue. As not all cells are in mitosis at any one point in time, it is helpful to have multiple treatments, and this "fractionation" of the dose is also helpful in limiting the dose to healthy tissue to permit it to repair itself. Applying radiation from several different angles further helps to reduce the dose to any one volume of healthy tissue.

In modern therapy machines, x-rays are produced by means of a bremsstrahlung conversion target. The dosimetric properties of these beams in tissue are then assessed by reference to their properties in water, and over a transverse field that might be 40 cm × 40 cm at a "source skin distance" (SSD) of 1 m from the conversion target. Because of the sharply peaked angular pattern of the bremsstrahlung spectrum produced by MV beams [15], "flattening filters" are employed to provide preferential absorption of the center of the x-ray beam, so as to provide a dose distribution that is more uniform transversely, across the treatment field.

X-ray energy is not uniquely defined for a bremsstrahlung spectrum produced from a microwave linac beam, since the electron beam spectrum is of finite width, and the bremsstrahlung process itself produces a broad spectrum. Because a distribution of energies is involved, the usual linac concept of energy is not sufficiently precise for characterizing the beam. Rather than energy, one refers to "quality" Q, with units of MV [16]. To be precise as to the meaning of results, one refers to the protocol employed, and it is common to employ "BJR 11." [17] The primary properties characterizing an x-ray beam of quality Q are that x-ray dose deposition in water has a characteristic curve climbing steeply from the surface to a maximum at

$$D_{\max}(\text{cm}) \approx 3.775 \log Q(\text{MV}) - 1.378,$$

for $4 \leq Q(\text{MV}) \leq 45$, and then dropping gradually to a dose at 10 cm that is, as a percentage of the dose at $D_{\max}$,

$$d(\%) = 26.09 \log Q(\text{MV}) + 46.78.$$

For engineering purposes, accelerator output is characterized by the average beam current on target, I_{avg}, and by output O in units of R/min measured at the depth of dose maximum, $D_{\max}$, in a 10 cm×10 cm field at an SSD of 1 m. The yield is defined as

$$Y(R/\text{min}/\mu\text{A}) = \frac{O(\text{rad}/\text{min})}{I_{\text{avg}}(\mu\text{A})},$$

and scales roughly as $Y \approx kE^n$, where $n \approx 3$, $E \approx Q$ in MV, and the constant $k < 0.1$ depends on the target and filter. Some indication of the features and variety of bremsstrahlung conversion targets can be found in the literature [18, 19]. Haimson quotes $k \approx 7 \times 10^{-2}$, and $n \approx 3.0$ at 4 MeV, $n \approx 2.7$ at 10 MeV, and $n \approx 2.5$ at 25 MeV [20]. Goldie finds data for a 0.25" Au target fit well to $k \approx 2 \times 10^{-2}$ and $n \approx 3.4$.

Using the steady state result for the cavity voltage, or "load line," $V_c = V_{\text{NL}} - mI_b$, one can show that optimal conversion of microwave power to x-ray output occurs when an accelerator is operated at a loaded voltage of about $V_c \approx 0.75\, V_{\text{NL}}$. This is different from the figure for optimal energy conversion, $V_c \approx 0.5\, V_{\text{NL}}$.

2.2. *History*

Radiation therapy dates back to the observation in the 19th century that radiation could kill cancer cells. This was followed by high-voltage tubes employing 150–350 kV energies where $D_{\max}$ lies within the skin, resulting in skin-burning. Van de Graaff accelerators enabled external beam treatments with 1–2 MV beams. Cobalt-60 was in use for therapy in the 1920s, with an effective energy in the range of 4 MV, providing a more skin-sparing depth–dose curve than the 2 MV machines. After World War II over 200 betatrons were built and installed around the world [21]. The first treatment with a betatron was described in detail by Quastler *et al.*, in 1949 [22]. The higher energies available, over 18 MV, provided a significant advantage for skin-sparing, but the dose–rate output in a flattened field was low, in the range of 40 R/min at 22 MeV, even for a relatively small field of 12.5 cm × 12.5 cm.

Both the electrostatic and induction machines were large and expensive installations. The cobalt machines had the advantage of cost, but the disadvantage of low and decaying output, and penumbra due to the finite source size. This penumbra resulted in poor edge definition for the treatment field.

And, in this postwar period, suddenly microwave linacs appeared. World War II had driven the development of high-power magnetrons and in the postwar period Hansen's concept of the cavity-resonator was exploited by several groups to combine a megawatt class magnetron and pulse modulator (a "transmitter"), coupled cavities and a high-voltage electron gun, to produce megavoltage electron beams.

Two groups took the lead: one was Hansen's group at Stanford, and the other was D. W. Fry's group at the Telecommunications Research Establishment (TRE) in England. Fry's group completed a 45 cm, 0.5 MV accelerator and produced the first beam in late 1946 [23]. This was a 3.0 GHz, 40-cm-long traveling wave structure designed for constant-gradient with 45 kV injection. The peak beam current was 36 mA under nominal conditions for 0.5 MeV. Hansen's group completed a 90 cm accelerator and produced a 1.7 MV beam in 1947 [3], and within three years had reached 6 MeV with a 14-foot linac, operating at 2856 MHz, powered by a 0.9 MW tunable magnetron [24]. The report by Becker and

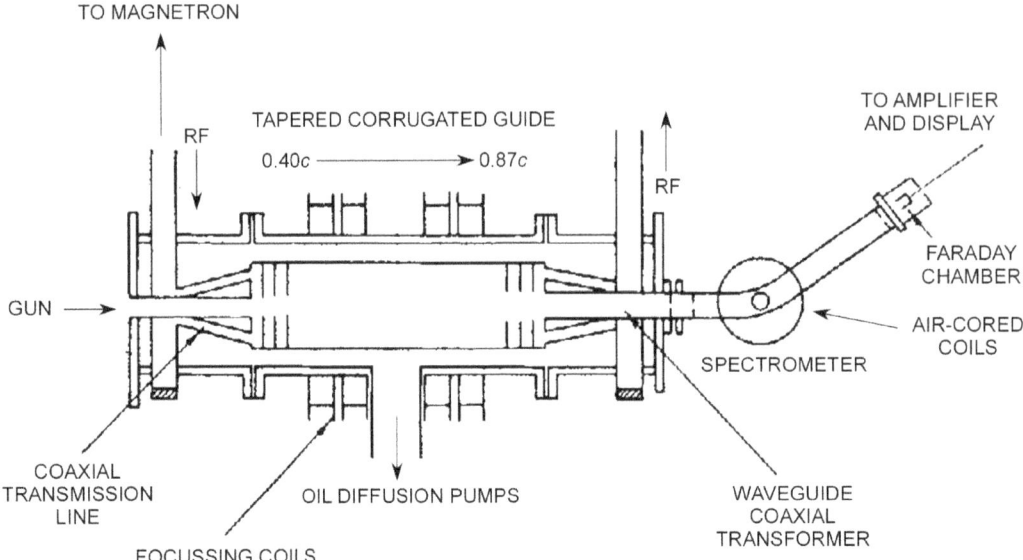

Fig. 14. The first megavoltage microwave linear accelerator or "guide" (from Fry *et al.*).

Caswell points to a number of issues that look very familiar to linac engineers 60 years later:

- Peak power was not as much as hoped for;
- A solenoidal field was needed, and coil alignment was tricky;
- The electron gun provided a largish beam diameter of 1/4";
- Magnetron frequency varied over time;
- Magnetron performance into the resonant structure was not the same as that into a load.

The first patient treatment with a microwave linac took place on August 19, 1953. This was at an installation at Hammersmith Hospital in London, developed by a collaboration sponsored by the British Ministry of Health. The accelerator team consisted of Fry's group, under the by-then-renamed Atomic Energy Research Establishment (AERE), working with the Metropolitan Vickers Electrical Company, later renamed Associated Electrical Industries (AEI), and with the Radiotherapeutic Research Unit of the Medical Research Council (MRC). This machine used a 2 MW magnetron and a 3 m accelerator on a stationary platform, producing an 8 MV beam and 100 R/min flattened over a 25-cm-diameter field. To provide for different treatment angles ("patient portals"), a 90° magnet at the output could be rotated, the treatment floor could be moved vertically, and the treatment table could be displaced laterally.

Meanwhile, in the US the high-power klystron was receiving more attention and over the period 1949–52 researchers at Stanford developed a 30 MW tube. This enabled the first large-scale electron linac for nuclear physics research — Mark III [5]. In 1954–55 three traveling wave linacs were built for electron therapy at energies of 45–70 MV, with a stationary linac.

Back at Hammersmith, the MRC was pursuing compact 4 MeV linacs on a rotatable gantry mount. The first installed was a double-gantry unit at Newcastle upon Tyne in 1953 [25], designed and built by Mullard Research Laboratories on behalf of Philips Electrical Ltd. The accelerator operated at 3.0 GHz and employed a traveling wave structure with the phase velocity tapered to match the electron velocity. Recirculation from the output to the input was employed via a rat race hybrid with a phase-shifter on the recirculation arm.

The power source was a 2.0 MW magnetron operated at a 45 kV, 1.7 μs pulse and up to 400 Hz maximum. The beam current was 250 mA, with an effective pulse length of 1.2 μs. The dose rate with buildup cap, at 1 m, with collimation and flattening was 250 R/min at 400 Hz. The vacuum pump was operated in a stationary position, through a rotary joint, to avoid getting oil in the system.

The first orientable single-gantry system was installed in Manchester in 1954. It employed a traveling wave accelerator structure powered by a

1.5 MW magnetron and was capable of producing a 4 MV electron beam and rotating through 120° about the isocenter. The length of the accelerator and the absence of a bend magnet resulted in a machine of considerable size for which a full, 360° rotation was not practical. It could rotate over 120°. This Metropolitan Vickers Orthotron is seen in Fig. 15, with a section of flooring removed to permit a −30° orientation. The output was 100 R/min in a 20 cm × 20 cm field at 1 m. Nominally 200 R/min could be achieved at the maximum PRF of 500 Hz. Additional details are provided by C. W. Miller [26, 27].

The first orientable machine in the US was developed by Ed Ginzton working with his faculty colleague from the medical school, Dr. Henry Kaplan. It was installed in 1954 in the radiology department, then in San Francisco. Ginzton's team included his graduate students Arnold Eldredge, Ken Mallory and Karl Brown. The system consisted of a 1.6 m traveling wave linac powered by a 1 MW klystron, providing a 6 MV beam with 110 R/min output, unflattened, in a 15 cm × 15 cm field at 1 m. Beam's-eye-view position verification was provided with an insertable 100 kVp x-ray tube. The accelerator was oriented in-line on a trunnion mount, permitting 5 feet of vertical travel and more than 90° of orientation, from vertical to horizontal. The first patient was treated in 1956, for retinoblastoma, and survived with vision intact.

The subsequent work of Kaplan, Ginzton and Varian Associates led in 1959 to the introduction of a "bent beam" machine, relying on a dipole magnet to produce a net 90° bend. This more compact "C-arm" configuration, seen in Fig. 17, permitted 360° rotation, and this "Clinac-6" was the first such 360° radiotherapy linac. The accelerator was 1.5 m long, traveling wave, and driven by a 2 MW magnetron. The 105° bend magnet system was not achromatic and consequently energy variation could translate into beam spot motion on the target, and symmetry variations. The source–axis distance (SAD) was 100 cm, unlike the prevalent 80 cm for cobalt-60 machines. The ram couch relied on a lead screw.

A subtle but important advance in technology in the Clinac-6 was the incorporation of a new invention, the sputter ion pump [28]. The "egg crate" anode "vacion" pump came out of a tube research program at Varian that commenced in 1956 with the goal of eliminating oil from the vacuum system for power tubes. By 1959 a separate "Vacuum Division" had been established and pumps were being shipped, including a 5000 L/s pump, to Oak Ridge National Laboratory, for fusion research. This oil-free pumping system facilitated stable performance of accelerators with rotation. Prior to this time, orientable accelerators such as the Orthotron had to deal with the problem of rotation with an oil diffusion pump. Oil can contaminate an accelerator and poison the gun, leading to loss of emission and output. An interim approach to this problem with Kaplan and Ginzton's first machine was to build the accelerator as a sealed tube, with no pump.

A product brochure from Vickers Research (not the same company as Metropolitan Vickers) in the early 1960s depicts a commercial 6 MV version as in Fig. 18 with a dose rate of up to 200 R/min, and 230° gantry rotation on a flat floor, with a 48° depression below the horizontal with the pit. Other characteristics include: a 1 m accelerator length, a 1 MW Ferranti VF10 magnetron operating at 9.25 GHz, 35 kV, with a 0.001 duty. The average beam power is 200 W, the average current 33 μA, and the beam spot 2 mm. The total floor load of gantry and couch is 3 tons. According to a product brochure, the economics assumed 1000 treatment hours per year, servicing by a technician at £3 per hour, and the £350 cost of replacement of the accelerating tube. Vickers Research went out of business after a few units had been installed.

The commercial goal with the linacs of this period, particularly the Clinac-6, had been to beat cobalt machines on performance, at a reasonable price. Unfortunately for linacs, AECL and Picker Cobalt machines were priced at US$75,000 *circa* 1958. A Clinac-6 sold for about twice this. The first two Clinac-6's were shipped in 1962, and eight more orders were received following that. Then the market dried up. But the year 1967 saw a radical change — the introduction of a new compact accelerator design.

From the load line equation one can appreciate that subject to the constraint of available peak power, a minimum energy requirement imposes a lower limit on shunt impedance, which then requires a certain length of accelerator. The great length evident in Fig. 15, oriented directly at the patient, imposed limits on treatment delivery, viz. limited gantry angles, and setup time to remove a piece of

D. H. Whittum

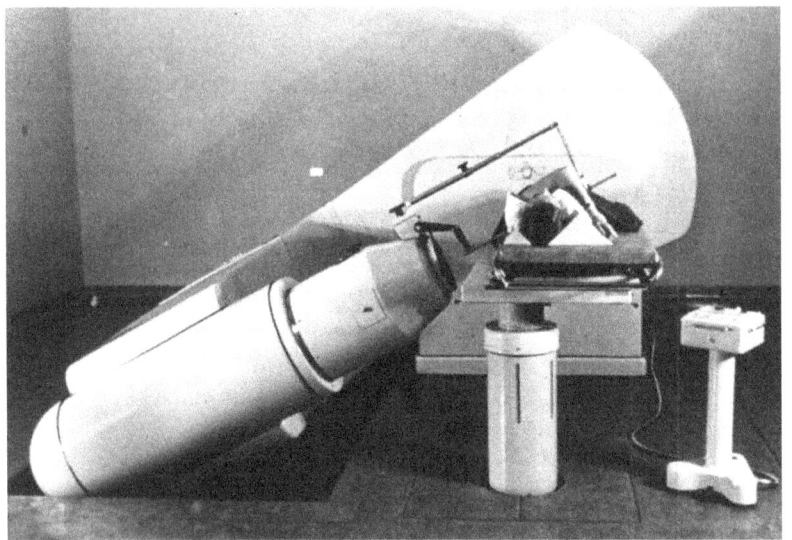

Fig. 15(a).   The Metropolitan Vickers Orthotron, *circa* 1954.

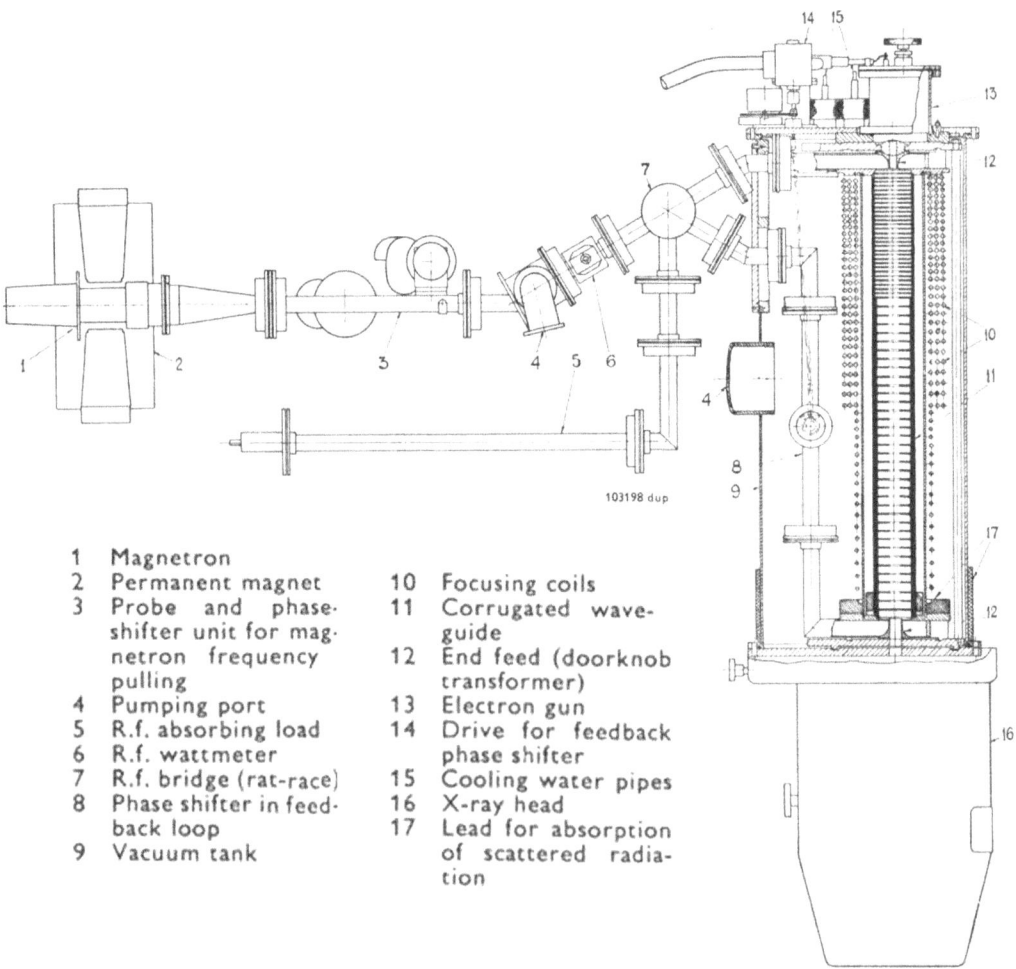

1   Magnetron
2   Permanent magnet
3   Probe and phase-
    shifter unit for mag-
    netron frequency
    pulling
4   Pumping port
5   R.f. absorbing load
6   R.f. wattmeter
7   R.f. bridge (rat-race)
8   Phase shifter in feed-
    back loop
9   Vacuum tank

10 Focusing coils
11 Corrugated wave-
    guide
12 End feed (doorknob
    transformer)
13 Electron gun
14 Drive for feedback
    phase shifter
15 Cooling water pipes
16 X-ray head
17 Lead for absorption
    of scattered radia-
    tion

Fig. 15(b).   The Orthotron accelerator system from C. W. Miller, Metropolitan Vickers.

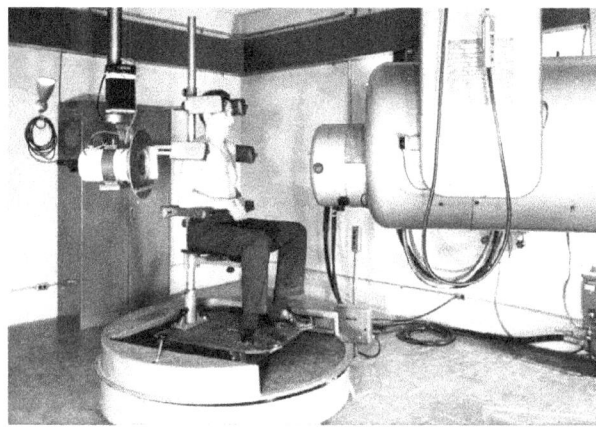

Fig. 16. Dr. Henry Kaplan's first machine, trunnion-mounted, with an isocentric rotating chair, and a fluoroscope behind the patient, to provide real-time imaging. A kV tube could be inserted at the head to provide beam's-eye imaging and image guidance.

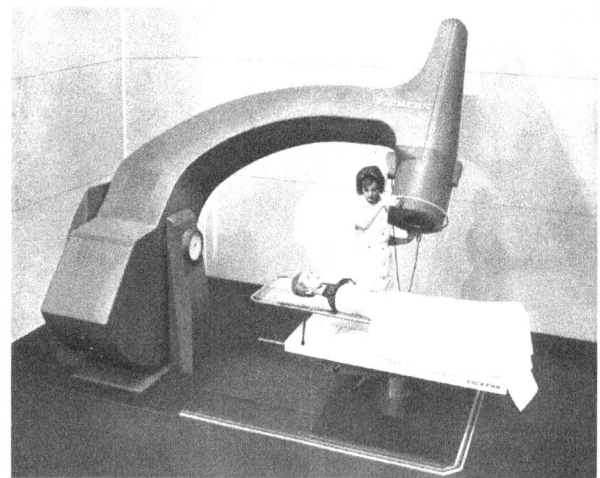

Fig. 18. The Vickers Research 6 MV X band system (from the product brochure).

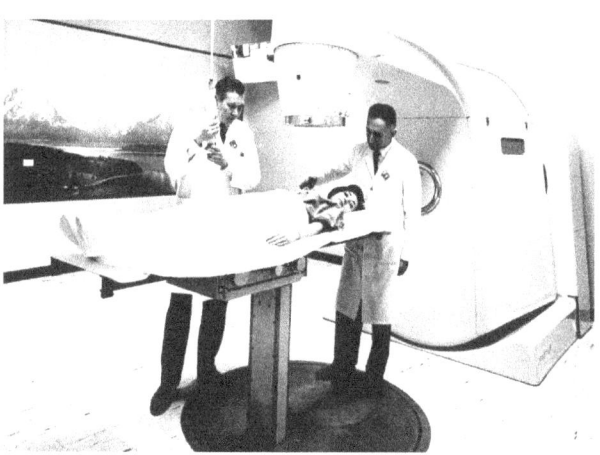

Fig. 17. Clinac-6 *circa* 1962, with Dr. Henry Kaplan at right. Ultimately 39 machines were sold by Varian, and a comparable number by the NEC under license.

flooring. In Fig. 17 the length issue has been resolved to some degree by the use of the bend magnet. However, in either case, the structure itself was limited by the lower shunt impedance per unit length typical of center-coupled structures. The Orthotron corresponded to $r \approx 47\,M\Omega/m$ and Clinac-6 to $r \approx 56\,M\Omega/m$, both commensurate with what had been achieved in the state of the art in high-energy electron accelerators of the time, Mark III and the Stanford Two-Mile Accelerator.

It was E. A. Knapp and colleagues at Los Alamos who conceived of and built a kind of accelerator capable of *twice* the shunt impedance per unit length of the center-coupled type; this was the side-coupled

standing wave accelerator [29–31]. With help from Craig Nunan, under the direction of Karl Brown, this high-shunt impedance design migrated from LANL at the L band (805 MHz) to Palo Alto at the S band (3 GHz) and made its first appearance as a medical linac in the form of a 4 MV straight-ahead machine, as seen in Fig. 19. Thanks to the compactness of the accelerator, and with the additional compromise of an SAD of 80 cm, this machine could fit into a cobalt-60 treatment room. In terms of performance, the machine had a better penumbra and a comparable depth–dose. It could rotate through 360°, and it had no sensitivity of position or symmetry to energy variations, since it had no bend magnet system. And it was affordable.

While in the 1960s the radiation field had been defined by the primary collimators and so had been rectangular in shape, the 1970s saw the advent of early *beam-shaping*, by which blocks and wedges were used to spare healthy tissue. These blocks were changed by hand for each beam angle, requiring multiple entries to the treatment room by the therapist. The typical number of beam angles employed was four or fewer.

The next great advance in medical linacs combined three advances in the field: an achromatic 270° bend magnet system, a long (1.5 m) side-coupled accelerator, and a reliable high-power klystron.

While an early 270° achromatic bend had been invented by Enge [32], manufacturability favored a segmented magnet invented by Karl Brown and William Turnbull in 1974 [33], not to be confused

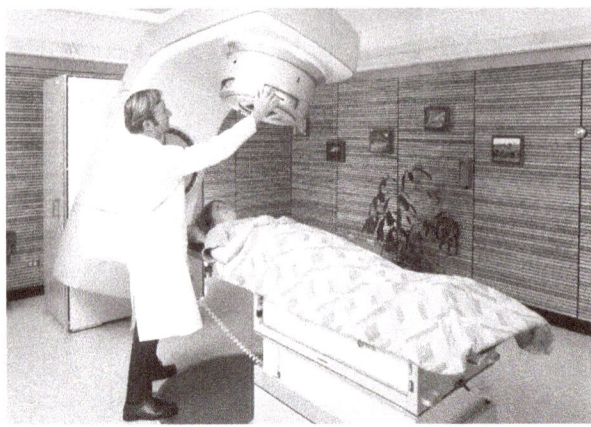

Fig. 19(a). Clinac-4, *circa* 1968, was based on a side-coupled standing wave accelerator, half the length of its predecessors, requiring no dipole bend magnet, and competitive in form, fit and function with cobalt machines. It was a bit more expensive, but superior performance, including the penumbra, eventually put the cobalt machines out of business.

Fig. 19(b). The biperiodic side-coupled accelerator design was the enabling technology for the first compact machines. The small beam aperture permits high electric fields confined near the beam axis, but then requires that microwave coupling from cavity to cavity be provided via side-mounted "coupling" cavities.

with a stepped field magnet they invented ten years later [34]. This segmented magnet system permitted a horizontal accelerator layout, while keeping beam focal spot motion and symmetry insensitive to energy variation.

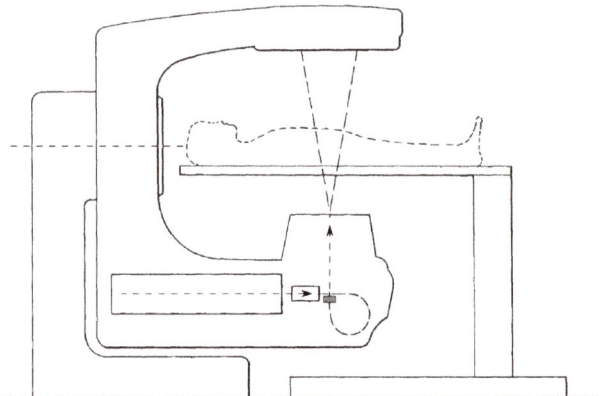

Fig. 20. The 270° deflection system invented by Brown and Turnbull achieved achromatic point-to-point imaging in both planes, providing unique stability of beam position and angle with respect to microwave system variations. This compact 270° bend, *circa* 1975, enabled the modern klystron-powered high-energy linac with energies ranging up to 22 MV, and down to 4 MV. The installed base of these machines is now over 4000 worldwide. (After Brown and Turnbull, US Patent No. 3,867,635.)

The horizontal layout, together with the high-shunt impedance of the side-coupled design, and the 5 MW capability of the klystron permitted energies as high as 22 MV to be accessed. Output in excess of 10,000 R/min was possible, and over 600 R/min flattened.

The late 1980s saw the use of custom-molded blocks to conform the dose to the tumor profile from each angle. Blocks were still changed by hand — a time-consuming process. Treatments used 4–6 beam angles. Single-beam isodose curves were synthesized to form an isodose curve conforming to the planning target volume in two dimensions.

A major breakthrough in accelerator design was the concept of the "energy switch," a water-cooled, mechanically adjustable microwave element that could effectuate a sharp electron energy spectrum at an energy determined by the operationally controlled switch setting. An early energy switch can be found described *circa* 1980 in US Patent No. 4,286,192 by Tanabe and Vaguine [35]. The probe shown in Fig. 22 is fully retracted, or fully inserted. When it is retracted, the phase advance between the two adjoining accelerating cavities is π, providing in-phase acceleration at the downstream cavity. When it is fully inserted, the phase advance is 2π, and the downstream cavity is decelerating. The mechanical action of this switch permits operation in

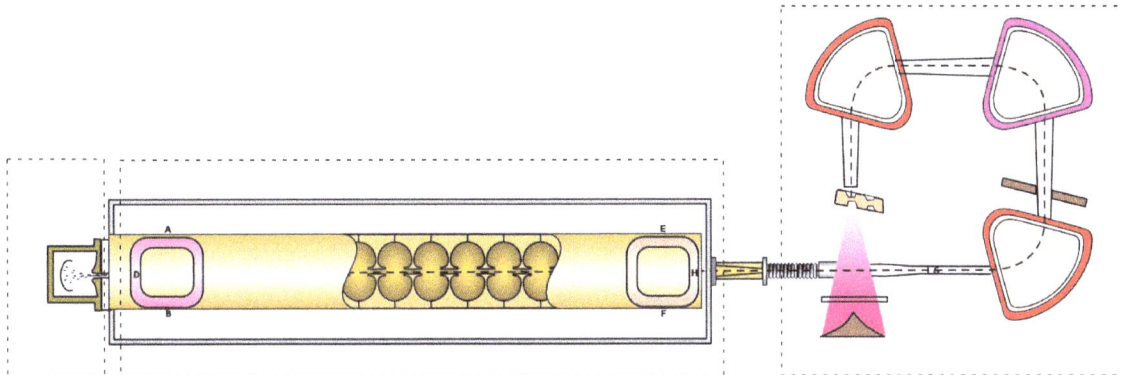

Fig. 21(a). The modern oncology linac system includes a triode gun, a high-shunt impedance accelerator structure, and an achromatic transport line with energy selection.

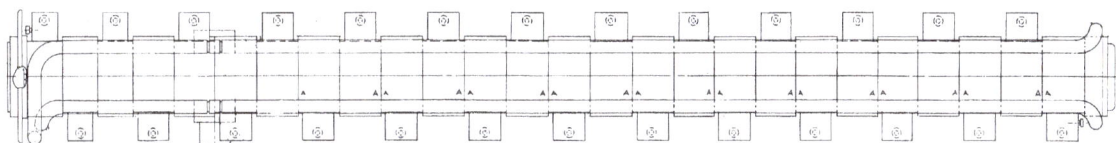

Fig. 21(b). Clinac-18 was the first of a generation of long, side-coupled medical guides. It consisted of $26\frac{1}{2}$ accelerating cells and was designed to work with a triode gun. It generated 6–18 MeV electrons and 10 MV photons.

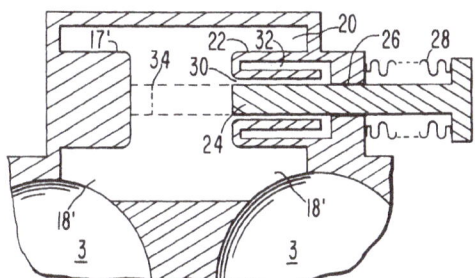

Fig. 22. The Tanabe–Vaguine switch, from US Patent No. 4,286,192.

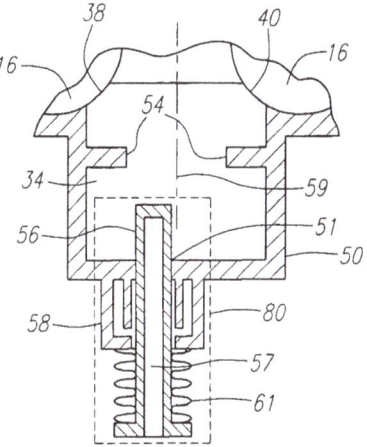

Fig. 23. The field step energy switch consists of a water-cooled, bellows-mounted sliding probe that perturbs a side cavity, resulting in a field step between the centerline cavities coupled through it. (From Ref. 37.)

two modes — a high-energy mode and a low-energy mode.

In the ensuing years a different kind of switch was also developed, one which effectuates an amplitude "step" between cells, rather than a phase flip [36]. The concept of the field step energy switch can be seen in Fig. 23. The action of the switch is to produce a field step α varying with probe position. Remarkably, the switch action permits the fields in the buncher end to be held fixed, by means of a two-step sequence: (1) adjust switch; (2) adjust forward power. With the buncher fields fixed, trapping and phasing of the beam is accomplished in a stable manner. The downstream field profile is completely adjustable in amplitude, and accepts the

beam "injected" into it from the "buncher." More elaborate switches are possible to conceive [37–41].

The 1990s saw the development of the multileaf collimator (MLC), permitting conformal dose shaping in three dimensions without custom blocks or therapist entry into the treatment room. With this technology treatments could expand to nine or more beam angles, helping greatly to spare healthy tissue, and to permit dose escalation. In addition, at every angle, the treatment field could be divided into 500 segments, so that it literally became possible to "paint" or conform the dose distribution with high precision, at the level of 1 mm at 1 m SSD in 3D.

The last decade has seen the rise of physiological gating, particularly respiratory gating, first reported by O'Hara *et al.* in 1989 [42]. One method of gating employs a reflective marker placed on the patient's chest as they lie on the treatment couch [43]. An infrared tracking camera system provides a wave form analog to the patient's respiratory pattern. Combined with treatment-specific trigger thresholds, the treatment beam can be gated on and off with the respiratory cycle via control of the grid pulse to the triode gun. For example, a lung nodule might be several mm in size, but between full expiration and full inspiration it may move by several centimeters. Without gating, the entire volume through which the nodule moves is treated. With gating, just a few millimeters around the nodule may be adequate. The result is reduced lung fibrosis over the six-week treatment, an escalated dose, and improved tumor control probability (TCP).

2.3. *Requirements for radiotherapy*

It was a significant advance for the therapist and the patient when the electronic portal imaging device (EPID) was introduced. A key aspect of treatment is confirmation of patient position. Commonly this was and is done using indelible ink or tattoo reference fiducials on the patient. The treatment room may have as many as five laser beams to provide registration of the patient fiducials to the isocenter. In addition, prior to the introduction of the EPID, every five treatments (once a week) the therapist would place an x-ray film behind the patient, exposing it with the MV treatment beam. This portal film was time-consuming, taking perhaps 4 min for the therapist to develop, while the patient had to remain in position, within the desired mm-level registration accuracy. This film was critical to informing a couch adjustment for offsets of 5 mm or larger. The EPID freed the therapist, and the patient, and introduced a new era of *image-guided radiation therapy* (IGRT).

In generic form IGRT refers to per-fraction analysis of setup error with patient repositioning to correct setup error. IGRT permits adjustment for intrafraction motion as with port films, and also interfraction motion, as between treatments patient anatomy may shift. Real-time motion includes respiration, peristalsis, and circulatory motion.

In recent years a kV x-ray tube with a digital panel has been incorporated in the state-of-the-art radiotherapy machine, providing 2D radiographic, fluoroscopic, or 3D cone beam computed tomography (CT) imaging. Imaging with the kV beam results in a lower dose and a better soft tissue contrast than with the MV beam.

The essential counterpart to imaging is rapid field-shaping enabled by the 120-leaf MLC covering a 40 cm × 40 cm field, permitting segments as small as 2.5 mm × 5 mm.

An excellent review of the geometric aspects of treatment has been provided by Yorke *et al.* [44]. They point out that since the early 1970s there has been evidence that dose increments of 10% can improve local control, or affect toxicity to normal tissue. Modeling from the same period suggested that regions of an underdose of 3–5 mm could greatly decrease the tumor control probability (TCP). In 1993 the International Commission on Radiation Units and Measurements (ICRU) compiled a framework for prescribing, reporting, and recording radiotherapy. This included concepts of the gross tumor volume (GTV), the clinical target volume (CTV), and the planning target volume (PTV). The largest of these, the PTV, includes a margin arising from uncertainties in treatment planning and delivery, including systematic (offsets to the treatment field) and random errors (blurring of the treatment field). Van Herk's result for the CTV-to-PTV margin is $M = 2.5\,\Sigma + 0.7\,\sigma$, where Σ and σ are one standard deviation of the systematic and random errors [45]. A margin is also incorporated to protect an organ at risk (OAR). Treatment protocols aim to reduce M, thus sparing healthy tissue, while permitting dose escalation to the lesion.

Where the PTV is affected by respiratory motion as with lung and liver cancer, respiratory gating is essential to the margin. Synthesis of a precise volumetric dose distribution, oftentimes using respiratory gating, is referred to as 3D conformal radiotherapy, to be distinguished from the use of 2D isodose curves. In 2006, Fang *et al.* reported a statistically significant 27% five-year survival for 3D radiotherapy, versus 6% for 2D [46].

Prior to a course of treatment, treatment planning relies on and exploits the capabilities of the hardware for treatment through multiple angles, including non-coplanar treatments, with a field rapidly adjustable and deformable thanks to the multileaf collimator. These capabilities permit a significant reduction in the dose to healthy tissue. For cases where the treatment volume is affected by respiratory motion, treatment planning via 4DCT can be performed, with subsequent treatment employing respiratory gating. 4DCT refers to CT imaging where frames are tagged by one of ten respiratory phases, as determined using an infrared marker and tracking system. These phases permit a 3D movie from which the oncologist selects not just a treatment volume, but a treatment phase or phases, such as full expiration. The treatment plan is developed to employ beam during the designated phases, and in this way to further spare healthy tissue. Fluoroscopy is also available as a verification tool.

In addition, positron emission tomography (PET) may be used for staging, and possibly a second PET for registration and comparison with the CT image. Other research focuses on real-time orthogonal imaging, as well as efforts to improve soft tissue contrast, and registration in areas with low contrast, including use of implanted fiducials [47, 48].

During preparation for treatment, patient position is verified using digital portal films, and can be verified in real time using orthogonal imaging with the kV tube and digital panel. During treatment, imaging may employ kV onboard orthogonal imaging either for kVCT or for real-time imaging. The megavoltage (MV) beam is routinely used for portal imaging to verify positioning of the treatment couch, and can also be employed for MVCT.

At the forefront today, researchers are endeavoring to find ever more practical and accurate means to assess changes in the tumor and its environment during the course of a six-week fractionated treatment and to adjust their treatment planning within the course of treatment [49].

2.4. *Modern linacs*

Thanks to the side-coupled biperiodic structure, the triode gun, the energy switch, the high stability of the klystron drive train, and the 120-leaf MLC, treatment outcomes have improved dramatically over the 50 years since the inception of microwave linacs for radiotherapy. As seen in Fig. 24, the modern C-arm gantry carries an arsenal of components to serve

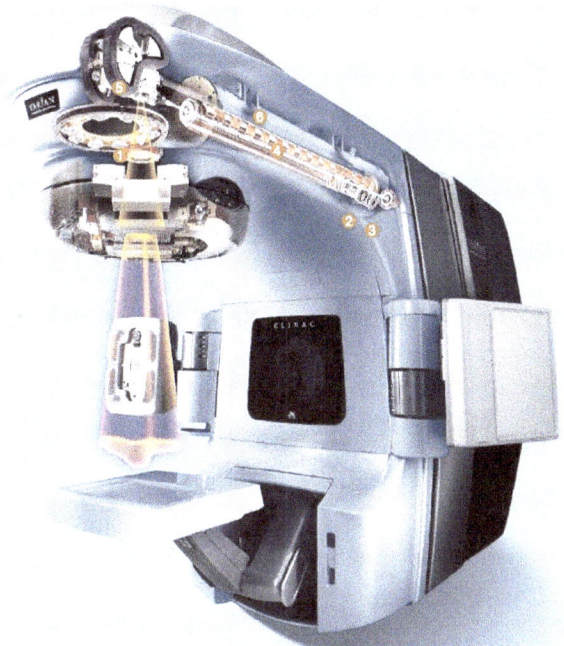

Fig. 24. The modern clinac provides real-time imaging via an orthogonal kV tube and digital panel, real-time position management based on gating via a gridded triode gun, and energy selection via a switched, multienergy accelerator structure. The tight spectrum permits a high output even through 6% energy slits, enabling short treatment sessions. The subsystems include: (1) flattening, dosimetry and collimation, (2) energy switch, (3) gantry, (4) accelerator structure, (5) 270° achromat with energy selection slits, and (6) solenoidal beam transport.

the therapy team. Imaging tools continue to reduce tumor margins, thus reducing normal tissue exposure and reducing ever more complex treatments to fit within the 15 min time frame regarded as "clinically practical." Today, the five-year survival rate post-treatment is over 60%, versus half that figure in the early 1960s.

The intervening years also have seen developments in the accelerator platform away from the C-arm gantry, to include a ring gantry [50] and a robotic arm [51].

The Tomotherapy, Inc. Hi-Art slip-ring powered gantry employs a magnetron-powered inline 6 MV S-band accelerator [52]. It uses a 64-leaf air-actuated binary (open or shut) MLC and a xenon image detector array. No flattening filter is employed. The maximum field size is 40 cm × 5 cm at the 85 cm isocenter. The rated output is 960 rad/min. While it gives up certain degrees of freedom for treatment, such as non-coplanar fields, it includes the ability to perform CT imaging and is well suited to treating small tumors.

The Accuray robotic arm employs a magnetron-powered inline 6 MV X band accelerator and offers the promise of tracking, i.e. moving the treatment head to accommodate respiratory motion. It is limited in not being able to perform posterior fields, and setup time tends to be long.

The Siemens mobetron [53] is a compact X band accelerator system designed for intraoperative radiotherapy in electron mode with 4, 6, 9, and 12 MV. Two structures — a buncher and a linac — are employed, with separate feeds. The buncher injects a 4 MV beam into the linac which receives a phase- and amplitude-adjusted feed, and provides an additional 0–8 MV acceleration.

Performance issues of interest in all configurations are output, stability, and precision. Stability concerns dose variation with beam-on cold [54–56], as well as focal spot motion during beam-on over the first 1–3 s [57]. Precision refers to registration of the treatment volume to the x-ray field at the 1 mm level. Stability of output on rotation is critical [58]. Consistency between the gated and nongated modes has been studied [59]. Output is limited by accelerator design, target life and, ultimately, average rf power. This provides an advantage to klystron-powered machines, albeit one that is offset for some by the lower cost of magnetrons.

To aid performance a number of feedbacks are employed. Dose output can be regulated by beam pulse width, pulse repetition frequency, or pulse-dropping. Energy can be regulated via the forward power, controlled either by the modulator voltage or the microwave power input ("rf driver") in the case of a klystron-based system. The control may dither within the energy acceptance of the slits, or servo on yield.

Energy can be changed with the help of an energy switch. Some linac systems are not equipped with a switch and employ more expedient methods of control, such as detuning of the rf drive from resonance, or use of beam-loading to vary energy — techniques that sacrifice spectrum, and thus output.

Beam position and angle on target affect symmetry and can be monitored using a segmented ion chamber. Four chambers covering four quadrants provide difference signals on which steering coil supplies servo. Some stability features are built into the hardware. Where a solenoid is employed, careful

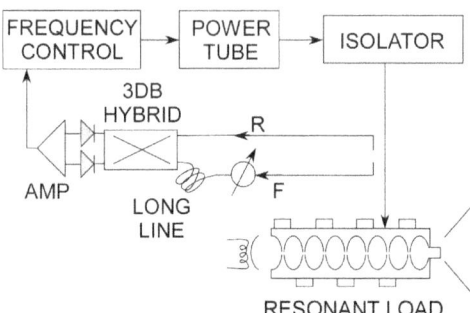

Fig. 25. Automatic frequency control, as conceived by Meddaugh, employs a phase comparison of reverse (R) and forward (F) wave forms, after addition of a long-line discriminator and a phase correction to the forward signal. The resulting error signal drives a motor that tunes the magnetron back to resonance.

minimizing of the transverse magnetic field helps to decouple focusing and steering. This also insures plenty of margin on the steering coil supplies for use during normal operation, for position feedback.

A bend magnet system that is achromatic insures that spectrum fluctuations within the energy acceptance do not translate into spot motion. If in addition a physical limiting aperture (a "scraper") is placed so as to image onto the target, then beam-size deviations are controlled.

An automatic frequency control (AFC) circuit is essential for standalone linac systems [60]. Unlike high-energy accelerators, medical linacs do not have a long time available after beam-on to reach equilibrium. Instead, with small commercial linacs, beam-on may last just a few seconds. Inlet water temperature is held at 30°–40°C, depending on the system, but may not be tightly regulated. During beam-on the accelerator frequency shifts at the rate of 50 kHz/°C, and the temperature may climb by 5°C or so, due to microwave power dissipation in the copper. Consequently the accelerator operating mode frequency, nominally 2856 MHz or 2998 MHz, moves freely with beam-on, by as much as 400 kHz in 1 min. During this time, the AFC circuit provides an error signal to either the magnetron tuner or the klystron rf drive, keeping the microwave system frequency locked on the moving accelerator resonance. The AFC module provides an analog of reverse voltage phase on the input line, relative to forward voltage. Suitably zero-adjusted with a one-time phase-shifter adjustment, this analog output provides an error signal adequate to track resonance.

Beyond the hardware level, a number of operational aspects are critical to treatment planning. In the factory, x-ray field characteristics are verified in a water tank to assess dose distribution in the transverse plane, in depth. On-site the medical physicist performs daily and weekly quality assurance of similar dosimetric properties of the beam in the various electron and x-ray modes.

Providing an accurate well-targeted dose is aided by the highest beam output possible from the accelerator, gated on the respiratory phase, and stable over the first seconds of beam-on. Sparing healthy tissue requires exquisite accelerator system stability, and peripherals integrated into the control system working with the imaging tools, to manage respiratory motion as well as planned motion of the treatment head. The key hardware engineering and physics challenges are for high output and stability from the microwave linac system as a whole.

3. Biperiodic Accelerator Structures

In this section we focus on biperiodic standing wave structures. First, we look at the side-coupled version — the highest shunt impedance structure. After that, we look at the on-axis coupled biperiodic structure, perhaps the most easily manufactured.

3.1. *Side-coupled biperiodic*

The side-coupled biperiodic structure seen in Fig. 26, originally devised by Knapp [61], consists of high-Q on-axis cavities, and side cavities made smaller by the use of nose cones, at the expense of cavity Q.

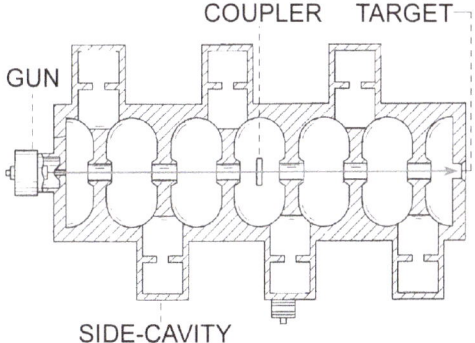

Fig. 26. Cutaway view of a biperiodic side-cavity-coupled structure, consisting of on-axis accelerating cavities, and off-axis side cavities which couple adjoining centerlines. Electrons are injected from the gun into the "buncher" cavity and travel to the right, striking the target. (From Ref. 36.)

Fig. 27. Two periods of a biperiodic side-cavity-coupled structure, shown excited in the $\pi/2$ mode. The five mode frequencies in this geometry can be employed for Brillouin-based tuning. The geometry depicted here is the *vacuum* portion of the structure, not the copper.

The accelerating mode ($\pi/2$ mode) of the structure employs field nulls in the coupling cavities, with the consequence that the wall Q for the mode is roughly that for the on-axis cavities.

A cutaway of two periods is seen in Fig. 27, showing the small beam tube that permits the high shunt impedance of this structure. Because this tube is so small in diameter, the microwave signal is evanescent within the beam tunnel, and cell-to-cell coupling through this tunnel is small. Instead, the microwave power flows through the side cavities in traveling from one on-axis cavity to the next. The $\pi/2$ mode is illustrated in Fig. 27.

3.2. *Tuning*

The side cavities in biperiodic accelerators are correctly tuned when a magnetic boundary applies at their midplane, with the system excited in the $\pi/2$ mode with resonance at the design frequency f_{des}, e.g. 2856 MHz. This definition does not in itself dictate a method of tuning, as it is not possible to apply a magnetic boundary at the side cavity midplane. Moreover, it is not possible to tune centerline cavities independently of side cavities. Thus any tuning procedure must take account of indirect observations, i.e. something other than simply a peak or dip indicating a resonance at f_{des}.

It is possible, for example, to insert a conducting rod through the beam tunnel, and in this way detune the system. We may think of this rod as detuning all centerline cavities; of course, it is perturbing the entire system of cavities. A probe inserted in a side cavity tuning port then permits observation of a frequency, f_{sc}, predominantly influenced by that side cavity geometry. The result obtained depends in principle on the rod diameter, and on the state of the tune of adjacent side cavities; the extent of this dependence, whether noticeable or negligible, depends on other features of the problem, particularly the coupling slot dimensions.

In this section we illustrate the analysis and measurements needed to establish the value to be employed as a target for f_{sc}. The desired value is produced by mechanically deforming the side cavity.

Tuning requires reference to signal propagation in the guide, either via measurement, and observation of a stop band, or via interpretation of mode frequency data, through a model dispersion relation. One can try to take mode frequencies alone and make progress, tuning to center the 0 and π modes around the $\pi/2$ mode, or tuning to center the adjacent modes around the $\pi/2$ mode. However, these approaches applied to a two-cell or three-cell test stack may result in a noticeable error for the tuning target needed to close the stop band. A long structure offers a great length over which cutoff signals can evanesce, and the consequences of an error in the tuning target are clear after the fact. It is preferable to have an algorithm capable of taking two-period data and using them to determine the precise tuning target.

For illustration, we consider the simplest biperiodic structure model dispersion relation, the result obtained by neglecting next-nearest neighbor coupling[29]:

$$k_1^2 \cos^2 \theta = \left(1 - \frac{f_1^2}{f^2}\right)\left(1 - \frac{f_2^2}{f^2}\right).$$

Here f is the frequency of excitation and θ is the phase advance per cell at this frequency. The copper geometry determines the structure constants, i.e. the coupling constant k_1 and the characteristic frequencies f_1, f_2. By tuning of the structure we affect the values f_1, f_2. The major issue for tuning is to insure that the dispersion curve does not have a "stop band," as seen in Fig. 28.

A five-mode structure provides five modes suitable to perform a least squares fit and determine the corresponding Brillouin curve constants, k_1, f_1, and f_2, and the stop band $\Delta f = f_-(\frac{\pi}{2}) - f_+(\frac{\pi}{2}) =$

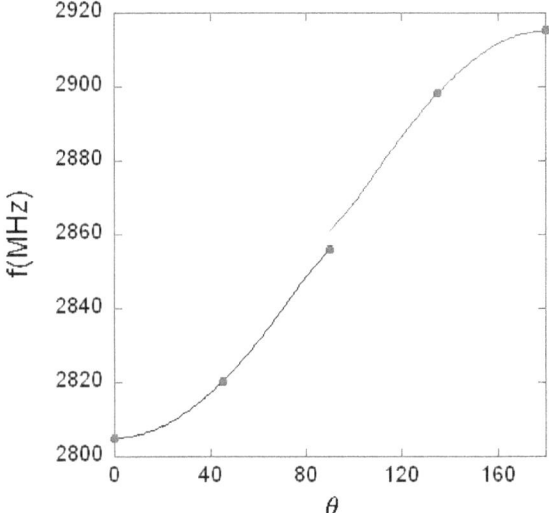

Fig. 28.   Brillouin curve for an example with a 5 MHz stop band, showing the corresponding frequencies in a five-mode structure.

$|f_1 - f_2|$. One tuning procedure consists of iterating to adjust these parameters so as to close the stop band.

The physical features that are tuned may be the side cavity nose height and the centerline bowl diameter. In an EM CAD model this is straightforward. On a prototype on the bench, a C-clamp is employed to gently dent the centerline bowl. A small clamp can be applied to the side cavity to either dent the bowl or push the noses closer, depending on which direction frequency is to be tuned. In the course of this procedure, whether on the bench or in an EM CAD model, one determines the Brillouin curve parameters by fitting the data to the biperiodic dispersion relation. To generate the dispersion curve one solves a quadratic equation for f in terms of $\cos \theta$. There are two branches to the dispersion curve — an upper branch, $f_+(\theta)$ and a lower branch, $f_-(\theta)$, with

$$f_\pm = \sqrt{f_1 f_2 / \alpha_\mp},$$

where

$$\alpha_\pm = \frac{b}{2} \pm \left(\frac{b^2}{4} - 1 + k_1^2 \cos^2 \theta\right)^{1/2},$$

$$b = \frac{f_1^2 + f_2^2}{f_1 f_2}.$$

If at $\theta = \pi/2$ there is a nonzero gap, Δf, the frequencies in this gap correspond to imaginary θ, and evanescent propagation. For these frequencies the periodic line acts like a cutoff waveguide.

The presence of such a stop band is undesirable, as it causes enhanced sensitivity to perturbations. Normal in-tolerance mechanical variations may result in out-of-spec field imbalance or "field slope." With a bit of algebra one can show that closing of the stop band $\Delta f = 0$ results in

$$f(\theta) = \frac{f_1}{\sqrt{1 - k_1 \cos\theta}} \quad \text{(tuned)}.$$

For a properly tuned structure one can show that $f_-(\pi) - f_+(0) \approx k_1 f_{\pi/2}$, so that k_1 is the full fractional bandwidth.

The process of tuning the side cavities tunes the side-cavity-centered $\pi/2$ mode approximately linearly. This side-cavity-centered $\pi/2$ mode corresponds to an E boundary centered in the side cavity, and an H boundary centered in the centerline. Tuning brings this mode frequency up to that of the centerline-centered $\pi/2$ mode (E boundary centered in the centerline, and H boundary centered in the side cavity). The closing of the stop band corresponds to the confluence of the side-cavity-centered and the accelerating-cavity-centered $\pi/2$ modes.

3.3. *Coupler design*

In the following we review coupler design using a particular accelerator for illustration — a biperiodic standing wave structure employing electric coupling through the beam tunnel, as seen in Fig. 29 [62, 63]. While center coupling does result in lower shunt impedance, the larger beam aperture facilitates reduced "guide glow," i.e. undesired radiation due to beam loss. In addition, this design simplifies manufacture.

Design is informed by a desired value for the external Q for the finished N-cell structure. This desired figure is first determined from the expected wall Q, and the desired coupling parameter β. The optimum coupling–parameter can be derived from the beam-cavity-waveguide equation and is $\beta_{\text{opt}} \approx 1 + P_b/P_d$. The figure P_b is the beam power required by the application, and this can be inferred from the specified energy and output. The power dissipated in wall losses is P_d and can be determined from the specified energy and a good working figure for shunt impedance. The single-cell external Q is determined from the N-cell value by dividing by N or, more precisely, by the ratio of energy stored in the full structure to that stored in the coupler cell. The result of all this is a desired number Q_e corresponding to a single coupler cavity coupled to the waveguide, as seen in Fig. 30.

Starting from the design for the basic period, a coupling hole is introduced into the side wall, tending to lower the frequency, and a bowl radius is reduced to raise the frequency back. With the remaining coupler geometry extending into the waveguide, the external Q is computed using the Slater method, as we describe next. Comparing the result to the target value, the coupling hole is adjusted, and the process is repeated (tuning adjustment, Q_e calculation). Here we focus on providing a straightforward description of the Slater method [12].

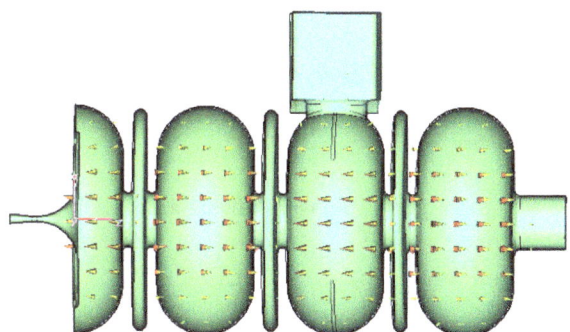

Fig. 29. Compact biperiodic accelerator structure, excited in the $\pi/2$ mode, as modeled in CST Microwave Studio, an electromagnetic computer-aided design program [64].

Fig. 30. Coupler cell EM CAD (CST) model. The solution shown corresponds to the shorting plane at a quarter-guide wavelength from the plane of the detuned short.

The essentials of the geometry are seen in Fig. 6, consisting of a cavity coupled by some means with a waveguide. Also shown are two planes — one located at coordinate z_0 and one at z_1. We define z_0 at the "plane of the detuned short," the location in the waveguide where, were the cavity driven off-resonance, one would find a node in the electric field. On resonance, this plane corresponds to an open circuit. We employ z_0 to mark the fixed but as-yet-unknown location of this plane. We will find it helpful to place a sliding short on the guide, and this location we mark with z_1. This system has generically at least two modes — one predominantly in the cavity and one predominantly in the waveguide. These modes interact by means of the coupling hole, and this interaction may be perturbed by means of the shorting plane. To see how this takes place, we need to do some calculations.

To analyze this system, we employ the beam-cavity-waveguide equation of Subsec. 1.4. We suppose that the system is driven at angular frequency ω, that the cavity resonance angular frequency is ω_0, and we neglect losses — since we will be applying this analysis to results of frequency domain simulations with lossless boundary conditions. The cavity response to a forward wave in the guide is described by

$$\left(\omega_0^2 - \omega^2 + j\frac{\omega\omega_0}{Q_e}\right)\tilde{V}_c = \frac{2j\omega\omega_0}{Q_e}\tilde{V}_F,$$

where $\tilde{V}_c$ is the cavity voltage, and $\tilde{V}_F$ is the forward voltage in the guide (incident on the cavity), defined with reference to the plane of the detuned short. The reverse voltage in the guide is $\tilde{V}_R = \tilde{V}_c - \tilde{V}_F$. The quantity ω_0 is the natural resonance frequency of the cavity, when there is an open circuit at the plane of the detuned short. The conducting boundary condition at the shorting plane takes the form

$$\tilde{V}_F e^{-j\beta D} + \tilde{V}_R e^{j\beta D} = 0,$$

where $D = z_1 - z_0$, and $\beta = (\omega^2 - \omega_c^2)^{1/2}/c$, with c the speed of light and ω_c the cutoff angular frequency in the waveguide. For a rectangular guide with long dimension a, $\omega_c = \pi/a$ for the TE_{10} mode. Combining these two results, one may show that

$$\tan(\beta D) = \frac{1/Q_e}{\omega_0/\omega - \omega/\omega_0}.$$

A more accurate form would add an unknown constant to the right side corresponding to a sum over all other cavity modes. This result is telling us that the system resonates at an angular frequency ω, different from the natural resonance frequency ω_0, i.e. it has been perturbed by the interaction with the shorting plane in the guide.

With this analysis in hand, we proceed as follows to determine Q_e. First, we set z_1 to a value a bit larger than the one-half guide wavelength, and perform an eigenmode simulation on the geometry of Fig. 30. Results must be inspected, including field plots, to identify the two modes of interest, one residing mostly in the cavity and one in the guide, i.e. the line resonance. The detuned short position may be determined from

$$z_0 = z_{1a} - \frac{\pi}{\beta(\omega_{\text{line}}(z_{1a}))}.$$

This defines a reference plane in the waveguide, one that enjoys a field null at the line resonance frequency, a frequency that should be well separated from the guide resonance. If one sees a significant field in both the cavity and the waveguide, one should try a different value of z_1. We also record the cavity resonance frequency from the same simulation output, and with this determine the next position for the shorting plane:

$$z_{1b} = z_0 + \frac{\pi}{2\beta(\omega_{\text{cav}}(z_{1a}))}.$$

This location is a quarter-guide wavelength from the plane of the detuned short. It will enjoy a field null, when the cavity is driven near resonance, and thus it is the place to locate a shorting plane in order to determine ω_0. We perform then a second eigenmode calculation and inspect the results to determine the frequency of the cavity mode. (Note that the cavity mode frequency is detuned due to the movement of the short.) In this calculation, we have placed the shorting plane a quarter-guide wavelength from the plane of the detuned short. Thus $\omega_0 = \omega_{\text{cav}}(z_{1b})$. We then compute

$$Q_e = \frac{\cot\{\beta(\omega_{\text{cav}}(z_{1a}))(z_{1a} - z_0)\}}{\frac{\omega_{\text{cav}}(z_{1a})}{\omega_0} - \frac{\omega_0}{\omega_{\text{cav}}(z_{1a})}}.$$

With the coupler dimensions fixed and the plane of the detuned short located, one may model the field in the full structure, as seen in Fig. 29.

4. Beam Dynamics

At the front of any electron linac is an electron gun. Guns may be classified according to the emission mechanism and the kind of electric field employed. To leave a surface, electrons must pass through a potential barrier of height given by the work function. This can be accomplished by tunneling in the presence of an electric field (field emission), absorption of a photon (photoemission), or by means of thermal motion (thermionic emission), in the case of a hot cathode. In addition, back-bombardment of a cathode by electrons may result in secondary emission. And heating from back-bombardment ("back-heating") may affect cathode temperature and thus thermionic emission.

4.1. *Thermionic emission*

The ancestry of thermionic guns is traced to the incandescent lamp *circa* 1801, Edison's light-bulb-testing lab, and the discovery of negative charge emission *circa* 1882. In 1911, Owen Willans Richardson proved that filaments, and not the ambient gas, emit electrons, and described the emission process as extraction from a Maxwell–Boltzmann distribution of electrons, over the surface potential barrier. Saul Dushman, working at the General Electric Research Laboratory, improved on this model taking into account the Fermi–Dirac electron distribution, deriving the so-called Richardson–Dushman law of thermionic emission. Richardson received the Nobel Prize for this work in 1928.

Thermionic emission relies on a heater filament to bring the cathode surface to a high temperature, $T \approx 10^3\,^\circ C$, such that electrons at the tail of the distribution may escape. The emitted current density,

$$ J \approx \eta A T^2 \exp\left(-\frac{\phi}{k_B T}\right), $$

with $A \approx 120\,\mathrm{A/cm^2 K^2}$ and $\eta < 1$, a transmission probability. This result corresponds to a zero applied electric field. Evidently, high current density requires high temperature, and the work function is critical. A low work function is important for emission, but low-work-function compounds melt and evaporate easily, and are prone to poisoning. The evolution of cathodes from tungsten to thoriated tungsten and to oxide cathodes and dispenser cathodes is described from a power tube perspective by Gilmour [65], and by Cronin, focusing on dispenser cathodes [66]. Unlike high-power microwave tubes, medical linacs do not require high cathode loading,

and end of life typically does not result from cathode depletion, provided that an adequate operating and engineering margin is planned into the system. Accelerators have run for over a decade in the field. For medical linacs, the primary limitation deriving from the gun is the need to maintain a pristine vacuum environment. This favors a welded, baked, and pinched-off assembly.

4.2. *Space charge limit*

Once electrons are flowing steadily through a gap, their negative charge distributed through the gap ("space charge") reduces the field seen at the cathode. The ultimate limit on emitted current density corresponds to zeroing of the field at the cathode. The applied field $E \approx V/L$, with L the gap length, while from Gauss's law the space charge field is $\propto \rho$, the charge density in the gap. Thus, on general grounds the space-charge limiting charge density $\rho \propto V$. Velocity in the gap $\propto V^{1/2}$, so that current density $\rho V \propto V^{3/2}$. In this way, combining energy and charge conservation with Gauss's law one can show that, in the limit $V \ll mc^2/e$, the space-charge-limited current takes the form $I = JA = KV^{3/2}$, a result first noted by Child, and later by Langmuir and Blodgett. The constant K is called the *perveance*, and is a function only of the diode geometry. For example, in the case of a planar gap of length L and plate area A, $K \approx 2.3A/L^2\,\mu\mathrm{perv}$, where $1\,\mu\mathrm{perv} = 10^{-6}\,A/V^{3/2}$.

The transition between temperature-limited and space-charge-limited operation is determined by the work function and can be assessed via the "rolloff" curve. This depicts gun current versus an analog of cathode temperature, such as gun filament current. On a semilog plot the two limits permit one to compare a relative measure of the work function within a lot.

Given the limits on cathode current density, and demand for high current, it is common to design the gun geometry to provide spatial convergence of the electron flow, as in a spherical diode, from the cathode through the beam tube aperture in the anode plane, and additional convergence after the anode plane, aided by magnetic focusing. It is desirable to have a degree of laminarity in the electron flow to aid in beam confinement through the rest of the device, and this requires self-consistent electrode design.

EGUN is a popular gun design program begun during the development of the XK-5 klystron gun, and benefits from over 30 years of development and benchmarking [67]. Addition of a grid over the cathode permits application of a control voltage and production of a shorter electron pulse. A second major advantage of a gridded gun is the ability to control current independent of gun high voltage. The high-voltage figure affects the trapping and phasing process in the linac, and thus the final beam spectrum. The grid does influence the beam waist and effective focal spot due to the change in the beam opening angle at the grid and the change in beam perveance.

4.3. *Envelope equation*

The space-charge-limited current may differ from the actual beam current; for example, if one forgets to turn on the power supply for the cathode heater. For this reason we distinguish the geometrical perveance of the gun from the beam perveance, computed as $I/V^{3/2}$, with I the actual beam current. Beam perveance determines the universal spreading curve governing envelope expansion when perveance is large and emittance is not. For accelerator modeling perveance is typically less than 1 μperv, and emittance should be included.

Here we consider what happens without a confining solenoidal field. A uniform, symmetric, centered, and aligned beam in a drift satisfies

$$\frac{d^2\sigma_x}{ds^2} + \hat{K}\sigma_x = \frac{\varepsilon^2}{\sigma_x^3},$$

where $\sigma_x^2 = \langle x^2 \rangle$, $\varepsilon^2 = \sigma_x^2\sigma_{x'}^2 - \sigma_{xx'}^2$, with $\sigma_{xx'} = \langle xx' \rangle$ and $\sigma_{x'}^2 = \langle x'^2 \rangle$. The constant $\hat{K}$ is due to space charge, causing defocusing of individual trajectories according to $x'' = -\hat{K}x$. For a nonrelativistic beam one can show that

$$\hat{K} = -\frac{K_b}{K_0}\frac{1}{a^2},$$

where $K_b = I_b/V_b^{3/2}$ is the beam perveance, $K_0 \approx 65.8\,\mu$perv, and $a = 2\sigma_x$ is the beam edge radius. In this way one arrives at a simplified but intuitively helpful description of beam expansion in a drift, under the influence of space charge and emittance:

$$\frac{d^2\sigma_x}{ds^2} = \frac{1}{4}\frac{K_b}{K_0}\frac{1}{\sigma_x} + \frac{\varepsilon^2}{\sigma_x^3}.$$

For a given emittance, perveance, and beam waist, this equation describes a curve unique up to translation.

4.4. *Simulation*

The injected phase space provided from the envelope equation permits assessment of capture, output spectrum, beam spot, and guide glow. In the paraxial limit, modeling of beam focusing and capture by rf fields is straightforward from a table of values for s, E_s evaluated on-axis. Results of a simulation for the biperiodic structure in Subsec. 3.3 are seen in Fig. 31.

For an accelerator on which the primary energy diagnostic will be attenuation, the characterization of the macroparticle distribution proceeds from a fit to data on a half-value layer (HVL) for monoenergetic beams. With the outgoing macroparticle distribution, one computes a figure representative of the dose-rate after a depth z. The effective HVL for the "numerical" beam is then determined, and a lookup table can be used to determine the monoenergetic beam HVL-equivalent energy.

5. High-Power Test

In this section we review basic aspects of the high-power test, focusing on a magnetron-based system and low-energy operation.

5.1. *Magnetron*

A comprehensive introduction to microwave tubes has been provided by Gilmour [66]. High-power tubes include a cathode, an anode, a high voltage pulsed across the gap acting to extract electrons from the cathode and, for high-current beams, a magnetic field

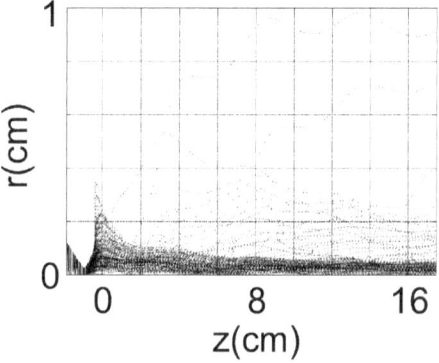

Fig. 31. Macroparticle trajectories at the design settings within the 1-cm-radius machine aperture. Beam loss, except to the anode plate, corresponds to 2 trajectories out of 360, or 0.6% of the injected beam.

confining the electrons. The electron stream develops an rf modulation either by growth of signal from noise, as in an oscillator, or from an injected rf signal, as in an amplifier. The magnetron oscillator is a crossed field device, by which it is meant that the electric field due to the pulsed high voltage producing the electron beam is orthogonal to the magnetic field confining the electron beam.

Due to the problem of oscillation in undesired modes ("moding"), the conventional magnetron employs straps across the vane tips to detune nearby modes. The closed geometry results in high-power density and limits due to breakdown. To forego the straps and reduce the power density, the coaxial magnetron has the anode resonator embedded in an outer coaxial cavity that stabilizes the oscillation in the design mode.

Magnetrons for medical linacs are preferably tunable and the tuner drive motor is the actuator closing the automatic frequency control loop. The yield servo may employ magnetron current as actuator. The dose servo will typically employ PRF.

The magnets are of two types — permanent magnets and electromagnets. Magnetrons operating with an electromagnet have a characteristic load line usually expressed in terms of the magnetic field and current, and operate most stably when kept on the load line. A permanent magnet is less expensive, but tends to provide less flexibility in terms of operating point and stability.

There are two prevalent practices for operating a magnetron into an accelerator. One approach is isolation, employing a three-port circulator with a load on one arm to keep reflections from the accelerator from impinging on the magnetron. This approach works with coaxial magnetrons. The second approach, invented by Howard Jory, is to employ an adjustable mismatch in front of the same load, so as to provide some reflection back to the magnetron, as seen in Fig. 32 [68]. The mismatch, or "phase wand," provides a factory-set phase and amplitude adjustment to the reflection. Qualitatively, the action of this phase wand is to couple the accelerator (a resonator) with the magnetron oscillator, and stabilize the output frequency near resonance. This effect can be analyzed with reference to the Rieke diagram. This diagram displays in the Smith chart format the variation in magnetron power output and frequency into loads of different VSWR and at different electrical distances.

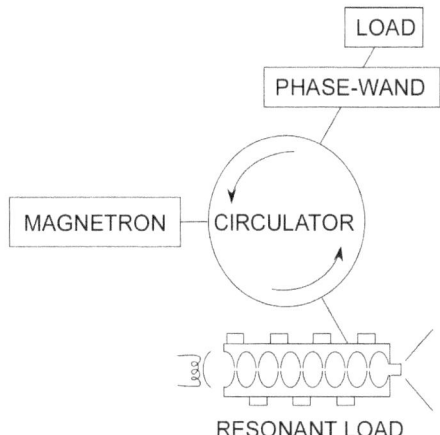

Fig. 32. RF network implementing the phase wand. Typically power to the load is monitored with a nondirectional coupler, and forward power to the accelerator is monitored with a directional coupler. A reverse and a forward power coupler are needed to provide signals to the AFC system.

Reliable commercial magnetrons may be found operating at the S band (3 GHz) up to 3 MW and 4 kW, and at the X band up to 1 MW and 1 kW. Higher peak and average powers are being explored but fundamental limitations arise from the small vane tip features, including cyclic stress from single-pulse heating and removal of average dissipated power.

5.2. *Accelerator characterization*

To illustrate the process we consider the biperiodic accelerator discussed in the previous examples. After cavity design, engineering design, machining, and several braze steps, the accelerator may be tuned. The weld assembly is completed and baked out at 450°C, gun-processed, pinched off, potted, and hi-pot-tested. The accelerator is then installed in a shielded vault as seen in Fig. 33, on an SF_6-pressurized three-port circulator network as in Fig. 32, with an AFC circuit as in Fig. 25. The system was powered by an e2V MG5349 magnetron, operated at 4.5 μs pulse length and 200 Hz PRF. The accelerator processed in minutes to less than 1 μA pump current ($<, 6 \times 10^{-8}$ torr). Characterization of the new accelerator included assessment of output, energy, and beam size at a variety of gun high-voltage settings and magnetron output power levels.

Ideally, accelerator output and beam quality are assessed with an ion chamber in a tank of water with a precise 3D translation mechanism. It is simpler for prototype studies to assess output with an ion

Fig. 33. Biperiodic structure of Subsec. 3.3; final assembly installed in a shielded vault for testing.

chamber placed at a depth matched to the appropriate D_{max} in a fixture of water-equivalent density material. The present example employed 1 cm of build up, in a 10 cm × 10 cm field at 100 cm SSD, with no flattening filter. Beam quality was assessed by observation of attenuation through several inches of steel and calculation of the HVL. For inference of equivalent energy from the HVL, use was made of data by Buechner *et al.* [18] covering 0.5–2.5 MV, and by Goldie *et al.* [69] covering 2–6 MV.

Testing was also performed to assess guide glow. In Table 1, the column marked "forward" corresponds to the machine output at the various settings and the corresponding HVL. The collimator was not blocked. The column marked "side output" corresponds to the dose rate output measured at 1 m and 90°. A figure for the "indirect" dose rate was obtained by placing several centimeters of lead on the guide. The difference is listed in the column labeled "side–indirect."

This accelerator provides low leakage at angles away from the design field — one of the primary goals of this design. Low leakage becomes important

to weight reduction as linacs become more compact. Weight meanwhile affects speed and the precision of the mounting system for the x-ray head package. Size and weight both affect future accelerator implementations, particularly configurations which achieve combined functions. For example, there is interest in marrying a magnetic resonance imaging (MRI) machine with a radiotherapy machine. Two major challenges with such an effort are noise due to the accelerator system and perturbations to the linac beam from the large magnetic field of the MRI. More compact and well-shielded accelerators are more versatile.

6. Summary

The development of electron linacs for cancer therapy has been a long series of evolutionary improvements, with some remarkable, and pivotal, hardware developments along the way. The basic requirements for linacs for oncology are high output, stability, precision, and versatility. Hardware development remains a critical path to enabling more effective real-time imaging, dosimetry, and verification. The larger challenge is to discern those hardware advances needed to enable the next great machine concept.

A natural direction in which to look for future machines is a higher gradient — a regime historically limited by field emission, trapping, and breakdown. Because of these phenomena, commercial linacs tend to operate at a limited surface field. Meanwhile, the problem of shielding results in diminishing returns as structures become more compact. One does not have to go far in terms of the gradient, or wavelength, before the accelerator structure ceases to be a significant weight or space constraint in the system. However, guide glow, the spurious radiation resulting from unintended beam loss, is a constraint, setting a lower limit on the weight of shielding required.

The goal of new concepts is to achieve improved outcomes not just in a study, but for a population of patients numbering in the hundreds of millions. Such a new machine must be manufacturable in large quantities and, since machines do not design and build themselves — they must be profitable.

Acknowledgments

This work was made possible by the support of Varian Medical Systems and the collaboration, advice, instruction, insights, and historical perspectives of

Table 1. Assessment of guide glow.

Ibeam (mA)	Egun (kV)	Forward output (R/min)	Forward HVL (inch)	Side output (R/min)	Side HVL (inch)	Side–indirect (R/min)
15	7.1	2	0.77	0.04	0.60	0.03
33	10.0	13	0.80	0.05	0.58	0.04
75	15.3	62	0.84	0.06	0.58	0.05
150	22.6	145	0.83	0.09	0.60	0.07

my colleagues Gard Meddaugh, George Merdinian, Art Salop, Mark Trail, Ray McIntyre, Greg Kalkanis, Bill Leong, Frank Gordon, and many more who work every day to reduce cancer from a life-threatening to a chronic ailment.

Work on the low-energy biperiodic accelerator discussed here was made possible by the expert technical assistance of Nick Cortese, Frank Gordon, Chris Patane, Lindsey Cramer, Wayne Biser, Julius Ng, and Stan Johnsen.

Special thanks are due to Steve Vanderet, Mike Kauffman, Rudy Potter, and Mark Trail for comments on the manuscript.

References

[1] W. W. Hansen, *J. Appl. Phys.* **9**, 654 (1938).

[2] E. L. Ginzton, *IEEE Trans. Electron. Devices* **ED-23**, 714 (1976).

[3] E. L. Ginzton, W. W. Hansen and W. R. Kennedy, *Rev. Sci. Instrum.* **19**, 89 (1948).

[4] G. Caryotakis, *IEEE Trans. Plasma Sci.* **22**, 683 (1994).

[5] M. Chodorow, E. L. Ginzton, W. W. Hansen, R. L. Kyhl, R. B. Neal, W. K. H. Panofsky and the staff, *Rev. Sci. Instrum.* **26**, 134 (1955).

[6] *The Stanford Two-Mile Accelerator*, ed. R. B. Neal (W. A. Benjamin, New York, 1968).

[7] E. L. Ginzton and C. S. Nunan, *Int. J. Radiat. Oncol. Biol. Phys.* **11**, 205 (1985).

[8] C. J. Karzmark, C. S. Nunan and E. Tanabe, *Medical Electron Accelerators* (McGraw-Hill, New York, 1993).

[9] D. I. Thwaites and J. B. Tuohy, *Phys. Med. Biol.* **51**, R343 (2006).

[10] D. Whittum, in *Frontiers of Accelerator Technology*, eds. S. I. Kurokawa, M. Month and S. Turner (World Scientific, Singapore, 1999), pp. 1–135.

[11] D. Whittum, in *Techniques and Concepts of High-Energy Physics*, ed. T. Ferbel (Kluwer, Amsterdam, 1999), Vol. X, pp. 387–486.

[12] J. C. Slater, *Microwave Electronics* (D. Van Nostrand, New York, 1950).

[13] H. Eickhoff and U. Linz, *Reviews of Accelerator Science and Technology* (World Scientific, Singapore, 2008), Vol. 1, pp. 143–161.

[14] *Quality Assurance for Clinical Trials: A Primer for Physicists*, AAPM Report No. 86 (AAPM, College Park, 2004), p. 7.

[15] A. Brynjolfsson and T. G. Martin III, *Int. J. Appl. Rad. Isotopes* **22**, 29 (Pergamon, Northern Ireland, 1971).

[16] P. D. LaRiviere, *Br. J. Radiol.* **62**, 473 (1989).

[17] Central axis depth dose data for use in radiotherapy, *Br. J. Radiol. Suppl.* **11** (British Institute of Radiology, London).

[18] W. W. Buechner, R. J. Van de Graaff, H. Feshbach, E. A. Burrill, A. Sperduto and L. R. McIntosh, *ASTM Bull.* **TP262**, 54 (1948).

[19] W. W. Buechner, R. J. Van de Graaff, E. A. Burrill and A. Sperduto, *Phys. Rev.* **74**, 1348 (1948).

[20] J. Haimson, *IRE Trans. Nucl. Sci.* **NS-9**(2), 32.

[21] A. Sessler and E. Wilson, *Engines of Discovery* (World Scientific, Singapore, 2007), pp. 49–54.

[22] H. Quastler *et al.*, *Am. J. Roentgenol. Radium Ther.* **61**, 2 (1949).

[23] D. W. Fry, R. B. Harvie, L. B. Mullett and W. Walkinshaw, *Nature* **160**, 351 (1947).

[24] G. E. Becker and D. A. Caswell, *Rev. Sci. Instrum.* **22**, 402 (1951).

[25] M. J. Day and F. T. Farmer, *Brit. J. Radiol.* **31**, 669 (1958).

[26] C. W. Miller, M.Sc., F. Inst. P., Metropolitan Vickers, Research Series No. 33 (1956), *Eight Int. Cong. Radiology* (Mexico City, July 1956), p. 3.

[27] *The Engineer*, Nov. 19, 1954.

[28] A. Roth, *Vacuum Technology* (North-Holland, New York, 1982).

[29] E. A. Knapp, B. C. Knapp and J. M. Potter, *Rev. Sci. Instrum.* **39**, 979 (1968).

[30] E. A. Knapp, *IEEE Trans. Nucl. Sci.* **NS-12**, 118 (1965).

[31] D. E. Nagle, E. A. Knapp and B. C. Knapp, *Rev. Sci. Instrum.* **38**, 1583 (1967).

[32] H. A. Enge, *Rev. Sci. Instrum.* **34**, 385 (1963).

[33] K. L. Brown and W. G. Turnbull, US Patent No. 3,867,635 (1975).

[34] K. L. Brown, W. G. Turnbull and P. T. Jones, US Patent No. 4,425,506 (1984).

[35] E. Tanabe and V. A. Vaguine, US Patent No. 4,286,192 (1981).

[36] G. E. Meddaugh, M. E. Trail and D. Whittum, US Patent No. 7,339,320 (2008).

[37] G. E. Meddaugh and G. Kalkanis, US Patent No. 6,366,021 (2002).

[38] J. Allen *et al.*, US Patent No. 6,376,990 (2002).

[39] S. M. Hanna, US Patent No. 7,112,924 (2006).

[40] G. Meddaugh, E. Tanabe and V. Vaguine, US Patent No. 4,382,208 (1983).

[41] I. Uetomi, M. Kimura and K. Ogura, US Patent No. 4,651,057 (1987).

[42] K. Ohara *et al.*, *Int. J. Rad. Oncol. Biol. Phys.* **17**, 853 (1989).

[43] M. L. Riaziat, S. Mansfield and H. Mostafavi, US Patent No. 6,690,965 (2004).

[44] E. D. York, P. Keall and F. Verhagen, *Med. Phys.* **35**, 828 (2008).

[45] M. van Herk *et al.*, *Int. J. Rad. Oncol. Biol. Phys.* **47**, 1121 (2000).

[46] L. C. Fang *et al.*, *Int. J. Rad. Oncol. Biol. Phys.* **66**, 108 (2006).

[47] R. D. Wiersma, W. Mao and L. Xing, *Med. Phys.* **35**, 1191 (2008).

[48] Y. Ma, L. Lee, O. Keshet, P. Keall and L. Xing, *Med. Phys.* **36**, 2215 (2009).

[49] L. Xiong *et al.*, *Med. Phys.* **33**, 1848 (2006).

[50] T. R. Mackie *et al.*, *Med. Phys.* **20**, 1709 (1993).

[51] A. Schweikard, H. Shiomi and J. Adler, *Med. Phys.* **31**, 2738 (2004).

[52] S. L. Mahan, D. J. Chase and C. R. Ramsey, *Med. Phys.* **31**, 2119 (2004).

[53] S. M. Hanna, in *Proc. 1999 Particle Accelerator Conference* (IEEE, New York, 1999).

[54] M. Partridge, P. M. Evans and M. A. Mosleh-Shirazi, *Med. Phys.* **25**(8), 1443 (1998).

[55] M. Buchgeister and F. Nusslin, *Med. Phys.* **25**, 493 (1998).

[56] R. Rajapakshe and S. Shalev, *Med. Phys.* **23**, 517 (1996).

[57] J.-J. Sonke, B. Brand and M. van Herk, *Med. Phys.* **30**, 1067 (2003).

[58] P. Francois and A. Mazal, *Med. Phys.* **36**, 816 (2009).

[59] C. R. Ramsey, I. L. Cordrey and A. L. Oliver, *Med. Phys.* **26**, 2086 (1999).

[60] G. E. Meddaugh, US Patent No. 3,820,035 (1974).

[61] E. A. Knapp, in *Linear Accelerators*, eds. P. M. Lapostolle and A. L. Septier (North-Holland, Amsterdam; Elsevier, New York, 1970), p. 601.

[62] J. McKeown and S. O. Schriber, *IEEE Trans. Nucl. Sci.* **NS-28**, 2755 (1981).

[63] S. O. Schriber, S. B. Hodge and L. W. Funk, US Patent No. 4,155,027 (1979).

[64] T. Weiland, M. Timm and I. Munteanu, *Microwave Mag.* **9**, 62 (2008).

[65] A. S. Gilmour, Jr., *Microwave Tubes* (Artech, Norwood, 1986).

[66] J. L. Cronin, *IEE Proc.* **128**, 19 (1981).

[67] W. B. Herrmannsfeldt, presented at Workshop on Pulsed Radio Frequency Sources for Linear Colliders (Montauk, Long Island, New York; Oct. 2–7, 1994), SLAC-PUB-6726 and "EGUN," SLAC-331 (1988).

[68] H. R. Jory, US Patent No. 3,714,592 (1973).

[69] C. H. Goldie, K. A. Wright, J. H. Anson, R. W. Cloud and J. G. Trump, *ASTM Bull.*, Oct. 1954, pp. 49–54 (TP211-216).

David H. Whittum is Manager of Microwave Applied Research, within Oncology Systems at Varian Medical Systems. Prior to joining Varian he served on the High Energy Physics Faculty at Stanford, and as a *joshu* at KEK (High Energy Accelerator Research Organization), in Tsukuba, Japan. He is a graduate of the University of California–Berkeley (Ph.D., Physics) and Harvard (A.B., combined Physics and Mathematics). He is a licensed engineer in the State of California, a member of the American Association of Physicists in Medicine, a Senior Member of the Institute of Electrical and Electronics Engineers, and a Life Member and Fellow of the American Physical Society.

Reviews of Accelerator Science and Technology
Vol. 2 (2009) 93–110
© World Scientific Publishing Company

Heavy-Particle Radiotherapy: System Design and Application

Hirohiko Tsujii*, Shinichi Minohara† and Koji Noda‡

National Institute of Radiological Sciences,
4-9-1 Anagawa Inage, Chiba-shi, Chiba 263-8555, Japan
*tsujii@nirs.go.jp
†minohara@nirs.go.jp
‡noda_k@nirs.go.jp

The requirements for an accelerator and a beam-delivery system for medical application are described (pointing out terms of heavy-particle radiotherapy), based on 15 years of experience with carbon-ion radiotherapy at HIMAC. The present heavy-particle radiotherapy and system are reviewed here. With a view to further development of carbon-ion radiotherapy, recent progress in heavy-particle radiotherapy and a new system are also described. Finally, the clinical applications are summarized.

Keywords: Heavy-particle radiotherapy; beam delivery system; heavy-particle radiotherapy facility; radiation quality; adaptive radiotherapy.

1. Introduction

The foundations of heavy-charged-particle therapy were laid in 1930 with the invention of the cyclotron by Ernest Lawrence, and in 1946 Robert Wilson proposed the clinical application of the cyclotron, advocating the use of protons and heavier ions in treating human cancer [1]. The fundamental physical features of the heavy-charged particle beams are their capability of depositing only relatively low doses as the beam enters the body en route to the target (plateau region), the release of the greatest amount of energy at the end of the beam range (Bragg peak), and the deposition of a very low dose in the tail region beyond the Bragg peak. So far, more than 70,000 patients have been treated with charged-particle beams around the world, with more than 87% of these treatments being delivered with proton radiotherapy (RT) and about 8% with carbon ions. Currently, there are 26 operating proton-therapy facilities, while carbon-ion RT is provided at four facilities. More than 20 hospital-based facilities are under construction or are being planned to be built within the next 10 years; most of them already have long-standing experience with modern photon RT.

This article describes the state-of-the-art technology as well as the intensification of preclinical and clinical research in this emerging field. Firstly, we describe the medical requirements for an accelerator and a beam-delivery system, based on our 15-year experience of cancer treatment with carbon-ion RT at HIMAC (Heavy-Ion Medical Accelerator, in Chiba) [2]. Secondly, we review the present heavy-particle RT and the RT-system. Thirdly, in consideration of the continuing development of carbon-ion RT, we present the current progress in heavy-particle RT and in RT-system development. Finally, the clinical applications of RT are summarized.

2. Medical Requirements for the Accelerator and Beam-Delivery System

2.1. *Treatment overview*

Heavy-particle RT utilizes complex and intelligent equipment to deliver therapeutic radiation for the maximum benefit of cancer patients. There are three processes in patient handling that determine the quality of the medical care.

2.1.1. *Treatment planning*

Treatment planning is the process of designing an optimum radiation field for a tumor and predicting its clinical effects on the patient.

2.1.2. *Patient positioning*

Patient positioning is the process of placing the patient exactly in the planned position and orientation with respect to the beam, typically at 1 mm precision, for beam delivery.

2.1.3. *Beam delivery*

The beam-delivery system modifies the extracted beam according to the treatment plan for administering the planned dose distribution in the patient.

2.2. **Range, field size, and SOBP size in Japan**

The statistics of the 15-year treatment period with HIMAC are presented, which may indicate reasonable clinical requirements of treatment beam properties for heavy-particle therapy.

2.2.1. *Residual range*

The residual range is defined as the water-equivalent depth that the beam can penetrate after it has passed through the necessary beam-modifying devices. Figure 1 shows the residual-range distribution for the delivered beams. It seems that a residual range of 250 mm in water may cover most of the patients. The residual range depends not only on the beam energy but also on the field-formation method. In the broad-beam methods such as the beam-wobbling and double-scatterer methods, range loss is caused

mainly by the scatterer. The maximum residual range is estimated to be 260 mm. In the pencil-beam scanning method, on the other hand, range loss can be minimized, and for HIMAC the residual range is typically 270 mm in water for a 400 MeV/n carbon beam.

2.2.2. *Field size and SOBP*

As shown in Fig. 2, a lateral-field diameter of 220 mm and a longitudinal-field extent or spread-out Bragg peak (SOBP) size of 150 mm can cover almost all types of patients treated with HIMAC. A larger field size of more than 200 mm is required mainly for the treatment of oblong tumors. In such cases, it is important to maintain the field length rather than the diameter. The field-patching method has been employed for a target size of more than 220 mm in diameter. The SOBP size should range from 40 to 150 mm.

2.3. **Dose rate**

For HIMAC, the irradiation dose rate is required to be 5 GyE/min/l so as to complete the fractional irradiation course within a tolerable time [2]. When one gives a certain biological dose, the ion number delivered to the surface of a patient is inversely proportional to the surface physical dose (LET) and the biological dose on the mid-SOBP over the physical dose to the surface (R). For example, the

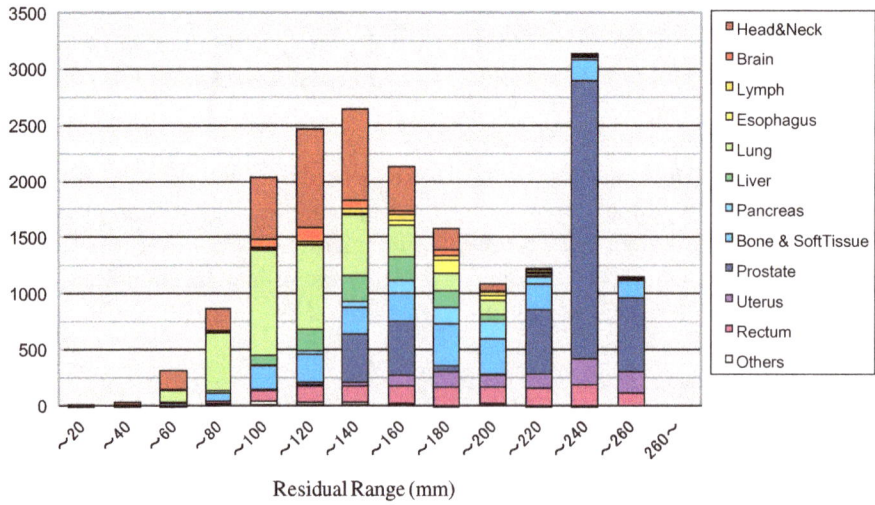

Fig. 1. Histogram of the number of irradiation shots as a function of the required residual range in mm.

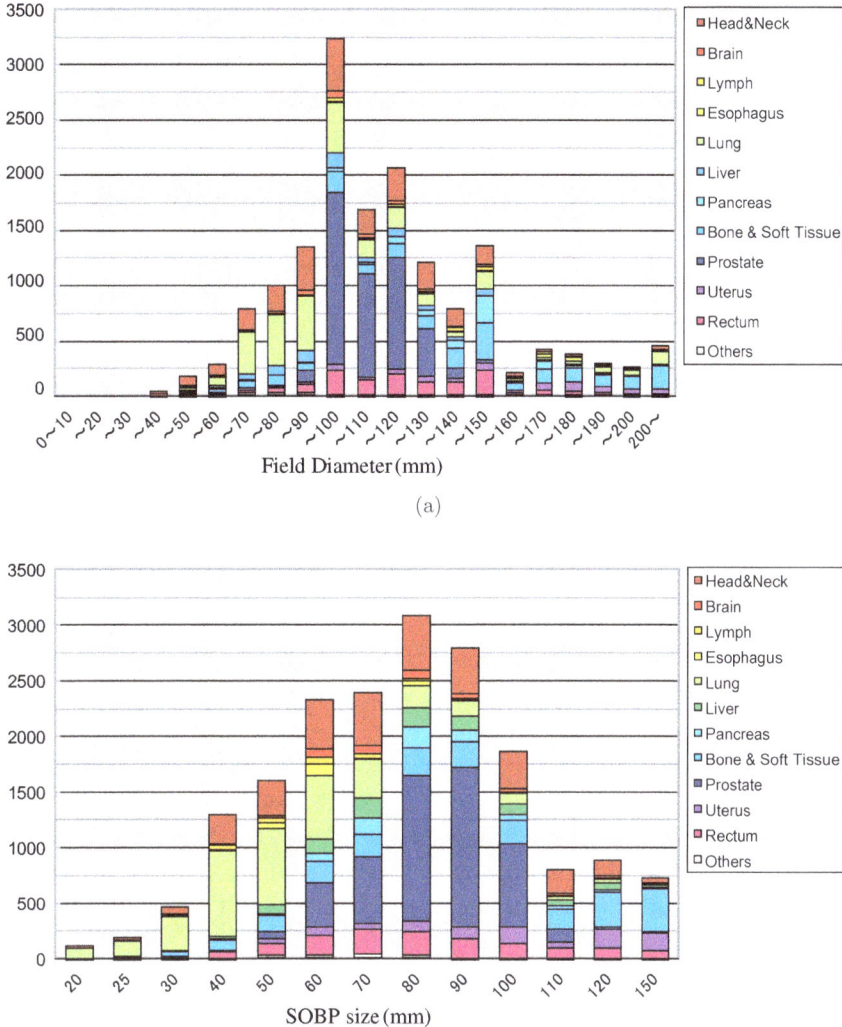

Fig. 2. Histogram of the session numbers as a function of the required field diameter in mm (a) and as a function of the required SOBP in mm (b).

LET of a 200 MeV proton is around 0.5 keV/μm while that of a 400 MeV/n carbon ion is around 10 keV/μm, and the R of a proton is estimated to be 1.2–1.3 while that of a carbon ion is 2.2–2.3. Therefore, the proton number is around 40 times larger than the carbon-ion one when one is delivering the same biological dose on the mid-SOBP. In carbon-ion RT using broad-beam methods, the dose rate of 5 GyE/min/l (10 cm × 10 cm × 10 cm) corresponds to around 10^9 particles per second (pps) of a beam intensity required at the entrance of a beam-delivery system, under a beam-utilization efficiency of around 20% in the beam-delivery system. In proton RT, on the other hand, an intensity of around

5×10^{10} pps is required at the entrance of the beam-delivery system under the same conditions as those used in carbon-ion RT. In the pencil-beam scanning method, the beam-utilization efficiency increases to almost 100%. The beam intensity is thus estimated to decrease by a factor of 3–5, compared with the broad-beam methods in both carbon-ion and proton RT. However, for scanning beams, the dose-rate limitation is determined by considering the following: (1) the quantity of an extra dose due to the finite time to turn off the beam-delivery in the spot-scanning method and (2) the amount of the extra dose delivered when the beam is moving between positions in the raster-scanning method, in which the beam is not turned off. In both cases, the beam is proportional to the beam intensity delivered.

2.4. *Number of treatment rooms*

When 1000 patients (pts) are treated per year, the required number of treatment rooms is estimated as follows. Since the number of treatment fractions (session number) is around 12 on average at HIMAC, the total session number is estimated to be 12,000 sessions/year for a treatment number of 1000 pts/year. Assuming the time required for one session to be 25 min, including patient positioning, one treatment room suffices to carry out 4600 sessions/year under the following working schedule: $8 \, (\text{h/day}) \times 5 \, (\text{days/week}) \times 48 \, (\text{weeks/year})$. From this, it is clear that the carbon-ion RT facility requires three treatment rooms. Further, the ratio of the treatment frequency using a horizontal irradiation port (H port) to that with a vertical one (V port) is around 5:4. Therefore, the three treatment rooms should be equipped with an H port, a V port, and H and V ports, respectively.

In proton RT, on the other hand, the average fraction number is usually more than twice that of carbon-ion RT. It appears that a working time of $16 \, (\text{h}) \times 240 \, (\text{days/year})$ is required for the treatment of 1000 pts or additional rooms within practical limitations.

2.5. *Safety*

2.5.1. *Safety considerations*

All systems in the medical facility should be designed with the safety of the patients and medical staff being of primary importance. One of the crucial considerations is radiation protection for the medical staff in the high-energy accelerator facilities [3]. For ensuring irradiation safety, each system needs to be designed to fail-safe standards. The system must be designed so that in any appropriate nonnominal situation, the instrumentation is designed to detect such a situation quickly, and the interlock system can be activated with sufficient promptness to turn off the beam fast enough to ensure patient and personnel safety. If this occurs in one treatment room, it must be possible to allow clinical operation in the other treatment rooms, provided that it is safe to do so.

2.5.2. *Secondary neutrons*

For curative treatments of young patients, risk of secondary cancer formation after radiotherapy has recently emerged as a matter of concern [4, 5]. Compared to photon beams, heavy-ion beams can minimize undesired exposure in normal tissues adjacent to the target volume because of their physical and/or biological advantages. It is also necessary to investigate the range of undesired exposure in normal tissues remote from the target, due mainly to secondary neutrons that are inevitably produced in patients and beam-line devices [5, 6], although the risk of low-dose exposure is controversial [7].

2.6. *Patient positioning*

Patient positioning or setup at times of beam deliveries based on reproducing the position in the planning x-ray computer tomography (CT) determines the therapeutic accuracy in radiotherapy in general. In addition, precise range control is a specific requirement for heavy-particle RT, which may require volumetric alignment considering tissue density. Reproducible immobilization of the patient, high-resolution detection of displacement between the *in-situ* and planned patient positions, and fine and precise adjustment of the position and orientation are therefore essential in patient positioning.

The precision required for the physical distribution of the dose and for the patient couch adjustment may be less than 1 mm. For example, the positioning accuracy of fixed targets such as a skull-based tumor is around 1 mm, and that of ocular-melanoma treatment of the eyeball is at the submillimeter level. It is not easy, however, to keep the precision of patient positioning over the course of the treatment. Especially in the case of organ motion during each delivery, the positioning precision is much greater than the 1 mm level. In clinical practice, the accuracy of the patient positioning depends on the skill of the radiotherapist, and its significance ultimately depends on target delineation by the radiation oncologist.

2.7. *Treatment planning*

The essential requirement for treatment planning of heavy-ion therapy is to maximally utilize the physical and biological advantages of heavy ions for cancer treatment. This involves precise delineation of a tumor volume for a given target, design of a customized beam and setting of the beam control parameters, prediction of dose distribution in the patient, and clinical assessment of the treatment plan for individual patients. In general, the sharper the dose distribution, the better the precision. For heavy ions, precision is generally expected to be in the order of 1 mm.

The physical interactions of heavy ions with matter are complicated, especially as beam modification and medium inhomogeneity are different from beam to beam and from patient to patient. Progress of the Monte-Carlo-simulation technique to more precisely calculate this will ultimately be feasible [8, 9], while at present faster empirical and approximate beam models are employed for dose-distribution calculation in clinical practice [10–15]. Relative biological effectiveness (RBE) is a specific feature that needs to be considered for heavy-particles [16, 17]. It relates the absorbed dose or energy per unit mass to the absorbed dose of the reference radiation for the same effect. The complexity of RBE depends on the incident particle type dose, beam customization and interactions in a body, tissue and tumor types, and the clinical endpoint.

2.8. *Quality assurance*

Quality assurance (QA) is used to warrant the correct use of radiations for therapeutic purposes. QA activities primarily focus on detection of any possible abnormal behavior with the highest sensitivity practically achievable. In case of any failure, corrective action and estimation of the significance have to be carried out. The quality management of clinical systems demands great efforts in the endless sequence of plan, do, check, and action, including routine (daily, weekly, monthly, annually) activities for inspection, measurement, recording, and reporting. Daily QA is usually performed prior to the first treatment of the day, typically within 30 min.

In spite of the QA work, however, subtle faults, errors such as software and human errors, are still possible, although safety is still assured. At HIMAC, information on the above difficulties has been qualitatively recorded since 1999, while prior to that only partial information had been recorded. Analysis of the types of information, in particular the number of events occurring due to hardware, has increased since 2005, around 10 years since the HIMAC facility was put into operation. The total number of serious

events that have occurred since April 2007 stands at 30. During this period, the number of sessions done in the three treatment rooms totaled around 16,000, meaning that the incidence is less than 0.2%. It should be noted that serious failure is defined as follows: Interruption of any treatment session must not exceed 1 h until resumption.

3. Particle-Therapy System

3.1. *Beam-delivery system*

The ion beam is extracted from the accelerator and transported to the recipient treatment room. The beam-delivery system of heavy-ion therapy consists of beam modifying and monitoring devices to deliver the prescribed dose distribution. There have been two approaches to forming the treatment beams: the broad-beam and pencil-beam scanning methods. Both approaches have advantages and disadvantages.

3.1.1. *Broad-beam irradiation system*

Figure 3 shows an example of a broad-beam irradiation system [18], which is the beam-delivery system with the beam-wobbling method at HIMAC. A pair of beam-wobbling magnets (wobbler) rotates the beam in a circular orbit at high frequency so as to generate a pseudostationary broad beam in conjunction with a heavy-metal scatterer. An exchangeable ridge filter modulates the beam range in the field to spread out the Bragg peak longitudinally. A range-shifter system inserts variable-thickness energy absorbers to adjust the beam range. Either a multileaf collimator (MLC) with movable

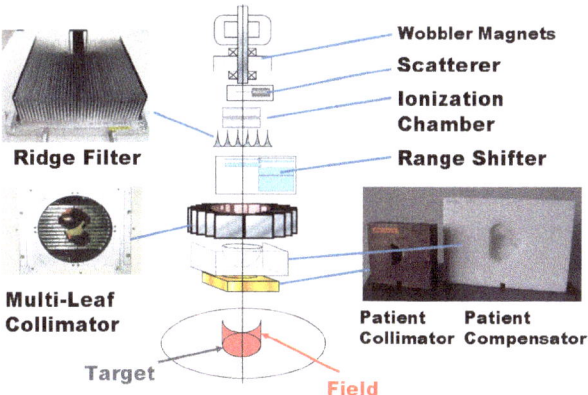

Fig. 3. Typical beam-delivery system with the beam-wobbling method as one of the broad-beam methods.

metal elements or a customized patient collimator defines the field aperture. A bolus — a sculptured plastic device — compensates for the beam ranges so that the beam end of the range conforms with the distal part of the target volume in the field.

The beam-wobbling method has advantages over the double-scatterer method in minimizing the material that might shorten the beam range. The AC power supplies are usually composed of LC resonant circuits at a constant frequency of 56.4 Hz to form uniform fields up to 22 cm in diameter in the HIMAC system.

In the HIMAC beam-delivery system, a ridge filter consisting of identical aluminum bar ridges spreads the beam range to give a uniform biological dose to the SOBP region.

3.1.2. Layer-stacking method

In the broad-beam method with a ridge filter, a constant SOBP over the field area results in an undesirable dose to the normal tissue proximal to the target. In order to avoid such unwanted doses, the layer-stacking method was proposed [19] and has been routinely used for carbon therapy at HIMAC [20, 21]. In this method, a mini-SOBP is produced with a ridge filter. The full widths at a 60% dose level of the SOBP layer are about 11.9, 12.8, and 15.9 mm for 290, 350, and 400 MeV/n, respectively. During irradiation, a mini-SOBP centroid is sequentially shifted in the longitudinal direction and in steps of 2.5 mm by changing the beam energy using a range shifter. The compensator and the MLC shown in Fig. 3 are also used as in the ordinary broad-beam method, while the aperture shape of the MLC conforms to each slice of the target. Once the planned dose is delivered to each slice, beam extraction is quickly cut off during the transition time to set the range shifter and the MLC for the next slice. Beam-ON/OFF is controlled by using the radiofrequency knockout (RF-KO) slow-extraction method [22]. Currently, the mechanical movement of the range shifter limits the transition time to within a fraction of a second.

3.1.3. Pencil-beam scanning system

Pencil-beam scanning is an irradiation method for painting the dose distribution with a small beam and narrow Bragg peak, which allows us to take full advantage of the heavy-particles. The pencil beam is laterally scanned so as to form a lateral irradiation field with orthogonal scanning dipole magnets, and is then longitudinally scanned by either a range shifter or stepwise energy change from the accelerator.

The beam-scanning path and the number of particles per location have been precisely determined in treatment planning to deliver the planned dose distribution. The scanning magnets therefore need to be controlled as a function of the number of particles detected by the dose monitor system. Fast and synchronous control of the dose monitor, magnetic scanning, and beam-extraction systems with precision and resolution better than 1% and 1 millisecond is normally required for clinically practical beam deliveries at HIMAC. There are two approaches to lateral beam scanning: the spot- and raster-scanning methods.

In the spot-scanning method, the scan path is quantized into spots and beam extraction is activated on a spot-by-spot basis. This method is sometimes referred to as a "dose-driven scanning" method, since the dose at any given spot is determined fully by the dose delivered at that spot and the beam is turned off after the dose is fully delivered. In this method, the switching speed and precision of beam extraction are crucial, and it is generally difficult to form fine dose distribution with a large number of spots within a tolerable duration.

In the raster-scanning method, the pencil beam is extracted continuously during magnetic scanning, which is controlled as a function of the number of particles delivered. This method is a variation of the dose-driven spot-scanning method, in that the beam is not turned off between spots, but the beam that is on, while it is moving from one spot to another, is included in the dose to each spot. This will greatly ease the requirements for the beam-extraction system. In treatment planning and beam control, the concept of discrete spots may still be used, but the extra dose delivered in transitions between spots must be mitigated or explicitly counted in plan optimization.

3.1.4. Rotating-gantry system

A rotating-gantry system allows wide choices of beam orientation, compared with a fixed port irradiation system. In our clinical practice with HIMAC, since the beam can be delivered from either the horizontal or vertical direction, the patient is fixed in supine, prone, and often rolled positions by typically 10°–20° from the horizontal plane in order to achieve a better combination of beams. This situation often adds to the patient's load, complicates the treatment planning, and makes precise positioning difficult. A rotating-gantry system, which allows 360° rotation around the patient, will resolve many of these problems, and it is the standard for conventional x-ray teletheraphy systems. A rotating gantry for ion RT, on the other hand, is much larger, as its size is typically 10 m in diameter in the commercialized proton RT systems. For a carbon-ion beam, the requirement for beam bending is a few times stronger. The HIT (Heidelberg Ion Therapy) facility [23] in Germany has the only heavy-ion rotating-gantry system existing today.

3.1.5. *Respiratory-gating system*

Organ motion during patient positioning and beam-delivery degrades targeting precision. In particular, breathing causes movement up to a few centimeters in the lung and liver regions, which may also influence the whole body when the patient is in the prone position. Respiratory-gated irradiation effectively mitigates such motion by controlling the beam-extraction timing synchronously with respiration.

Breathing can be detected with an infrared light spot and a position-sensitive detector, which gives a respiration-wave-form signal. The organs are normally most stable at the end of expiration, for which gating for beam extraction is set [24]. The respiration pattern and its reproducibility are patient-dependent. Real-time detection of the respiration wave form, fast and robust gating logic, and responsiveness of the beam-extraction system are therefore essential for the respiratory-gating system. In the case of HIMAC, beam extraction is turned on and off within sub-milliseconds owing to the RF-KO slow-extraction method.

3.1.6. *Patient-positioning system*

Patients need to be immobilized rigidly and reproducibly on the treatment couch for high-precision therapy. For this reason, patient-specific fixtures such as polyurethane foam and thermoplastic shells are made for x-ray CT simulation and used to keep the

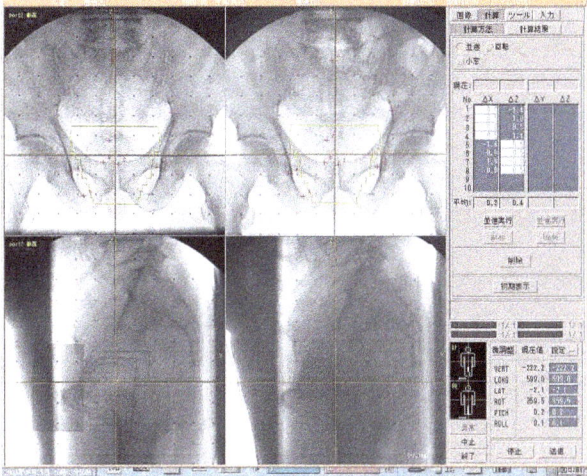

Fig. 4. Patient positioning with orthogonal x-ray images.

patient always in the planning position during the entire course of the treatment.

The patient-positioning system of HIMAC consists of a treatment couch, orthogonal x-ray imagers, and an image-registration system. The treatment couch is finely adjustable around three translational axes and about three rotational axes. Two sets of x-ray imagers utilizing either image-intensifiers or flat-panel detectors acquire images of the bony structure of the patient. As shown in Fig. 4, an image-registration system compares the on-site images with the reference images that represent the planning position and quantifies any misalignment to correct the couch position.

The reference images should be either digitally reconstructed radiographs (DRRs) from the planning CT images or from their approximate x-ray images taken in the simulation, or from past treatment sessions. Due to image-quality differences, DRRs are often not very suitable for such comparison between images. Patient positioning is a time-consuming process in the daily treatment procedure, typically taking about 15–20 min [18].

3.1.7. *Dosimetry system*

Clinical RT systems are required to have two independent dose monitors to securely turn the beam off and to estimate the delivered dose in case of any failure with either monitor. These dose monitors are generally not very sensitive to field formation. A flatness monitor with two-dimensionally arranged electrodes is thus installed in the beam line.

For QA of the treatment beam, not only consistency among these beam monitors, but also constancy of the dose output measured at the irradiation site, are routinely verified. For ion beams, it is essential to verify depth–dose curves, and daily measurements have to be done quickly and easily. For this purpose, a multilayer ionization chamber (MLIC) with multiple electrodes layered in the longitudinal direction is useful, as this allows the depth–dose distribution to be measured in a single measurement. The MLIC is calibrated by a Markus-type parallel-plate ionization chamber once annually, in accordance with IAEA TRS-398. The Markus chamber is also calibrated once annually according to the standard.

Due to complication of the beam-delivery system for lateral and longitudinal beam modifications, the relation between the dose-monitor reading and the dose output at the treatment site is complex in heavy-particle therapy. The monitor setting for the absorbed dose prescribed at the reference mid-SOBP depth is therefore usually determined experimentally for each planned beam prior to the first treatment session. For QA of the measurement, the dose output per monitor reading is compared to a calculation or an empirical value for a similar beam in the database.

3.1.8. *Control system and safety interlock system*

At HIMAC, each treatment room has an identical set of control systems. There is a server computer that interfaces the irradiation control systems with the accelerator control system and the hospital network system. A number of programmable logic controllers (PLCs) are in use to interface the computers with the devices.

The beam-monitor system directly closes the beam shutter when the measured dose reaches the prescribed dose. In addition, the abort-signal generated by any system in which serious problems occur is transmitted to a central interlock system via one of the PLCs through an optical-fiber cable and a coaxial metallic cable. At the same time, the control system transmits the command to close the beam shutter to the accelerator control system. The central interlock

system also commands the beam shutter to be closed by software.

3.1.9. *Secondary neutrons*

Dosimetry studies on secondary neutrons in heavy-particle RT are being actively pursued [25–39]. Allowance of secondary neutrons is more important in broad-beam irradiation, because the dominant neutron sources are the beam-modifying devices. The secondary-neutron ambient dose equivalents for patients were measured using a rem meter at two carbon-ion RT facilities: HIMAC and HIBMC (Hyogo Ion Beam Medical Center) and four proton RT facilities: PMRC (Proton Medical Research Center in Tsukuba University), NCCHE (National Cancer Center Hospital East), SCC (Shizuoka Cancer Center), and HIBMC in Japan [36]. All of these facilities utilize the broad-beam method.

Figure 5 shows one example of our results, where RBEs of protons and carbon ions were assumed to be 1.1 and 2.3, respectively: (1) the neutron ambient dose equivalent in carbon-ion RT was lower than that in proton RT; (2) the differences in the measured neutron doses among the facilities were within a factor of 3; (3) the neutron ambient dose equivalents for the broad-beam proton and carbon-ion RT modalities were equal to or less than those in the case of photon RT.

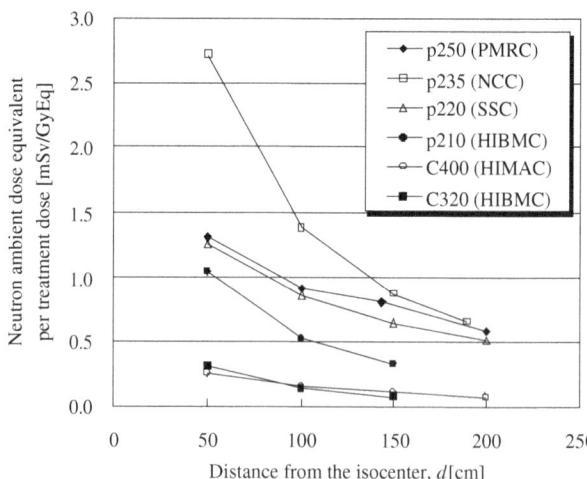

Fig. 5. Measured ambient dose equivalents for broad beams of protons and carbon ions with maximum beam energy at each beam line. The legend shows the beam species, the energy, and the facility. "p" and "C" indicate protons and carbon ions, respectively. The values following p and C indicate the beam energy in MeV/n.

The deviation appearing in the results of proton RT facilities is considered due to the difference in geometrical configuration in the irradiation room or in the setting of beam-delivery devices. These factors also influence in part the lesser production of secondary neutrons by carbon beam; however, it should be noted that the production of neutrons is in principle less in the carbon beam than in the proton in terms of the absorbed dose to the target, as the number of particles required to carry the same amount of energy to the target is more than 10 times larger in the proton. This is not tolerated by a smaller production cross section of protons.

3.2. *Accelerator system*

3.2.1. *Carbon-ion radiotherapy facility*

In Asia, the existing carbon-ion RT facilities are HIMAC, HIBMC, and IMP (Institute of Modern Physics in China). At HIBMC, treatments have been carried out using both proton and carbon. At IMP, a cyclotron is utilized as an injector for the main cooler synchrotron, and the intensity is increased by the cool-stacking method. In Europe, GSI (Gesellschaft für Schwerionenforschung) has carried out carbon-ion RT since 1997. Based on the GSI results, the HIT facility was constructed at Heidelberg, Germany, and beam commissioning has been fully completed.

As an example of a heavy-ion RT facility, the HIMAC accelerator facility is described here. The HIMAC accelerator complex consists of two electron cyclotron resonance (ECR) ion sources (10 GHz and 18 GHz) and a Penning ionization gauge (PIG) ion source, an injector cascade [100 MHz radiofrequency quadrupole (RFQ) and Alvarez linacs], two identical synchrotron rings, high-energy beam lines and beam-delivery systems. The layout of HIMAC is shown in Fig. 6. At present, the HIMAC accelerator complex can deliver ion species from proton to Xe, and maximum energy is determined so as to obtain a 30 cm range in water. Carbon-ion RT has been performed during daytime hours from Tuesday to Friday with a carbon beam of 290, 350, and 400 MeV/n energy. The beam-wobbling method has been used as the beam-delivery method. The respiratory-gated irradiation and layer-stacking irradiation methods were developed and have been routinely used. During nighttime hours and on weekends, basic-research studies have been carried out using various ions.

3.2.2. *Proton radiotherapy facility*

LLUMC (Loma Linda University Medical Center) was the first proton synchrotron facility dedicated to proton RT [40]. Many proton RT facilities have since been constructed and have been in service in the world. For example, MGH (Massachusetts General Hospital) and PSI (Paul Scherrer Institute) have employed an AVF cyclotron manufactured by Ion Beam Application (IBA) and a superconducting cyclotron by ACCEL Company, respectively, and MDACC (MD Anderson Cancer Center) has

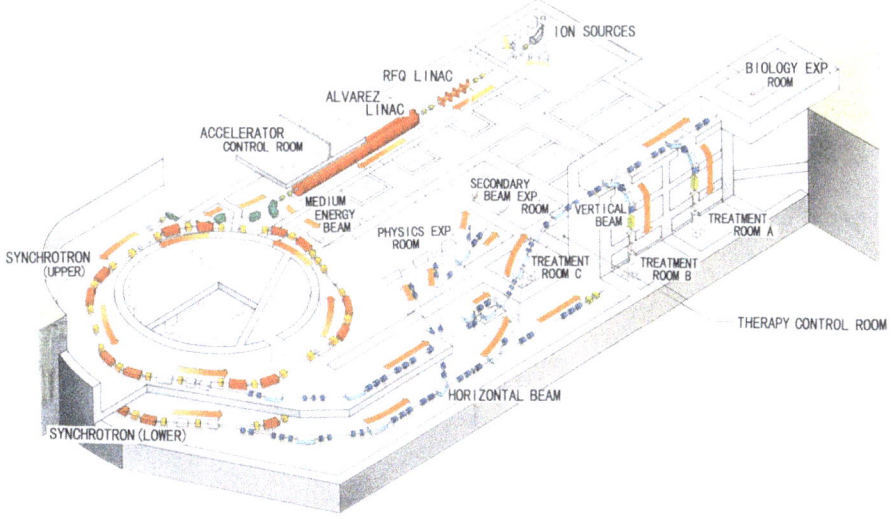

Fig. 6.   Bird's eye view of HIMAC with the new treatment facility.

Fig. 7.   235 MeV fixed-energy AVF cyclotron for proton therapy at NCCHE.

Fig. 8.   250 MeV synchrotron at the University of Tsukuba. The injection line can be seen on this side and the extraction line is on the opposite side.

employed a compact synchrotron by Hitachi. Of them, we herein describe the cyclotron and synchrotron facilities in Japan.

NCCHE has carried out proton RT since 1998. The main accelerator of this facility is an AVF cyclotron (shown in Fig. 7), which was manufactured by IBA and SHI (Sumitomo Heavy Industry). The cyclotron is operated at a fixed energy of 235 MeV for protons. The energy-separator system (ESS) consisting of energy degraders and a momentum separator, which is installed downstream of the extraction point of the cyclotron, can be used to vary the proton energy. Two rotational gantries and one horizontal beam port are installed in three different treatment rooms. The double-scattering method is employed for one rotational gantry, while the beam-wobbling method is used for the other. The facility treated 607 patients between 1998 and December 2008.

Since 1983, the University of Tsukuba has accumulated considerable experience of proton RT using a 500 MeV booster synchrotron at KEK (High Energy Accelerator Research Organization in Japan). The number of patients treated has meanwhile reached 700. Because of serious limitations on the use of the accelerator for physical research, a new, medically dedicated facility was constructed. In the new facility, the number of patients treated between 2001 and December 2008 reached 1367. The accelerator and beam-delivery systems were manufactured by Hitachi. The main accelerator is a compact synchrotron with a strong focusing function, as shown in Fig. 8. The synchrotron is diamond-shaped,

with a circumference of 23 m. Output-beam energy varies in a range of 70–250 MeV for protons. The rise and fall timing of the main magnet current can be triggered by external signals generated from the patient's respiration curve. Two rotational gantries are routinely used for cancer treatment and one horizontal beam line for basic experiments. Based on this synchrotron design, the proton RT facility at MDACC was developed.

The accelerator system at SCC, manufactured by Mitsubishi Electric Corporation, is also very compact. A weak-focusing synchrotron with edge focusing was chosen as the main accelerator in order to suppress the space-charge effect, and its diameter is around 6 m. Clinical trials were commenced in 2003, and 692 patients have been treated up to December 2008. There are three treatment rooms — two equipped with rotating gantries and the third with a horizontal beam line.

4. Progress of Heavy-Particle Radiotherapy and the Radiotherapy System

4.1. Adaptive radiotherapy

It is very difficult to accurately predict the temporal variations of the target position and shape in the treatment-planning phase. The usual practice is to add unpredictable temporal variations of the target to the internal margin of the planning target volume.

In addition, the distribution of biological radiation response on the target will change in the course of fractionated irradiation. For example, as the volume of some uterine tumors decreases to about a tenth during fractionated irradiation, we need to revise the treatment planning a few times during the treatment.

One of the ideal approaches to these variations of the target would be adaptive irradiation. In this approach, the treatment plan is dynamically modified at each fraction to optimize the dose distribution in the process of patient positioning. Then revised parameters for the irradiation field are calculated and set on the beam-delivery system. To realize this process, it is necessary to obtain volumetric images more often. One way to do this is to install a CT scanner in the treatment room and use a scanning beam-delivery system that does not require a patient-specific collimator or compensator. In addition, it is essential to register the target and critical organs using prior planning information and to optimize the revised dose distribution in a very short time. Such a system has not yet been implemented for clinical use.

4.2. *Radiation quality*

The mechanism underlying the clinical effect of carbon ions has still not been fully understood. In addition, the complicated physical interactions in a patient are hard to predict. In order to fill this gap in our knowledge, some extent of approximation has been introduced in the treatment-planning system (TPS) [16]. The beam model to be used in the TPS is being improved through studies. When heavy ions travel in a patient's body, some of them suffer fragmentation reactions with target nuclei, and various species of fragments are generated and become widely distributed. The biological effectiveness of heavy ions is affected not only by the deposited energy but also by the particle species [41]. Thus, precise calculation of the spatial distribution of radiation quality such as dose, dose-averaged LET, fluence and energy distributions for each species of particles plays a crucial role.

In general, the deflection of primary particles in a thick medium has been well described by Moliere's multiple-scattering theory [42]. On the other hand, multiple scattering alone is not sufficient to account for the distribution of fragment particles. Our recent study revealed that the large deflection of fragment particles in a substance could be accounted for in the multiple-scattering formula by considering an additional term representing a lateral "kick" at the production point of the fragment [43]. This additional term can be explained as a transfer of the intranucleus Fermi momentum of a projectile to the fragment, and its extent obeys the expectation derived from the Goldhaber model [44]. The angular distribution of fragment particles was measured with a monoenergetic $290\,\mathrm{MeV}/n$ $^{12}\mathrm{C}$ beam through a nuclear reaction in a thick water target [43]. They determined a parameter describing the extent of transferred momentum in the Goldhaber model so as to reproduce the observed angular distributions.

Based on these studies, a semianalytical beam transportation code was developed for energetic heavy-ion beams in which the three-dimensional distribution of radiation quality can be calculated for each species of particles [45]. The production of secondary and tertiary fragments is considered, and the effects of Fermi momentum transfer are taken into account at their production point. Despite its simplicity, the developed code was able to reproduce the experimental result well. Another approach to deriving the spatial distribution of radiation quality in matter is the utilization of Monte Carlo codes. Although Monte Carlo codes are time-consuming and currently not practical for implementation in a TPS of heavy-ion therapy, some articles reported that they can provide precise estimates of radiation quality in matter [46, 47].

The last issue to be mentioned is fractionation. Hypofractionation is in general not preferable for conventional photon RT, as it increases the risk of normal-tissue complication, though a shorter therapy period by this modality is considered to be beneficial for patients both physically and financially. The superior dose conformation property of carbon ions is considered to benefit hypofractionated RT while not causing any severe side effects to surrounding normal tissues. Even though the dose distribution of carbon ions is preferable, multiportal irradiation should be the choice in order to reduce the dose to normal tissues. Here, the time gap between consecutive deliveries in a day should be as short as possible in order to avoid repair of sublethal damage in tumor cell nuclei. A rotating gantry will contribute to shortening the time gap. It will also help radiation technologists by reducing the workload of patient repositioning.

4.3. *Compact carbon radiotherapy facility*

For the purpose of achieving a more widespread use of carbon-ion RT in Japan, NIRS (National Institute of Radiological Sciences) designed a standard type of carbon-ion RT system [48] to reduce construction costs. Based on the design study, Gunma University has been constructing a standard-type facility since 2006, and the first patient will be treated in March 2010. This facility has an ECR ion source, an RFQ and an APF-IH linac [49], a synchrotron ring, three treatment rooms and one experimental room for basic research. In this facility, a C^{4+} beam, which is generated by a compact 10 GHz ERC source, is accelerated to $4\,\mathrm{MeV}/n$ through the injector linac cascade. After the C^{4+} beam is fully stripped by a thin carbon foil, the C^{6+} beam is injected into the synchrotron by the multiturn injection scheme and is accelerated to a maximum of $400\,\mathrm{MeV}/n$. All magnets in the beam transport lines are made of laminated steel in order to permit a change in the beam line within 1 min. The beam-delivery system employs

Table 1. Main specifications of the compact carbon-ion RT system.

Ion species	Carbon ions only
Range	25 cm max. in water (400 MeV/u)
Field size	15 cm square
Dose rate	5 GyE/min (1.2×10^9 pps)
Treatment rooms	3 (H, V, H&V), No rotational gantries
Fourth room	Prepared for future developments
Irradiation technique	(1) Respiratory-gated method
	(2) Layer-stacking method

a spiral beam-wobbling method [50] for forming uniform lateral dose distribution with a relatively thin scatterer.

The facility is downsized to one-third of the HIMAC facility. The specifications are summarized in Table 1, and an image view of the Gunma University facility with a full complement of equipment is shown in Fig. 9.

4.4. *New treatment facility with at HIMAC*

At HIMAC, a new treatment facility was proposed for further development of heavy-particle RT toward adaptive cancer therapy.

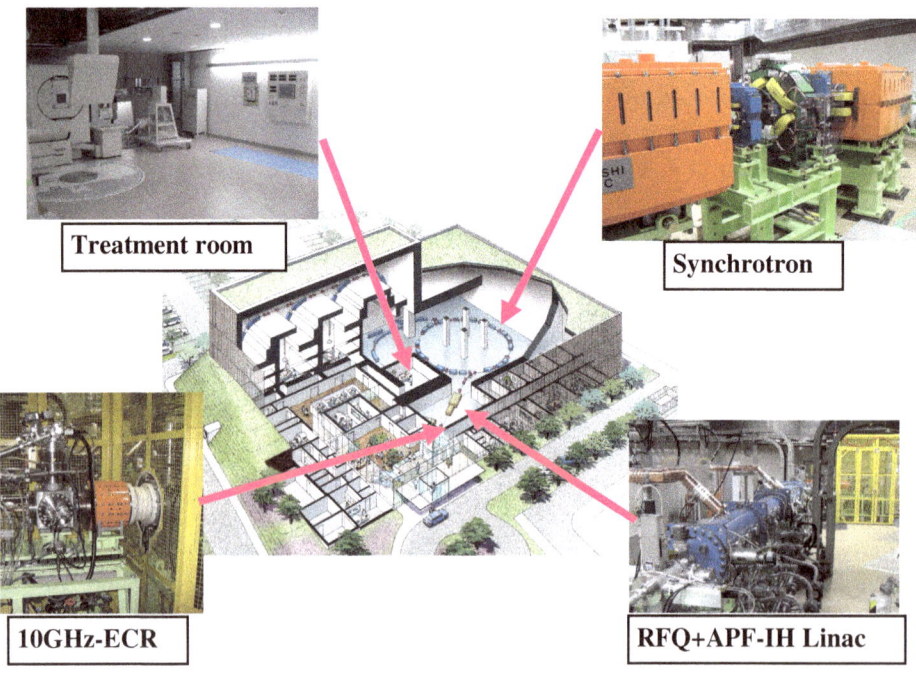

Fig. 9.   Image view of the Gunma facility with installed equipment.

4.4.1. *Phase-controlled rescanning method*

The new facility should be designed to employ a pencil-beam scanning method for a fixed target, a moving target, and/or a target near critical organs, toward the target of the implementation of adaptive cancer therapy. For this purpose, we have proposed the phase-controlled rescanning (PCR) method with a pencil beam [51]. In the PCR method, rescanning completes the irradiation of one slice during a single gated period corresponding to the phase between the end of expiration and the beginning of inspiration, because the organs are most stable during this gated period. Further, since the average displacement of the target over a single gated period is close to "zero," we can obtain uniform dose distribution even under irradiation of a moving target. The PCR method requires two main technologies: (1) the intensity-modulation technique, for a constant irradiation time on each slice having a different cross-section; and (2) the fast pencil-beam scanning technique, for completing several-time rescanning within a tolerable time.

4.4.1.1. Intensity modulation

We have developed a spill-control system [52] in order to deliver the beam with intensity modulation, based on improvement of the RF-KO slow-extraction method. The core part of this system requires the following functions: (1) calculation and output of an AM signal according to request signals from an irradiation system, (2) real-time processing with a time resolution less than 1 ms, and (3) feedforward and feedback controls to realize the extracted intensity as requested. This system allows us to dynamically control the beam intensity almost as required, as shown in Fig. 10.

4.4.1.2. Fast pencil-beam scanning

For the fast pencil-beam scanning, we have developed three key technologies: (1) new treatment planning for raster scanning, (2) extended flat-top operation of the synchrotron, and (3) high-speed scanning magnet.

Raster scanning has been chosen, instead of spot scanning, in order to save the beam-off period during spot-position movement. With the raster-scanning method, however, it is inevitably necessary to deliver an extra dose to the position between

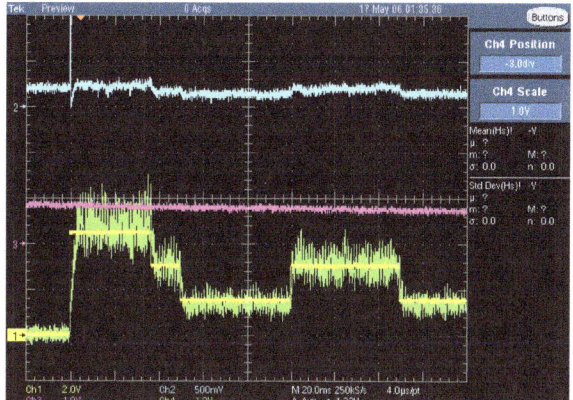

Fig. 10.   Time structure of extracted beam obtained by the spill-control system. Spill time structure (green modulated) by request signal (yellow).

the spot positions. It should be noted that the extra dose is proportional to the delivered intensity. Owing to the high reproducibility and uniformities in the time structure of the extracted beam through the spill-control system, we can predict the extra dose and incorporate its contribution in the treatment planning. Consequently, we can increase the beam intensity and shorten the irradiation time.

On account of a high beam-utilization efficiency of around 100% in the scanning method and an intensity upgrade to $2 \cdot 10^{10}$ carbon ions, we can complete single-fractional irradiation of almost all treatment procedures in a single-operation cycle of the synchrotron. This single-cycle operation, which can be realized by using a clock-stop technique in the flat-top period, can increase the treatment efficiency, especially for respiratory-gated irradiation. Thus we have proposed the extended flat-top operation of the synchrotron. In this operation mode, the stability of the beam was tested, and it was verified that the position- and profile-stability was less than ~ 0.5 mm at the isocenter in 100 s of extended flat-top operation. This extended flat-top operation can shorten the irradiation time by a factor of 2.

Scanning speed is designed to be 100 mm/ms and 50 mm/ms in the horizontal and vertical directions, respectively, faster by around one order than that in the conventional way [53]. In order to increase the scanning speed, we designed a scanning magnet with slits in both ends of the magnetic poles, according to thermal analysis, including an eddy-current loss and a hysteresis loss. The power supply of the scanning magnet was designed for fast scanning, and this consists of two-stage circuits: the first stage for voltage forcing by IGBT switching elements and the second stage for the flat-top-current control by FET switching elements. Testing showed a maximum temperature rise of around 30°, which was consistent with our thermal analysis.

4.4.1.3. Experimental verification

In the first stage, we carried out a fast raster-scanning experiment by using the HIMAC spot-scanning test line [53]. The irradiation control system was modified so as to be capable of raster-scanning irradiation instead of spot-scanning. In the experiment, we adapted the measured dose response of the pencil beam with an energy of $350\,\text{MeV}/n$, corresponding to a 22 cm range in water. The beam size at the entrance was adjusted to 3.5 mm at one standard deviation. Using a miniridge filter, the Bragg peak was slightly spread out to Gaussian shape with a width of 4 mm at one standard deviation. The validity of the beam model and the optimization calculation had already been verified experimentally [54]. In the experiment, the extraction beam rate was highly stabilized during the extended flat-top operation, owing to the spill-control system, and we were able to successfully carry out the pencil-beam raster-scanning experiment.

We designed and constructed a test irradiation port for the fast raster-scanning experiment in order to verify the design goal, which was the same configuration as the fixed beam-delivery system adapted to the new treatment facility. In a preliminary test, we delivered irradiation to a target with a spherical shape, 6 cm in diameter, and obtained uniform 3D dose distribution within 10 s even with 10-time rescanning.

4.4.2. Facility design

The ^{12}C beam will mainly be used for treatments that have been carried out in the existing HIMAC facility. Different ion species will also be employed for the further development of particle therapy at NIRS. In addition, positron-emission beams, such as ^{11}C and ^{15}O, will be used to verify the irradiation area and their ranges in a patient's body. Thus, an R&D study has been carried out in order to

obtain positron-emission beams accelerated directly through the HIMAC accelerator [55], instead of using the projectile-fragmentation method. In order to carry out this study in a manner identical to the existing HIMAC treatment, the residual range required has to be more than 25 cm. Thus, the maximum ion energy is designed to be $430\,\text{MeV}/n$ in the fixed beam-delivery system, corresponding to a residual range of 30 cm in a ^{12}C beam and 22 cm in a ^{16}O beam. The maximum lateral-field and SOBP sizes are 20 cm × 20 cm and 15 cm, respectively, in order to cover almost all treatments with HIMAC. On the other hand, the rotating-gantry system employs a maximum energy of 400MeV/n, a maximum lateral-field of 15 cm × 15 cm, and a maximum SOBP size of 15 cm in order to be able to downsize the gantry.

The new treatment facility is connected to the upper synchrotron of HIMAC. In the treatment hall, placed beneath the facility, three treatment rooms are prepared in order to treat more than 800 patients per year. Two of them are equipped with fixed beam-delivery systems in both the horizontal and vertical directions, while the other is equipped with a rotating gantry. Two treatment-simulation rooms are also prepared for patient positioning as a rehearsal place, and for observing any changes of target size and shape with x-ray CT during the entire treatment. Furthermore, six rooms are devoted to patient preparation before irradiation. A bird's-eye view of the new treatment facility at HIMAC is shown in Fig. 11.

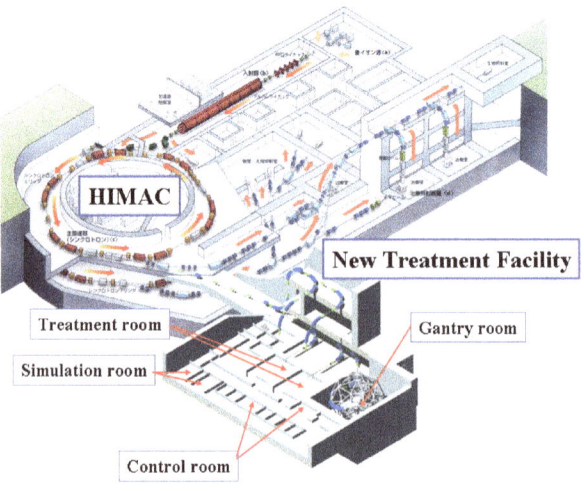

Fig. 11.　Bird's-eye view of the existing HIMAC and new treatment facility.

5. Clinical Application

The clinical indications in the early phase of proton RT included difficult-to-treat anatomic sites and those in immediate proximity to critical normal structures. For patients with skull-base chordoma and chondrosarcoma, large clinical series and long-term outcomes are available. Furthermore, hypofractionated proton therapy has been extensively used for uveal melanomas, in which local tumor control in excess of 95% has been accomplished [56].

The next clinical phase of proton therapy emphasized the reduction of treatment-related side effects by protons, due to the reduction of the integral dose to normal tissues. Prominent examples are the treatment of prostate cancer, pediatric tumors, hepatoma, and early-stage lung cancers. For prostate cancer, cumulative experiences worldwide with protons indicate a reduction of long-term rectal and genitourinary toxicities. For pediatric patients, abundant publications on treatment-planning comparisons are available, and early clinical data have convincingly demonstrated the superiority of proton RT in sparing normal, developing organs. The concept of reduced long-term damage to the surviving child, by reduction of the integral dose, has been accepted quickly in the world.

The present phase comprises the evaluation and use of proton RT for nearly all solid malignancies requiring RT — specifically the most-frequently-occurring malignancies for which RT is part of multimodality treatment. An initial study in Japan indicated excellent local control for hepatomas, which frequently develop in patients with poor liver function due to cirrhosis, as well as for early-stage non-small-cell lung cancer in medically inoperable patients.

Currently, there is no evidence that protons would be specifically contraindicated for any disease that is treated with external beam photon irradiation. Since there is no conceivable advantage for normal tissues receiving radiation, and with increasing awareness and understanding of the long-term side effects of normal-tissue radiation exposure not only in children but also in adults, it is predictable that not only proton RT, but also carbon-ion therapy, will continue on their respective paths of success.

Regarding higher-LET heavy-charged particles, experiences in Japan (NIRS and HIBMC) and Germany (GSI) have consistently demonstrated improved local control with carbon ions in the clinical setting for a large number of patients, now totaling more than 5000. Other facilities (Heidelberg, Gunma, Pavia, Marburg, and Kiel) are under construction or consideration at this time.

Carbon-ion therapy has so far proven very effective in treating many kinds of tumors that have also been very effectively treated by low-LET particle beams, including eye, head and neck, skull base, lung, liver, prostate, pelvic recurrences of rectal cancer, and sacral chordomas and osteosarcoma [57, 58]. Thus far, a lot of emphasis has been placed on demonstrating carbon's efficiency in delivering hypofractionated RT without increased toxicity, thereby allowing more patients to be treated over the same period of time than can be accomplished with other forms of RT. For example, stage I lung cancer and liver cancer can now be treated with only one or two fractions. Even for prostate cancer and bone and soft-tissue tumors requiring a longer course of RT, only 16–20 sessions has been sufficient, roughly half the number of fractions presently in general use in the case of low-LET heavy-charged-particles or conventional RT.

For tumors that are in contact with or infiltrate the gastrointestinal tract, high-dose high-LET RT may cause severe side effects, such as gastrointestinal ulcer or perforation. After the early experience with these problems following carbon-ion treatment at NIRS, it was often decided to precede RT by first performing a colostomy or inserting a spacer (or both) to displace the gastrointestinal tract; this approach resulted in greatly fewer complications.

Excellent local control, which is required to cure cancer, has been achieved with carbon-ion RT for some otherwise-intractable cancers, such as inoperable bone and soft-tissue sarcomas (Fig. 12).

Depending on the type of tumor, however, improving local control does not always prolong survival. In locally advanced malignant melanoma, for example, long-term survival was not always obtained although local control had been significantly improved with carbon-ion RT, because concurrent primary tumors or distant metastasis had developed. Thus, carbon-ion RT must sometimes be used in combination with chemotherapy, since cancer can be a systemic disease.

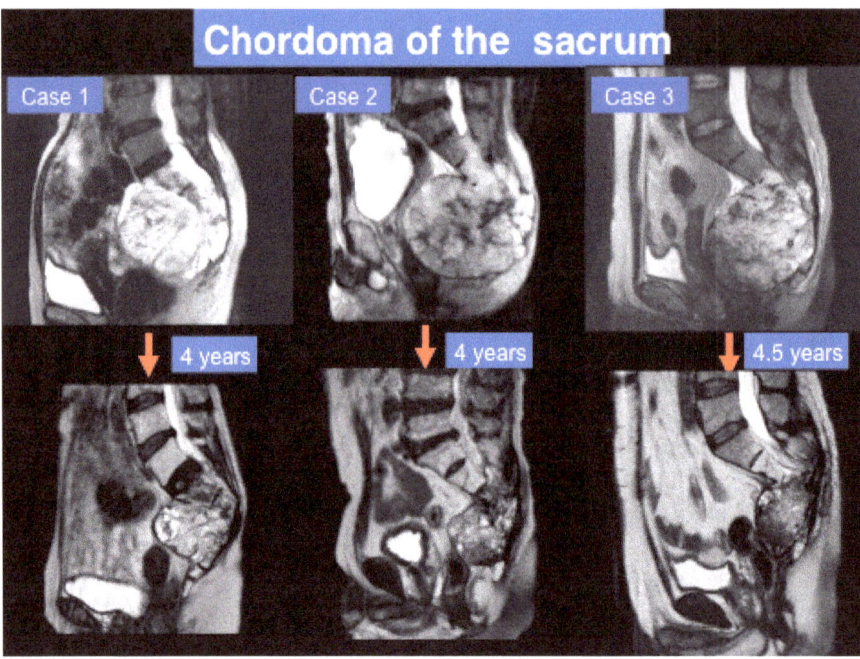

Fig. 12. Unresectable chordoma of the sacrum treated with carbon-ion therapy. Remarkable shrinkage of tumors is obtained.

Although controlled clinical trials comparing proton therapy with carbon-ion therapy have never been performed, a number of prospective and retrospective trials provide evidence that proton therapy is effective against such tumors as uveal melanoma, skull base and intracranial tumors, early-stage lung cancer, hepatocellular carcinoma, and low-risk prostate cancer. Pediatric tumors are considered a clear indication for proton therapy, since a reduction in dose deposition in nontarget tissue is believed to reduce the risk of secondary malignancies. The efficacy of carbon-ion therapy may be similar or superior to that of proton therapy in these tumors except for pediatric tumors. The types of tumors that have been preferably treated with carbon ions include bone and soft-tissue sarcoma in the pelvis and paraspinal regions, head and neck tumors with pathological types of adenocarcinoma, adenoid cystic carcinoma and malignant melanoma, high-risk prostate cancer, and pelvic recurrent rectosigmoid cancer. Furthermore, in carbon-ion therapy having a superior dose localization and greater biological effectiveness, a significant reduction in overall treatment time and fractions has been accomplished without increasing toxicities. This means that the carbon-ion therapy facility can be operated more efficiently, enabling the treatment of a larger number of patients than is possible with other modalities over the same time period.

6. Conclusion

Heavy-charged-particle beams have attracted growing interest for cancer radiotherapy based on their high dose localization. Recently, therefore, heavy-particle RT has been successfully carried out at 26 proton and 4 carbon-ion RT facilities in the world. Several construction projects for facilities providing heavy-particle RT have also been progressing. Stimulated by the good results obtained with heavy-particle RT, many more projects will be promoted. For future facility planning, therefore, medical requirements for the accelerator and beam-delivery systems have been summarized, based on our 15-year experience of cancer treatment with carbon-ion RT at HIMAC. For further development of heavy-particle RT, NIRS has developed fast pencil-beam scanning methods for treatment of both fixed and moving targets, and is constructing a new treatment facility. This new facility holds the promise of achieving more reliable and safer cancer therapy for even more treatment sites in the near future.

Acknowledgments

We would like to express our thanks to Drs. S. Fukuda, T. Furukawa, T. Inaniwa, N. Kanematsu, N. Matsufuji, S. Mori, M. Torikoshi, and S. Yonai at the Department of Accelerator and Medical Physics, NIRS, for their useful discussions and advice. We are also grateful to Dr. T. Kamada and the other members of the Research Center for Charged Particle Therapy, NIRS, for their useful advice and warm support.

References

[1] R. R. Wilson, *Radiology* **47**, 487 (1946).

[2] Y. Hirao *et al.*, *Nucl. Phys. A* **538**, 541c (1992).

[3] H. Tujii *et al.*, *Igakubuturi* **28**, 172 (2009).

[4] D. Brenner *et al.*, *Cancer* **88**, 398 (2000).

[5] E. Hall, *Int. J. Radiat. Oncol. Biol. Phys.* **65**, 1 (2006).

[6] D. Brenner and E. Hall, *Radiother. Oncol.* **86**, 165 (2008).

[7] M. Tubiana, *Radiother. Oncol.* **91**, 4 (2009).

[8] H. Paganetti *et al.*, *Phys. Med. Biol.* **53**(17), 4825 (2008).

[9] Y. Kase, N. Kanematsu, T. Kanai and N. Matsufuji, *Phys. Med. Biol.* **51**(24), N467 (2006).

[10] P. L. Petti, *Med. Phys.* **19**, 137 (1992).

[11] L. Hong *et al.*, *Phys. Med. Biol.* **41**, 1305 (1996).

[12] M. Endo *et al.*, *J. Jpn. Soc. Ther. Radiol. Oncol.* **8**, 231 (1996).

[13] N. Kanematsu *et al.*, *Jpn. J. Med. Phys.* **18**, 88 (1998).

[14] B. Scaffner *et al.*, *Phys. Med. Biol.* **44**, 27 (1999).

[15] M. Kraemer *et al.*, *Phys. Med. Biol.* **45**, 3299 (2000).

[16] T. Kanai *et al.*, *Int. J. Radiat. Oncol. Biol. Phys.* **44**(1), 201 (1999).

[17] M. Kraemer and M. Scholz, *Phys. Med. Biol.* **45**(11), 3319 (2000).

[18] M. Torikoshi *et al.*, *J. Radiat. Res.* **48**, A15 (2007).

[19] T. Kanai *et al.*, *Med. Phys.* **10**, 344 (1983).

[20] Y. Futami *et al.*, *Nucl. Instrum. Methods A* **430**, 143 (1999).

[21] T. Kanai *et al.*, *Med. Phys.* **33**, 2989 (2006).

[22] K. Noda *et al.*, *Nucl. Instrum. Methods A* **374**, 269 (1996).

[23] H. Eickhoff *et al.*, *Proc. EPAC04*, pp. 290–294.

[24] S. Minohara *et al.*, *Int. J. Radiat. Oncol. Biol. Phys.* **47**, 1097 (2000).

[25] P. Binns and J. Hough, *Radiat. Prot. Dosimetry* **70**, 441 (1997).

[26] X. Yan *et al.*, *Nucl. Instrum. Methods A* **476**, 429 (2002).

[27] G. Mesoloras *et al.*, *Med. Phys.* **33**, 2479 (2006).

[28] R. Tayama *et al.*, *Nucl. Instrum. Methods A* **564**, 532 (2006).

[29] M. Moyers *et al.*, *Med. Phys.* **35**, 128 (2008).

[30] U. Schneider *et al.*, *Int. J. Radiat. Oncol. Biol. Phys.* **53**, 244 (2002).

[31] Y. Zheng *et al.*, *Phys. Med. Biol.* **52**, 4481 (2007).

[32] J. Polf and W. Newhauser, *Phys Med Biol.* **50**, 3859 (2005).

[33] S. Agosteo *et al.*, *Radiother. Oncol.* **48**, 293 (1998).

[34] H. Jiang *et al.*, *Phys. Med. Biol.* **50**, 4337 (2005).

[35] C. Zacharatou-Jarlskog *et al.*, *Phys. Med. Biol.* **53**, 693 (2008).

[36] S. Yonai *et al.*, *Med. Phys.* **35**, 4782 (2008).

[37] H. Iwase *et al.*, *Radiat. Prot. Dosimetry* **126**, 615 (2007) .

[38] K. Gunzert-Marx *et al.*, *Radiother. Oncol.* **73**, S92 (2004).

[39] U. Schneider *et al.*, *Int. J. Radiat. Oncol. Biol. Phys.* **68**, 892 (2007).

[40] S. Vatnitsky *et al.*, *Phys. Med. Biol.* **44**, 2789 (1999).

[41] E. A. Blakely *et al.*, *Adv. Radiat. Biol.* **11**, 295–390.

[42] G. Molière, *Z. Naturforsch.* **3a**, 78 (1948).

[43] N. Matsufuji *et al.*, *Phys. Med. Biol.* **50**, 3393 (2005).

[44] A. S. Goldhaber, *Phys. Lett. B* **53**, 306 (1974).

[45] T. Inaniwa *et al.*, *Phys. Med. Biol.* **52**, 7261 (2007).

[46] S. Kameoka *et al.*, *Radiol. Phys. Tech.* **1**, 183 (2008).

[47] H. Nose *et al.*, *Med. Phys.* **36**, 870 (2009).

[48] K. Noda *et al.*, *J. Rad. Res.* **48**, A43 (2007).

[49] Y. Iwata *et al.*, *Nucl. Instrum. Methods A* **572**, 1007 (2007).

[50] M. Komori *et al.*, *Jpn. J. Appl. Phys.* **43**, 6463 (2004).

[51] T. Furukawa *et al.*, *Med. Phys.* **34**, 1085 (2007).

[52] S. Sato *et al.*, *Nucl. Instrum. Methods A* **574**, 226 (2007).

[53] E. Urakabe *et al.*, *Jpn. J. Appl. Phys.* **40**, 254 (2001).

[54] T. Inaniwa *et al.*, *Nucl. Instrum. Methods B* **266**, 2194 (2008).

[55] S. Hojo, T. Honma, Y. Sakamoto and S. Yamada, *Nucl. Instrum. Methods B* **240**, 75 (2005).

[56] D. Schulz-Ertner and H. Tsujii, *J. Clin. Oncol.* **25**, 953 (2007).

[57] H. Tsujii *et al.*, *J. Radiat. Res.* **48**, A1 (2007).

[58] H. Tsujii *et al.*, *New J. Phys.* **10**, 075009 (2008).

Hirohiko Tsujii, M.D. is an Executive Director, National Institute of Radiological Sciences (NIRS) and Professor of Chiba University and Gunma University. He graduated from Hokkaido University School of Medicine, Japan and was trained at Hokkaido University Hospital and St. Vincent Hospital and Medical Center of New York, USA. He is involved in Pi-meson Therapy at Los Alamos in 1978 and PSI in 1982, Proton Therapy at Tsukuba University in 1988–94, and Heavy-ion Therapy at NIRS since 1994. He has been majored in radiation oncology, particularly in charged particle radiotherapy. He received Scientific Award, Princess Takamatsu Cancer Research Fund in 2005 and NISTEP Award in 2006. He is Chairman of PTCOG since 2006.

Shinichi Minohara, PhD is a medical physicist at National Institute of Radiological Sciences since 1990, working on heavy ion radiotherapy in HIMAC. He is a head of therapy system and medical physics sections in Department of Accelerator and Medical Physics. He originated the respiratory gated irradiation system for carbon radiotherapy together with K. Noda and T. Kanai *et al.* He is primarily concerning patient positioning and treatment planning.

Koji Noda, PhD is an accelerator physicist at National Institute of Radiological Sciences (NIRS). He contributed to the experimental study of the slow-extraction and the electron cooling of the cooler-synchrotron ring "TARN 2" at Institute of Nuclear Science, Tokyo University. In 1989, He joined the HIMAC project at NIRS. He and his group designed a compact accelerator facility for a carbon-ion cancer therapy, which has been constructed at Gunma University. They have also developed a next-generation irradiation system at NIRS. Currently he is Director of Department of Accelerator and Medical Physics at NIRS.

Reviews of Accelerator Science and Technology
Vol. 2 (2009) 111–131
© World Scientific Publishing Company

High Frequency Linacs for Hadrontherapy*

Ugo Amaldi

*University Milano-Bicocca and TERA Foundation,
Via Puccini 11, I-28100 Novara, Italy*
ugo.amaldi@cem.ch

Saverio Braccini

*Albert Einstein Center for Fundamental Physics and
Laboratory for High Energy Physics
University of Bern
Sidlerstrasse 5, CH-3012 Bern, Switzerland*
saverio.braccini@cem.ch

Paolo Puggioni

*ADAM SA, Rue de Lyon 62,
CH-1211 Geneva, Switzerland*
paolo.puggioni@cem.ch

The use of radiofrequency linacs for hadrontherapy was proposed about 20 years ago, but only recently has it been understood that the high repetition rate together with the possibility of very rapid energy variations offers an optimal solution to the present challenge of hadrontherapy: "paint" a moving tumor target in three dimensions with a pencil beam. Moreover, the fact that the energy, and thus the particle range, can be electronically adjusted implies that no absorber-based energy selection system is needed, which, in the case of cyclotron-based centers, is the cause of material activation. On the other side, a linac consumes less power than a synchrotron. The first part of this article describes the main advantages of high frequency linacs in hadrontherapy, the early design studies, and the construction and test of the first high-gradient prototype which accelerated protons. The second part illustrates some technical issues relevant to the design of copper standing wave accelerators, the present developments, and two designs of linac-based proton and carbon ion facilities. Superconductive linacs are not discussed, since nanoampere currents are sufficient for therapy. In the last two sections, a comparison with circular accelerators and an overview of future projects are presented.

Keywords: Carbon ion therapy; cyclinac; dose delivery; hadrontherapy; linac; medical accelerators; particle therapy; proton therapy.

1. The Challenges Confronting Hadrontherapy

Hadrontherapy, the treatment of tumors with hadron beams, is a new frontier in cancer radiation therapy which is nowadays undergoing rapid development. Since its beginnings, more than 60,000 patients have been treated with protons and light ions in the world [1]. However, about one third of all the patients treated with proton therapy have been irradiated in nuclear and particle physics laboratories by means of nondedicated accelerators. Moreover, less than 2% of all these patients have been treated with pencil beam delivery systems in which

the tumor target is uniformly painted with a large number of successive spots, thus making the best possible use of the properties of charged hadron beams. This fundamental technical advance took place at the end of the last century in two physics laboratories: the Paul Scherrer Institute (PSI; in Villigen, Switzerland), where the spot scanning technique was developed for protons [2], and the Gesellschaft für Schwerionenforschung (GSI; in Darmstadt, Germany), where the raster scanning technique was developed for carbon ions [3]. In 2009 almost all hospital-based centers are still using passive dose delivery systems in which the beam is

---

*In memory of Mario Weiss, who led the developments of linacs at TERA from 1993 to 2003.

scattered in successive targets and flattened and/or shaped with appropriate filters and collimators [4]. In some centers, the more advanced semiactive "layer stacking" technique is used [5].

In the next few years, hadrontherapy centers must use new approaches to the delivery of the dose if they want to keep pace with the competition of conventional radiotherapy — mainly performed with x-rays produced by electron linacs. Indeed, new techniques have been introduced in the last ten years to conformally cover moving tumors with many crossed beams and spare more and more the surrounding healthy tissues. Many hospitals routinely employ intensity-modulated radiation therapy (IMRT) [6] and are starting to use image-guided radiation therapy (IGRT) [7, 8]. Further improvements have recently been brought by Tomotherapy [9, 10] and rapid arc technologies [11]. Hadron dose delivery systems have to become more sophisticated in order to bring to full fruition the intrinsic advantages of the dose distribution due to a single narrow ion beam characterized, at the end of its range in matter, by the well-known Bragg peak.

Proton beams of energy between 200 and 250 MeV (and very low currents, about 1 nA on target) and carbon ion beams of energy between 3500 and 4500 MeV (and currents of about 0.1 nA on target) are advantageous in the treatment of deep-seated tumors because of four physical properties [12]. Firstly, they deposit their maximum energy density abruptly at the end of their range. Secondly, they penetrate the patient with limited diffusion and range straggling (from this point of view carbon ion beams are about three times better than proton beams). Thirdly, being charged, they can easily be formed as narrow-focused and scanned pencil beams of variable penetration depth, so that any part of a tumor can be accurately irradiated. The fourth physical property is linked to radiobiology and pertains to ions, particularly carbon ions: since each ion leaves in a traversed cell about 24 times more energy than a proton having the same range, the damage produced in crossing the DNA of a cell nucleus is different and includes a large proportion of multiple close-by double strand breaks. This damage cannot be repaired by the usual cell repair mechanisms, so that the effects are qualitatively different from the ones produced by the other radiations; for this reason, carbon ions can control tumors, which are otherwise radioresistant to both protons and x-rays [13].

The first property is the main reason for using charged hadrons in radiotherapy, since the single beam dose distribution is in all cases superior to that of x-rays, which has an almost exponential energy deposition in matter after a maximum dose delivered only a few centimeters inside the patient's body. Thus beams of charged hadrons allow in principle a more conformal treatment of deep-seated tumors than beams of x-rays; they give minimal doses to the surrounding tissues, and — in the case of carbon ions — open the way to the control of radioresistant tumors.

The challenge of hadrontherapy is in making full use of the above four properties, especially when the tumor moves, mostly because of the breathing of the patient. The fact that protons and ions have an electric charge, the third property, is the key to any further development but, surprisingly enough, till now practically all therapy beams have been shaped by collimators and absorbers as if hadrons had no electric charge.

In the GSI active "raster scanning" technique, a pencil beam of 4–10 mm width (FWHM) is moved in the transverse plane almost continuously (without switching off the beam) by two bending magnets located about 10 m upstream of the patient. After painting a section of the tumor, the energy of the beam extracted from the carbon ion synchrotron is reduced to paint a less deep layer. In practice, to obtain a variable speed the beam is moved in steps three times smaller than the FWHM of the spot and the next small step is triggered when a predetermined integral of the fluency has been recorded by the ionization chambers placed just before the patient. In this approach the beam is always on.

In the PSI active "spot scanning" technique (which is also called "hold and shoot"), the 8–10 mm (FWHM) spot is moved (switching off the beam) by much larger steps (of the order of 75% of the FWHM of the spot) and, as in the previous case, the transverse movement — which takes about 2 ms — is triggered by ionization chambers measuring the fluence. During the movement of the spot the proton beam extracted from the cyclotron is interrupted for 5 ms by means of a fast kicker.

In both cases the tumor target is painted only once and this is an inconvenience in the case of moving organs, since any movement can cause important

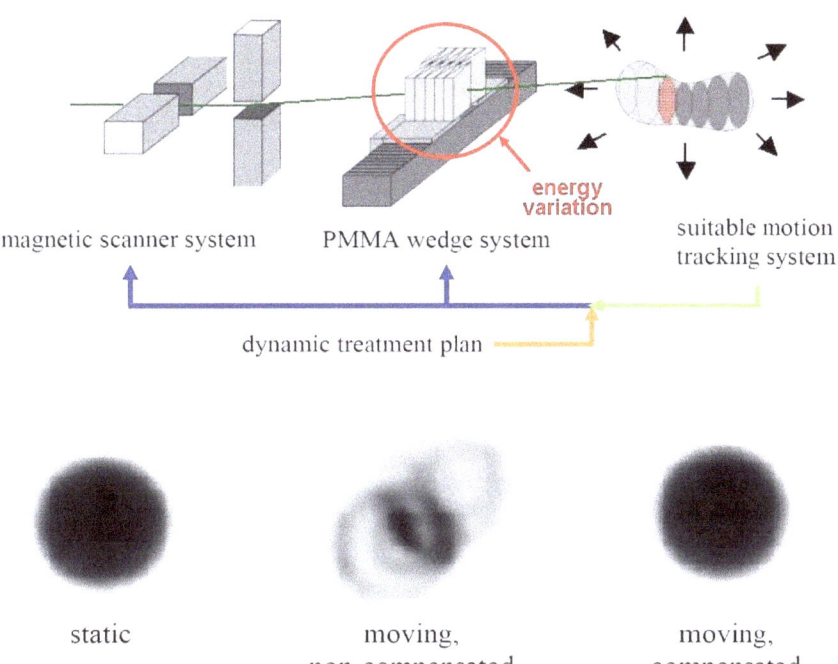

Fig. 1.   The feedback system — numerically and experimentally studied at GSI — compensates for the movements of the organs acting, with two bending magnets, to correct the transverse movements and, with absorbers of variable thickness, to compensate for longitudinal movements [14]. (*Courtesy of GSI.*)

local under- or overdosages. Three strategies have been considered to reduce such effects. In order of increasing complexity, they are:

(1) In the irradiation of the thorax and the abdominal region, the dose delivery is synchronized with the patient expiration phase in a process called "respiratory gating," so that the effects on the distribution of the dose due to the movements of the organs are reduced to a minimum (this technique is also used in conventional radiotherapy);

(2) The tumor is painted many times in three dimensions so that the movements of the organs (if not too large) can cause only small ($\leq$ 3%) overdosages and/or underdosages;

(3) The movement is detected by a suitable system, which outputs in real time the 3D position of the tumor, and a set of feedback loops compensates for the predicted position in the dose delivery plan with on-line adjustments of the transverse and longitudinal locations of the following spots, as shown in Fig. 1 [14].

An optimal delivery mechanism should be such as to allow the use of any combination of these three

approaches: respiratory gating, multipainting and active angular/energy feedback.

To face these challenges, innovative technological solutions are developed. In this framework, linacs, which are fast-cycling accelerators, offer several advantages and are particularly suited to the multipainting of moving organs, as discussed in Subsecs. 5.2 and 6.1.

2. Linacs Enter Hadrontherapy

This section describes the early design studies of the linacs for proton therapy in a chronological order, from the first proposals in 1989 to the Top-project in 1995.

The focus is on linacs which produce beams directly employed for treating patients, so the developments in the design of hadron low energy linacs used as injectors of medical synchrotrons are not discussed. The reader is referred to the recent papers by U. Ratzinger and collaborators [15, 16].

2.1. *The first proton linac for therapy designed at FNAL*

The first design of a proton linac for therapy dates back to 1989 [17–19], when at FNAL J. Lennox *et al.*

proposed a hospital-based accelerator for (i) eye treatment with 66 MeV protons, (ii) fast neutron therapy, (iii) boron neutron capture therapy and (iv) isotope production. This multipurpose 24-m-long accelerator had a duoplasmatron H^+ source, a low energy beam transport (LEBT) system, a radiofrequency quadrupole linac (RFQ) and a drift tube linac (DTL) that could deliver up to a $180\,\mu A$ average current. The advertised advantages, with respect to the usual approach based on cyclotrons, were the higher dose rate, the limited power costs and the operation in a safer radioactive area.

The RFQ [20, 21] is efficient for very low beta particles ($\beta < 0.06$). The 3 MeV protons were injected into a DTL (consisting of four independent modules) operating at 425 MHz with a low repetition rate (30 Hz) and relatively long pulses (315 μs). The protons, focused by a system of permanent magnetic quadrupoles (PMQs), could be accelerated at five different energies (3, 7, 27, 47 and 66 MeV) by switching off a certain number of DTL modules. The energy modulation was considered important for obtaining a beam suitable for the applications requiring different proton energies.

2.2. A 3 GHz high repetition rate solution

In 1991, R. Hamm, K. Crandall and J. Potter [22] of Accsys Technology proposed a linac solution composed of three sections. The system is made up of an RFQ–DTL operating at 499.5 MHz, followed by a 3 GHz side-coupled cavity linac (SCL, now called CCL) that accelerates protons from 70 to 250 MeV (Fig. 2). The energy modulation could be achieved by switching off the modules and by using degrading foils. This design was based on a higher frequency (3 GHz), a higher repetition rate (100–300 Hz) and shorter beam pulses (1–3 μs) than that of Lennox *et al.*

The high frequency enhances the shunt impedance ($Z \sim f^{1/2}$ [23]) and, for the same power consumption, the total length of the accelerator could be reduced by increasing the mean electric field.

Note that the high repetition rate favors beam scanning while the small output beam size and emittance allow a compact gantry design. The position of the beam can be moved fast (up to 100–300 times in a second) to cover all the area of the treatment. Moreover, the short beam pulses mean an affordable cost of the wall-plug power, because the duty cycle of the RF system (i.e. the repetition rate times the RF effective pulse length) is always smaller than 10^{-3}.

2.3. A 1.28 GHz linac as booster of an existing cyclotron

In 1992, M. P. S. Nightingale *et al.* proposed linear accelerators as boosters of existing hospital cyclotrons, so as to have a cost-effective machine [24]. The 1.28 GHz CCL was designed to boost protons from 62.5 MeV to 200 MeV in about 20 m. The main problem of this structure is the matching with the cyclotron, which usually produces a beam of 50–300 μA with large emittance. The Scanditronix MC60 cyclotron of the Clatterbridge hospital, considered in this first study, could be modified to produce a 100 μA pulsed beam of about 20 μs with a transverse rms emittance of $9.3\,\pi$ mm mrad, as was demonstrated in 1998 in a study conducted for the TERA Foundation [25].

The design synchronous phase was $\varphi_s = -30°$, so that the longitudinal capture efficiency ($3\varphi_s/360$ [26]) was about 25%. The duty cycle of the RF was set at 0.1%, so that the accelerated average current was about 4×10^3 times smaller than the one injected in the linac.

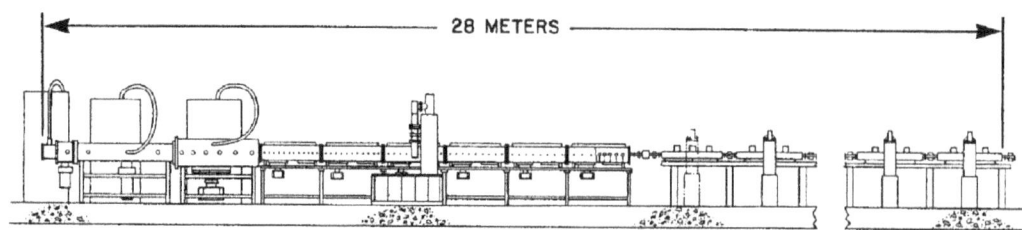

Fig. 2.   Schematic layout of the model PL-250 proton therapy linac designed in 1991 by R. Hamm, K. Crandall and J. Potter [22].

The bore radius was calculated so that the FODO structure of the series of PMQs had twice the acceptance of the input emittance ε; the 70° transverse phase advance guaranteed a minimum β Twiss parameter in each quadrupole [27], so that the transverse physical dimension of the beam ($\sim \sqrt{\varepsilon\beta}$) was smaller than the linac beam hole.

2.4. *A traveling wave solution*

An innovative approach was proposed by D. Tronc in 1993 [28, 29], when he designed an H-coupled 3 GHz traveling wave (TW) structure. The claim was that this TW linac has higher shunt impedance and a higher quality factor than the classical CCL. By removing the side-coupling cavities, the accelerator has a smaller diameter, so that simultaneous acceleration and focusing become feasible with the introduction of a special external helical focusing [30–32].

In order to get a large Q value and high shunt impedance, the length of the cavities should be as large as possible. This is even more effective at high frequencies (small wavelength λ) and low beta values, when the lengths naturally shrink to maintain the synchronism between the particle and the RF wave. The formula that determines the distance d between the midplanes of two accelerating cavities is

$$d = \frac{\beta\lambda}{2\pi}\Delta\phi, \qquad (1)$$

where $\Delta\phi$ is the phase shift between two adjacent cells.

Tronc chose a forward TW linac working in the $-3/4\pi$ mode, which means that $\Delta\phi = (2\pi - 3/4\pi) = 5/4\pi$. Thus, the length of the cavities of this TW linac is larger than that of a CCL that works in the $\pi/2$ mode and has $\Delta\phi = \pi$. According to Tronc's calculations, for $\beta = 0.25$ (30 MeV protons), the shunt impedance of a $-3/4\pi$ TW linac is about 50% higher than for an equivalent CCL structure.

So far, this has been the only attempt to design a TW linac for proton therapy.

The main characteristics of the four approaches described above are listed in Table 1.

2.5. *Further designs based on standing wave structures*

From 1993 on, and in parallel with the work done for the hadrontherapy center now in construction

Table 1.   Characteristics of the four proposals.

Subsection	Type	Freq. (MHz)	Energy (MeV)	Length (m)
2.1	SW	425	0–66	24
2.2	SW	2998	0–250	28
2.3	SW	1280	62–200	20
2.4	TW	2998	0–250	25

in Pavia, the CNAO (Centro Nazionale di Adroterapia Oncologica, Italy [33]), one of us (U. A.) proposed [34, 35] and the TERA group developed a novel type of high frequency and high repetition rate accelerator — a "cyclinac" — which produces charged hadron beams, fulfilling the clinical requirements better than cyclotrons and synchrotrons, as explained in Sec. 8. A cyclinac is an accelerator complex which makes use of a linac as booster of a cyclotron that could be used also for other medical purposes. The study soon branched into two approaches described in the "Green Book [36]."

2.5.1. *The cyclinac approach of the TERA foundation*

The initial proposal concerned a 30 MeV cyclotron used as injector of a 3 GHz proton linac (Fig. 3). This, as explained above, would imply high gradients and thus a relatively short accelerator.

The choice of the cyclotron energy of the first complete study was dictated by the fact that at 30 MeV the accelerating cells of the first module ($\beta = 0.25$) have very thin separating walls so that the mechanical tolerances and the cooling could be critical. Thus, it was decided that the first CCL would be designed for a 62 MeV input energy, having in mind in particular the cyclotron which is used for eye proton therapy at the Clatterbridge center for Oncology (Liverpool). In 1994 the results of the optimization were presented by M. Weiss and K. Crandall [37], who completed the first design of the linac which in 1998 was dubbed LIBO (LInac BOoster). The developments which followed are described in Secs. 3, 5 and 6.

2.5.2. *The all-linac approach*

An all-linac solution was studied by L. Picardi *et al.* for the Top project of ENEA and Istituto Superiore di Sanità (ISS–Rome) [38]. This machine is made up

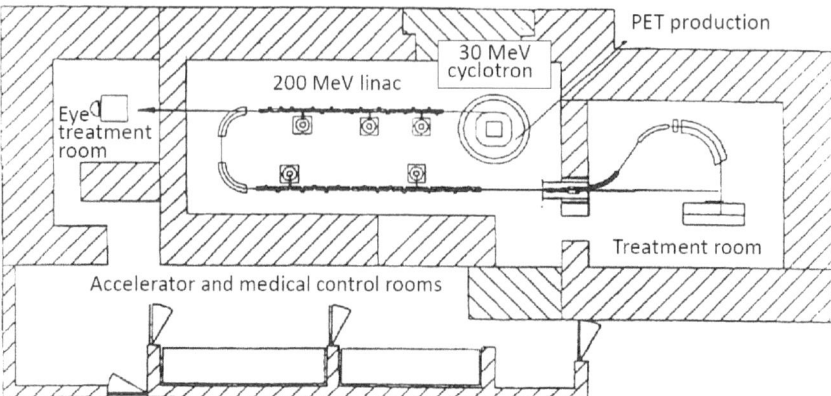

Fig. 3. The first sketch of what was later called a "cyclinac" was based on a 30 MeV commercial cyclotron used also for the production of radiopharmaceuticals [36].

of three sections: (i) an injector (RFQ + DTL) that accelerates protons up to 7 MeV, and (ii) a 3 GHz side-coupled drift tube linac (SCDTL) that injects 65 MeV protons into (iii) a 3 GHz CCL of the LIBO type.

This solution is similar to the one proposed by Hamm *et al.* (Subsec. 2.2), but in the range between 7 and 65 MeV the DTL is replaced by the innovative 3 GHz SCDTL patented in 1995 [39]. In this new structure, a certain number of DTL cavities form a "tank." The tanks are then coupled by off-axis coupling cavities and oscillate at 3 GHz working in the $\pi/2$ mode.

At low β, this structure has the same high shunt impedance of a DTL (at $\beta = 0.25$ about three times the corresponding one of the CCL) because of the considerable length of the cavities. Moreover, while in a DTL at 3 GHz the gaps between the tubes are so small that there is no space for the PMQs, in the SCDTL the PMQs can be placed on-axis at the location of the coupling cells. At last, the $\pi/2$ operating mode gives great field stability and insensitiveness to tuning errors of the cavities (see Subsec. 3.3). A prototype to accelerate protons from 7 to 11 MeV has been built.

For $\beta \sim 0.34$ (65 MeV protons) the SCDTL shunt impedance decreases and a CCL is the most efficient (see Fig. 16). In the first Top project design, a linear CCL booster accelerated protons from 65 to 200 MeV.

At present the Top IMPLART facility (Intensity-Modulated Proton Linear Accelerator for Radiation Therapy) has been financed for construction at IFO (Istituto di Fisioterapia Ospedaliera, Rome). In this case the SCDTL structure accelerates protons from 7 to 40 MeV and is followed by the CCL structure described in Sec. 5.

3. Testing of the LIBO Prototype and Recent Developments

For a cyclinac, the fraction of the transmitted beam is in the range 10^{-5}–10^{-4}. In the case of hadrontherapy, such a minute overall acceptance does not pose any problem because — as remarked above — tumor therapy with protons and carbon ion beams requires beam currents of only 1 nA and 0.1 nA on target, respectively. These very small currents are easily obtained if the linac is placed downstream of a commercial cyclotron capable of producing without problems 10^6–10^7 times larger currents. This solution has the added advantage that, if so desired, these high currents can produce in parallel radioisotopes for diagnostics, pain palliation and tumor therapy or be used for research purposes.

Based on these ideas, the 62–200 MeV linac of Ref. 37 was designed in detail and LIBO has been the first prototype of a linac for proton therapy ever built and tested (Fig. 4). This section describes this experience and the ongoing developments.

3.1. *The LIBO prototype*

In 1998, a collaboration was set up among TERA, CERN (E. Rosso *et al.*), the University and INFN of Milan (C. De Martinis *et al.*) and the University and INFN of Naples (V. Vaccaro *et al.*), with the aim of building and testing the first high frequency proton linac.

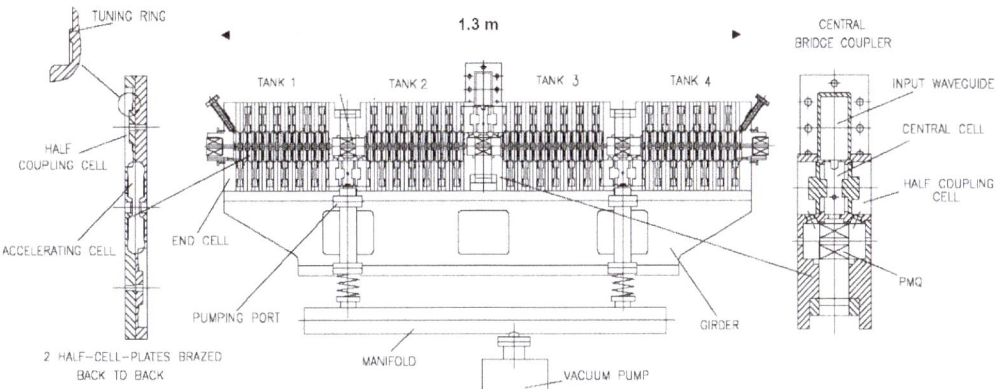

Fig. 4. Mechanical design of the four "tank" of the LIBO protype, forming one "unit" made up of two "modules." Each tank is made up of a number of basic units machined with high accuracy in copper and called "half-cell plates." Permanent magnetic quadrupoles (PMQs) are located between two successive tanks to focus the accelerated proton beam [40].

The LIBO prototype is a 3 GHz side-coupled linac with a design gradient of 15.7 MV/m. As shown in Fig. 4, it is composed of four accelerating tanks, each made one of 23 half-cell plates brazed together. The unit, 1.3 m long, is powered through a single central bridge coupler connected to a klystron. During the power tests, performed in the LIL tunnel at CERN, the design gradient was easily reached by injecting the nominal peak power of 4 MW. With the maximum available power from the klystron, a gradient of up to 27 MV/m was reached without discharges [40].

In 2001, the beam acceleration test was performed at the Laboratori Nazionali del Sud of INFN in Catania, by using the LNS Superconducting Cyclotron as injector of LIBO. Protons were accelerated from 62 to 73 MeV, well in agreement with the simulations [41]. The spectrum of the accelerated particles is shown in Fig. 5. Hence, the working principle of a linac as a booster of a cyclotron was completely demonstrated. A paper detailing the tests made and the measurements of the longitudinal acceptance is being completed [41].

3.2. *A new design of proton linacs starting from 30 MeV*

After the success of the LIBO beam acceleration test at 62 MeV, it was possible to reconsider the initial idea of a 3 GHz proton linac starting from 30 MeV. At this energy the proton speed is about 1.4 times smaller than at 62 MeV and the longitudinal dimensions of the cavities ($d = \beta\lambda/2$, where λ the wavelength of the RF pulse) shrink by the same factor.

In the case of very short cavities ($d = 12$ mm) the cooling, as already said, is more demanding and the machining and the tuning are particularly delicate. Moreover, mechanical tolerances are very tight (better than 10–20 μm) and the measurements of second order coupling effects between the cavities, which could be neglected for higher β and lower frequencies, become critical [42].

Thanks to the use of powerful software for 3D electromagnetic field calculations and the introduction of innovative design procedures [42], the technical problems have been solved and an accelerating module, made up of accelerating cells similar to the ones tested at larger energies, could be built and tested at low power (Fig. 6). These developments are the basis of the linac design which is at present pursued by ADAM SA [43], a CERN spinoff company which is building, for the end of 2009, the

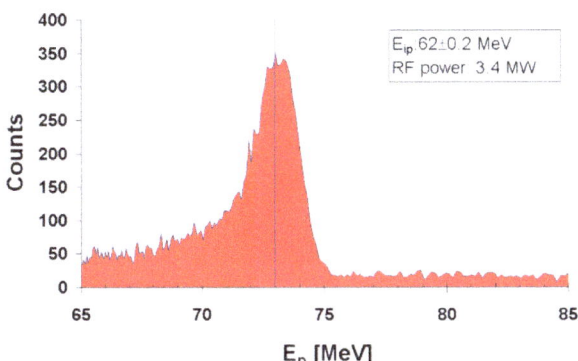

Fig. 5. Proton energy spectrum observed with a NaI crystal located downstream of the LIBO module [41].

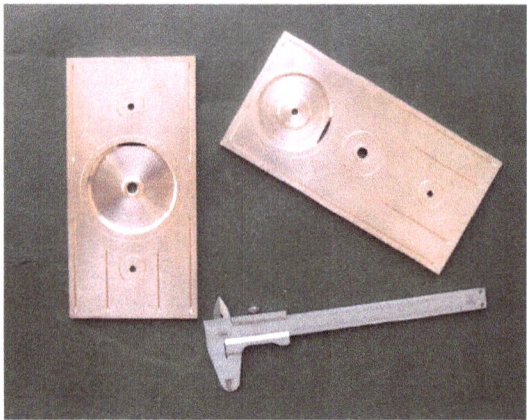

Fig. 6.   Two half-cells (left) and the bridge coupler (right) of the 50-cm-long module — made up of two tanks — which accelerates protons from 30 to 35 MeV.

first two modules that accelerate protons from 30 to 41 MeV.

In the last five years the groups led by V. Vaccaro and C. De Martinis have developed a new patented design of the linac plates called a back-to-back accelerating cavity (BBAC) [44]. In the "standard" design of Fig. 6 a tank is made up of identical half-cell plates which exhibit a half coupling cavity on one face and a half accelerating cavity on the other face. The BBAC design foresees instead a portion of an accelerating cavity on one face and the complementary part on the opposite one. The same applies to the coupling cavity. The cutting plane is such as to divide one of the two coupling slots so that the cavities exhibit an asymmetric cut. Therefore one new tile is equivalent to two half-cell plates of the standard design. The main advantages of this solution are:

• The septum between two adjacent cavities is no longer obtained by setting two tiles back to back so that its thickness can be reduced with an increase of the volume/surface ratio and thus of the shunt impedance;
• The reduced number of tiles required to build a tank entails a reduction of the machining and brazing costs.

This design was implemented in the first module of ACLIP, a 3 GHz linac intended to accelerate protons from 30 to 62 MeV. The linac consists of 5 different modules for a total length of 3.1 m [45]. Its first module is madeup of 26 accelerating cells arranged

in two tanks. This module was built [46] and power-tested [47] with a 4 MW magnetron/modulator on the premises of the e2v Company (UK) without any indication that the limit of the field gradient had been reached. In autumn 2009, beam acceleration tests will be performed at the Catania INFN-LNS superconducting cyclotron.

These two lines of activities are pursued in Italy in collaboration with CERN, while the studies described in Subsecs. 2.1–2.4 have been discontinued.

4. Standing Wave Linacs for Hadrons

To clarify the most important technical issues, only standing wave (SW) linacs are considered in this section since, as discussed above, among all the design studies of linacs for hadrontherapy which have been performed so far, only one prefigures the use of a traveling wave (TW) structure. TW linacs for electrons have been discussed in Vol. 1 of *Reviews of Accelerator Science and Technology* by P. Wilson [48].

This section is devoted to a short collection of the most important facts and formulae needed in the design of low β SW linacs, with a particular focus on CCL structures.

4.1. *RF figures of merit and scaling laws*

• *Transit time factor T*. This measures the reduction in energy gain caused by the sinusoidal time variation of the field while the particle is transiting in the gap. It approaches 1 if the gap between the "noses" of the accelerating cavities is small with respect to $\beta\lambda/2$:

$$T = \frac{\int E(0,z)\cos\omega t(z)dz}{\int E(0,z)dz}. \quad (2)$$

- *Effective shunt impedance per unit of length ZTT.* This measures the efficiency of producing an effective axial voltage $V_0 T$ for a given dissipated power P per unit of length L:

$$ZT^2 = \frac{(V_0 T)^2}{P_0 L}. \quad (3)$$

- *Internal quality factor Q_0.* This takes into account the lossy behavior of the resonator and is proportional to the number of oscillation periods needed to dissipate the energy stored in the cavity:

$$Q_0 = \frac{\omega U}{P_0}, \quad (4)$$

where ω is the resonant frequency, U the stored energy and P_0 the dissipated power. Q_0 is also related to the width of the resonance peak. For a critically coupled cavity [49]:

$$\Delta_H = \frac{2\omega}{Q_0}, \quad (5)$$

where Δ_H is the FWHM of the resonant peak and ω is the resonant frequency.

The shunt impedance scales as $f^{1/2}$, and the quality factor as $f^{-1/2}$. Thus higher frequencies linacs can have the same accelerating gradient consuming less power.

4.2. *Figures of merit of the field distribution*

- *Field nonuniformity F_{nu}.* It is the relative standard deviation of the fields X stored in the accelerating cavities of a tank:

$$F_{\mathrm{nu}} = \left\langle \frac{\Delta X}{X} \right\rangle_{\mathrm{rms}}. \quad (6)$$

According to the studies of Ref. 50, this parameter is not critical for linac operation. Errors up to $\pm 10\%$ can be accepted without affecting significantly the beam dynamics, provided that the average tank fields, which are determined by the RF power level, are within $\pm 1\%$ of the correct value. However, the requirements for therapy are more stringent. For example, in order to have a precision

of ± 1 mm in the 32 cm water range of 230 MeV protons, the mean energy of the beam must be correct within $\pm 0.2\%$.

- *Power efficiency ε_{p}.* It is the ratio between the sum of the energy stored in all the accelerating cavities (effective for the acceleration) and the total energy stored in the whole structure:

$$\varepsilon_{\mathrm{p}} = \frac{U_{\mathrm{AC}}}{U_{\mathrm{AC}} + U_{\mathrm{CC}} + U_{\mathrm{BC}}}, \quad (7)$$

where $U_{\mathrm{AC}}, U_{\mathrm{CC}}$ and U_{BC} are the sum of the energies stored in the accelerating cells (ACs), coupling cells (CCs) and in the bridge coupler (BC), if present, respectively.

4.3. *The choice of the $\pi/2$ mode and the stop band*

In 1967, Knapp *et al.* [51,52] demonstrated that the $\pi/2$ mode has many advantages as far as the performance and the stability of the accelerator are concerned:

- Frequency errors of the single cavities affect the frequency and the field distribution of the whole system only through second order effects;
- The losses do not produce any phase shift of the oscillations in the different cavities;
- The spacing between the working frequency and its neighbor modes is larger than in any other mode.

Nowadays, all CCLs work in the $\pi/2$ mode, and also new types of accelerators take advantage of this special mode. For example, structures like SCDTL (discussed in Subsec. 2.5.2) and CLUSTER (discussed in Sec. 7 and in Ref. 53) can accelerate low β particles with greater efficiency and stability than the classical DTL.

In the $\pi/2$ mode, half of the cavities are excited (accelerating cavities, ACs) and half are not (off-axis coupling cavities, CCs). The chain is thus biperiodic, made up of cells with two different geometries and resonant frequencies: ACs and CCs, resonating respectively at ω_a and ω_c. The stop band is the region of frequencies of the dispersion curve (see Fig. 7) in which the structure cannot be excited. It arises when the resonant frequencies of the ACs and CCs do not match.

The stop band is closed only if the following relation is satisfied:

$$\frac{\omega_a}{\sqrt{1-k_a}} = \frac{\omega_c}{\sqrt{1-k_c}}, \quad (8)$$

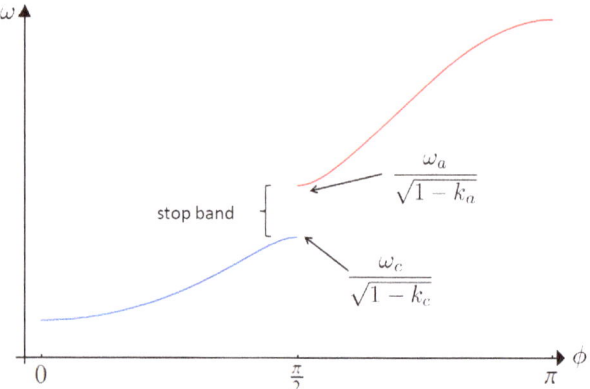

Fig. 7.   Dispersion relation of an infinite biperiodic chain (the vertical axis is in arbitrary units). In the stop band no excitation of the structure is possible.

where k_a and k_c are the second order coupling coefficient of ACs and CCs, respectively. As explained in Refs. 51 and 52, in a circuit representation they are proportional to the mutual inductance coefficient between two second neighbor cells. It can be proven that the sensitivity of the system to frequency errors in single cavities is proportional to the amplitude of the stop band. If the stop band is opened, all the advantages of the $\pi/2$ mode vanish.

4.4. *Constraints on the number of cavities per tank*

In order to minimize the length of the accelerator, to reduce the number of bridge couplers and to lower the power consumption, it is advantageous to have a maximum of accelerating cavities in the same tank.

The energy gain ΔW of a tank is

$$\Delta W = N_c L_c E_0 T \cos \phi, \qquad (9)$$

where ϕ is the stable phase [26] and N_c and L_c are the number and the length of the cavities in the tank, respectively. The total power consumption P is given by

$$P = \frac{(E_0 T)^2 N_c L_c}{ZT^2}. \qquad (10)$$

By combining Eqs. (9) and (10), the energy gain in a tank of length $N_c L_c$ can be written in the form

$$\Delta W = \sqrt{N_c L_c Z T^2 P} \cos \phi. \qquad (11)$$

Thus, for a fixed tank power consumption P, the energy gain is proportional to $N_c^{1/2}$.

However, there are constraints that have to be considered during the design and that limit the

number of cavities per module:

- A structure with N coupled cavities has N resonant modes on the dispersion curve. As N increases, the distance between the $\pi/2$ mode and its neighbors ($\delta\Omega$) decreases [54] as

$$\frac{\delta\Omega}{\omega_{\pi/2}} = k_1 \frac{\pi}{2N}, \qquad (12)$$

where k_1 is the first order coupling coefficient, which is the mutual inductance coefficient between two neighbor cavities. Mode-mixing problems may arise if the half width at half maximum Δ_H is approximately as large as $\delta\Omega$. Typical values of the parameters in a 3 GHz CCL for $\beta = 0.25$ are $Q \approx 5000, \Delta_H \approx 1.5\,\text{MHz}, k_1 \approx 0.05, N \approx 65$, and thus $\delta\Omega \approx 3.5\,\text{MHz}$.

- The field nonuniformity and the power efficiency deteriorate with increasing N. In Refs. 51 and 52, Knapp *et al.* demonstrate that the field nonuniformity F_{nu} and the ratio $U_{\text{CC}}/U_{\text{AC}}$ are both proportional to N.

4.5. *Effects of tuning errors of the ACs and the CCs*

Tuning errors of the ACs and the CCs affect the field distribution figures of merit (defined in Subsec. 4.2). The surfaces in Fig. 8 show the values of F_{nu} and ε_{p}, on the left and on the right respectively, for a given pair of rms errors of ω_a and ω_c.

It is seen that requirements on the precision of ω_a are more critical than those on the precision of ω_c. The power efficiency ε_{p} is independent of the errors of the CCs, while it is linear in the errors of the ACs. On the other hand, the field nonuniformity F_{nu} depends on the errors of both the ACs and the CCs. However, if the rms error of the ACs is zero, even large errors of the CCs do not change the field distribution.

An error in the resonant frequency of a CC causes the redistribution of the energy stored in the neighbor ACs (affecting F_{nu}) but does not increase the amount of energy stored in the CC itself (ε_{p} is not affected).

On the other hand, an error on the resonant frequency of an AC increases the field in the neighbor CCs (affecting ε_{p}) and, at the same time, redistributes the energy stored in that AC and the two neighbor ACs (affecting F_{nu}).

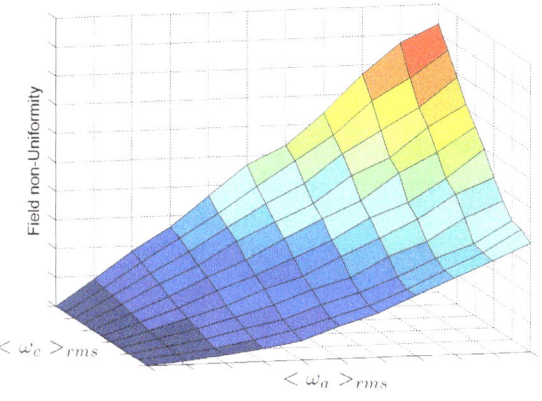 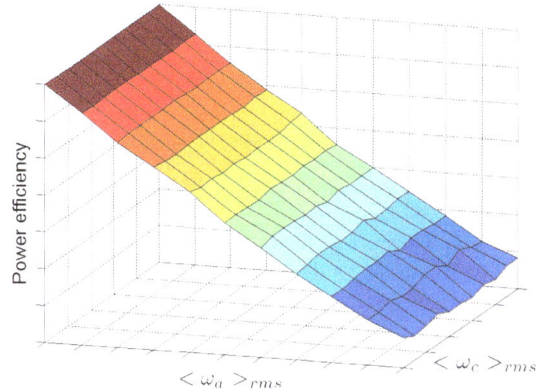

Fig. 8. Qualitative effect of tuning errors on the figures of merit of the field distribution (for the definitions, see Subsec. 4.2). "field nonuniformity" F_{nu} (left) and "power efficiency" ε_{p} (right). Given a pair of rms errors on ω_a and ω_c, the surface shows the values of F_{nu} and ε_{p}. All the quantities are in arbitrary units.

The reason for these different behaviors is that, in the $\pi/2$ mode, a very low field is stored in the CCs with respect to the one stored in the ACs.

Relative frequency errors of about 10^{-4} for the ACs (and errors 2–3 times larger for the CCs) are typical requirements for SW linacs.

5. A Linac-Based Facility for Proton Therapy

In 2001, TERA proposed the cyclinac as the heart of a fully fledged multidisciplinary center, named IDRA (Institute for Diagnostics and Radiotherapy) [55]. The main idea of IDRA is to combine on the same site four activities in cancer treatment and research [56]:

- Radioisotope production for diagnostics with PET (positron emission tomography) and SPECT (single photon emission computed tomography),
- Radioisotope production for endotherapy to treat metastasis and systemic tumors,
- proton therapy,
- Research in nuclear medicine and radiation therapy.

IDRA is a physical and cultural space where radiation oncologists, nuclear medical doctors and medical physicists can work together toward the common goal of diagnosing and curing solid tumors and their metastases with both teletherapy and endotherapy techniques.

The main features of IDRA are:

- A 30 MeV high current commercial proton cyclotron with several external beams,

- Various 30 MeV high current beams for isotope production and research,
- a high gradient side-coupled linac — based on the LIBO prototype — which accelerates protons from 30 to 230 MeV with a continuous range of energies,
- One or more treatment rooms equipped with fixed beams and/or rotating gantries for the treatment of deep-seated tumors.

5.1. *The linac of IDRA*

The parameters of the linac are summarized in Table 2. An artist's view of IDRA featuring an eye therapy beam and three gantries is shown in Fig. 9 [57, 58]. In only 18 m, 30 MeV protons are accelerated up to 230 MeV. The high repetition rate (100–200 Hz) makes this linac particularly suitable for the spot scanning technique (Subsec. 5.2).

The small effective duration of each RF pulse (3.2 μs) determines the 150 kW total plug power. The difference between the effective duration of the RF pulse and the duration of the proton pulse (1.5 μs) is due to the filling time of the structure: $Q_0/2\omega$.

The effective shunt impedance per unit of length is low for the first modules (about 30 MΩ/m), as the CCL is not efficient for low-β particles, but then rises to 90 MΩ/m at the end of the linac. With such impedances, the needed overall RF peak power is 60 MW, which can be provided by 10 compact modulator/klystron systems similar to the one shown in Fig. 10. These modulators are robust commercial solid state devices which, in case of failure, can within 2–3 h be easily exchanged as a single unit with their klystron.

Table 2.   Main parameters of LIBO [58].

Accelerated particles	p^{+1}
Type of linac	CCL
RF frequency (MHz)	2998.5
Input energy (MeV)	30
Output energy (MeV)	230
Total length of the linac (m)	18.5
Cells per tank/tanks per module	16/2
Number of accelerating modules	20
Thickness of a half cell in a tank (mm)	6.3–14.6
Diameter of the beam hole (mm)	7.0
Normalized transversal acceptance (mm mrad)	$1.8\,\pi$
Number of permanent magnetic quadrupoles	41
Length of each PMQ (mm)	30
PMQ gradients (T/m)	130–153
Synchronous phase (deg)	−15
Peak power per module (with 20% losses) (MW)	3.0
Effective shunt impedance ZT^2 (inj.-extr.) (MΩ/m)	30–90
Axial electric field (inj.-extr.) (MV/m)	15–17
Number of klystrons (peak power = 7.5 MW)	10
Total peak RF power for all the klystrons (MW)	60
Klystron RF efficiency	0.42
Repetition rate (Hz)	≤ 200
Duration of a proton pulse (μs)	1.5
Max. number of protons in 1.5 μs (2 Gy L^{-1} min^{-1})	$4 \cdot 10^7$
Effective duration of each RF pulse (μs)	3.2
RF duty cycle	$3.2 \cdot 10^{-4}$
Plug power at 100 Hz + 100 kW auxiliaries (kW)	150

This accelerator complex presents many advantages with respect to the currently used proton therapy machines (see Sec. 8). The dose delivery can naturally be performed by active methods in all three dimensions. The transversal coordinates of the beam are controlled by the use of bending magnets, while the longitudinal one is determined by continuously and rapidly varying the energy of the beam. If each module is powered by one klystron, the depth of the Bragg peak can be changed by selecting the number of active klystrons and by adjusting the power sent to the last active one. Thus, as shown in Fig. 11, a continuous range of energies is achieved and the penetration depth can be varied in only 2 milliseconds in steps of ±1 mm. This is obtained by rapidly adjusting only the low power signals of the drivers of the klystrons.

In the design of Table 2, to reduce the number of modulator/klystron systems, each of those powers two modules at the same time. This still allows one to rapidly vary the energy in the 90–230 MeV range.

5.2. *Dose delivery and multipainting techniques with protons*

In radiation therapy, a ±2.5% uniform dose has to be delivered to the tumor target. To obtain such uniformity using the spot scanning technique, the optimal distance between the spots is calculated from their natural FWHM. As already mentioned, in the PSI spot scanning technique [2] the distance is 75% of the FWHM so that the dose nonuniformity is smaller

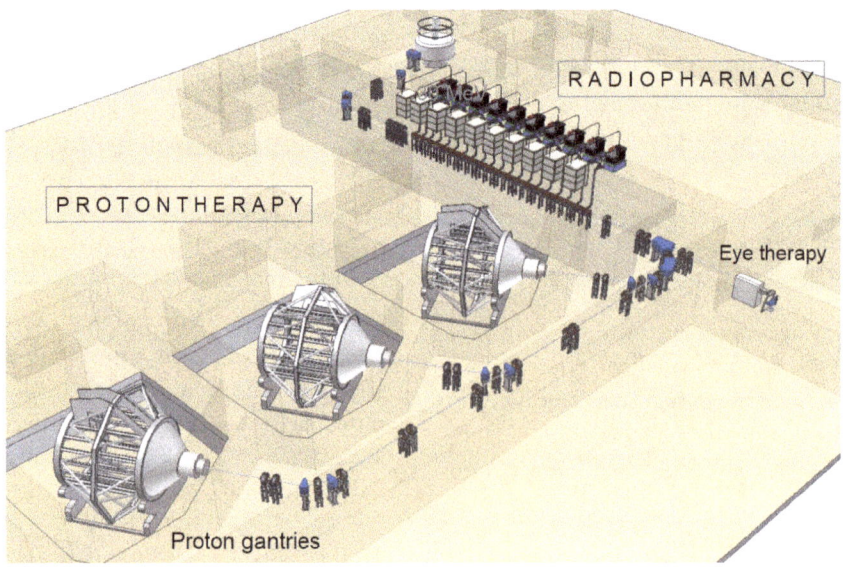

Fig. 9.   A typical layout of IDRA features a 30 MeV cyclotron, a linac of the LIBO type and three treatment rooms equipped with rotating gantries and a fixed beam line for the treatment of eye tumors [58].

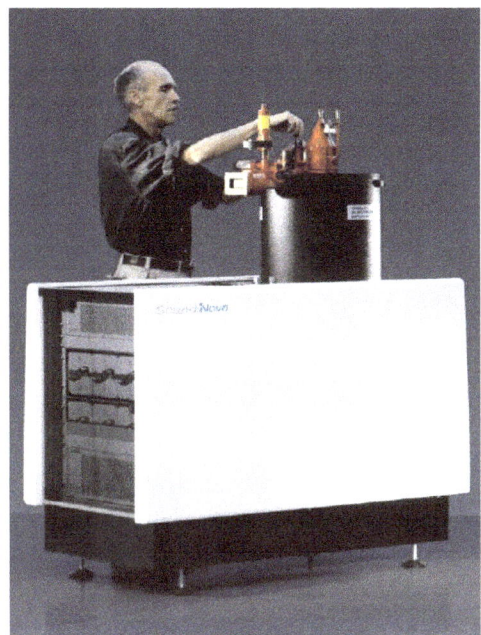

Fig. 10. The 7.5 MW klystron is powered by a solid state modulator commercialized by Scandinova Systems AB (Uppsala). LIBO employs 10 modulator/klystron sytems.

than ±1.25%. In the GSI raster scanning method the distance is 30% of the FWHM and the tumor is painted only once without switching off the beam in between the "visits" to the 2.5 denser voxel lattice. A pulsed cyclinac beam can be used both ways in conjunction with a 3D feedback system, but for the treatment of moving organs, as discussed at the end of Sec. 1, spot scanning with multipainting is preferred. The reasons are that both systematic errors in the delivered dose average out when the same voxel is visited more than 10 times and, if a spot is missing, which corresponds to a 3% drop of the local dose, the error can be corrected during the next paintings.

At a 200 mm water depth the natural lateral spread of the Bragg peak has an FWHM of 11.5 mm, which, combined with a proton pencil beam having an FWHM of 7 mm, gives an overall FWHM of 13.5 mm. This corresponds to a 6.4 mm lateral falloff (80%–20%), which has to be compared with the 5.5 mm "natural" value. Figure 12, taken from Ref. 58, shows the relative number of protons to be stopped in each voxel so as to uniformly irradiate from a single direction and with "almost round" spots a 1 L volume (diameter = 12.4 mm). The number of protons peaks at the distal edge, because the front slices are crossed by the beams reaching deeper voxels. The figure is just an example, since in a real treatment more directions will be used, in particular when employing the linac variable beam energy to implement the very effective "distal edge tracking" technique (DET) [59].

A 12-times painting of a moving organ with spots containing a number of protons (adjusted

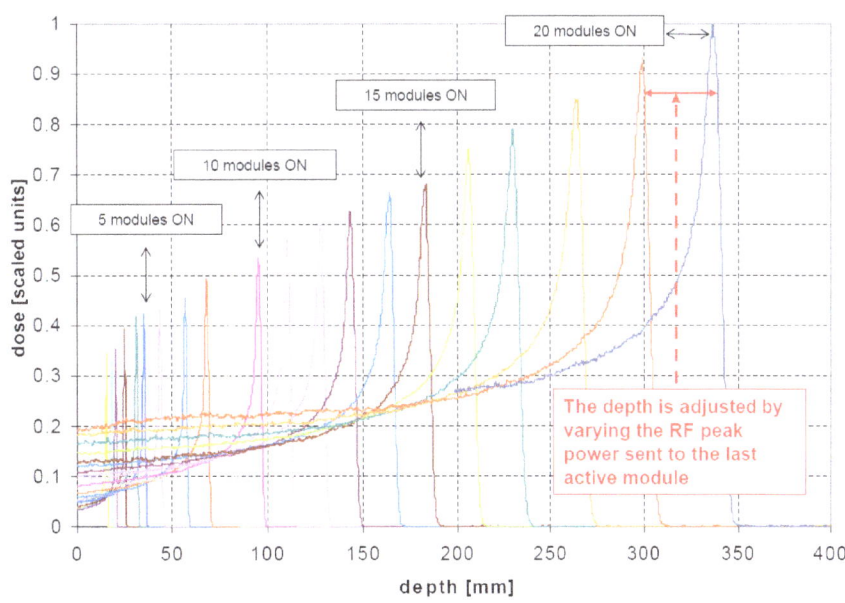

Fig. 11. Proton depth dose distribution when the number of the active accelerating modules is varied one by one. To avoid superpositions a different normalization is used for each curve [58].

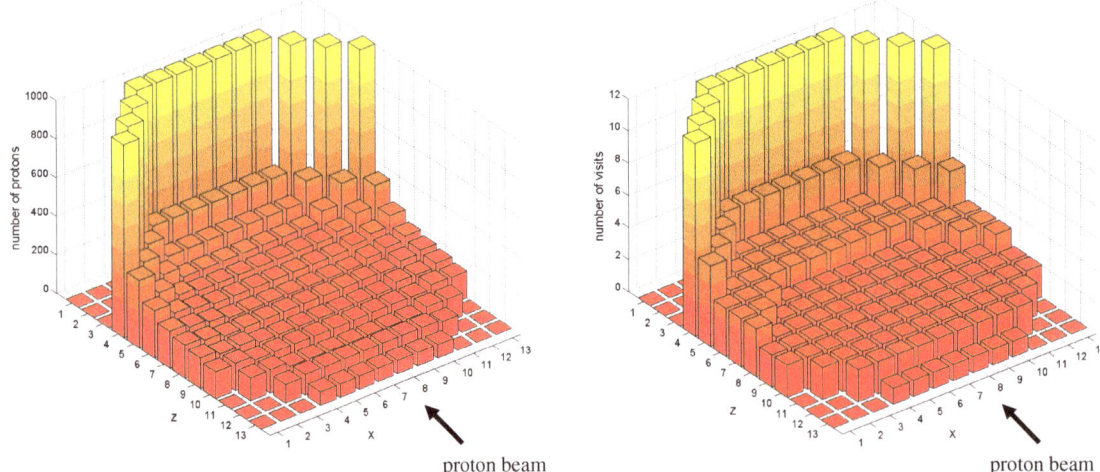

Fig. 12.   Number of protons (in arbitrary units) delivered in each voxel of the central transversal slice needed to obtain a $\pm 1.25\%$ uniform dose distribution to a 6.2-cm-radius spherical volume (1 L) centered at a 20 cm depth in water (left); number of "visits" needed to obtain a flat equivalent dose distribution with the condition with the condition that any missing visit dose not change the total local dose by more than 3% (right). The coordinates z and x are given as a number of voxels; z is the longitudinal and x the transversal coordinate [58].

by controlling the cyclotron source) which fluctuate from one visit to the next by $\pm 5\%$ implies a $\pm 1.5\%$ effect on the dose accuracy. The right panel of Fig. 12 shows that the proximal voxels need many less visits so that, on average, each spot is painted 3.5 times [58]. Table 2 shows that the maximum number of protons in a spot needed to deliver, at 100 Hz, the $2\,\mathrm{Gy\,L^{-1}\,min^{-1}}$ standard dose is $N_m = 4 \cdot 10^7$. By taking into account the linac overall transmissions, this corresponds to a $150\,\mu\mathrm{A}$ current from the cyclotron, which is 3–5 times smaller than the one routinely produced by commercial 30 MeV cyclotrons. Of course, when sending one of the cyclotron beams to the linac, the source will be chopped at the linac repetition rate so to minimize the activation of the components.

6. A Linac-Based Facility for Carbon Ion Therapy

In 2004, TERA designed a LIBO-like structure to postaccelerate carbon ions having $300\,\mathrm{MeV/u}$, such as those produced by the superconducting cyclotron designed by L. Calabretta *et al.* of the LNS-INFN laboratories in Catania and dubbed SCENT (Superconducting Cyclotron for Exotic Nuclei and Therapy) [60, 61]. The working principle of CABOTO (CArbon BOoster for Therapy in Oncology) is similar to that of LIBO. High frequency (3 GHz), high repetition rate ($\leq 400\,\mathrm{Hz}$) and short hadron

Table 3.   Parameters of the carbon ion Linac.

Accelerated particles	C^{+1}
Type of linac	CCL
RF frequency (MHz)	2998.5
Input energy (MeV/u)	300
Output energy (MeV/u)	430
Total length of the linac (m)	22
Cells per tank / tanks per module	15/2
Number of accelerating modules	16
Thickness of a half cell in a tank (mm)	15–18
Diameter of the beam hole (mm)	8
Normalized transversal acceptance (mm mrad)	$2.8\,\pi$
Number of permanent magnetic quadrupoles	33
Length of each PMQ (mm)	60
PMQ gradients (T/m)	140–170
Synchronous phase	$-15°$
Peak power per module (with 25% losses) (MW)	4.5
Effective shunt impedance ZT^2 (inj.-extr.) (MΩ/m]	100–110
Axial electric field (inj.-extr.) (MV/m)	25–23
Number of klystrons (peak power $= 7.5$ MW)	16
Total peak RF power for all the klystrons (MW)	75
Klystron RF efficiency	0.42
Repetition rate (Hz)	≤ 400
Duration of a carbon ions pulse (μs)	1.5
Max. number of C ions in $1.5\,\mu$s ($2\,\mathrm{Gy\,L^{-1}\,min^{-1}}$)	$1.6 \cdot 10^5$
Effective duration of each RF pulse (μs)	3.2
RF duty cycle	$1.3 \cdot 10^{-3}$
Plug power at 400 Hz $+$ 100 kW auxiliaries (kW)	330

pulses (1.5 μs) are the main characteristics of this 22-m-long linac for carbon ions, which is particularly suited for the spot scanning technique with multipainting [62].

The most relevant parameters of a recent version of CABOTO are collected in Table 3. It has to be underlined that in this case the ion source is a critical component since, to obtain the maximum number of carbon ions in a visit $N_m = 1.6 \cdot 10^5$, when the transmissions of the cyclotron and the linac are taken into account, the source has to deliver in $1.5\,\mu s$ about $1.6 \cdot 10^5$ fully stripped ions at $400\,\mathrm{Hz}$ [58]. Such intensity can be produced by the new superconducting Electron Beam Ionization Sources (EBIS) produced by DREEBIT GmbH (Dresden) [63].

Carbon ions can be accelerated from 300 up to $430\,\mathrm{MeV/u}$ in a continuous range of energies by selecting the number of "active" modules and modulating the energy by changing the input power in the last active module, as already discussed for IDRA.

A scheme of the dual carbon ion and proton center designed by G. Cuttone *et al.* is shown in Fig. 13. The installation of the 16 accelerating modules of CABOTO will be a second phase of the facility which is planned for the Cannizzaro Hospital in Catania [64]. In the first phase, the $17\,\mathrm{cm}$ water range of $300\,\mathrm{MeV/u}$ carbon ions will allow the treatment of 85% of all head and neck tumors and 80% of all lung and liver tumors [62].

It is worth noting that the carbon ion linac is shorter than the standard transport lines present in every center to bring the hadrons from the accelerator to the treatment rooms.

6.1. *Dose delivery and multipainting with carbon ions*

The dose delivery system is based on the spot scanning technique, used also for LIBO, but it has to take into account the different behavior of carbon ions with respect to protons. As a matter of fact, the Bragg peak produced by carbon ions is sharper and the lateral falloff is smaller than the proton one. For instance, the natural FWHM of the spot produced at $20\,\mathrm{cm}$ by a $330\,\mathrm{MeV/u}$ carbon beam is $3.1\,\mathrm{mm}$, almost 4 times narrower than that of protons having the same range. By using a pencil beam with an FWHM of $5\,\mathrm{mm}$, the overall transverse value of the FWHM is $5.9\,\mathrm{mm}$, corresponding to a $2.8\,\mathrm{mm}$ falloff, to be compared with the $1.5\,\mathrm{mm}$ natural one. Longitudinally the FWHM is intrinsically smaller than $5.9\,\mathrm{mm}$, but the unique property of the linac beam comes to the rescue: by slightly varying the proton energy, when visiting 12 times the same voxel, the Bragg peak can be widened as needed. With the same

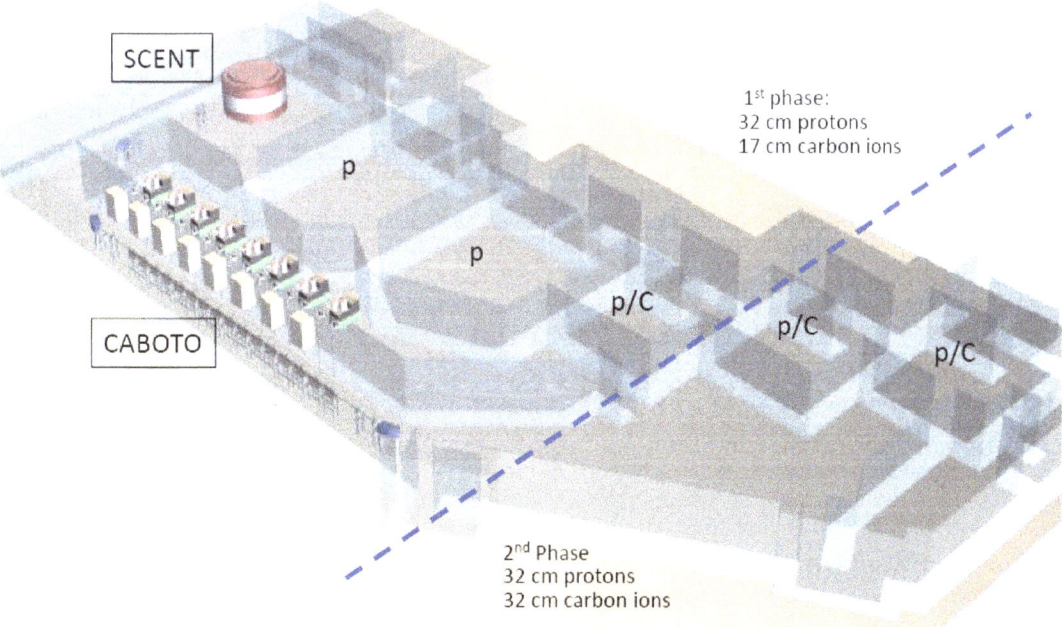

Fig. 13. The hadrontherapy center designed by the Catania group is the one schematically shown on the left of the blue line. The installation of the line will allow reaching with carbon ions a water depth of 32 cm in the rooms on the right of the blue line.

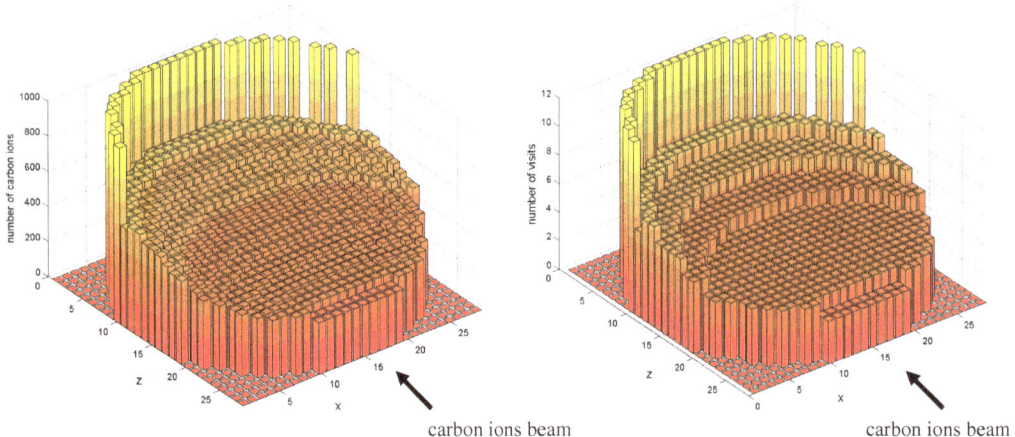

Fig. 14. Number of carbon ions (in arbitrary units) delivered in each voxel of the central transversal slice needed to obtain a ±1.25% uniform biological dose distribution to a 6.2 cm-radius spherical volume (1 L) centered at a 20 cm depth in water (left); number of "visits" needed to obtain a flat dose distribution with the condition with the condition that any missing visit dose not change the dose by more than 3% (right). The coordinates z and x are given as a number of voxels; z is the longitudinal and x the transversal coordinate. With respect to protons, due to the smallet FWHM of the beam, the number of spots for each dimension is double [58].

PSI criterion adopted for proton scanning, the distance between the spots is set at 75% of the overall FWHM and the number of voxels needed to cover the 1 L sphere is easily obtained.

The two histograms of Fig. 14 and the value $N_m = 1.6 \cdot 10^5$ needed to deliver $2 \, \text{Gye} \, \text{L}^{-1} \, \text{mim}^{-1}$ (Table 3) have been computed by taking into account the fact that the "physical dose" is different from the "equivalent dose," which is calculated by multiplying the physical dose by the effective local RBE (relative biological effectiveness) [65]. This semiempirical parameter takes into account the relative effectiveness (with respect to the x-rays) of the carbon ions in causing lethal damage to the cells. Since for carbon ions the RBE is typically 1.5 at the beginning of the path inside the tissue and increases to about 3 at the very end of the range, the physical dose delivered to the distal slices of the tumor target has to be lower than the one delivered in the middle in order to obtain a "flat" equivalent dose.

7. CLUSTER, an Innovative Low β H-Type Structure

If the linac has to accelerate carbon ions having an energy definitely smaller than 100 MeV/u, the relatively low shunt impedance of CCL structures implies a further increase of the power consumption.

The need for high power efficiency in the low β range (0.05–0.3) leads to the choice of H-mode

accelerating cavities, also called TE cavities because the electric field is naturally directed transversally with respect to the structure axis. These structures have been studied since 1950 [66, 67] and are nowadays used at low frequencies (100–200 MHz) at GSI [68] and in Linac3 at CERN [69].

H-mode cavities are drift tube cavities operating in the $H_{n1(0)}$ mode, where the index n is usually 1 (IH cavities; already existing) or 2 (CH cavities, under development). These cavities are very attractive because of the high shunt impedance for low β particles due to the fact that the generally transverse electric field is made parallel to the axis and concentrated in the accelerating gaps by the metallic drift tubes. Moreover, they are π-mode structures, i.e. the RF accelerating field is phase-shifted by 180° between successive gaps, a feature allowing higher average gradients, which in the present case are further increased by the choice of a large frequency (3 GHz).

In 2003, the TERA Foundation designed and patented a new type of H-mode accelerator that is particularly suitable for high frequencies and low β. The concept of CLUSTER (Coupled-cavity Linac USing Transverse Electric Radial field) is to connect a certain number of H-mode tanks, by using special bridge couplers, in a single resonant structure operating in the π/2 mode, as shown in Fig. 15. This choice is the novelty of this design and gives great stability to the field at these high frequencies (see

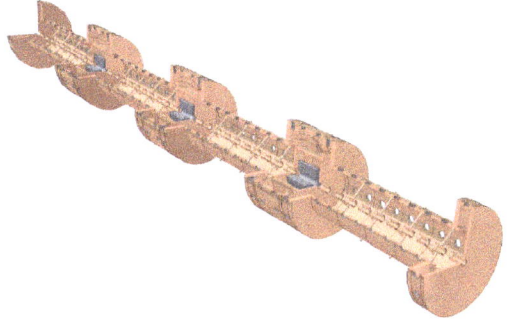

Fig. 15. Module of CLUSTER, the Couple-cavity Linac USing Transverse Electric Radial field. The accelerating tank consists of a sequence identical (constant β) accelerating units, each formed by an accelerating gap and two half drift tubes. The accelerated beam is focused by PMQs [53].

Subsec. 4.3). In order to further increase the shunt impedance, at 3 GHz the tanks consist of CH cavities, while, at lower frequencies, classical IH cavities could also be adopted. The coupling cell of the bridge couplers resonates in the TEM_{011} mode and their geometrical dimensions have been chosen so that the PMQs can be positioned on axis [53, 70].

In Fig. 16, the efficiency of this structure is compared with the approaches discussed in the previous sections. This interesting low β, high frequency and high shunt impedance structure can be adapted to many applications:

(1) High current proton acceleration at 500–700 MHz for radioisotope production using a linac system;

(2) Low current booster for proton therapy, to be used, for instance, in an IDRA center (see Sec. 5) that features an 18 MeV cyclotron and needs a linac capable of accelerating $\beta = 0.2$ protons;

(3) Low current booster for carbon ions, in a center having, for instance, a 60 MeV/u cyclotron ($k = 250$) as injector of the linac.

8. Linacs and Circular Accelerators: A Comparison

At present, all the hadrontherapy centers in operation or under construction are based on circular accelerators: cyclotrons and synchrotrons. For proton therapy both solutions are in use and commercial companies offer complete centers based on one or the other technology. On the other hand, due to the larger energy and magnetic rigidity, synchrotrons are employed to accelerate carbon ions. Only recently has it been announced that the first prototype of a superconducting cyclotron for protons and carbon ions will be built by the company IBA [71].

As far as the size is concerned, proton cyclotrons — normal or superconducting — have 4–5 m diameters while proton synchrotrons have 6–8 m diameters. For carbon ions the diameters of the synchrotrons are in the range 19–25 m.

The beam produced by cyclotrons is characterized by a fixed energy — usually in the range from 230 to 250 MeV for protons — and a 30–100 MHz pulsed beam which can be considered continuous when compared with the human respiration period. This kind of beam is surely suited for coping with the organ motion problem but needs a quite long special device installed in the beam line — usually called ESS, for "energy selection system" — which varies the beam energy by mechanically moving absorbers in times of the order of 100 ms and, downstream of the absorbers, requires a set of quadrupoles, bending magnets and slits to select the energy and "clean" the lower energy beam. The ESS hall becomes a radioactive area due to beam losses — especially if 60–70 MeV energies are used for eye treatments. Due to fragmentation, this system represents an even more critical issue for carbon ions.

The beam produced by synchrotrons is characterized by a spill time of about 1 s, during which the beam is extracted for therapy, and by a filling and accelerating time of about 1–1.5 s in which the beam is not available. From spill to spill the energy can be varied as one wishes even if, in case of passive scattering, only a few energies are usually commissioned and used. It has to be noted that the beam periodicity is similar to that of the respiration cycle,

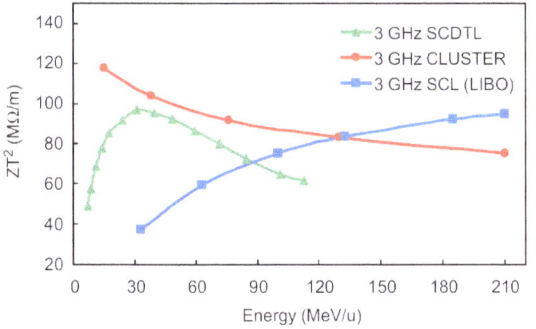

Fig. 16. Effective shunt impedance for three 3 GHz linacs, with a 2.5 nm iris radius: LIBO, SCDTL, CLUSTER [53].

Table 4.   Properties of the beam of various accelerators.

Accelerator	The beam is always present?	The energy is electronically adjusted?	Which is the approx. time (in ms) to vary E_{max}?
Cyclotron	Yes	No	100
Synchrotron	No	Yes	1000
Linac	Yes	Yes	1

which represents a disadvantage for the irradiation of moving organs with the "gating" technique.

As shown in Table 4, the beam produced by linacs presents several advantages with respect to both cyclotrons and synchrotrons and it can be considered as optimal for applications in hadrontherapy. Linacs are in fact completely flexible in their capability of varying both the energy and the intensity of the beam in 1–2 ms.

In a cyclinac, the energy can be varied between the cyclotron output value and the maximum possible for the linac, but this feature will never be used because of the finite momentum acceptance of the beam transport channel. However, a $\pm 1.5\%$ momentum acceptance is sufficient to obtain a very fast adjustment ΔR of the particle range: $\Delta R/R \approx \pm 5\%$. This corresponds to a longitudinal fast adjustment of ± 10 mm for an $R = 200$ mm. For deep-seated tumors, this is more than enough to compensate for the longitudinal variation of the particle path in the patient's body due to organ movements.

For tumors located at a 50–70 mm depth, the ± 3 mm fast adjustment may not be enough, but the range variation can be more than doubled by using larger energies and a 10 cm absorber located very close to the patient.

This possibility can be combined with the standard use of two transverse magnetic fields and allows the use of a fast and electronically controlled 3D feedback system. This system acts on the power levels of the last active klystron to vary the energy, and on the intensity of the cyclotron source to adjust the number of particles delivered in the next spot. The absence of passive absorbers and mechanical devices is surely advantageous in terms of reliability, maintenance and radiation protection.

Particle beams accelerated by linacs have many features in common with the ones produced by (nonscaling) fixed field alternating gradients accelerators (FFAGs), which are, typically, high current accelerators but have recently been designed for producing the nanoampere proton [72] and carbon ion beams [73] needed in radiation oncology. It has to be noted that nonscaling FFAGs have not yet been built, their RF systems are complicated and the extraction of a variable energy beam is difficult. On the contrary, high frequency linacs are very common, their RF systems are commercial items and beam extraction poses no problem.

9. Very High Gradient Linac Structures and Future Developments

The natural yardstick for measuring a medical linac is the 15–20 m length of the ESS needed for reducing the energy of the proton and carbon ion beams extracted from cyclotrons. The designs of Tables 2 and 3 have these lengths, and new approaches to shortening them are certainly worthwhile. Moreover, if shorter linacs could be produced, one could build "single room facilities" in which a proton linac rotates around the patient, as described in the patent of Ref. 74 under the name TULIP, which stands for "TUrning LInac for Proton therapy."

The first limitation on the miniaturization of hadron linacs is power consumption, which — for a given total acceleration voltage — increases proportionally to the electric field and — fixing also the field — is inversely proportional to the length [Eqs. (9)–(11)]. A second limitation comes from electron field emission (FE) with the consequent breakdown phenomena — which can locally destroy the metal surface.

In the 1950s, Kilpatrick assumed that destructive breakdowns happen when FE is enhanced by a cascade of secondary electrons ejected from the cathode by ion bombardment [75]. A simple calculation led to the Kilpatrick criterion, which states that the limiting surface electric field increases roughly as the square root of the RF frequency. With the data available at the time, the Kilpatrick field at 3 GHz was computed to be $E_{max} = 49$ MV/m. In the following years, structures were built in which the maximum surface field was twice the Kilpatrick field.

In the last 20 years, in connection with the design of normal conducting electron–positron colliders in the 10–30 GHz range, many more data have been collected which show that (i) the phenomena are complicated and ions do not play an important role [48], (ii) at 3 GHz the limit is definitely larger

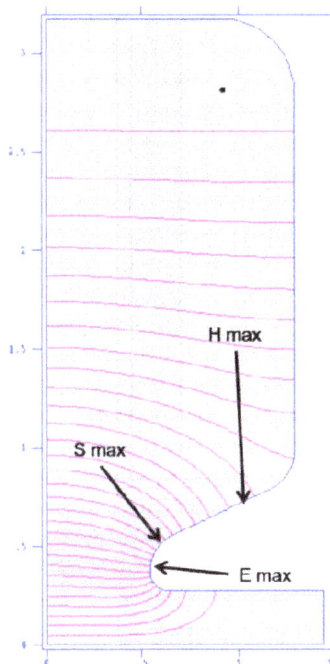

Fig. 17. The red curves represent the electric field lines of the accelerating mode and the arrows indicate the regions of a typical CCL accelerating cavity where the Pointing vector S and the electric and magnetic fields (E, H) are maximal.

than $150\,\mathrm{MV/m}$ [76], and (iii) $E_{\max}$ is roughly constant above about $15\,\mathrm{GHz}$ [77]. Recently, at CERN, a new quantity has been introduced — the "modified Poynting vector," [78] which has been shown to determine the breakdown rate. This new understanding has opened the way to the design of shorter high frequency linacs for hadrontherapy.

In an SW cavity such as the one in Fig. 17, the ratio between the maximum field $E_{\max}$ and the accelerating field in the gap can be varied in the range 5–8, so that at $3\,\mathrm{GHz}$ accelerating gradients as large as $30\,\mathrm{MV/m}$ can be obtained. At larger frequencies the gradient can be further increased, so since 2008 TERA and the CLIC RF structure group at CERN led by W. Wuensch have been collaborating on the design of new 9–$12\,\mathrm{GHz}$ structures.

The development of larger gradient structures finds its limit in the power consumption, which, for a given repetition rate, is proportional to the duration of the RF pulse. In the case of SW linacs this duration cannot be reduced below a couple of microseconds because of the filling time of the structure, which at $3\,\mathrm{GHz}$ is about $1.5\,\mu\mathrm{s}$ (Subsec. 5.1). TW linacs do not have this limitation and are thus good candidates for short hadron linacs running at frequencies larger than $3\,\mathrm{GHz}$.

Acknowledgments

The financial support of the Monzino Foundation (Milano), the Price Foundation (Geneva), the Associazione per lo Sviluppo del Piemonte (Torino) and Accelerators and Detectors for Medical Applications — ADAM SA, Geneva — is gratefully acknowledged.

References

[1] Particle Therapy Cooperative Group (PTCOG), http://ptcog.web.psi.ch/ptcenters.html

[2] E. Pedroni, R. Bacher, H. Blattmann, T. Böhringer, A. Coray, A. Lomax, S. Lin, G. Munkel, S. Scheib, U. Schneider and A. Tourosvsky, *Med. Phys.* **22**, 37 (1995).

[3] T. Haberer, W. Becher, D. Schardt and G. Kraft, *Nucl. Instrum. Methods A* **330**, 296 (1993).

[4] A. M. Koelher *et al.*, *Med. Phys.* **4**, 297 (1977).

[5] N. Kaematsu *et al.*, *Med. Phys.* **29**, 2823 (2002).

[6] S. Webb, *Intensity-Modulated Radiation Therapy* (Institute of Physics Publishing, Bristol and Philadelphia, 2001).

[7] D. A. Jaffray, *Semin. Radiat. Oncol.* **15**, 208 (2005).

[8] G. Baroni, M. Riboldi, M. F. Spadea, B. Tagaste, C. Garibaldi, R. Orecchia and A. Pedotti, *J. Radiat. Res.* **48**, A61 (2007).

[9] T. R. Mackie *et al.*, *Med. Phys.* **20**, 1709 (1993).

[10] www.tomotherapy.com

[11] F. Lagerwaard, W. Verbakel, E. van der Hoorn, B. Slotman and S. Senan, *Int. J. Radiat. Onc. Biol. Phy.* **72**, S530 (2008).

[12] U. Amaldi and G. Kraft, *Rep. Prog. Phys.* **68**, 1861 (2005).

[13] H. Tsujii *et al.*, *J. Radiat. Res.* **48**(Suppl.), A1 (2007).

[14] S. Groetzingen *et al.*, *Phys. Med. Biol.* **3**, 34 (2008).

[15] H. Vormann, B. Schlitt, G. Clemente, C. Kleffner, A. Reiter and U. Ratzinger, Status of the linac components for the Italian hadrontherapy center CNAO, in *Proc. EPAC08* (2008), pp. 1833–1835, and references therein.

[16] U. Ratzinger, G. Clemente, C. Commenda, H. Liebermann, H. Podlech, R. Tiede, W. Barth and L. Groening, A 70 MeV proton linac for the FAIR facility based on CH-cavities, in *Proc. LINAC 2006* (2006), pp. 526–530.

[17] A. J. Lennox, F. R. Hendrickson, D. A. Swenson, R. A. Winje and D. E. Young, Proton linac for hospital-based fast neutron therapy and radioisotope production. Fermi National Accelerator Laboratory, TM-1622 (1989).

[18] A. J. Lennox, *Nucl. Instrum. Methods B* **56/57**, 1197 (1991).

[19] A. J. Lennox and R. W. Hamm, A compact proton linac for fast neutron cancer therapy, in *Proc. Acc. Tech.* (Long Beach, California, 1999), pp. 33–35.

[20] T. P. Wangler, *Principles of RF Linear Accelerators* (John Wiley and Sons, 1998), pp. 225–257.

[21] I. M. Kaochinskiy and V. A. Tepliakov, *Prib. Tekh. Eksp.* **2**, 12 (1970).

[22] R. W. Hamm, K. R. Crandall and J. M. Potter, Preliminary design of a dedicated proton therapy linac, in *Proc. PAC90*, Vol. 4 (San Francisco, 1991), pp. 2583–2585.

[23] Ref. 20, p. 50.

[24] M. P. S. Nightingale, A. J. T. Holmes and N. Griffiths, Booster linear accelerator for proton therapy, in *Proc. LINAC92* (Ottawa, 1992), pp. 398–401.

[25] J. A. Clarcke *et al.*, Assessing the suitability of a medical cyclotron as an injector for an energy upgrade in *Proc. EPAC98* (Stockholm, 1998), pp. 2374–2376.

[26] P. Lapostolle and M. Weiss, Formulae and procedures useful for the design of linear accelerators, CERN-PS-2000-001 (2000), available at http://preprints.cern.ch

[27] Ref. 20, pp. 209–210.

[28] D. Tronc, *Nucl. Instrum. Methods A* **327**, 253 (1993).

[29] D. Tronc, Compact protontherapy unit design, in *PAC93* (1993), pp. 1768–1770.

[30] D. Tronc, Patent F 91 09292.

[31] D. Tronc, Patent F 92 06290.

[32] D. Tronc, Patent F 93 03152.

[33] U. Amaldi and G. Magrin (eds.), *The Path to the Italian National Center for Ion Therapy* (Mercurio, Vercelli, 2005).

[34] U. Amaldi, The Italian hadrontherapy project, in *Hadron Therapy in Oncology*, eds. U. Amaldi and B. Larsson (Elsevier, 1994), p. 45.

[35] U. Amaldi and G. Tosi, The hadron therapy project three years later. TERA 94/13, GEN 11.

[36] M. Weiss *et al.*, *The RITA Network and the Design of Compact Proton Accelerators*, eds. U. Amaldi, M. Grandolfo and L. Picardi (INFN, Frascati, 1996), Chap. 9.

[37] K. Crandall and M. Weiss, Preliminary design of a compact linac for TERA. TERA 94/34, ACC **20**, (1994).

[38] L. Picardi *et al.*, Progetto del TOP Linac. ENEA-CR, Frascati (1997), RT/INN/97-17.

[39] L. Picardi, C. Ronsivalle and A. Vignati, Struttura SCDTL. Patent No. RM95-A000564.

[40] U. Amaldi, P. Berra, K. Crandall, D. Toet, M. Weiss, R. Zennaro, E. Rosso, B. Szeless, M. Vretenar, C. Cicardi, C. De Martinis, D. Giove, D. Davino, M. R. Masullo and V. Vaccaro, *Nucl. Instrum. Methods A* **521**, 512 (2004).

[41] C. De Martinis *et al.*, Acceleration tests of a 3 GHz proton linear accelerator (LIBO) for hadron therapy. To be submitted to *Nucl. Instrum. Methods A.*

[42] P. Puggioni, Radiofrequency design and measurements of a linear accelerator for hadrontherapy, Thesis, Milano-Bicocca University (2008).

[43] www.adam-geneva.com

[44] V. G. Vaccaro, Patent No 2008 A25.

[45] V. G. Vaccaro *et al.*, ACLIP: a 3 GHz side coupled linac to be used as a booster for 30 MeV Cyclotrons, in *Proc. Cycl.* (2007), pp. 172–174.

[46] V. G. Vaccaro *et al.*, Design, construction and low power RF tests of the first module of the ACLIP linac, in *Proc. EPAC08* (2008), pp. 1836–1838.

[47] V. G. Vaccaro *et al.*, RF high power tests on the first module of the ACLIP linac, in *Proc. PAC09* (2009).

[48] P. B. Wilson, Electron linac for high energy physics, in *Reviews of Accelerator Science and Technology*, Vol. 1, 2008, pp. 7–42.

[49] S. Turner (ed.), *CAS* (CERN accelerator school), Ch. 2. CERN Yellow Reports (1992).

[50] G. R. Swain, R. A. Jameson, E. A. Knapp, D. J. Liska, J. M. Potter and J. D. Wallace, *IEEE Trans. Nucl. Sci.* 614 (1971).

[51] E. A. Knapp, B. C. Knapp and D. E. Neagle, *Rev. Sci. Instrum.* **38–11**, 1583 (1967).

[52] E. A. Knapp, B. C. Knapp and J. M. Potter, *Rev. Sci. Instrum.* **39-7**, 979 (1968).

[53] U. Amaldi, A. Citterio, M. Crescenti, A. Giuliacci, C. Tronci and R. Zennaro, *Nucl. Instrum. Methods A* **579**, 924 (2007) and arXiv:physics/0612213.

[54] Ref. 20, p. 112.

[55] U. Amaldi *et al.*, Institute for Advanced Diagnostics and Radiotherapy — IDRA. TERA note, 2001/6 GEN 31 (July 2001).

[56] R. Zennaro, *ICFA Beam Dynam. Newslett.* **36**, 62 (2005).

[57] U. Amaldi, S. Braccini, A. Citterio, K. Crandall, M. Crescenti, G. Magrin, C. Mellace, P. Pearce, G. Pitta, E. Rosso, M. Weiss and R. Zennaro, Cyclinacs: fast-cycling accelerators for hadron therapy, [arXiv:0902.3533v1 (2009)].

[58] U. Amaldi, S. Braccini, M. Crescenti, G. Magrin, C. Mellace, P. Pearce, G. Pitta, P. Puggioni, E. Rosso, M. Weiss and R. Zennaro, Accelerators for hadron therapy: from Lawrence cyclotrons to linacs. To be published in *Nucl. Instrum. Methods A* (2009).

[59] U. Oelfke and T. Bortfeld, *Technol. Cancer Res. Treat.* **2-5**, 401 (2003).

[60] L. Calabretta and M. Maggiore, Study of a new superconducting cyclotron to produce a 250 MeV — 50 kW light ion beams, in *Proc. EPAC02* (Paris, France, 2002), pp. 614–617.

[61] L. Calabretta, G. Cuttonea, M. Maggiorea, M. Rea and D. Rifuggiato, *Nucl. Instrum. Methods A* **562**, 1009 (2006).

[62] U. Amaldi, Cyclinacs: novel fast-cycling accelerators for hadron therapy, in *Proc. Cycl. 2007* (2008), pp. 166–168.

[63] G. Zschornack, F. Grossmann, V. P. Ovsyannikov and E. Griesmayer, Dresden EBIS-SC: a new generation of powerful ion sources for the medical particle therapy, in *Proc. Cycl.* (2007), pp. 298–299.

[64] G. Cuttone, private communication.

[65] M. Scholz and G. Kraft, *Radiat. Prot. Dosimetry* **52**, 29 (1994).

[66] P. Blewett, Linear accelerator injector for proton synchrotrons, in *Proc. High Energy Accelarators and Pion Physics* (Geneva, 1956).

[67] P. M. Zeidlitis and V. A. Yamnitskii, *J. Nucl. Energy, Part C* **4**, (1962).

[68] U. Ratzinger, The new GSI prestripper linac for high current heavy ion beams, in *Proc. LINAC96* (CERN, Geneva, 1996), pp. 288–292.

[69] N. Angert, W. Bleuel, H. Gaiser, G. Hutter, E. Malwitz, R. Popescu, M. Rau, U. Ratzinger, Y. Bylinski, H. Haseroth, H. Kugler, R. Scrivens, E. Tankle and D. Warner, The IH linac of the CERN lead injector, in *Proc. LINAC94* (Tsukuba, Japan, 1994), pp. 743–746.

[70] U. Amaldi, M. Crescenti and R. Zennaro, Linac for ion beam acceleration. US Patent 6888326.

[71] Y. Jongen, presented at PTCOG 47 (Jacksonville, USA, May 2008).

[72] D. Trbojevic, A. G. Ruggiero, E. Keil, N. Neskovic and A. Sessler, Design of a non-scaling FFAG accelerator for proton therapy, in *Proc. Cycl. 2004* (2005), pp. 246–248.

[73] E. Keil, A. M. Sessler and D. Trbojevic, *Phys. Rev. ST Accel. Beams* **10**, 054701 (2007).

[74] U. Amaldi, S. Braccini, G. Magrin, P. Pierce and R. Zennaro, Ion acceleration system for medical and/or other applications. Patent WO 2008/081480 A1.

[75] W. P. Kilpatrick, *Rev. Sci. Instrum.* **28**, 824 (1957).

[76] J. W. Wang and G. A. Loew, Field emission and RF breakdown in high-gradient room-temperature linac structures. SLAC PUB 7684 (Oct. 1997).

[77] W. Wuensch, High-gradient breakdown in normalconducting RF cavities. EPAC02 (2002), pp. 134–138.

[78] A. Grudiev and W. Wuensch, A new local field quantity describing the high gradient limit of accelerating structures, in *Proc. LINAC08* (Victoria, Canada, 2008), pp. 936–938.

Ugo Amaldi as staff of Italian National Health Institute (ISS, Rome) in the 60s worked in radiation physics and opened two lines of research: (e,e′p) in nuclei and (e,2e) in atoms. In 1973 he moved to CERN where he co-discovered the rise of the hadronic cross-sections with energy, published the first paper proposing a high-energy superconducting linear collider, founded and directed for 13 years the DELPHI collaboration at LEP and published a paper about the supersymmetric unification of the fundamental forces, which has more than 1200 citations. From the end of the 80s he has taught particle and medical physics in the two Milan Universities. In 1992 he created the TERA Foundation to design the Italian national carbon ion facility CNAO, which is being commissioned in Pavia, and to apply linac technologies to hadrontherapy. More than one third of the Italian high-school pupils study physics on his textbooks.

Saverio Braccini is a senior physicist at the Laboratory for High Energy Physics of the University of Bern where he leads the research activities on medical applications of particle physics. He was formerly Technical Director of the Foundation for Oncological Hadrontherapy TERA, where he contributed to the development of innovative accelerators and detectors for the treatment of tumours with hadron beams. Previously, he has been active in particle physics at the Large Electron Positron Collider (LEP) and at the Large Hadron Collider (LHC) at Cern, giving important contributions to low energy QCD and to the construction of high precision particle detectors.

Paolo Puggioni is a young physicist graduated in Milano-Bicocca University with Prof. U. Amaldi. In the last two years his research activity focused on radiofrequency design, measurements and beam dynamics of low current linacs for hadrontherapy. He is a collaborator of the TERA Foundation and of ADAM SA, a CERN spin-off company in the medical accelerator business. At the moment he is completing his postgraduate studies in Neuroinformatics at the Edinburgh University.

Reviews of Accelerator Science and Technology
Vol. 2 (2009) 133–156
ⓒ World Scientific Publishing Company

Medical Cyclotrons

D. L. Friesel*

*Parttec, Ltd., 2620 N. Walnut St., Ste. 805,
Bloomington, IN 47404, USA
dennis.friesel@parttec.com*

T. A. Antaya

*Massachusetts Institute of Technology,
77 Massachusetts Avenue,
Cambridge, MA 02139, USA
tantaya@mit.edu*

Particle accelerators were initially developed to address specific scientific research goals, yet they were used for practical applications, particularly medical applications, within a few years of their invention. The cyclotron's potential for producing beams for cancer therapy and medical radioisotope production was realized with the early Lawrence cyclotrons and has continued with their more technically advanced successors — synchrocyclotrons, sector-focused cyclotrons and superconducting cyclotrons. While a variety of other accelerator technologies were developed to achieve today's high energy particles, this article will chronicle the development of one type of accelerator — the cyclotron, and its medical applications. These medical and industrial applications eventually led to the commercial manufacture of both small and large cyclotrons and facilities specifically designed for applications other than scientific research.

Keywords: Radioisotopes; irradiation; nuclear medicine; specific activity; accelerators; cyclotron; imaging; therapy.

1. Introduction

The discovery by Rutherford in 1919 that nuclear disintegration could be caused by bombarding nitrogen with alpha particles from a naturally occurring radioactive substance precipitated an intensive effort to produce ever-more-energetic nuclear particles to probe and understand matter. The result has been an exciting period of development of a variety of machines (particle accelerators) to produce charged particles of increasingly high energies to probe the nucleus. This development, which started with simple electrostatic linear accelerators in 1924 and the Lawrence cyclotron in the 1930s, continues today with the commissioning of the LHC at CERN [1]. Particle energies have increased over the last 80 years to nearly 10^7 times that available from naturally decaying elements, and have allowed a rich, if not yet complete, understanding of the makeup of matter and the Universe.

Many different methods for producing high energy nuclear particles were proposed and developed since Rutherford's call for "a copious supply" of more energetic particles in 1919. They are discussed in previous review articles in Vol. 1 of this publication [2]. The present contribution will focus on one method of acceleration, the cyclotron, particularly in relation to the precipitant development of the medical applications spawned by it — applications which continue to expand today.

From its inception in 1930 by Ernest. O. Lawrence [3] through its many design variations to increased particle energy and intensity for research, the cyclotron has been used for a variety of biological, medical and industrial applications. Soon after the first experimental demonstration of the cyclotron resonance principle by Lawrence and Stanley Livingston [4, 5], new radioisotopes produced by high energy particles were discovered and used for

*Work partially supported by grant 2-4570.5 of the Swiss National Science Foundation.

the study of both biological processes and chemical reactions. Lawrence developed the cyclotron for nuclear physics research, yet he was very much aware of its possible applications in medicine. An interesting and comprehensive historical review of the development and growth of medical applications for cyclotron-produced beams and radioisotopes initiated at Berkeley by Lawrence and his colleagues can be found in Ref. 6.

The earliest medical applications of cyclotron beams began at Berkeley when Lawrence brought his brother John to join his group in 1935 [7]. John Hundale Lawrence, a physicist with an M.D. from the Harvard Medical School (1930), quickly demonstrated the worth of cyclotron-produced radioisotopes in disease research [8]. He became the Director of the Division of Medical Physics at the University of California at Berkeley. In 1936 he opened the Donner Laboratory to treat leukemia and polycythemia patients with radioactive phosphorus (^{32}P) [9]. These were the first therapeutic applications of artificially produced radioisotopes on human patients. By 1938, the Berkeley 27-inch (later upgraded to 37-inch) cyclotron had produced ^{14}C, ^{24}Na, ^{32}P, ^{59}Fe and ^{131}I radioisotopes, among many others that were used for medical research [6].

Lawrence believed in the promise of accelerator-produced radioisotopes as a possible weapon against disease and hailed these biomedical applications to appeal to local philanthropists to fund his accelerator development programs. Philanthropists at the time donated more money to medicine, public health, and biology than to physics. His appeals attracted funding for the 200-ton multiuse Crocker 60-inch (magnet pole) diameter cyclotron in 1938, which was commissioned in 1939, ostensibly as a "medical" cyclotron. One might consider the Crocker machine to be the first cyclotron built specifically for medical research and applications. A photo of the completed cyclotron is shown in Fig. 1.

John Lawrence and Cornelius Tobias, another student of Ernest Lawrence, used this cyclotron to research one of the earliest biomedical uses of radioactive isotopes. They used radioactive nitrogen, argon, krypton, and xenon gases to provide diagnostic information about the functioning of specific human organs. Their research discovered the nature of decompression sickness, known as the "the bends," experienced by many military aviators when flying

Fig. 1. Donald Cooksey and Ernest Lawrence (right) at the completed 60-inch cyclotron in the Crocker Laboratory.

Fig. 2. E. O. Lawrence (seated), the inventor of the cyclotron, and his brother, J. H. Lawrence, the "father of nuclear medicine," at the console of the Crocker 60-inch cyclotron.

at high altitudes without pressurized suits [10]. A photo of the Lawrence brothers at the console of this cyclotron is shown in Fig. 2.

In other activities, Drs. John Lawrence and Robert Stone were the first to employ hadron therapy to treat cancer using the Crocker 60" cyclotron. They began clinical trials treating cancer with neutrons in 1938, just six years after the discovery of the neutron by Chadwick in 1932. Neutron radiation damage is done primarily by nuclear interactions — interactions that are now known to have a high linear energy transfer (LET). With high LET

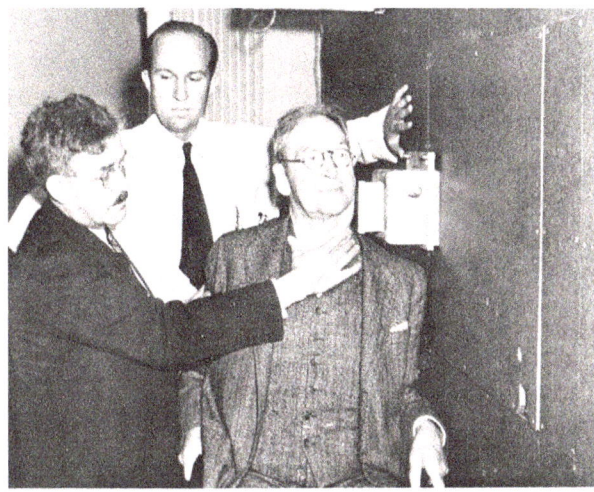

Fig. 3. John Lawrence and Robert Stone preparing a patient for neutron irradiation at the 60" Crocker cyclotron in 1938.

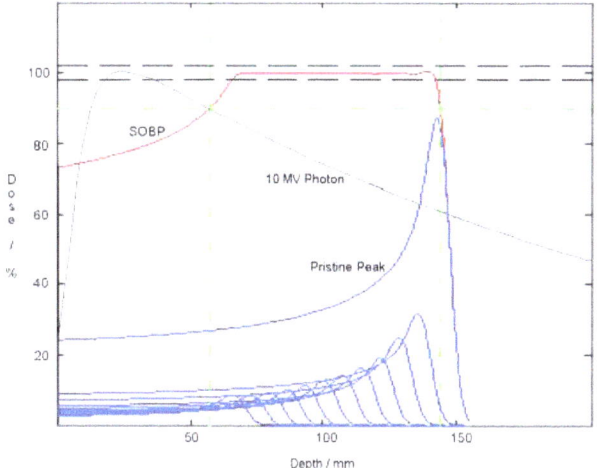

Fig. 4. Energy loss profiles for a 160 MeV proton beam compared with a 10 MeV photon. A 160 MeV proton pristine Bragg peak (blue curve) and a combined energy 9 mm spread-out Bragg peak (SOBP) madeup of a series of lower energy proton beams (red curve) are shown.

radiation such as neutron radiation, the chance for a damaged tumor cell to repair itself is very small [11]. A photo of a patient being prepared for a neutron treatment is shown in Fig. 3. This initial work was terminated, because the cyclotron was needed for the war effort during World War II. However, after the war, renewed interest in neutron therapy precipitated clinical trials at several facilities around the world in the 1970s. Unlike the early trials at Berkeley, most of the later trials were conducted using accelerators other than cyclotrons. By the 1980s, neutron therapy was no longer used for routine cancer treatment.

Yet another graduate student of Lawrence, Robert Wilson, realizing the advantages of the hadron Bragg peak, proposed the use of high energy protons and other charged ions to treat deep-seated tumors in the human body [12]. The basic physics that makes hadron therapy so attractive is the manner in which high energy ions lose energy while passing through matter. Energetic ionizing (charged) particles lose energy slowly through atomic interactions as they penetrate matter until near the end of their range, where they give up the last 85% of their energy. This energy loss profile is illustrated in Fig. 4 for a 160 MeV proton beam. The large energy loss peak at the end point (blue curve) at 15 cm into the body is called the pristine Bragg peak. Protons are known today as low LET radiation. However, the depth that an energetic proton will penetrate into the human body varies predictably with energy,

such that most of the proton energy can be delivered precisely in a tumor at any depth by selecting the appropriate energy. Beams of lower energies and intensities can be directed at the tumor so as to cover the whole depth of the tumor. This is shown in Fig. 4 for a 9 cm spread-out Bragg peak, SOBP (red curve). The radiation damage to the body during entry is significantly less than in the tumor. Since the beam stops in the tumor, very little damage is done behind it. An energy deposition profile for a 10 MeV photon (gray curve) is also shown on the same scale in this figure for comparison, illustrating higher entry and exit doses to the patient together with a smaller energy deposition in the target (tumor) area. The energy loss profile for a neutron is similar in shape to that for a photon, and hence these particles damage both cancerous and health tissue. The proton, though a low LET particle, has the advantage that it can concentrate its energy in the tumor.

Wilson's proposal led to the routine use of high energy ion beams for the direct treatment of localized (cancerous) tumors within the human body. Today there are over 30 operating ion beam therapy (IBT) facilities around the world, many of them designed and built by commercial vendors, with several more planned or under construction. Over 60% of these facilities use one commercial cyclotron design as the source of energetic ions required for the treatment [13].

Through these and other pioneering work, John Lawrence became known as the "Father of Nuclear Medicine and the Donner laboratory is recognized as its birthplace" [14]. The cyclotron development activities at the Berkeley Radiation Laboratory became the crucible for the growth of nuclear medicine and hadron therapy into an indispensable part of modern health care. Accelerator-produced radioisotopes are now routinely used for imaging diagnostics or to treat diseases. Human organs can be readily imaged, and disorders in their function revealed. "It is estimated that 15 to 20 million nuclear medicine imaging and therapeutic procedures are performed every year around the world, and demand for radioisotopes is increasing rapidly. In developed countries (about a quarter of world population) the frequency of diagnostic nuclear medicine procedures performed is approximately two per 100 persons per year, and the frequency of therapy with radioisotopes is about one tenth of this" [6].

In the sections that follow, the development of the cyclotron from the classical design invented in 1930 by Lawrence to the relativistic superconducting machines of today will be briefly reviewed and the medical applications each new capability inspired will be presented. While virtually all the advances in cyclotron design and performance were accomplished in the pursuit of scientific research, there was a persistent demand for access to them for applications in medicine and industry. The rapid advances in cyclotron performance quickly made earlier designs obsolete for research, and many of these were given over to medical research and applications work. This growing demand eventually (~1970) made the commercial manufacture of cyclotrons dedicated to these applications viable. These commercially developed and manufactured machines and their medical applications will be given special attention.

2. A Brief Review of Cyclotron Development

2.1. *The classical cyclotron*

The classical cyclotron (also called the "conventional" or "Lawrence" cyclotron), invented by Lawrence in 1930, was quite simple in concept and construction. The underlying physics principles are that charged ions (protons, electrons, etc.) are accelerated with electric fields and contained or focused

by magnetic fields. Lawrence's brilliant insight was that the orbit period of a particle of charge q, mass m and velocity v traveling in a circle in a uniform magnetic field B normal to the particle velocity is constant; only the radius R of the orbit increases with the particle momentum (mv): [15, 16]

$$R = \frac{mv}{qB}.$$ (1)

The particle orbit period τ is given by

$$\tau = \frac{2\pi m}{qB}.$$ (2)

Hence, a constant frequency sinusoidal oscillating voltage on the accelerating cavities — called dees because of their shape — matching the cyclotron resonance condition ($\omega = qB/m$) accelerates the particles twice per revolution, causing them to increase their orbit radius as they gain energy. The repetitive dee gap crossing of the recirculating beam allows it to be accelerated to high energies with relatively low dee voltages, thus eliminating the need for the high voltages used in the competing technologies of the time — the Van de Graaff [17] and Cockcroft–Walton linear accelerators [18].

The ideal kinetic energy gain per revolution in a cyclotron for a synchronous particle of charge q is given by

$$T = 4qV_0.$$ (3)

For a peak dee voltage V_0 of $\pm 2\,\mathrm{kV}$, a particle of charge q receives an $8\,\mathrm{keV}$ kinetic energy gain per revolution, i.e. each time it crosses the gap between the dees (every $180°$ of rotation) it receives an energy gain of $4\,\mathrm{keV}$. A proton traversing 300 orbits would gain a maximum of $2.4\,\mathrm{MeV}$. A classic schematic showing the major electromagnetic components of a conventional cyclotron is shown in Fig. 5.

2.1.1. *Orbital stability*

A critical design issue for all particle accelerators is the orbital stability of the circulating beam during acceleration. The particles must remain focused into small bunches in all three spatial dimensions, and orbit oscillations about the magnet midplane and equilibrium orbit must be small enough to keep the beam from getting lost on the magnet poles or dee structures. Electric and magnetic restoring forces must be built into the accelerator to keep the beam centered in the orbit. Also, the magnetic field must

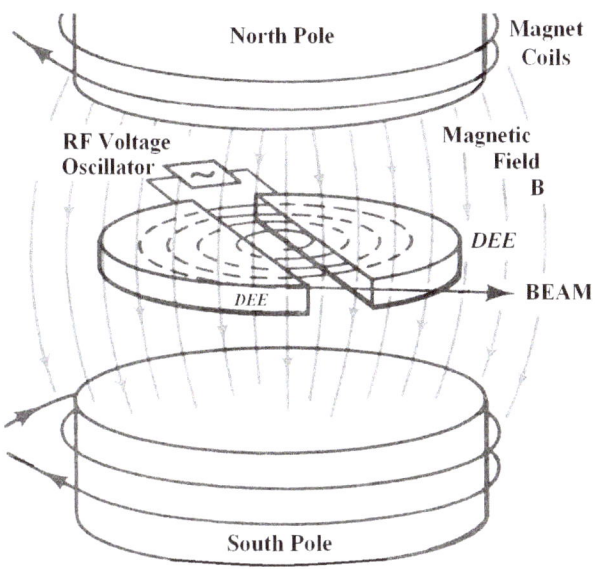

Fig. 5. Three-dimensional exploded view of the major cyclotron electromagnetic components as invented by Ernest Lawrence in 1930.

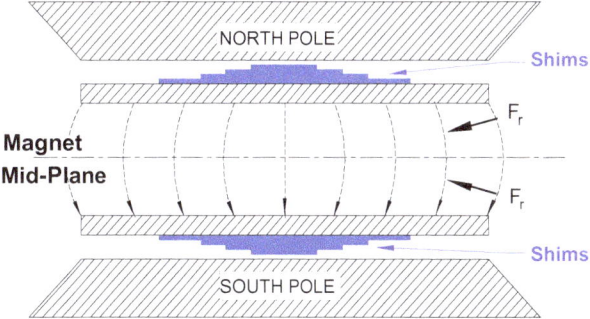

Fig. 6. Cross-sectional view of a cyclotron magnet pole gap, showing the field-shaping shims (blue) installed to provide a stable circulating beam during acceleration.

be constant to a high precision to maintain a constant orbit frequency that matches the constant rf electric field frequency throughout the many revolutions of the acceleration cycle. This later condition, called "isochronism," insures that the particles arrive at the acceleration gap when the RF voltage is near its peak value, V_0. The two requirements of beam focusing and isochronism compete with one another in the classical cyclotron and ultimately limit the maximum energy of this initial design.

2.1.2. *Focusing*

Beam focusing and orbit stability in a cyclotron require a small restoring force to push a divergent circulating beam back into the midplane equilibrium orbit. The magnetic field of a classical cyclotron (shown in Fig. 5) tends to bulge out and decrease slightly with the radius because of leakage near the pole outer edges. The resulting magnetic field thus has a small radial component (B_r) that applies weak axial and radial forces to the circulating beam. The slight field decrease with the radius is too small to provide the necessary focusing forces to keep the beam in the machine throughout the acceleration cycle. Lawrence's team added iron "shims" to the magnet pole tips to produce a more rapid field falloff with the radius so as to provide the required focusing forces. The shims, shown in Fig. 6, increase the

pole gap from the center outward with the radius to reduce the field in a controlled way. The field decrease with the radius for a classical cyclotron is

$$B_r = \frac{-nB_0 z}{R}, \qquad (4)$$

where n is a constant defined as the "field index." The resulting radial and axial focusing forces are

$$Fr = -m\omega^2(1-n)r, \qquad (5)$$

$$Fz = -m\omega^2 nz. \qquad (6)$$

For values of $0 < n < 1$, these forces keep the beam focused and cause it to make small oscillations about the magnet midplane during acceleration. The orbit oscillation frequencies are called "Betatron tunes" and are given by

$$\nu_r = \frac{\omega(1-n)^{1/2}}{2\pi}, \qquad (7)$$

$$\nu_z = \frac{\omega(n)^{1/2}}{2\pi}, \qquad (8)$$

where again $\omega/2\pi$ is the particle orbit frequency [19]. The resulting oscillation period of a deviant particle about the equilibrium orbit in both the axial and radial directions is smaller than the orbit frequency — the definition of a weak focusing accelerator. The mathematical formalism for the above equations can be found in Ref. 16.

2.1.3. *Isochronism*

For a constant sinusoidal rf accelerating voltage $\pm V_0$, a synchronous particle arriving at the dee gap at the maximum voltage receives a kinetic energy gain per revolution given in Eq. (3), as shown for three consecutive orbits (blue dots) in Fig. 7. The only

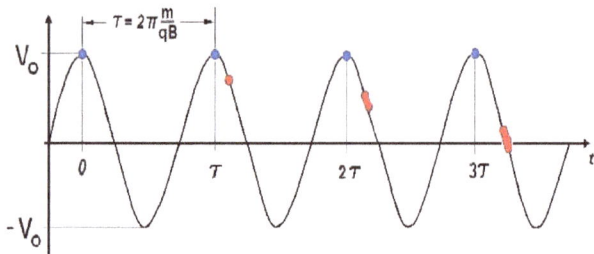

Fig. 7. Sinusoidal rf accelerating voltage, illustrating isochronous acceleration (blue) and a "phase slip" due to a decreasing magnetic field with the radius (red) for three revolutions.

force maintaining the particle in synchronism with the accelerating voltage is the magnetic field, which must be maintained to a very high precision ($\sim 0.1\%$) for particles making hundreds of turns. Variation in the magnet gap or the magnet excitation current will cause the particle orbit period to deviate from the synchronous value. For a field constrained to decrease with the radius as required for focusing, the particle orbit periods will be longer than the synchronous orbit, and will hence arrive at the dee gap at increasingly later times (red dots) relative to the rf period, causing the particles to become increasingly out of phase with the rf electric field. This is referred to as a "phase slip," i.e. the particles slowly slip out of phase with the rf accelerating voltage with each passing turn. Two things happen when this occurs. The accelerating voltage experienced by the particle is less than V_0, i.e.

$$V = V_0 \cos\theta, \qquad (9)$$

where θ is the rf phase angle traversed during a single particle orbit. The resulting kinetic energy gain per turn would then be

$$T = 4qV_0 \cos\theta. \qquad (10)$$

The lower energy gain per turn causes the particles to make a larger number of orbits in the cyclotron to reach the maximum design energy. In the worst-case scenario, the particles will eventually arrive late enough after many turns to receive no acceleration or even deceleration. Another effect of a phase slip is an increase in the particle bunch spatial size and time width during acceleration, also shown in Fig. 7. Both effects cause beam intensity loss during the acceleration process.

A third effect of acceleration in any cyclotron that causes the particles to lose synchronism with

the fixed frequency rf electric field is that the particle mass $m(t)$ increases with velocity according to Einstein's theory of relativity. As the particle velocity $v(t)$ increases with time, the particle mass $m(t)$ increases according to

$$m(t) = \frac{m_0}{(1 - \beta(t)^2)^{1/2}}, \qquad (11)$$

where m_0 is the particle mass at rest, $\beta(t) = v(t)/c$, with c being the speed of light. A 20 MeV proton's mass is 2% higher than one at rest, according to Eq. (10). This mass increase further increases the orbit period [Eq. (2)], adding to the loss of synchronism. To compensate for the relativistic mass increase with energy, the field must increase with the radius in proportion to the particle mass increase, exactly the opposite of what is required for focusing. Using high dee voltages to reduce the number of turns required to achieve the maximum energy can mitigate the competing requirements of relativity, focusing and isochronism. Even with this, the maximum proton energy capability of the classical cyclotron originally invented by Lawrence can be shown to be approximately 20 MeV. This situation lasted until about 1958 for the classical cyclotron design.

In the years following the development of the first practical cyclotron (11″, 1.2 MeV p) in 1932, several larger cyclotrons were constructed at Berkeley and many machines were subsequently built around the United States and the world. At Berkeley, Lawrence built the 27″ (3.6 MeV p, 1932), 37″ (8 MeV d, 1937) and 60″ Crocker (16 MeV d, 1939) cyclotrons and started the construction of the 184″ cyclotron. Designed for 340 MeV protons, this machine would not have worked as a classical cyclotron for the reasons just given, but was modified to operate as a synchrocyclotron. Lawrence's graduate students and other scientists from Berkeley started the construction of many research cyclotrons at other institutions based on the classical Berkeley designs. By the late 1930s, machines were built at Cornell (S. Livingston), Columbia, Indiana University (Franz Kurie and L. J. Laslett), Princeton, MIT (S. Livingston), Rochester and Yale, just to name a few of the approximately 24 cyclotrons operating or under construction at academic institutions around the United States by the mid-1940s [20]. Cyclotrons were also built in Sweden (Stockholm, 32″), Japan (Reiken, 26″) and other countries in Europe. Some of these machines were used for medical and scientific studies well into the 1960s.

2.2. *The synchrocyclotron*

One obvious solution to the classical cyclotron energy limit is to reduce the frequency of the rf accelerating voltage with time in synchronism with the increase in the particle orbit period caused by the effects, primarily relativity, described above, i.e.

$$\omega_{\mathrm{rf}}(t) = \frac{qB}{m(t)}. \tag{12}$$

This "frequency-modulated" (fm) operation requires a single beam bunch to be accelerated with the phase of the accelerating particles shifted to be between $40°$ and $60°$ after the voltage peak, as shown in Fig. 8. The synchronous accelerating voltage V_s is then less than V_0. When a particle is out of phase with $\omega_{\mathrm{rf}}(t)$, i.e. has an orbit period either smaller ($\theta_{\min}$) or larger ($\theta_{\max}$) than the synchronous value (θ_s), it receives an energy gain less or greater than $4qV_0\cos\theta_s$. This causes the particle orbit period to move back toward the synchronous value over several orbits, acting like a feedback loop. Consequently, the particle orbit period will oscillate about the synchronous value during the acceleration cycle, a process called phase stable acceleration. The phase angle range ($\theta_{\min} - \theta_{\max}$) of the particle motion is referred to as an "rf bucket." The principle of "phase stability" was postulated simultaneously by Veksler [21] and McMillan [22],

and is also the fundamental principle for acceleration in a synchrotron. Phase stable acceleration also allows the particles to accelerate through small magnetic field anomalies in the cyclotron caused by pole tip machining errors and poor magnet excitation current regulation. One drawback of the synchrocyclotron is that once a beam bunch is captured and accelerating, the next bunch cannot be accelerated until the first is accelerated to full energy and the rf frequency reset to the injection value. The resulting extracted beam has a pulse period several thousand times the rf accelerating frequency, compared to the classical cyclotron pulse period of twice the accelerating rf frequency, significantly reducing the average extracted beam intensity.

The principle of phase stable acceleration was first demonstrated when the 37″ Berkeley cyclotron was converted to a synchrocyclotron in 1946 [23]. The 184″ cyclotron at Berkeley, shown in Fig. 9, was then quickly redesigned to operate as a synchrocyclotron capable of accelerating protons to 350 MeV. Following these successes at Berkeley, 14 large synchrocyclotrons were constructed around the world that produced proton beams of up to 1 GeV (Gatchina) in energy [24].

2.2.1. *Early proton therapy medical developments*

The higher beam energies (>100 MeV protons) of the synchrocyclotron gave medical researchers the opportunity to study and test proton beam therapy as originally proposed by Robert Wilson in 1946 [12].

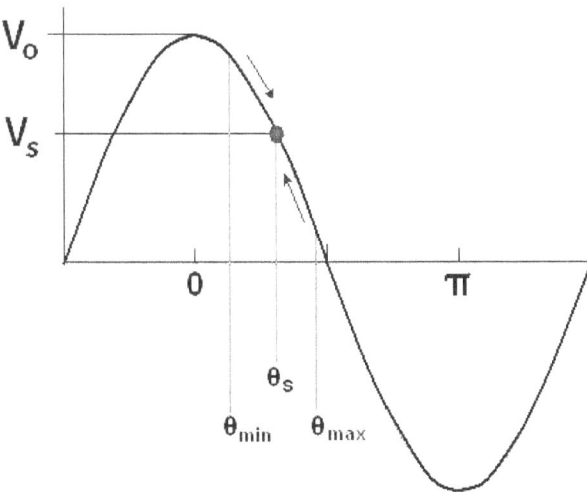

Fig. 8.   RF acceleration cycle, illustrating the principle of phase stable acceleration in a synchrocyclotron. The area between $\theta_{\min}$ and $\theta_{\max}$ is the "rf bucket."

Fig. 9.   The Berkeley 184″ synchrocyclotron, 1946.

Fig. 10. The Harvard 160 MeV synchrocyclotron just prior to its commissioning on June 15, 1949, with Profs. Norman Ramsey (cyclotron laboratory director; right) and Lee Davenport (associate director). This machine was shut down on June 20, 2002.

John Lawrence was the first to treat cancer patients with high energy protons in 1954 using the Berkeley 184″ synchrocyclotron, by irradiating the pituitaries of 30 patients with metastatic breast cancer [25].

Another of the early synchrocyclotrons, the Harvard synchrocyclotron commissioned in 1949 (160 MeV p), shown in Fig. 10, would go on to become a dedicated proton source for a pioneering program of ion beam radiation therapy for cancer from the 1960s until its decommissioning in 2002. Robert Wilson established many of the design parameters for this machine while he was an associate professor at Harvard, during which time he published the famous paper proposing the use of high energy protons for the treatment of deep-seated tumors in the human body [12]. After a distinguished career in nuclear and medical research, the Harvard accelerator was used to begin a study of the treatment of various neurological lesions with high energy protons in the 1960s in collaboration with the Massachusetts General Hospital (MGH) neurosurgery department. This collaboration resulted in the development of several important beam manipulation techniques (double scattering, range modulation), treatment protocols and patient immobilization techniques for proton radiotherapy that are still in routine use today [26]. A total of 9115 patients were treated using the Harvard synchrocyclotron from 1961 until its decommissioning in 2002.

Another of these early synchrocyclotrons performed preliminary studies of proton beam therapy from 1957 to 1968. At the Gustaf Werner Institute (GWI) in Uppsala, Sweden, a synchrocyclotron with a radius of 90″ and a maximum energy of 200 MeV for protons was commissioned on December 9, 1951, with Ernest Lawrence in attendance. This machine was used for radiotherapeutic research and clinical tests with 185 MeV proton beams from 1957 to 1968. These are but a few examples of how early physics-driven accelerators were used for medical research and applications, a trend that has continued to this day.

2.3. *The Thomas (isochronous) cyclotron*

The major disadvantage of the synchrocyclotron, i.e. low average intensity pulsed beams, was overcome by the development a third type of cyclotron, known as the isochronous cyclotron, which is capable of accelerating a continuous stream of particle bunches at a constant orbit frequency to high energies. The approach to addressing the relativistic mass increase of Eq. (10) is to allow the field to increase radially at the same rate as the relativistic mass increases during acceleration. The radial field variation of the axial component of the magnetic field must be

$$B_z = \frac{B_0}{[1 - (Z/A)^2(r/a)^2]^{1/2}} \qquad (13)$$

where $a = E_0/ecB_0$. With this radial field variation, the cyclotron frequency is constant and independent of energy.

2.3.1. *The radial ridge cyclotron*

The high energy isochronous cyclotron was not considered in the early days of cyclotrons because the increasing field violated the conditions for axial stability $(0 < n < 1)$ of the classical and synchrocyclotrons discussed earlier. A method to overcome the weak focusing properties of the required radially increasing field was proposed in 1938 by Llewellyn Thomas [27]. He suggested using an azimuthally varying magnetic field to provide edge focusing for particles entering and exiting the high and low field regions of the magnet. This was accomplished by dividing the cyclotron magnet pole faces into regions of high fields, called "hills" (H), and low fields, called

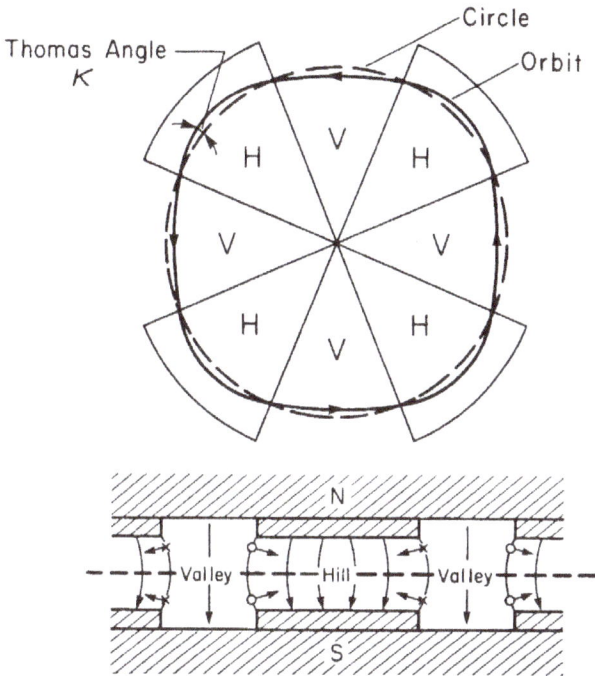

Fig. 11. Schematic of a four-sector radial ridge cyclotron magnet, showing the noncircular obits, the Thomas angle and the regions of high (H) and low (V) fields.

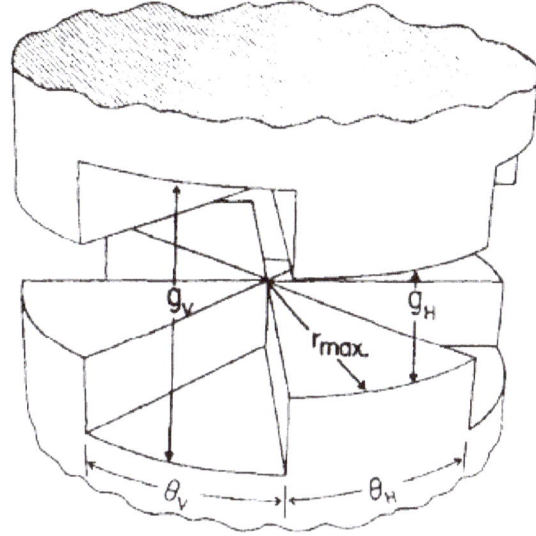

Fig. 12. Isometric view of the pole tip geometry for a four-sector radial pole tip gap variation needed for a Thomas cyclotron. The "valley" and "hill" pole tip gaps, g_V and g_H, are shown.

Fig. 13. The lower pole tip of an Advanced Cyclotron Systems TR 24 radial ridge cyclotron. This cyclotron delivers up to 1 amp of 15–24 MeV protons to produce PET or SPECT isotopes.

"valleys" (V), such that the average radial field of the cyclotron increases with the energy according to Eq. (13) to maintain a constant orbit period. The simplest example of this is the four-sector radial ridge cyclotron magnet pole face depicted schematically in Figs. 11 and 12.

The azimuthally varying magnetic field makes the protons travel in noncircular orbits, causing them to pass through the interface between the high and low fields at an angle k, referred to as the "Thomas" angle. The radial components of the fields at the interface can be made strong enough to produce adequate radial and axial focusing forces to maintain beam stability throughout the acceleration cycle. These forces are proportional to the Thomas angle k and the ratio of the high and low field values, which must be calculated during the design of the accelerator. An ion traversing a pole gap with an axial variation in pole height sees a net axial focusing force back toward the cyclotron median plane, as shown in Fig. 10. A photo of a commercially available isochronous 24 MeV proton cyclotron using radial ridge pole tips is shown in Fig. 13 [28].

While the benefits of fixed frequency and CW (continuous dc beam) operation were clear, it took a while for this "Thomas" cyclotron to be adopted. The first operational isochronous cyclotron was a three-sector radial ridge electron machine constructed and operated at Berkeley during the 1950s. This major development was deemed classified and was therefore not made public until 1956 [29].

The first isochronous proton cyclotron, a three radial-sector machine with a pole diameter of 86 inches, was built at DELFT in 1958. This machine accelerated protons to 12.7 MeV [30]. Another demonstration of such focusing was the Thomas shimmed classical cyclotron at Los Alamos [31].

The slow adoption of Thomas's 1938 proposal seems to have been due both to the fact that the orbit dynamics of this machine needed to be absorbed by the cyclotron builders, and the fact that the existing cyclotrons of the 1930–40s were simply adequate for the nuclear research in progress. The priorities of WWII once again had a role in the slow development of this concept. Eventually the demand for higher energies and higher intensities for nuclear science forced the development of the isochronous cyclotron. By 1959 there were two isochronous cyclotrons in operation and 13 more in various stages of design and assembly.

The pole field variation required to provide the Thomas focusing may be done with a sine wave or square wave pole gap variation, and with radial or spiral ridge pole shapes. A distinguishing feature of isochronous cyclotrons from the beginning was that a quantitative beam dynamics simulation was required to verify the field design, and indeed one of the first uses of the new digital computers in the 1950s was for the design of an isochronous cyclotron. The complete theoretical basis of the isochronous cyclotron, in which it is seen that Thomas focusing is an example of the general principle of alternating gradient focusing, is shared with another accelerator type, the fixed field alternating gradient accelerator (FFAG), and was developed principally by members of the MURA Group in the mid-1950s, led by Symon [32]. From the late 1950s onward most new cyclotron projects were based upon this isochronous cyclotron type.

2.3.2. *The spiral ridge cyclotron*

The energy capability of the radial ridge design is limited to about 45 MeV by the small Thomas angle that can be achieved, which limits the strength of the axial focusing forces that can be obtained. This constraint was removed by the introduction of spiral, rather than radial, ridge pole tip sectors. The spiral angle magnet pole sectors caused the circulating particles to cross the pole edges at an angle greater than the Thomas angle, producing a stronger

axial focusing force. The spiral pole tip shape can be adjusted during the design process to select the strength of the focusing required for orbit stability. This process could not be done empirically, but required the use of digital computers, which became available to scientists in the late 1950s. The radial and spiral ridge cyclotrons belong to a cyclotron group referred to as isochronous, azimuthally varying field, and sector-focused cyclotrons. One of the largest spiral ridge cyclotrons, TRIUMF, was made in Vancouver, B.C. This accelerator, a six-sector cyclotron (shown in Fig. 14), accelerated H⁻ ions to 520 MeV and is the physically largest cyclotron ever built (pole diameter 17.17 m), because the maximum field was limited to 6 kG to prevent magnetic striping of the H⁻ ions during the acceleration process [33].

The development of the sector-focused cyclotron required sophisticated machining and fabrication techniques, and was initially available only for scientific research. However, the efficiency and compactness of the design made the cyclotron ideal for the production of medical isotopes for SPECT and PET. Today, with the omnipresence of accelerator design computer codes, the sector-focused cyclotron has become an immensely practical high energy, efficient and relatively low cost machine that has made the applications of high energy particle beams a common commercial commodity used for the production of a large number of medical imaging, diagnostic and therapeutic applications. One example of a commercial cyclotron employing the spiral ridge

Fig. 14.   TRIUMF spiral ridge cyclotron pole magnets shown during construction.

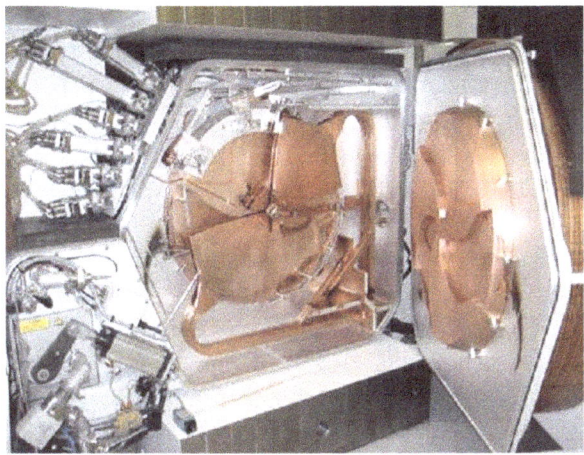

Fig. 15. Four-sector isochronous cyclotron magnet pole tips with spiral edges.

Fig. 16. PSI K600 eight-sector isochronous cyclotron, four accelerating structures located between the spiral ridge magnets that produce an energy gain of 2.9 MeV per revolution. This machine holds the world record for beam intensity extracted from a cyclotron (2 mA protons).

pole design is the General Electric 14 MeV proton "PETrace" isotope production cyclotron [34], shown in Fig. 15. These commercial machines will be discussed in Sec. 3.

2.3.3. *The separated sector ring cyclotron*

The separated sector cyclotron is the logical extension of the sector-focused design where the valleys are eliminated and the cyclotron is constructed from individual magnets spaced in the form of a ring. This design was proposed in the 1960s by Willax [35]. The first two machines of this design were constructed at PSI in Zurich, Switzerland (590 MeV protons) [36], and at Indiana University (IUCF) in Bloomington, Indiana (200 MeV protons) in the 1970s [37]. Each cyclotron was designed for a different particle and energy range, and the machines have different numbers of dipole magnets (eight and four, respectively) and employ different pole shape designs (spiral and radial pole edges), as shown in Figs. 16 and 17, illustrating the flexibility of the ring cyclotron design. The PSI and IUCF cyclotrons, the first of their type, became operational for the first time in 1975.

The separated sector cyclotron has several practical advantages over the sector-focused designs. The rf accelerating dee structures can be removed from the magnet gap and located between the individual magnets, making their design more efficient and powerful. Several high power accelerating dee structures can be located symmetrically between the magnets that accelerate the beam twice per dee passage, or twice the number of dee structures used,

Fig. 17. The Indiana University K220 four-sector cyclotron is shown during construction. Two of the four gaps between the magnets contained accelerator structures that give up to 0.8 MeV energy gain per turn.

resulting in multi-MeV energy gains per revolution. Four accelerating structures located between the spiral edge magnet sectors, as seen in Fig. 15 for the PSI eight-sector cyclotron, produce an energy gain of 2.9 MeV per revolution throughout the acceleration cycle. Beam extraction efficiencies approaching 100% can be achieved. The beam injection and extraction devices can also be removed from the magnet gaps and placed in the spaces between the magnets. With all this hardware removed from the magnets, the magnet gap can be reduced to no more than that required for beam orbit containment. Since the electrical power required to generate the resonant magnetic field increases as the square of the magnet gap, the magnet operating efficiency can be made very high. It also produces a sharper field edge between the high and low field regions of the ring, providing increased axial focusing needed for the higher energy beams. Another example of the versatility of the separated sector design is the IUCF cyclotron. This accelerator was originally constructed as a variable energy (20–370 MeV) and multiparticle (p, d, ^{3,4}He, ^{6,7}Li) research machine. To achieve this wide range of particles and energies, the magnet pole faces were augmented with a set of 21 trim coils that could shape the magnetic field as a function of the radius to meet the relativistic requirements of each particle and energy selected. The field increase from injection to extraction required for 220 MeV protons was 25%, while the field could essentially be constant with the radius for low energy helium and lithium beams. The trim coils required as much power to operate for high energy proton beams as the main magnets themselves. This facility is yet another example of a research cyclotron completing its scientific mission and then being dedicated to medical and industrial applications. The IUCF facility was converted into a proton therapy cancer treatment facility (MPRI) in the years between 2000 and 2006, and the cyclotron was reconfigured to operate as a 205 MeV fixed energy proton source [38]. This cancer treatment facility, which continues to operate today, will be discussed in more detail in Sec. 3.

The separated sector cyclotron is neither compact nor generally suited for practical medical applications because of its size and complexity. These high energy and intensity machines were designed for scientific research, and have higher performance capabilities than required for medical use. Nevertheless, most of the 13 large separated sector ring cyclotrons [24] devote some of their time to the development or application of hadron therapy. Indeed, places such as PSI, IUCF and NAC have helped pioneer some of the more advanced applications of hadron radiation therapy.

2.4. *The superconducting cyclotron*

In the late 1960s and early 1970s, large scale superconducting magnets, based upon NbTi conductors operating at liquid helium temperatures, began to appear in advanced scientific applications, including bubble chamber magnets, fusion experiments and magnetohydrodynamic (MHD) devices. As the set of technical developments that made possible these new superconducting magnetic devices, operating at fields as high as 3–4 T, they became more widely known. Cyclotron designers soon began to consider their possible role in isochronous cyclotrons. One of the vexing problems of isochronous cyclotron design had been that, at the practical limit of magnetic excitation levels of resistive copper windings, typically less than 2 T, the iron in the cyclotron had a wide range of saturation magnetizations that could not be modeled with the tools available, and hence model magnets were needed to verify and optimize the orbit properties of new cyclotron designs. Fraser and colleagues at Chalk River Nuclear Laboratories were the first to realize that for operations above 3 T, there could be a simultaneous improvement in magnetic field accuracy due to the likelihood of full saturation of the iron poles, while significantly reducing size and overall cost for a given energy [39]. Cyclotrons have a unique scaling of the final energy (T), radius (r) and field (B), for a given ion of charge $Q = Ze$ and mass Am_0:

$$T = \left(\frac{Q^2}{A}\right)\frac{e^2}{2m_0}r^2B^2. \tag{14}$$

This equation shows that for a given final energy and ion, increasing the field results in a decrease in the size of the cyclotron, and this scaling holds for all three cyclotron types. A significant argument for the development of the superconducting cyclotron proposed by Blosser and Johnson was to construct a compact hadron therapy cyclotron suitable for hospital use [40]. In the 1980s, superconducting magnets were used to exploit this scaling law to realize compact heavy ion isochronous cyclotrons at fields around 5 T [41]. The first superconducting cyclotron, a K500 heavy ion accelerator (shown schematically in Fig. 18), was successfully fabricated at Michigan State University (MSU). This accelerator is still in service for research today as an injector for a larger K1200 cyclotron built at MSU.

A total of nine superconducting isochronous cyclotrons operating at fields in the range of 3–5 T were built for heavy ion nuclear science from 1981 through 1992. All of these machines are still

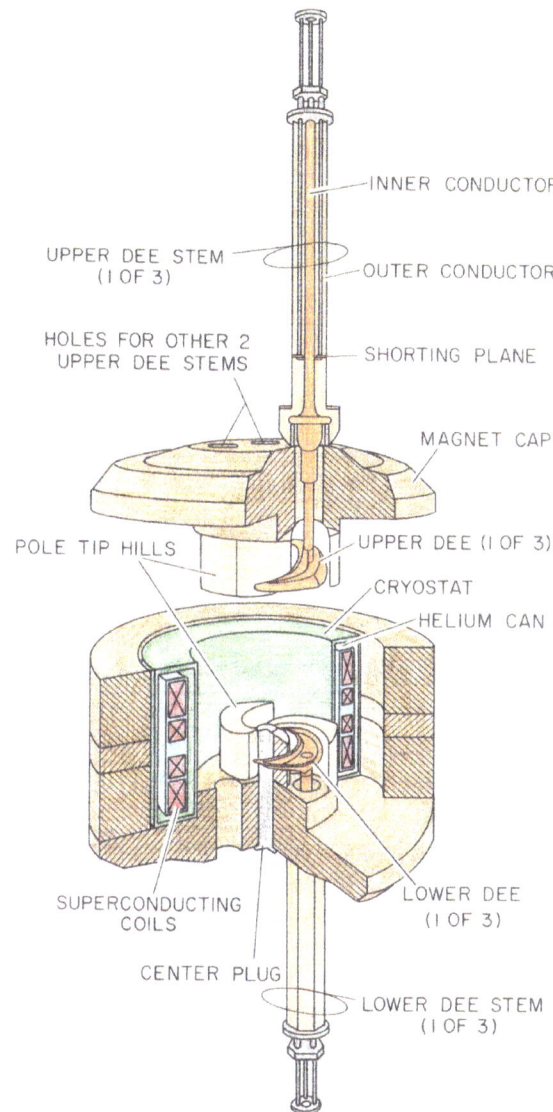

INNER CONDUCTOR

UPPER DEE STEM
(1 OF 3)

OUTER CONDUCTOR

HOLES FOR OTHER 2
UPPER DEE STEMS

SHORTING PLANE

MAGNET CAP

POLE TIP HILLS

UPPER DEE (1 OF 3)

CRYOSTAT

HELIUM CAN

SUPERCONDUCTING
COILS

LOWER DEE
(1 OF 3)

CENTER PLUG

LOWER DEE STEM
(1 OF 3)

Fig. 18. The MSU K500 cyclotron is a three-sector, spiral hill superconducting isochronous cyclotron. It has a peak field of 5.5 T. It is a variable energy heavy ion accelerator, and was the first operating accelerator of any kind to employ a superconducting magnet.

was viewed as limiting, as was the engineering complexity of higher field isochronous machines. One of the major difficulties with the very high field magnets was the strength of the materials required to stably contain the coils and related structure. Successful studies were performed for a superconducting synchrocyclotron [42], and for an 8 T cyclotron magnet [43]. However, through the late 1990s, resistive cyclotron magnet designs remained at 1–3 T field levels and superconducting magnet designs remained around 3–5 T.

While dedicated medical cyclotrons are discussed in detail in Sec. 3, we note one unique superconducting cyclotron designed and fabricated at MSU for clinical medical use there. Studies were initiated at MSU in 1984 for a 50 MeV compact superconducting deuteron cyclotron for neutron beam radiotherapy [44]. This cyclotron, shown in Fig. 19, was mounted on a pair of gantry rings in an available spectrometer pit at MSU National Superconducting Cyclotron Laboratory (NSCL) for testing and had an internal beryllium target that yielded a neutron spectrum peak at 25 MeV with an order of magnitude more neutrons per deuteron

Fig. 19. The 50 MeV Harper Grace deuteron cyclotron for neutron beam radiotherapy. The 4.6 m O.D. gantry rings allowed this cyclotron to rotate +/− 180° around the reclining patient with the liquid-helium-cooled magnet fully energized but the beam stopped during rotation.

operating, and are almost an order of magnitude more compact than equivalent energy conventional resistive magnet cyclotrons. Since cyclotron subsystems and facility overall size scale with the cyclotron size, this also had a dramatic effect on the overall facility costs. Further, since the magnets are superconducting and the rf systems are now more compact, overall power and cooling requirements are again significantly decreased.

In the 1980s, studies of cyclotrons with fields beyond 5 T were made, but the magnet technology than a similar energy proton beam [45]. It was built jointly by NSCL and the MedCyc Corporation, commissioned at NSCL in April 1989, and installed the next year at Harper Grace Hospital in Detroit, Michigan [46]. Neutron radiotherapy never caught on as a mainstream oncology technique, but this machine was in clinical operation until 2008, when the neutron radiotherapy program at Harper Grace was discontinued.

3. Commercial Medical Cyclotrons

The original cyclotron concept, invented by Lawrence in 1931, has been developed over the last eight decades into machines that can provide any ion and energy desired for research or applications given the practical limit of cost. The applications of cyclotron beams in medicine and industry have grown from the first investigations of John Lawrence in the 1930s to the point where commercial cyclotrons are designed and built to specifications to meet a large array of user applications, including industry, national security and medicine.

In general, most of the commercially available isotope production machines are room temperature sector-focused cyclotrons employing either radial (< 30 MeV) or spiral ridge sector magnets. The hadron therapy facilities in operation today use a variety of decommissioned high energy research accelerators (IUCF, for example), a few specially built accelerators and an ever-growing assortment of commercially available treatment facilities. Indeed, manufacturers are now marketing complete isotope production and hadron therapy facilities to hospitals around the world. In this section, an attempt will be made to highlight some of the medical applications and the commercial manufacture of the medical cyclotrons. Given the large and increasing number of these machines operating today around the world, the authors will undoubtedly miss some of these.

3.1. *Medical isotope production*

Phillips introduced the first commercial cyclotron for medical isotope production in 1966 [47]. This 28 MeV proton cyclotron demonstrated that cyclotrons could be a cost-effective source of medical isotopes. Cyclotrons have been the main accelerator source of medical isotopes ever since, with hundreds of units deployed by many companies. The Philips cyclotron produced 15 MeV deuteron and 38 MeV ^{3}He beams. Prior to the commercial isotope production cyclotron, most radioisotopes were produced in a research nuclear reactor.

Cyclotron-produced medical isotopes are used in planar (2D) imaging studies with the gamma camera, and computed tomography (3D) such as single photon emission computed tomography (SPECT), positron emission tomography (PET) and PET/CT. Generally a compound labeled with a radioactive tracer, prepared in a modular chemistry unit from an irradiated target material, is introduced *in vivo*. The tracer element, a gamma ray emitter, travels through the body, and collects in specific parts of the body, depending upon the chemistry of the compound, which can then be imaged for clinical diagnostic purposes or treatment. For instance, iodine collects in the thyroid glands and tumors. In the simplest case, planar imaging, single emitted gamma rays are imaged with a gamma camera, showing a 2D distribution of the tracer radioisotope at active sites in the body. SPECT is similar to planar imaging, in that multiple 2D images are obtained by rotating the gamma camera around the patient and then a 3D image is computer-generated. The primary SPECT isotopes used for medical imaging produced by cyclotrons are ^{99m}Tc for bone, myocardial and brain scans, ^{123}I for tumor scans and ^{111}In for white blood cells.

PET differs from SPECT in that the tracer elements are short-lived positron emitters, i.e. ^{11}C, ^{13}N, ^{15}O and ^{18}F, which are used to study brain physiology and pathology [8]. Positrons readily annihilate with any free electron in the body, yielding a pair of 511 keV photons. The two 511 keV photons are emitted at nearly 180° from each other. Timing can be used to determine the location of the positron annihilation event and thus 3D images can be constructed with computer analysis. The timing also improves the signal-to-noise ratio and fewer events are needed to construct the image. PET/CT and PET/MRI are used to coregister anatomic and metabolic information simultaneously from PET and CT or MRI data. The main PET tracer is FDG, a glucose analog molecule labeled with flourine-18 produced typically in an 18 MeV proton cyclotron.

A representative (i.e. very incomplete) list of the manufacturers of medical isotope production cyclotrons is provided in Table 1, which illustrates the variety of beams and energies available to the medical community for isotope production [48]. One of the earliest commercial cyclotrons was the CS-15 H^+/D^+ machine, marketed by the Cyclotron Corporation in Berkeley, California. Several of the accelerators listed were designed and manufactured by a single firm (TR-30 and CS-30, for example), and were originally designed by the accelerator group at TRIUMF but were marketed and sold under the names used by the companies listed in Table 1. From this list, it is obvious that in the years after 1975, the sale and use of these isotope production machines grew at a very rapid pace.

The majority of medical isotope production cyclotrons accelerate H^- ions. The reason for this is that a simple beam extraction process is used. Some of the H^- ions impinging on a thin internal target, called a "stripper foil," set at an internal radius, have their two electrons removed ("stripped" away) and the resulting H^+ ions follow a reverse curvature orbit (with respect to the H^- ions) and are directed out of the cyclotron. The remaining, unstripped ions continue to accelerate to a larger radius, where they can be stripped at a higher energy. Multiple thin stripper foils can be inserted at several radii within the cyclotron, making it possible to simultaneously extract several H^+ beams of different energies from a single cyclotron. H^- cyclotrons operate at magnetic fields of 1 T or less to avoid Lorentz stripping, i.e. magnetic removal of the extra electron attached to the proton [49].

The IBA Cyclone 30, introduced in 1986 — shown in Fig. 20 — was a big step forward, producing significantly higher H^- currents per unit wall plug power and having two independent stripper foils and isotope production targets [50]. Beam intensities up to 1 mA are now available from these accelerators. Today, isotope production cyclotrons are available commercially at energies from 5–90 MeV from EBCO, GE, Japan Steel Works, IBA and Siemens (formerly CTI). Most of these later generation cyclotrons are variable energy and particle ones, can have multiple extracted beams, and come complete with built-in target stations and computer control systems designed to permit operation by trained technical personnel who are not required to be expert accelerator designers or builders.

3.2. *Ion beam therapy cyclotrons*

The 1946 suggestion of Robert Wilson [12] to use high energy protons to kill deep-seated tumors in the human body became a reality beginning in the late 1950s and has grown into a well-established protocol for curing a host of otherwise untreatable cancers, as well as a preferred method of curing other cancers while reducing radiation damage side effects. This cancer-fighting technique is referred to as hadron therapy or ion beam therapy (IBT) and is most effective in eliminating well-localized cancerous tumors located within the human body, particularly in the head and neck areas. This section, like those above, is not intended to describe the full range of medical applications for IBT, but seeks to describe the required properties of the treatment beams and delivery systems, and list the present commercial suppliers of applicable cyclotron based IBT facilities.

The basic physics that makes hadron therapy so attractive is the manner in which high energy particles (protons, deuterons, pions and heavier ions) lose energy while passing through matter, as described in Sec. 1. The peak in the energy loss profile at the end of an energetic particles range, illustrated in Fig. 1, is called the Bragg peak. The depth of penetration of the ion beam into the body depends precisely upon the particle type and energy used for the IBT treatment, as well as the density of the area to be penetrated. This energy loss property allows the physician to precisely target a tumor located within the human body while sparing radiation damage to the healthy tissue around it. A 160 MeV proton beam will penetrate 15 cm into the human body. Successive beams of lower energies and intensities can be directed at the tumor so as to uniformly irradiate the whole depth of the tumor. Beam apertures and Lucite compensators are used to map the lateral shape and distal (rear) edge of the tumor. These devices are shown in Fig. 21. A contour map of the area radiated by a proton beam using this technique is also shown, with the red area being the highest concentration of radiation delivered to the body. The green circle represents a sensitive organ that should not receive damaging radiation, illustrating the capability of this treatment protocol to spare healthy tissue. These very basic techniques have been improved over the years with precise beam scanning and intensity

Table 1.   Representative list of commercial cyclotrons used for isotope production.

Manufacturer	Model	Particle	Energy (MeV)	Intensity (μA)	First available
Advanced Cyclotron Systems	TR 13	$H-$	13	100	1994
7851 Adlerbridge Way	TR 14/TR19	$H-$	14–19	300	
Richmond, BC,	TR 24	$H-$	15–24	300	
Canada (EBCO)	TR 30	$H-$	15–30	1000	1990
	TR 30/15	$H-$, $D-$	30/15	1000	1994
Ion Beam Applications, s.a.	Cyclone 3	$D-$	3.5	100	
Chemin du Cyclotron 3	Cyclone 10/5	$H-$, $D-$	10/5	100	1988
B-1348 Louvan-la-Neuve	Cyclone 18$^+$	H^+	18	2000	1994
Belgium	Cyclone 18/9	H^-, D^-	18/9	100	1992
	Cyclone 30	H^-	30	500	1986
Scanditronix Wellhofer, AB	MC 16	H^-, D^-	17/8.3	50	1990
Stalgatan 14		^{3}He, ^{4}He	12.4/16.5		
Uppsala, Sweden	MC 32 NI	H^-, D^-	16–32/8.5–16	60	1990
	MC 35	H^-, D^-	7.5–35/3.8–18	75	
		^{3}He, ^{4}He	5.6–47/7.5–16.5		
	MC 35, 40	H^+, D^+	8–40/20	75	1979
	MC 50	H^+, D^+	50/25	50	1989
	MC 60PF	H^+	60	35	1984
General Electric	MINItrace	H^-	9.6	1993	
www.GEhealthcare.com	PETtrace	H^-, D^-	16.5/8.4		1993
Seimens (CTI)	RDS eclipse	H^-	11.0	100	1987
www.Seimens.com/healthcare	RDS 112				
	RDS 111				
The Cyclotron Corporation*	CS-15	H^+, D^+	15/8	60	1967
950 Gilman St.Berkeley, CA,	CS-22	H^+, D^+	22/12	50	1970
USA	CS-28	H^+, D^+	24/14	50	1974
	CS-30	H^+, D^+	26/15	60	1973
	CP-42	H^-	42	200	1980
Sumitomo Heavy Industries,	HM-18	H^-, D^-	18/10	70	1991
Ltd. 2-1-1 Yato-cho	480 AVF	H^+	30	80	1985
Ranashi City, Japan	750 AVF	H^+	70	55	1985
	930 AVF	H^+	90	10	
Japan Steel Works, Ltd.	BC168	H^+, D^+	16/8	70	1982
Muroran, Japan	BC1710	H^+, D^+	17/10	70	1981
	BC2211	H^+, D^+	22/11	70	1989
	BC3015	H^+, D^+	30/15	70	1985

modulation techniques that are used to give a precise three-dimensional conformal map of the tumor with the treatment beam.

The majority of the existing hadron therapy facilities today use protons, but a few are able to use heavier ion beams such as helium, carbon and neon. IBT using pion beams has also been conducted. The ion beam property used to compare the radio-biological effectiveness of hadron treatment is its linear energy transfer (LET) to the tumor. Ionizing radiation destroys the ability of cells to divide and grow by damaging their DNA strands. Activated radicals produced from atomic interactions are the primarily cause of radiation damage by photons, electrons and protons. These types of radiation are called low linear-energy-transfer (low LET) radiation. Neutrons and heavy ions have a high LET compared with protons and photons. All these ions have been used for Hadron Therapy in the past. There is much debate among clinicians about the relative effectiveness of the various ion species in curing cancerous tumor diseases, although a few very general facts are agreed upon by most, namely:

- Most localized cancers can be effectively treated with low LET protons or with photons.

Fig. 20. The widely deployed IBA Cyclone 30 isotope production cyclotron accelerates protons from 18 to 30 MeV, also simultaneously accelerating deuterons (9–15 MeV), and features dual isotope production targets.

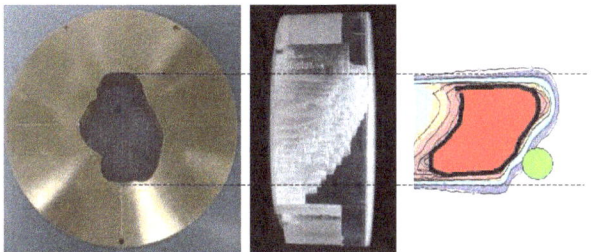

Fig. 21. Brass beam aperture and Lucite compensator used together with the SOBP to confine the maximum ionizing energy of a proton beam to a tumor.

- About 10% of tumors have a high resistance to low LET treatment and they can often be successfully treated with the higher LET heavy ion beams such as carbon.
- Carbon ions have better dose distributions (less lateral scattering and collateral damage to adjacent healthy tissue) than protons.
- More study is required to accurately determine the most effective treatment protocols.

The facility costs are the primary reason that there are thousands of photon cancer treatment facilities today. Commercially available photon radiation treatment facilities, based on low energy electron linear accelerators, are well within the budget of most hospital facilities. There are currently about 30 operating IBT facilities in the world, of which only a few have a heavy ion capability. Yet, regardless of the techniques used, all IBT facilities require

an ion beam source, a particle accelerator such as a cyclotron, which can deliver a precise beam energy and intensity.

3.2.1. *Proton therapy treatment facilities*

A 230 MeV proton beam will penetrate 32 cm into the human body, a depth large enough for most human applications; hence this has become the canonical energy for all proton therapy accelerators. The early trials in the development of IBT by John Lawrence and the scientists at the GWI, Sweden and the Harvard synchrocyclotron facilities were discussed in Subsec. 2.2. This work was done with research cyclotrons or cyclotrons converted for medical applications at the end of their useful physics research life. The first hospital-based proton therapy treatment facility in the United States, The John M. Slater Proton Treatment and Research Center, was installed at the Loma Linda University Medical Center in San Bernardino, California. The facility was the first specifically designed for use as a proton therapy treatment facility. The accelerator, a 370 MeV proton synchrotron, and beam delivery systems, including the 360° rotating gantries, were designed at the Fermi National Accelerator laboratory in Batavia, Illinois by Fred Mills and Frank Cole [51]. While this article is not discussing synchrotron accelerators, this facility is worth mentioning here, as it was a milestone for the acceptance of IBT as a superior treatment protocol for several cancers in the United States. Several proton therapy facilities around the world had been operating for several years, particularly in Europe and Japan, but the primary impediment to the use of proton therapy in the US is cost.

The second proton therapy facility to begin operations in the US was the Northeast Proton Treatment Center (NPTC), a new commercially built IBT facility installed at the Massachusetts General Hospital in Boston [52]. This facility, recently renamed the Francis H. Burr Proton Therapy Center, replaced the pioneering and aging Harvard synchrocyclotron proton therapy facility which had been in operation in conjunction with the Harvard Medical School since 1961. The new treatment facility was designed and built by Ion Beam Applications (IBA) in Belgium [53], the same company that manufactures the Cyclone series of isotope production cyclotrons listed in Table 1. This facility, consisting of a 230 MeV

Fig. 22. An IBA Cyclone 230 superconducting cyclotron manufactured for IBT. Compare the size of this 230 MeV proton cyclotron with that of the 220 MeV Indiana University room temperature proton cyclotron shown in Fig. 17, using the men in the respective pictures.

superconducting cyclotron, a beam delivery system and three treatment rooms, two of which house 360° rotating gantries, is the first commercially built, installed and operating hadron therapy facility in the world. A photo of the cyclotron, an IBA C235, and the layout of this treatment facility are shown in Figs. 22 and 23. The IBA facilities typically consist of 3–4 treatment rooms. Since the commissioning of this facility, IBA has installed a total of 13 additional facilities worldwide, 60% of the total number

of commercial proton therapy facilities (30) installed or under construction worldwide as of 2009.

Yet another company, Varian Medical Systems, Palo Alto, California, is marketing a proton therapy system based on an Accel 250 MeV proton superconducting synchrotron, shown in Fig. 24. This cyclotron was based on the design for a superconducting cyclotron by H. Blosser, the inventor of the superconducting cyclotron. Varian acquired Accel Corporation in 2007, and fabricated the machine now installed at PSI, Switzerland and being used for IBT [54].

3.2.2. *Heavy ion beam facilities*

There are currently four IBT facilities around the world that use heavy ions (carbon) for IBT [55]. Two of them are located in Japan, which has seven operating IBT facilities. There are six more IBT facilities under construction in Japan, and three of them will be heavy ion facilities. An additional three carbon IBT facilities are presently under construction in Europe and one is under construction in China.

All of the existing carbon IBT facilities use a synchrotron accelerator to produce the energy required for a carbon beam to penetrate deep into the human body. Carbon, with 6 protons and 6 neutrons, is approximately 12 times more massive than a proton,

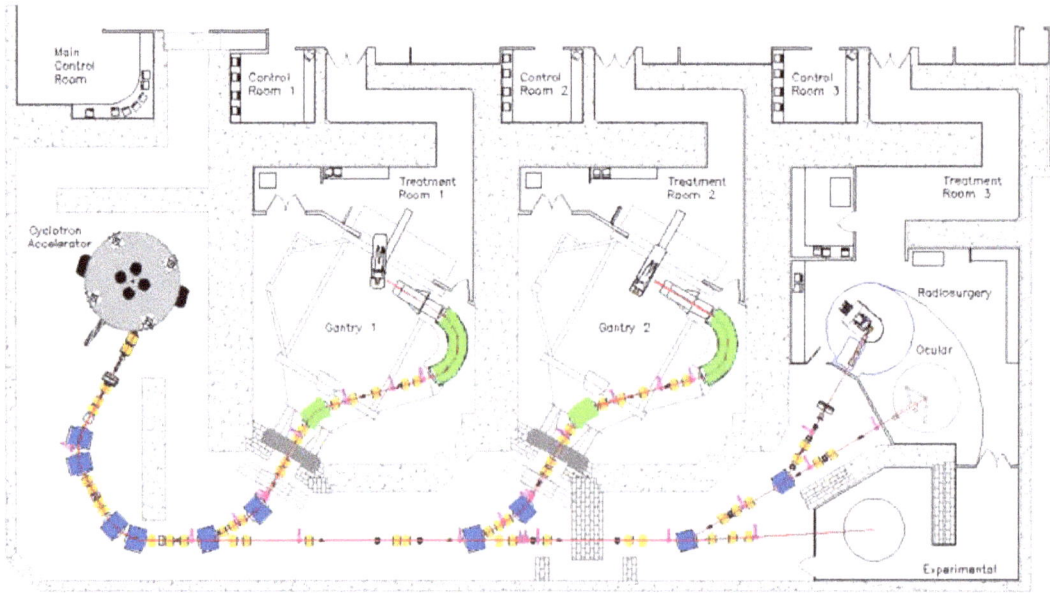

Fig. 23. Facility layout of the proton therapy treatment center (NPTC), showing the relative locations of the Cyclone 30, beam delivery systems and three treatment rooms.

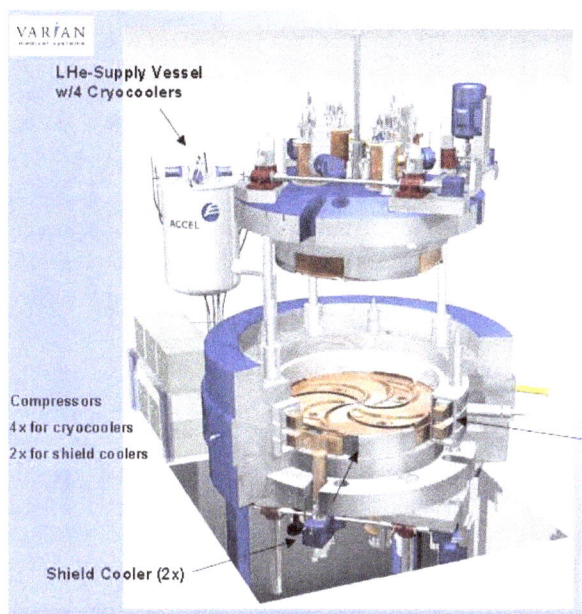

Fig. 24. 3D schematic of the Accel superconducting 250 MeV proton cyclotron, now being marketed by Varian Corporation.

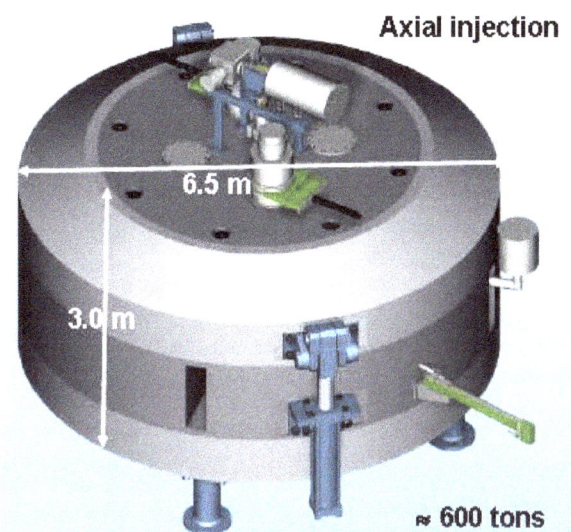

Fig. 25. Schematic of the IBA C400 superconducting heavy ion cyclotron, showing some dimensional characteristics.

and thus requires significantly higher energies to penetrate deep into the human body than protons. For a charge-to-mass ratio (q/a) of $\frac{1}{2}$, a fully stripped 320 MeV/amu carbon beam will penetrate 15 cm into the body. Consequently all of the present carbon facilities currently use high energy synchrotrons to produce the required energy carbon beams for IBT. They are large and complex facilities compared to the commercial proton IBT centers.

One company, IBA of Belgium, has designed a high field (4.5 T) superconducting cyclotron to accelerate ions with $q/a = \frac{1}{2}$ (H_2^+, He, Li, B, C, Ni, O, Ar) to 400 MeV/amu. This cyclotron, called the C400, is based on the IBT C235 design but with higher magnetic fields and a larger diameter (6.4 m vs. 4.7 m). The machine will be able to provide 265 MeV protons as well as 400 MeV/amu heavy ions, making it an all-purpose accelerator for IBT applications within a very small footprint [56] and a serious competitor to the synchrotron as a practical and affordable source of ions for hadron therapy. The first of these accelerators will be installed in Caen, France. A schematic of the C400 design is shown in Fig. 25.

3.2.3. Commercial hadron facilities and vendors

There are only a few vendors of completely designed and operational hadron therapy facilities, and they

Table 2. Commercial hadron therapy system vendors.

Manufacturer	Accelerator type	Maximum energy/particle
Ion Beam Applications, Belgium	SC cyclotron	235 MeV/P
Varian, Inc., Palo Alto, CA, USA	SC cyclotron	250 MeV/P
Still Rivers, Inc., Boston, MA, USA	SC synchrocyclotron	235 MeV/P
Optivus Proton Therapy, Loma Linda, CA, USA	Synchrotron	370 MeV/P
Mitsubishi Heavy Industries, Japan	Synchrotron	370 MeV/P
Hitachi, Japan	Synchrotron	250 MeV/PP
Sumitomo Heavy Industries, Japan	SC cyclotron	235 MeV/P

are listed in Table 2. Three of the vendors are the major heavy industrial corporations in Japan, Mitsubishi, Hitachi and Sumitomo. Of these, Sumitomo is the only company marketing cyclotrons for both PET isotope production (see Table 1) and hadron therapy. Three of the vendors listed market facilities based on superconducting cyclotrons.

It is risky to count the hadron facilities operating or under construction today, since the number changes rapidly. Nevertheless, there are presently

Table 3.　List of current IBA proton treatment facilities.

Facility name	Location, country
Northeast Proton Therapy Center (NPTC)	Massachusetts General Hospital, Boston, MA, USA
Midwest Proton Radiotherapy Inst. (MPRI)	Indiana University, Bloomington, Indiana USA
University of Florida Proton Therapy Center (UFPTCI)	University of Florida, Jacksonville, FL, USA
Procure 1 Proton Therapy Center	Oklahoma City, OK, USA
Procure 2 Proton Therapy Center	Warrenville, IL, USA
Roberts Proton Therapy Center	University of Pennsylvania, Philadelphia, PA, USA
Hampton Univ. Proton Therapy Inst.	Hampton University, VA, USA
The Center for Proton Therapy of Orsay	Curie Institute, Orsay, France
West German Proton Radiotherapy Center (WPE)	Essen, Germany
Wan Jie Proton Therapy Center (WPTC)	Wan Jie Hospital, Zi-Bo, China
Beijing Greatwall International Cancer Center	The Soni-Japanese Friendship Hospital, Beijing, China
The National Cancer Center	Seoul, South Korea
National Cancer Center Hospital East	Kashiwa City, Chiba Prefecture, Japan

30 operating proton therapy centers around the world, with 14 more facilities under construction or planned. Of the 30 existing facilities, 16 listed in Table 3 use cyclotrons as the high energy particle source. Of these, 13 were built and installed by IBA using the IBA C235 superconducting cyclotron. A 14th facility is marketed by Sumitomo Heavy Industries, in collaboration with IBA, and uses the Proteus C235 cyclotron design.

4. Future Accelerators for Medical Applications

4.1. *New high field compact medical cyclotrons*

In 2003–4, Antaya introduced a superconducting synchrocyclotron design with 9 T fields for proton beam radiotherapy [57]. The purpose of this effort was to use the field scaling of Eq. (12) to produce a compact cyclotron that would enable the development of a low cost single room proton beam

radiotherapy treatment (PBRT) system. To be feasible, the compact cyclotron would have to be gantry-mounted and have a final proton energy of at least 250 MeV. A synchrocyclotron type accelerator was chosen because: (a) the prior design study at 5 T [43] could be used as a guide, (b) it was possible to demonstrate quickly how the beam dynamics of synchrocyclotron would scale at a high field, (c) the intensity requirements for PBRT, namely a proton beam intensity of 20–40 enA, are easily achievable in a synchrocyclotron, (d) the anticipated simplicity and inherent robustness of a weak focusing cyclotron would be ideally suited to the requirements of a nonspecialist-operated clinical PBRT system, and (e) this field level, while high for accelerators, was far from the limits of the technology for high field superconducting magnets. In order to complete this 9 T synchrocyclotron design, it was necessary to develop a quantitative computational beam dynamics model for weak focusing phase-stable cyclotrons [58]. In addition, it was necessary to establish feasible magnet and rf engineering solutions for such high field cyclotrons [59]. Finally, advances in cryogenic engineering and components allowed the engineering of an orientation-independent superconducting cyclotron magnet that is cryogen-free [60]. Since the radius decreases in proportion to the field increase, this 9 T synchrocyclotron is more than 50 times less massive than a conventional machine of the same final proton energy, as shown in Fig. 26, and later in Fig. 28.

Many of the beam dynamics and technical challenges resolved for the introduction of the high field synchrocyclotron are shared in the design and engineering of all cyclotron types. Weak focusing cyclotrons favor low energy gain per turn (ΔT_1), and a 250 MeV proton machine requires about 15,000 turns ($\Delta T_1 \sim 15$ keV/turn), to reach the full energy. Since the turn spacing varies inversely with energy and $f(\gamma) \sim 1$, the turn spacing at full energy is of the order of 10 microns in this 9 T synchrocyclotron. High extraction efficieny is desirable for such a patient-in-the-room accelerator, and hence it was necessary to develop a new beam extraction technique, which required an exceptionally precise beam dynamics simulation, using highly accurate computed fields [62]:

$$\frac{dr}{dN} = r\frac{\Delta T_1}{T}f(\gamma). \tag{15}$$

Fig. 26. The 9 T superconducting synchrocyclotron for proton beam radiotherapy. This compact high field cyclotron enables single room treatment at a fraction of the cost of conventional facilities, and widespread adoption is expected.

Figure 27 shows a nonlinear growth of a radial oscillation that results in an exponential growth in turn separation from around 6 microns to 1 cm in 20 turns, followed by passive magnetic-element-guided ion extraction at 250 MeV.

4.2. *Compact IBT facilities*

While the number of proton therapy treatment centers is growing at a reasonable pace, they are generally placed in high population density areas because of their size and cost, approximately US$100 million for a three-treatment-room facility, with gantries. To make this treatment available to a larger fraction of the population, manufacturers are designing more modest and innovative one- or two-room facilities that would be more affordable (US$40 million), and hence be practical for installation in smaller population centers. The development of high field superconducting cyclotrons or synchrocyclotrons is making this option a reality. Still River Systems [63] is commissioning a one-room PBRT system, called the Monarch250TM, based on a gantry-mounted 9 T synchrocyclotron (shown in Fig. 28 and discussed in Subsec. 4.1). This 250 MeV synchrocyclotron is much smaller than the IBA C235 and Accel 250 MeV 3 T superconducting cyclotrons shown in Figs. 22 and 24, and is an excellent visual demonstration of the dimensional scaling laws with magnetic field for cyclotron magnets. This cyclotron is currently being commissioned.

The Still River Systems single-room PBRT system's conceptual design is shown in Fig. 29. The Monarch synchrocyclotron is seen mounted on a 180° rotating gantry within the treatment room, similar to what was done with the 50 MeV Harper Grace deuteron cyclotron in 1980. Barnes Jewish Hospital in St. Louis, Missouri and M. D. Anderson

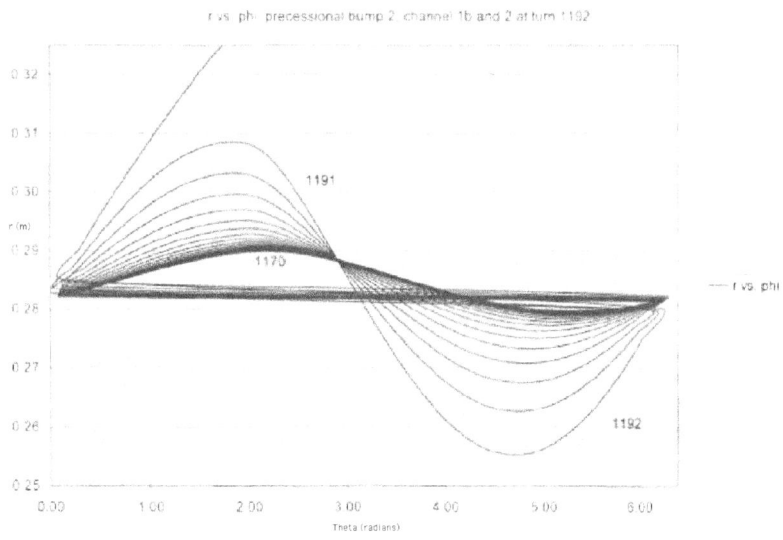

Fig. 27. An unfolded set of orbits near 250 MeV in a 9 T synchrocyclotron. The radial orbit separation moves exponentially from a few microns to millimeters in about 20 turns, and then is directed out of the edge of the cyclotron field, as a result of a set of small field perturbations.

Fig. 28. The Still River Systems Monarch$^{250\text{TM}}$ superconducting synchrocyclotron. The small size of this 250 MeV Proton machine is evident from the size of the man working on it.

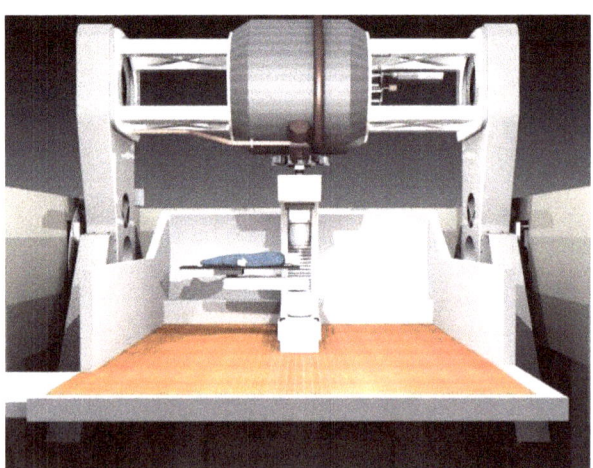

Fig. 29. The conceptual design for the Still Rivers Systems single-treatment-room PBRT system, showing the Monarch$^{250\text{TM}}$ mounted on a gantry within the treatment room.

Cancer Center in Orlando, Florida have made partial commitments to purchase this system, which is advertised to cost US\$20 million for the treatment equipment [64]. One of the advantages of the single-room concept is that the equipment could be

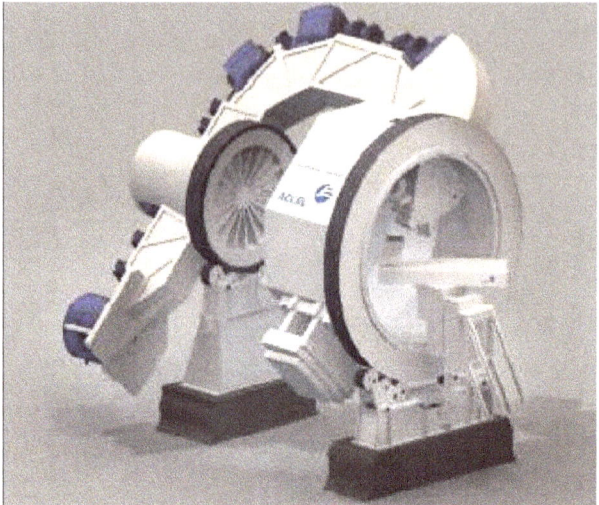

Fig. 30. The Accel Corp. conceptual design of a single-room PBRT system with a gantry-mounted cyclotron and beam delivery system.

installed in an existing space within a hospital or traditional photon radiation facility.

Accel has also made a tentative design, shown in Fig. 30, for a single-room PBRT system based on a high field cyclotron mounted on a gantry, somewhat like the Still Rivers Systems design [63]. Other commercial manufacturers, such as Procure and IBA, are also studying ways to reduce the high cost of hadron facilities.

With the recent advances in cyclotron design, the cost of the cyclotron ($\sim$ US\$10 million) is not the major expense for a hadron therapy facility. Larger treatment centers with 3–5 treatment rooms, most with gantries, and the infrastructure to house the facility and shield public areas from stray radiation are the major cost drivers. Thus, a small single-room facility has the potential to significantly reduce the overall cost of a hadron therapy treatment system.

5. Conclusions

The invention and development of the cyclotron over the last 80 years has played a major role in our ability to understand the nature of matter through scientific research and to control disease through medical applications. It is remarkable how this invention has grown from the 27-inch 3.7 MeV proton Berkeley classical cyclotron in 1932, to the massive variable energy, variable particle research machines like the 1 GeV proton synchrocyclotron at Gatchina in the 1970s and '80s, and, today, into very powerful commercially available 250 MeV compact proton superconducting applications cyclotrons. The Still River Systems/MIT 250 MeV superconducting proton synchrocyclotron is smaller than the 27-inch Berkeley cyclotron used for the initial isotope production and fledgling medical studies performed by John Lawrence in 1938. The development of the cyclotron into ever more powerful and compact machines is continuing at a rapid pace to provide smaller, more powerful, more flexible and less expensive isotope production and hadron therapy facilities for medical applications. Today, there are literally thousands of cyclotrons operating daily around the world for medical applications. More cyclotrons are developed, built and sold commercially than for scientific research, which has moved on to much higher energy synchrotron accelerators like the LHC for its studies. Who would have guessed in 1932 that the initial developments and studies of the Lawrence brothers would have such a widespread impact on today's society?

References

[1] http://lhc.web.cern.ch/lhc

[2] A. W. Chao and W. Chou (eds.), *Reviews of Accelerator Science and Technology*, Vol. 1 (World Scientific, New Jersey, 2008).

[3] E. O. Lawrence and N. E. Edlefsen, *Science* **72**, 376 (1930).

[4] E. O. Lawrence and M. S. Livingston, *Phys. Rev.* **37**, 1707 (1931).

[5] M. S. Livingston, Ph.D. thesis, University of California, Berkeley (1931).

[6] W. T. Chu, paper LBNL-59884, University of California, Berkeley (2005). http://en.wikipedia.org/wiki/John_H._Lawrence

[7] J. L. Heilbron and R. W. Seidel, *Lawrence and His Laboratory: A History of the Lawrence Berkeley Laboratory*, Vol. 1 (University of California Press, Berkeley, Los Angeles, Oxford, 1989). http://www.aip.org/history/lawrence/radlab.htm.

[8] *J. Nucl. Med.* **11**(6), 292 (1970). http://jnm.snmjournals.org/cgi/reprint/11/6/292.

[9] F. N. D. Kurie, *J. Appl. Phys.* **9**, 691 (1938).

[10] C. A. Tobias, H. B. Jones, J. H. Lawrence and J. G. Hamilton, *J. Clin. Inves.* **28**, 1375 (1949).

[11] NIU Institute for Neutron Therapy at Fermi Lab. http://www.neutrontherapy.niu.edu/neutrontherapy/therapy/index.shtml

[12] R. R. Wilson, *Radiology* **47**, 487 (1946).

[13] Y. Jongen, *Workshop on Hadron Beam Therapy of Cancer* (Erice, Italy, 2009). http://erice2009.na.infn.it/index.htm

[14] J. Kahn and J. B. Kahn, Science Articles Archive, LBL (1996).

[15] M. K. Craddock and K. R. Symon, in *Reviews of Accelerator Science and Technology*, Vol. 1 (World Scientific, 2008), p. 65.

[16] M. S. Livingston and J. P. Blewett, *Particle Accelerators* (McGraw-Hill, New York, 1962).

[17] Van de Graaff generator, US Patent #1,991,236 (1935).

[18] http://www.aip.org/history/lawrence/epa.htm

[19] D. W. Kerst and R. Serber, *Phys. Rev.* **60**, 53 (1941).

[20] M. K. Craddock and K. R. Symon, *Reviews of Accelerator Science and Technology*, **1**, 71 (2008).

[21] V. I. Veksler, *J. Phys. USSR* **9**, 153 (1945).

[22] E. M. McMillan, *Phys. Rev.* **68**, 143 (1945).

[23] J. R. Richardson, K. R. MacKenzie, E. K. Lofgren and B. T. Wright, *Phys. Rev.* **68**, 669 (1946).

[24] M. K. Craddock and K. R. Symon, in *Reviews of Accelerator Science and Technology*, Vol. 1 (World Scientific, 2008), p. 74.

[25] C. A. Tobias, J. H. Lawrence, J. L. Born *et al.*, *Cancer Res.* **18**, 121 (1958).

[26] R. Wilson, *A Brief History of the Harvard University Cyclotrons* (Harvard University Press, 2004). http://www.physics.harvard.edu/~wilson/cyclotron/history.html.

[27] L. H. Thomas, *Phys. Rev.* **54**, 580 (1938).

[28] Advanced Cyclotron Systems, Inc., 7851 Alderbridge Way, Richmond, B.C. V6X 2A4. info@advancedcyclotron.com.

[29] K. Boyer, *Proc. Sector-Focused Cyclotrons, NAS Nucl. Sci. Ser.* **26**, 656 (1959).

[30] E. L. Kelly, R. V. Pyle, R. L. Livingston, J. R. Richardson and B. T. Wright, *Rev. Sci. Instrum.* **24**, 492 (1956).

[31] F. A. Heyn and K. K. Tat, Design and performance of a 12-MeV isochronous cyclotron, in *Proc. Sector-Focused Cyclotrons, NAS Nucl. Sci. Ser.* **26, 656**, 29 (1959).

[32] K. R. Symon, D. W. Kerst, L. W. Jones, L. J. Laslett and K. M. Terwilliger, *Phys. Rev.* **103**, 1837 (1956).

[33] J. R. Richardson *et al.*, *Proc. PAC '75, IEEE Trans.* **NS-22**, 1402 (1975).

[34] http://www.gehealthcare.com/usen/about/ge_factsheet.html.

[35] H. Willax, *Proc. Int. Conf. Sector-Focused Cyclotrons & Meson Factories* (1963), CERN63-19, pp. 386–397.

[36] W. Joho, *Proc. PAC '75, IEEE Trans.* **NS-22**, 1397 (1975).

[37] R. E. Pollock, *Proc. PAC '77, IEEE Trans.* **NS-24**, 1505 (1977).

[38] D. Friesel, *AIP Conf. Proc.* **600**, 27 (2001).

[39] C. B. Bingham, J. S. Fraser and H. R. Schneider, Chalk River Nuclear Laboratories Report AECL-4654 (1973).

[40] H. Blosser and D. Johnson, *Nucl. Instrum. Methods* **121**, 301 (1974).

[41] P. S. Miller, Status report on the MSU K500 cyclotron, in *Proc. 9th Int. Conf. Cyclotrons and Their Applications* (Les Editions de Physique, Paris, 1982), p. 191.

[42] M. M. Gordon and X. Y. Wu, Extraction studies for a 250 MeV superconducting synchrocyclotron, in *PAC '87* (Washington, D.C., 1987), p. 1255.

[43] J. Kim, H. Blosser, S. Hickson, L. Lee, F. Marti, J. Schubert, G. Stork and A. Zeller, *IEEE Trans. Appl. Supercond.* **3**, 266 (1993).

[44] H. G. Blosser, *IEEE* **84CH1996-3**, 431 (1984).

[45] M. Yudelev, J. Burmeister, E. Blosser, R. L. Maughan and C. Kota, Hospital-based superconducting cyclotron for neutron therapy: Medical physics perspective, in *Proc. 16th Int. Conf. Cyclotrons and Their Applications* (American Institute of Physics, 2001), p. 40.

[46] G. Coutrakon, D. Miller, B. J. Kross, D. F. Anderson, P. Deluca and J. Siebers, *Med. Phys.* **18**, 817 (1991).

[47] A. A. van Kranenburg, D. Weirts and H. L. Hagedoorn, *1966 Int. Conf. Cyclotrons, IEEE* **NS-13**, 448 (1966).

[48] B. F. Milton, *Int. Phys. Conf.* TRI-PP-95-57 (Beijing, 1995).

[49] M. A. Furman, Lorentz stripping of H-ions, LBNL-42722/CBP Note-287 (1999).

[50] Y. Jongen and G. Ryckewaert, *IEEE* **NS-32**, 2703 (1985).

[51] F. T. Cole, P. V. Livdahl, F. E. Mills and L. C. Teng, *PAC '89, IEEE Trans.* **89CH2669-0**, 737 (1989).

[52] Y. Jongen *et al.*, *PAC '97* (Vancouver, B.C., 1998), p. 3816.

[53] Ion Beam Applications S.A., Chemin du Cyclotron 3, B-1348 Louvan-la-Neuve, Belgium. http://www.iba-worldwide.com.

[54] A. G. Geisler *et al.*, *Proc. EPAC '08*, (2008), p. 9.

[55] K. Noda, *Workshop on Hadron Beam Therapy of Cancer* (Erice, Sicily, Italy, 2009). http://erice2009.na.infn.it.

[56] Y. Jongen *et al.*, *Proc. EPAC '08* (2008), p. 1806.

[57] T. Antaya, High field superconducting synchrocyclotron, US Patent App. Publication US2007/017101 (2007).

[58] T. Antaya and J. Feng, VPAC variable frequency circular particle accelerator code, MIT PSFC/TED Report-061102 (2006).

[59] T. A. Antaya, A. L. Radovinsky, J. H. Schultz, P. H. Titus, B. A. Smith and L. Bromberg, Magnet structure for particle acceleration, Int. Patent App. Publication WO2007084701 (2007).

[60] A. L. Radovinsky, A. Zhukovsky and V. Fishman, Cryogenic vacuum break thermal coupler, US Patent App. (1996).

[61] M. Buntaine, Still River Systems Inc., private communication.

[62] T. A. Antaya and J. Feng, Baseline extraction deflection proof of principle demonstration with a small compensated processional bump. MIT PSFC/TED Report-080523 (2007).

[63] L. Calebretta, *Workshop on Hadron Beam Therapy of Cancer* (Erice, Sicily, Italy, 2009). http://erice2009.na.infn.it.

[64] http://www.barnesjewish.org/cancer/default.asp?NavID=3339

Dennis Friesel received his Ph.D. in Nuclear Physics at Notre Dame University in 1970. During his 35-year tenure as an accelerator physicist at Indiana University (IU), he was principally involved in the construction of the IUCF k220 separated sector cyclotron (1975), the k500 "Cooler" synchrotron storage ring using electron cooling and the k240 Cooler Injector Synchrotron (CIS). He became the project manager (2000) for the conversion of the IUCF cyclotron and research facility into the Midwest Proton Radiotherapy Institute for the treatment of cancer. Dr. Friesel retired from IU in 2006 and continues to do research and development work on particle physics related projects at PartTec Ltd., a small research company in Bloomington, Indiana.

Tim Antaya received his Ph.D. in Accelerator Physics from Michigan State University in 1984. His dissertation involved the beam formation processes at the center of the K500 cyclotron, the first superconducting cyclotron and the first accelerator of any type to employ a superconducting magnet. At MSU he contributed to the development of all the MSU sc cyclotrons and led the ECR Ion source Program. After 7 years at Babcock and Wilcox, he joined MIT in 2002, where he introduced a new class of compact high field sc cyclotrons. The first of these, a 9T synchrocyclotron, is now being commercialized for single treatment room Proton Beam Radiotherapy.

Reviews of Accelerator Science and Technology
Vol. 2 (2009) 157–178
© World Scientific Publishing Company

Synchrotrons for Hadrontherapy

Marco G. Pullia

CNAO Foundation, Strada Privata Campeggi,
27100, Pavia, Italy
marco.pullia@cern.ch

Since 1990, when the world's first hospital-based proton therapy center opened in Loma Linda, California, interest in dedicated proton and carbon ion therapy facilities has been growing steadily. Today, many proton therapy centers are in operation, but the number of centers offering carbon ion therapy is still very low. This difference reflects the fact that protons are well accepted by the medical community, whereas radiotherapy with carbon ions is still experimental. Furthermore, accelerators for carbon ions are larger, more complicated and more expensive than those for protons only. This article describes the accelerator performance required for hadrontherapy and how this is realized, with particular emphasis on carbon ion synchrotrons.

Keywords: Medical accelerator; synchrotron; hadrontherapy; particle therapy.

1. Introduction

The first proposal to use hadrons for radiotherapy dates back to 1946 [1]. The term "hadron" means "heavy particle," and hadrons are one of the two main groups into which particles are classified, the other group being leptons. Hadrons are made up of quarks and there are a large number of particles that fall under this classification. This article deals only with the hadrons used in radiotherapy. Many different particles have been used in this field but mainly neutrons, pions, protons and the nuclei of a few ions. Neutrons and pions are nowadays somehow being abandoned and this article will deal only with protons and ions up to oxygen.

Progress during the first four decades was confined to spinoff projects in high energy physics laboratories, but after the completion of the world's first dedicated, hospital-based, proton therapy facility at Loma Linda, California in 1990, interest and activity in radiotherapy grew rapidly.

About one third of the population have to deal with cancer in their lives. Approximately 42% of the tumors are metastatic or diffuse, while the remaining 58% are localized. Of the latter tumors, 40% are treated today by surgery, by radiotherapy (RT) or by a combination of the two, but for the remaining 18%, local control fails.

This report starts by describing how ionizing particles can be used to kill cancer cells and what functional features are needed in particle accelerators to deliver the appropriate beams.

These requirements are then converted into technical specifications and design solutions.

2. Radiotherapy and Hadrontherapy Generalities

As radiation traverses matter, it deposits energy along its path. The energy deposited per unit of mass in a body is called the "physical dose," and is expressed in gray ($1\,\mathrm{Gy} = 1\,\mathrm{J/kg}$).

The energy deposited can harm and eventually kill the cells. With conventional photon radiotherapy, the mechanism is generally *not* a direct interaction between the radiation and a critical structure of the cell (DNA), but rather a "poisoning" of the cell by creation of free radicals (mainly hydroxyl radical, OH^-, and superoxide, O_2^-). Of course, direct interaction is also possible, but this type of damage caused by photons is generally a single strand break which can be repaired relatively easily by the cells.

The probability of killing a cell increases with the dose, which suggests that as big a dose as possible should be given to the tumor. At the same time, irradiating the healthy tissue damages it as well and might even induce new cancers. It is therefore necessary to give large doses to the ill tissue while keeping the dose to the healthy parts around it as small as possible.

To obtain conformity of the dose to the tumor volume, two strategies are possible: (1) radiotherapy can be delivered from *inside* (or close to) the tumor with a rapid falloff of the dose around the tumor itself (brachytherapy) or (2) radiotherapy can be delivered from outside using a scheme that gives large doses to only the tumor (external radiotherapy). In both cases, safety margins around the lesion are always adopted when one is defining the target volume to be irradiated.

Brachytherapy is performed by surgical implantation of radioactive "seeds," but it is not the subject of this article and will not be considered in the following.

To understand external radiotherapy, it is necessary to consider the way in which energy is deposited in matter. In Fig. 1 a few examples of dose profiles against penetration depth are given (original from U. Weber, taken from Ref. 2). In this figure, the radiation is entering from the left hand side, the skin is at depth 0 cm and a tumor extends from 12 cm and 17 cm depth.

In comparison with photons, protons and carbon ions exhibit a much more favorable energy distribution with a low dose along the first part of the path, up to 12 cm, a large peak at the level of the tumor and almost no dose after the tumor. This energy deposition curve, typical of hadrons, is known as the *Bragg peak*.

In practice, Fig. 1 is a little too optimistic, inasmuch as for the treatment of a thick tumor one has to

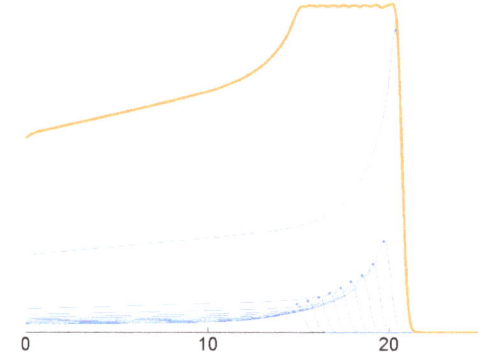

Fig. 2. To treat a real tumor with a finite thickness, a "spread-out Bragg peak" is needed, which is obtained as a sum of single peaks of different energy and intensity.

overlap many Bragg peaks, as shown in Fig. 2 [3], to obtain a *spread-out Bragg peak* (SOBP). Even with the additional dose in the entry channel, the SOBP still provides a far better dose profile than the photon curve.

Indeed, it is difficult to see from Fig. 1 how any reasonable dose conformity can be obtained with photons. The normal strategy is to irradiate the tumor from several directions (generally called "fields") so as to sum the dose inside the tumor and distribute the energy delivered outside the tumor over a large amount of healthy tissue to keep the dose low.

State-of-the-art photon machines and treatment plans (intensity-modulated radiation therapy, IMRT) typically use up to nine different fields. Of course the same strategy can be applied with hadrons, but the same coverage of the target can be achieved with a smaller number of fields (one or two typically).

Figure 3 shows a comparison of an IMRT treatment plan with nine fields and a single field proton therapy plan (figure taken from Ref. 4). The kidneys

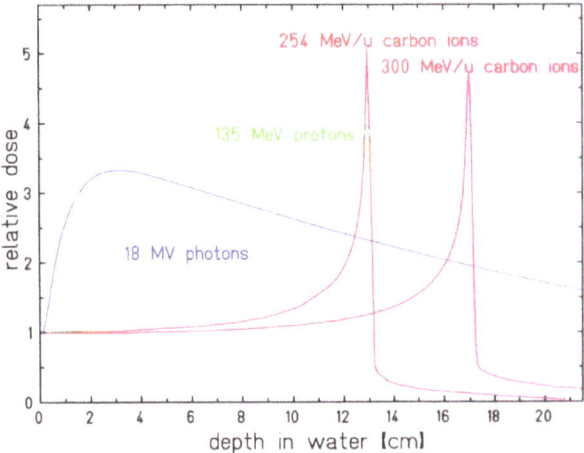

Fig. 1. Energy deposition in depth for photon and particle beams. In the case of photons, the dose decreases exponentially after a maximum a few cm below the skin.

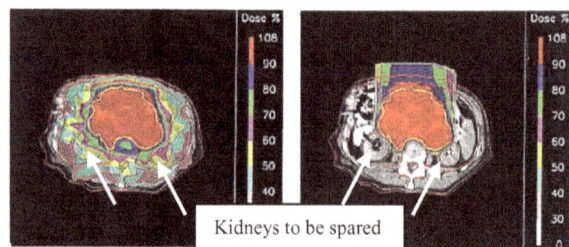

Fig. 3. The proton treatment plan, shown on the right, is definitely better than the IMRT treatment that uses nine x-ray beams, plotted on the left.

and the spinal cord receive significantly lower doses in the proton plan than they do in the photon plan.

So far, only the better conformation of the dose to the tumor has been considered. This improvement is true both for protons and for carbon ions. There are small differences, such as a different lateral scattering (improving with Z) or a different length of the trailing tail due to fragmentation (getting worse with Z), but the general advantage of a Bragg peak versus an exponentially decreasing dose is similar.

Within the hadron family, carbon ions have the additional advantage over protons of a higher *linear energy transfer* (LET). This means that in the same path length they deposit more energy. Qualitatively the energy deposited by carbon ions is more efficient, in terms of cell destruction, than the energy deposited by protons.

The reason for the difference is to be found in the density of ionization, as can be seen in Fig. 4, which shows the ionization at the passage of a particle on the scale of the DNA [5].

The higher density of ionization around the carbon track makes the direct interaction more effective, causing double and multiple strand breaks, which are more difficult to repair.

Compensating for the lower energy deposited per particle by using a larger number of particles, which is what happens in practice, does not yield exactly the same result.

Physically, the difference is that a large number of low ionizing particles causes a more or less homogeneous dose distribution on the size of a cell nucleus (micrometers), while delivering the same dose with high ionizing particles would correspond to a small number of very dense spots inside the same area.

This situation introduces the concept of a "radiobiological dose" and its relationship to the underlying "physical dose."

The higher efficiency in killing cells is expressed by the *relative biological effectiveness* (RBE), which is the ratio between the photon and the ion doses which are necessary for producing the same biological effect.

Depending on the ion LET, which changes along the path as shown by the Bragg curve, the RBE for carbon can be higher than 3 in the Bragg peak region and just slightly larger than 1 in the entry channel.

The variable RBE of carbon ions improves their efficiency, but requires an additional complication

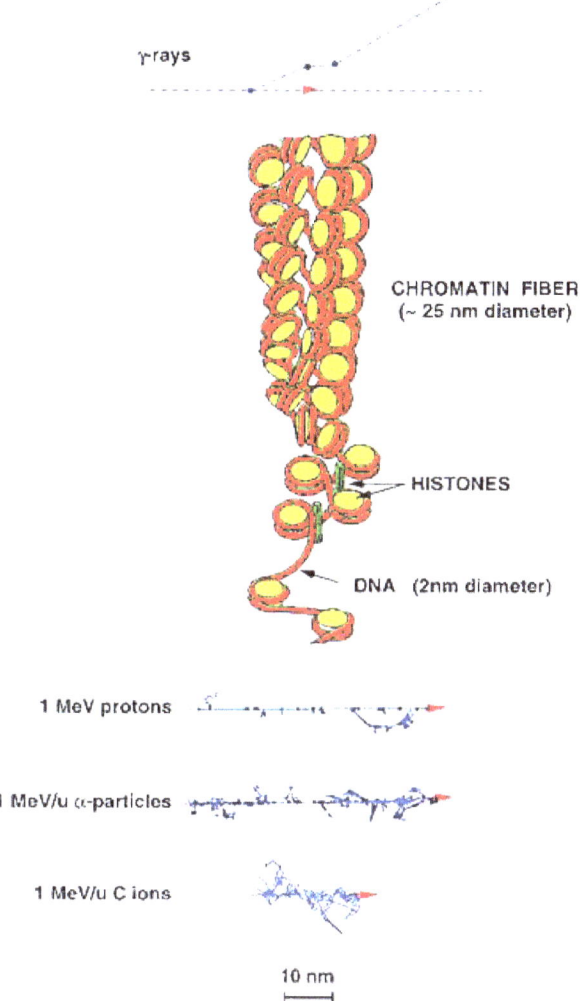

Fig. 4.   Ionization density on the DNA scale.

when one is planning the treatment since the physical dose is no more sufficient and the "biological dose" has to be considered.

Figure 5 shows the physical dose distribution necessary for obtaining a flat biological dose [6].

Another aspect has to be considered in radiotherapy, i.e. fractionation. Healthy cells repair themselves more efficiently than the cancerous cells. For this reason the dose is generally delivered in fractions administered on different days when the cancerous cells are progressively depleted while the majority of healthy cells survive.

This differential mode of cell destruction has given rise to the name "radiotherapy," rather than "radioablation." Typical fractionation schemes with photons and protons foresee 30 sessions of 2 Gy each, while with carbon the same dose is delivered in only

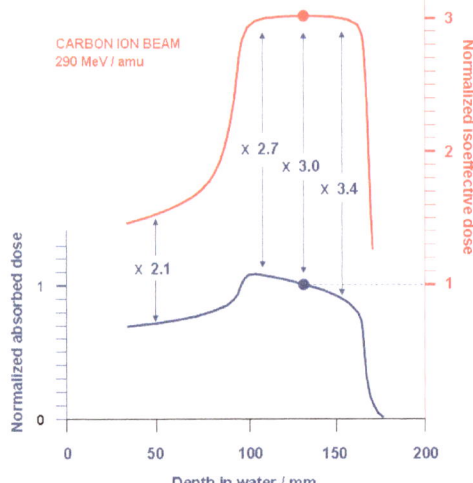

Fig. 5.   To obtain a flat biological dose, a nonuniform physical dose is needed to account for a variable RBE.

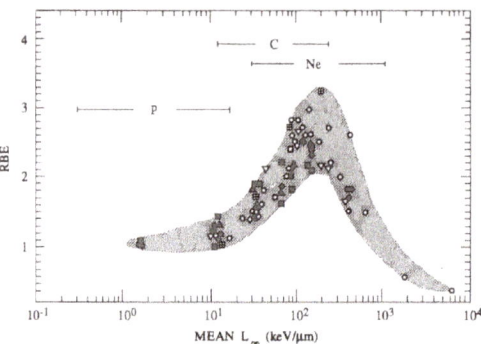

Fig. 6.   RBE dependence on LET. Different symbols correspond to different cell types.

15 fractions and hypofractionation schemes are under study.

Things are even more complicated when one considers that the effects do not depend only on the particle and its energy, but also on the dose, cell type, blood perfusion and many other parameters, not all well known.

One important result from the interaction of all these parameters is that the enhanced RBE of the heavier ions makes it possible to treat some of the radioresistant tumors, such as hypoxic ones, which are difficult to treat with photons or protons because the lack of oxygen inhibits the creation of free radicals.

In summary, it has been shown that hadrons can provide a much better dose conformity and that the heavier ions can have also a radiobiological advantage with respect to conventional photon radiotherapy.

Only protons and carbon ions have been cited so far, but in principle there are many other possible ions.

Therapy with protons is more established and less controversial than heavy ion therapy, because the radiation quality (LET) of protons is very similar to that of x-rays and because the 50 years of clinical x-ray experience in tumor control and toxicity can be carried over to protons. In spite of all the cell studies with carbon ions, the *in vivo* results are not as well established as they are for protons and x-rays. The end result is that we expect carbon ions to be better

for radiation therapy, but this is still considered to be experimental. This important clinical research can only be achieved with new carbon therapy centers.

Other ion species can be considered, but using ions heavier than carbon does not seem to be advantageous and even the higher LET does not correspond to a higher RBE, as shown in Fig. 6 [5].

For nuclei heavier than carbon, the RBE in the Bragg peak begins to decrease due to an "overkill" effect, while the RBE in the entrance region continues to grow. Carbon, therefore, gives the maximum ratio of RBE in the Bragg peak to RBE in the entrance region for light and heavy ions. In other words, carbon causes repairable damage in the input channel and irreparable damage at the Bragg peak. Lighter ions lack the punch at the peak, while heavier ions are lethal also in the entry channel.

Furthermore, since the same dose would be given with fewer particles, statistical problems in the dose homogeneity can arise. In the extreme, if the whole dose could be given with just one particle, it would be impossible to treat the whole tumor.

Up to now, carbon has been considered to be the optimal particle, in this respect.

3. Beam Delivery

As mentioned in the previous section, the dose has to be "shaped" with a contour closely surrounding the tumor. In conventional radiotherapy, this is achieved by "illuminating" the tumor from many directions. To avoid an unnecessary dose to the healthy tissue, the photon beam is "cut" transversally with a collimator that is a shielding block having the shape of the tumor projection in the irradiation direction.

Multileaf collimators exist that can be programmed to obtain either statically or dynamically the wanted shape without having to machine a dedicated collimator per field.

Exploiting the finite range and the low lateral scattering, with hadrons, further optimization is possible and necessary. While with photons the field is designed only according to the transverse shape of the tumor, with hadrons the field can be matched also to the longitudinal form by selecting the appropriate energy of the particles.

There are basically two ways to obtain the dose distribution: *passive* and *active* beam delivery (or beam spreading).

3.1. *Passive dose delivery*

Passive systems (see Fig. 7, are in some ways similar to conventional radiotherapy systems.

The beam is first enlarged by scattering; then a variable thickness device degrades the energy to

Fig. 8.   Range shifter wheel [7].

match the tumor depth. This is followed by a ridge filter, used to increase the energy spread to create an SOBP. The beam is then collimated to select the central uniform region. At this point a *bolus* (or compensator) is inserted. A bolus is a patient-specific device with a shape that conforms the distal Bragg peak surface to the distal contour of the tumor (the posterior surface of the tumor in the direction of the beam). It compensates for the shape of the distal surface of the target as well as any density variations in front of the volume. Finally, a multileaf collimator similar to the one used in conventional radiotherapy is employed to obtain the transverse shape.

Alternatives exist for creating the SOBP. To obtain a flat (physical) dose profile, it is necessary to overlap Bragg peaks at different depths and of different heights, as already shown in Fig. 2. One common way is to use a rotating wheel range modulator. The concept can be easily understood by looking at Fig. 8. As the wheel rotates, it changes the depth of the Bragg peak; the relative angular occupancy of the different thicknesses defines the fraction of particles given at the various longitudinal positions.

3.2. *Active dose delivery*

Active systems are in some ways similar to a cathode ray tube. Since protons and carbon ions are charged particles, a pencil beam can be deflected by a couple of magnets and can be used to "paint" an image. The principle, illustrated in Fig. 9, has been developed at PSI (first scan in 1989 [8]), LBL [9] and GSI [10].

The target volume is subdivided longitudinally into isoenergetic "slices," which are the equivalent to the different thicknesses in the range modulator for the passive system. The first slice (generally the

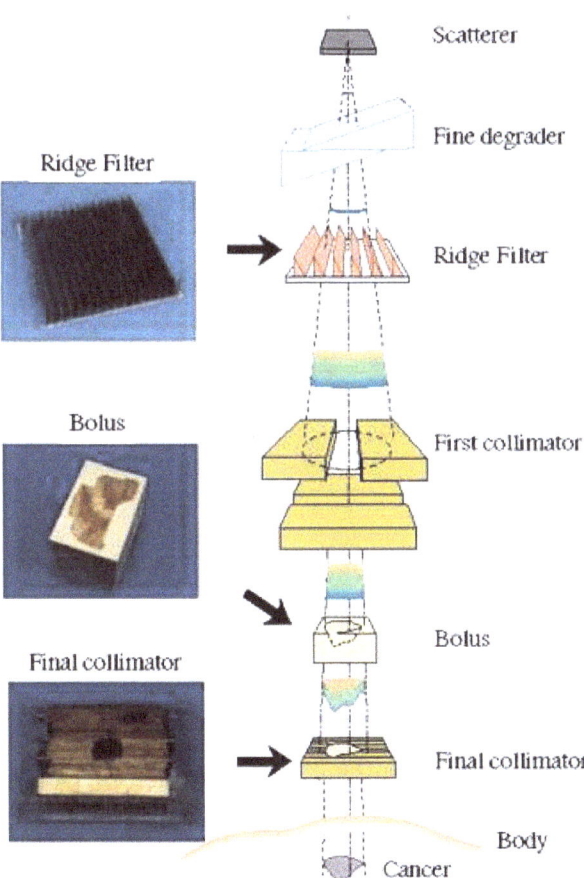

Fig. 7.   Passive beam delivery system.

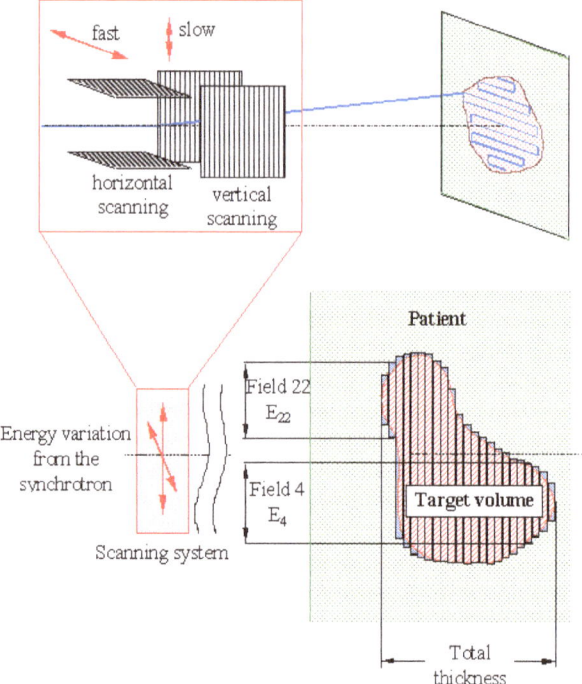

Fig. 9.   Active beam delivery system.

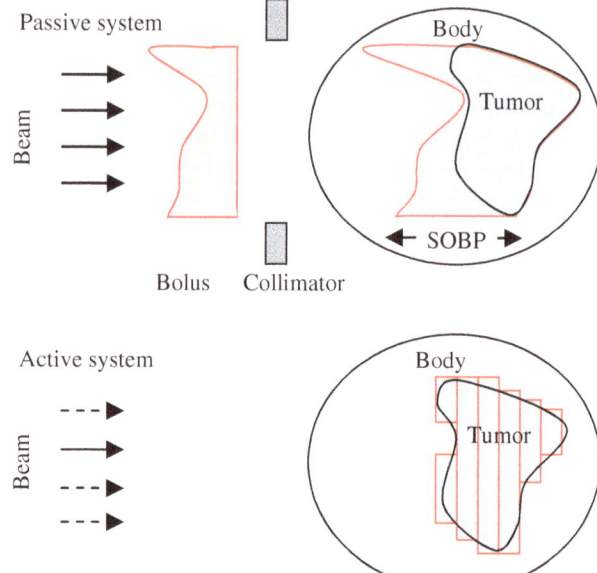

Fig. 10.   The constant thickness SOBP of passive systems gives the dose also to parts of healthy tissues that can be better spared with active beam delivery.

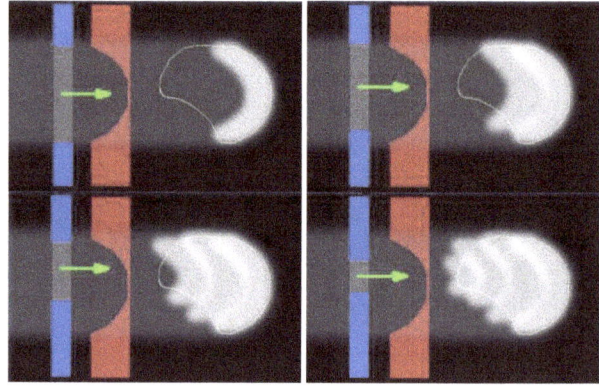

Fig. 11.   Layer-stacking method for dose delivery [11].

more distal one) is painted, then the beam energy is varied, the second-most-distant slice is painted, and so on.

Active scanning, irradiating layer by layer, can provide a better conformation of the dose, as shown in Fig. 10. The bolus used in the passive system shapes the distribution according to the distal part of the target volume, but the same shape is then reproduced on the proximal side. This implies that in the entrance channel some normal tissue receives the full dose.

An intermediate delivery system in which the volume is irradiated in slices that are passively filled and with a multileaf collimator that shapes the field slice by slice is also conceivable (*layer-stacking irradiation method*; see Fig. 11), but it requires bolus and collimator and works only if the tumor geometry is such that the multileaf collimator moves only in the closing direction during the treatment. If not, the part of a slice that was not irradiated with the previous one would receive a smaller dose.

With active scanning the dose is distributed in small volumes, often called *voxels* (volume pixels) or *spots*, which are treated individually. This means that each voxel of a slice can receive a different dose,

depending on how big a dose the voxel has already received by particles directed to deeper layers.

This overcomes the problem of the layer-stacking method, permits *intensity-modulated particle therapy* and allows the treatment of basically arbitrary tumor shapes.

A graphical demonstration of the method is shown in Fig. 12.

Active scanning can thus provide better dose shaping and does not require patient-specific hardware, but this is not for free: active systems are much more difficult in operation. It will be shown later that

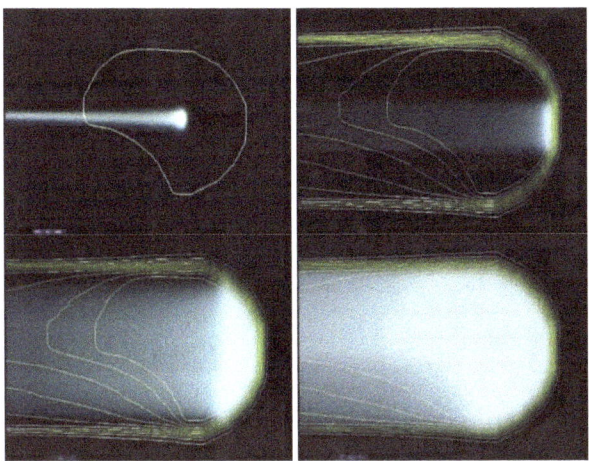

Fig. 12. Filling a target with an active beam delivery system [12].

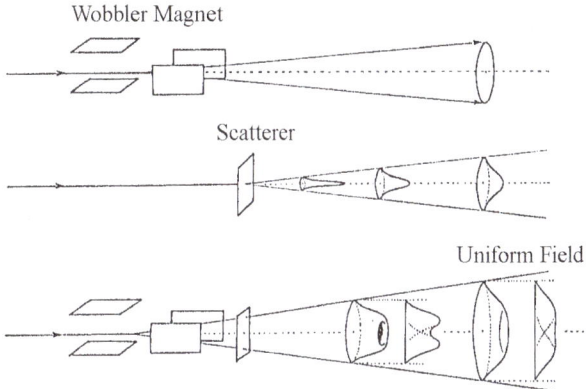

Fig. 13. Wobbling beam delivery system.

with active beam delivery, the beam has to be positioned with high precision.

Furthermore, although it has been shown that hadrontherapy is capable of delivering a precise dose distribution, to be consistent the target has to be known at least to the same precision. This means that both the target shape and the target position *inside the body* have to be known. Apart from the requirements for the medical imaging, this means that the target volume has to be "fixable." For a tumor in the head region this is possible, but for tumors that move with respiration or with the heartbeat it becomes impossible.

For a moving tumor, it is evident that passive spreading, irradiating the whole volume inside which the lesion can move, is more suitable or at least easier. Scanning systems, on the contrary, treat a small volume at a time and need time to treat the whole tumor. If during this time the target moves, there will be an uneven dose distribution.

Because of that, today nobody is treating moving tumors with pencil beams.

To treat moving organs with active scanning, synchronization with breathing, repainting, tumor tracking and following, and active energy compensation methods are under development worldwide [13].

3.3. *Wobbling*

An important variant of the passive beam delivery is the wobbling system used at HIMAC [14] and originally proposed at LBL [15]. In this technique two

magnets move the beam on a circle before a scatterer, as shown in Fig. 13. A proper combination of scatterer thickness and wobbling circle radius results in a flat region that can be used in a passive system with collimators and bolus.

The use of a completely passive system is not advisable with carbon ions, for several reasons. The first is that carbon ions scatter less than protons and to obtain a large treatment field requires the scatterer either to be placed at a large distance or to have a large thickness. Large thickness implies large energy loss and large beam loss, which in turn means higher energy and higher current for the primary beam to reach the desired range.

The second reason is that carbon ions, interacting with the scatterer, break down into fragments with a longer range (causing a bigger dose to be delivered after the tumor) and suffer more nuclear interactions with production of neutrons.

4. Requirements for a Synchrotron for Hadrontherapy

Considering what has been described above, the best performance can be achieved with a machine that accelerates *carbon* ions and uses an *active* beam delivery system. The beam spreading system will be ready to implement new techniques for the treatment of moving organs. Below, machines aimed at being at the leading edge and therefore compliant with this choice, will be considered.

4.1. *Particles and ranges*

To treat deep-seated tumors, a range of 27–30 cm of water is required. For carbon ions, this corresponds to a beam energy of 400–430 MeV/u and to a magnetic rigidity of 6.35–6.62 Tm.

To treat large tumors, a treatment field of 200 mm × 200 mm is a very common choice.

Considering that the energy loss per unit length dE/dx scales approximately as Z^2, the energy per nucleon required to reach a given depth increases with Z. This in turn implies that the necessary beam rigidity $B\rho$ increases as well with Z.

Thus, at least as far as magnets are concerned, a synchrotron capable of accelerating carbon ions to the necessary energy is capable of accelerating any ion between hydrogen and carbon as well.

Oxygen is an interesting ion too, but the rigidity necessary for therapy is higher than for carbon. Nevertheless, a synchrotron that can accelerate carbon to 400 MeV/u can also accelerate oxygen to the same energy. Thus, oxygen can also be used, though with a smaller range of 19 cm.

Most carbon machines, existing or planned, are designed to accelerate both protons and carbon ions. This will allow medical doctors to compare the merits of the two species under the same conditions. It is, indeed, generally difficult to compare results obtained by different teams with different accelerators in different places.

It must be realized that designing the machine to also use protons is not for free: it approximately doubles the range of extracted beam rigidities, which has consequences for the power supply specifications. It does not require more current or more power, but to fulfill the precision prescriptions also at lower rigidity is more demanding for the power supply.

Also, the injector has to be designed for dual species operation, which increases its complexity and requires one additional ion source.

4.2. *Particles per spill and dose rate*

The duration of the treatment is another important parameter. During each treatment session, the patient is indeed immobilized with masks, bite blocks and other restraints that cause him some stress. Most of the time that the patient spends immobilized, though, is not irradiation time: the tumor position has to be verified with x-rays and the patient has to be aligned with respect to the treatment line. These activities require something like 20–30 min.

The treatment time has then to be short with respect to half an hour, but it makes a small difference if it is 2 or 3 min.

Typical doses are in the order of 2 Gy/fraction; considering tumors up to 2 l, a dose rate of 2 Gy/min/l would provide the treatment in 2 min, which seems reasonable. In many projects dose rates as high as 5 Gy/min/l are specified, leaving margins for hypofractionation schemes or very large tumors.

To avoid confusion, in the following, only the physical dose (not the biological dose) will be considered.

To deliver these dose rates, a high beam current is not necessary. The fully active beam spreading systems considered here foresee changing the beam energy from the accelerator. In a synchrotron, this normally means that any beam eventually remaining in the accelerator is dumped, a new beam is injected and it is accelerated to the new energy. Thus, a concept of "particles per spill" is more interesting than the mere current. A normal duty cycle for operation is of the order of

1 s(slow extraction)/2.5 s(complete cycle) = 0.4.

As a first approximation, consider a "cubic" tumor 100 mm × 100 mm × 100 mm, and let us divide it into 3 mm slices (33 slices).

A carbon ion deposits 30 MeV/u (30 × 12 = 360 MeV total) in the last 3 mm of its range. In the meanwhile it deposits 14 MeV/u in the three mm upstream. Thus, the second slice needs only half of the particles of the first. The third, after treatment of the first two slices, has already received two thirds of the dose and therefore needs only one third, etc. Examples of energy–range correspondence and of the relative doses to be delivered for the first slices are summarized in Tables 1 and 2. Only the last 30 mm of the range of a carbon ion are shown in Table 1. A more energetic ion would have a longer range, but after losing the rest of its energy it would fall inside this table.

To deposit 2 Gy in the distal slice, one needs approximately 10^9 carbon ions. Thus, if the accelerator can provide 10^9 particles per spill (pps), 33 spills are necessary (one per slice). If the accelerator can provide only one third of the particles, 37 spills are needed, 3 for the distal slice, 2 for the second and the third, and then 1 as in the previous case.

Table 1. Energy–range correspondence for carbon ions in water calculated with SRIM 2008 [16].

Range (mm)	Energy (MeV/u)
3	29.8
6	43.9
9	55.7
12	65.8
15	74.6
18	82.4
21	89.4
24	96.0
27	102.3
30	108.2

Table 2. Relative doses in the slices of a cubic treatment volume. Slice #1 is the most distal.

Slice	R. dose
1	1.00
2	0.53
3	0.35
4	0.28
5	0.25
6	0.23
7	0.21
8	0.20
9	0.18

4.3. *Dose precision and beam position*

In the design of most of the existing hadrontherapy accelerators, the prescription for the dose uniformity is $\pm 2\%$ to $\pm 3\%$.

To reach such a tight specification is not an easy task, from the point of view of either transverse beam positioning or the temporal spill homogeneity.

To get the feel of the precision required, let us consider a one-dimensional Gaussian beam and sum a number of Gaussians to obtain a flat distribution, as shown in Fig. 14.

As long as the Gaussians are summed with regular spacing, a flat distribution is obtained independently of the step size until the distance between curves gets to approximately 2σ. With such spacing the inhomogeneity is $\pm 1.4\%$, as shown in Fig. 15.

On the contrary, if in a regular pattern one single beam is mispositioned, then the homogeneity is strongly affected.

Figure 16 shows the same situation as in Fig. 14, but with a small error of 0.1σ in the position of the

The difference becomes more visible when larger doses are needed, because in this case many spills may be necessary also for the more superficial layers.

To give a 6 Gy dose with 10^9 pps, 37 spills are required, while with one third of the particles 64 spills are needed.

The dose rate clearly depends on the tumor shape; for the simple case considered, the 33 spills at 2.5 s per cycle correspond to 2 Gy/l in 83 s = 1.5 Gy/min/l.

For a 2 l tumor with a double section (200 mm × 100 mm), only 35 spills are needed and the dose rate is 2.7 Gy/min/l.

For the sake of completeness, when the biological dose is considered, the number of particles required goes down more slowly than indicated in Table 2 because, as shown in Fig. 5, the physical dose has to increase, when moving backward with the Bragg peak, to compensate for the RBE.

On the other hand, the 2 Gy considered correspond to approximately 5–6 GyE.

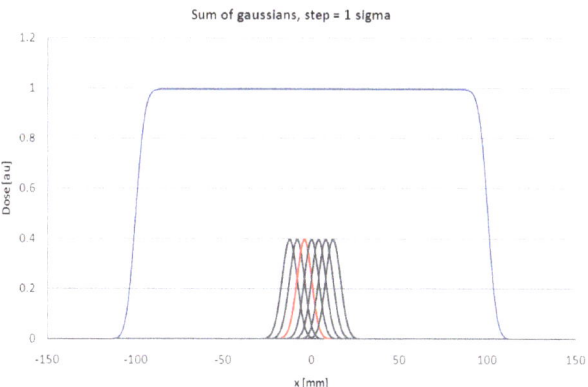

Fig. 14. Sum of Gaussians, step = 1σ.

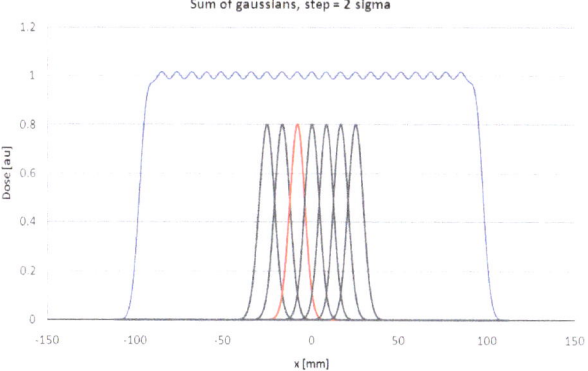

Fig. 15. Sum of Gaussians, step = 2σ.

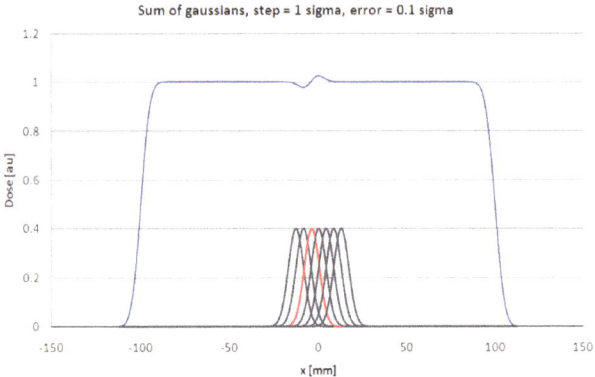

Fig. 16. Sum of Gaussians, step $= 1\sigma$. The red Gaussian is displaced by $0.1\,\sigma$.

Gaussian drawn in red. A dose error of $\pm 2.4\%$ is created.

If random errors between -0.1σ and $+0.1\sigma$ are applied to the position of the Gaussians, inhomogeneities in excess of $\pm 5\%$ are easily obtained, as shown in Fig. 17.

The sensitivity with respect to position errors increases with the distance between Gaussians. The same error of 0.1σ in the position of one Gaussian causes a dose error of $\pm 0.6\%$ for a step of 0.25σ, a dose error of $\pm 1.2\%$ for a step of 0.5σ and a dose error of $\pm 2.4\%$ for a step of 1σ.

Therefore the step size should be as small as possible.

However, to give the same dose with more beamlets means that each beamlet must be fainter, which in turn means that it is more difficult to measure accurately and more difficult to keep in place with feedback on the scanning magnets. Furthermore, it means more frequent beam displacements.

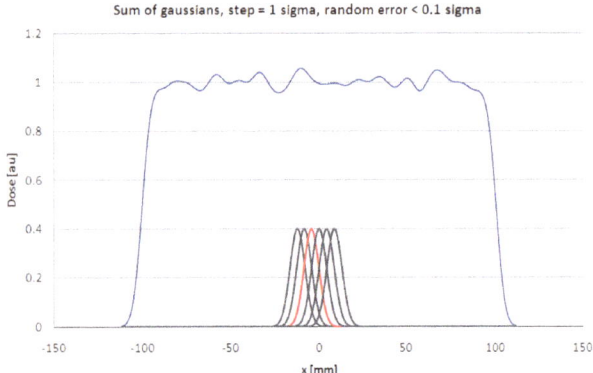

Fig. 17. Sum of gaussians, step $= 1\sigma$. Random errors on the position of all the Gaussians in the range $\pm 0.1\sigma$ are applied.

Fig. 18. A position error of 0.1σ in the position of one Gaussian yields a dose error of $\pm 0.85\%$.

A common[a] scheme is to use a step of 0.78σ, i.e. FWHM/3 (full width at half maximum) [17]. In the simple unidimensional scheme considered here, this corresponds to a height of the Gaussian of approximately 1/3 of the total dose to be delivered. In a 2D situation, overlapping at FWHM/3 in both the transverse directions would correspond to a ninth of the total dose in each position.

Summing 2D distributions yields smaller errors, because each beamlet contributes for a smaller fraction to the total dose. An error of 0.1σ in each direction (step 0.78σ) gives a dose error of $\pm 0.85\%$, as shown in Fig. 18.

As for the 1D model, for the 2D model random errors on the position of all the beamlets yield larger errors. A random error distribution between -0.1σ and $+0.1\sigma$ in the position of the beam yields errors in excess of $\pm 3\%$ (Fig. 19).

Despite the oversimplified model used, the results are more or less correct and yield tight tolerances in beam positioning.

For active scanning, beam size (FWHM) is to be selectable between 4 and 10 mm, according to the shape of the slice to be treated.

Since

$$\text{FWHM} = 2\sqrt{2\ln(2)}\sigma \approx 2.35\sigma, \qquad (1)$$

FWHM $= 4$ mm means that $0.1\sigma = 0.17$ mm. In the monodimensional model, a step of FWHM/3 with a position error on one Gaussian of 0.1σ gives a

---

[a]Since active systems are planned in only a few centers, and are used only at PSI and GSI, the word "common" has to be considered in relative terms.

Step = 0.78 sigma, random error < 0.1 sigma

Fig. 19.   A random position error of 0.1σ in the position of the Gaussians yields a dose error $> \pm 3\%$.

dose error of $\pm 1.9\%$, which is nonnegligible when the overall dose uniformity prescription is 2–3%.

The 0.17 mm are generally shared between measurement error and position error, resulting in a beam position precision specification of 0.1 mm.

Considering error distributions, larger errors are found which are to some extent compensated for by the 2D beam distribution, which averages over more beam positions.

4.4. *Dose precision and spill uniformity*

As already mentioned in the previous subsection, the dose has to be uniform to within $\pm$ 2–3%. When one is administering the treatment, it is therefore necessary to be able to measure and to react at least within the time in which, say, 2% of the dose is delivered to a voxel.

The time in which the full dose is given to a voxel depends, of course, on how big a dose has to be given and how quickly the beam is extracted from the accelerator.

Slow extraction is necessary; if single turn extraction were used, a beam which is a fraction of a microsecond long should be spread over many voxels, which means that only a passive system could be used.

Typical spill lengths with slow extraction are in the order of 1 s and the typical time necessary to deliver the full dose to a voxel is of the order of 5–10 ms.

To obtain a 2% precision on the dose, it is therefore necessary to work on a time scale of 0.02×5 ms =

100μs, which means that when one is measuring the extracted spill, the measurement frequency has to be at least 10 kHz.

The maximum acceptable amount of particles in a "reaction time," defined as the time needed to measure, decide and eventually steer the beam to the desired position or to move it to the next position (or to stop it for the last voxel of a slice), is twice the tolerance, i.e. $2 \times 2\% = 4\%$ of the "full voxel." If 98% of the dose has been reached, then the beam can be displaced; if just 97.9% of the dose has been delivered, then during the next "reaction time" the dose given will not cause an overdose.

If the maximum number of particles delivered in one "reaction time" is larger than the allowed 4%, then the average current has to be reduced until the specification is met. This means longer times to treat one voxel and consequently longer treatment times.

Reaction times of the order of 100–200 μs, and thus compatible with the 5–10 ms voxel filling time, are feasible.

From the accelerator point of view, the requirement is that, given the desired extracted current (average), the maximum intensity modulation is

$$\frac{I_{\max}}{I_{\text{av}}} < 2$$

for measurement frequencies up to 10 kHz.

The "handwaving" considerations given in this section yield a few specifications, summarized in Table 3.

5. Slow Extraction

One key point in the design of a synchrotron for hadrontherapy with active beam delivery is therefore the slow extraction scheme used.

Table 3.   General specifications for a hadrontherapy syncrotron.

Accelerated ion	p, C
Particle range (g/cm^2)	3–30
Energy range (MeV/u)	60–225 (p)
	120–430 (C)
Field size (mm $\times$ mm)	200 $\times$ 200
Particles per spill	4×10^{10} (p)
	10^9 (C)
Beam size (FWHM) (mm)	4–10
Beam position precision (mm)	0.1
Extraction	Slow
Dose uniformity	$\pm 2.5\%$

Slow extraction is performed with resonant schemes in which the machine is operated near to a resonance and, when the beam has to be extracted, the particles are brought out of a stable region such that their betatron oscillation amplitude grows until they reach the extraction septum and are finally sent to the treatment room.

The first hadrontherapy dedicated synchrotron was the one built by Fermilab for LLUMC. In this facility, a "half-integer" slow extraction is performed. The stable region has approximately an elliptical shape in phase space with two separatrices around, as shown schematically in Fig. 20 [18]. When the beam has to be extracted, the horizontal tune is changed toward the value 0.5, the stable region shrinks and the now unstable particles begin to jump from one separatrix to the other until they are extracted.

The extraction duration at Loma Linda is specified to be 0.05–5 s and the intensity stability is required to be within $\pm 2.5\%$ at a 1 kHz sampling rate [19].

All the other operating and planned synchrotrons for carbon therapy use a third integer resonant extraction. In this scheme, the tune is next to $N/3$ and a sextupolar field is used to drive a resonance.

The sextupolar field distorts the particle trajectories in phase space and creates a triangular stable region outside which the motion of the particles becomes unstable.

Also in this case, the amplitude of oscillation of a particle in the unstable region grows turn after turn until the particle enters a septum and is extracted.

The stable region in (normalized) phase space, and to the first order approximation described by the Kobayashi Hamiltonian [20], is schematically shown in Fig. 21.

The size of the stable region is parametrized by the apothem h,

$$h = \frac{4\pi}{S}\delta Q, \qquad (2)$$

where Q indicates the horizontal tune $\delta Q = (Q_{\text{part}} - Q_{\text{res}})$ and S is the normalized sextupole strength,

$$S = \frac{1}{2}\beta_x^{3/2}\frac{\ell_S}{|B\rho|}\left(\frac{d^2 B_z}{dx^2}\right)_0 = \frac{1}{2}\beta_x^{3/2}\ell_S k'. \qquad (3)$$

Equation (2) defines the stability limit. If the relation is represented in a Q amplitude plot, the Steinbach diagram is obtained, shown in Fig. 22.

The Steinbach diagram, using the relation $\Delta Q = Q'\,\Delta p/p$, is often drawn with $\Delta p/p$ as abscissa.

Depending on the stop band width with respect to the beam, the slow extraction type can be divided into three main categories, illustrated in Fig. 23 [21].

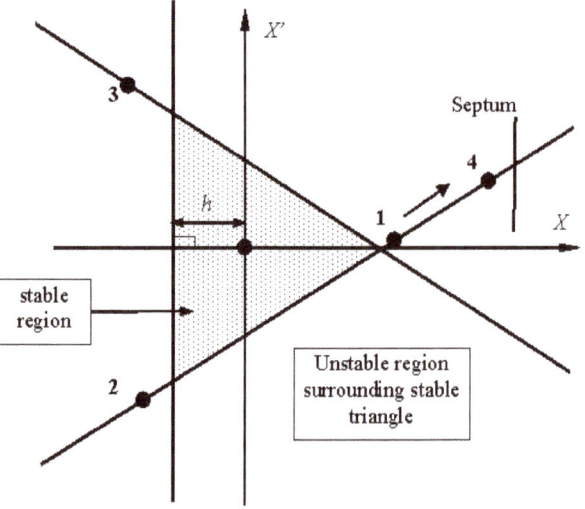

Fig. 21.   Third integer extraction.

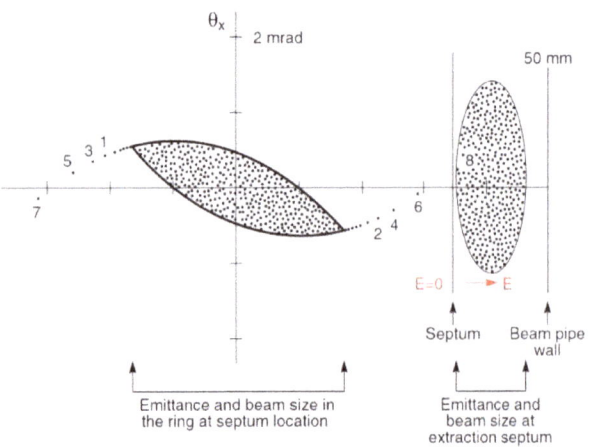

Fig. 20.   Half-integer extraction.

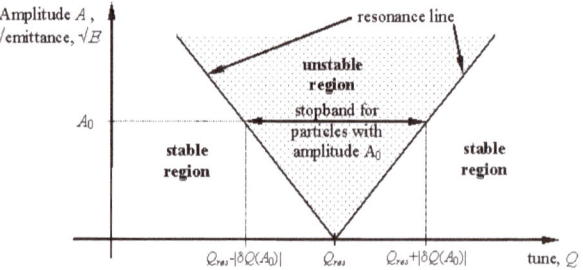

Fig. 22.   Steinbach diagram.

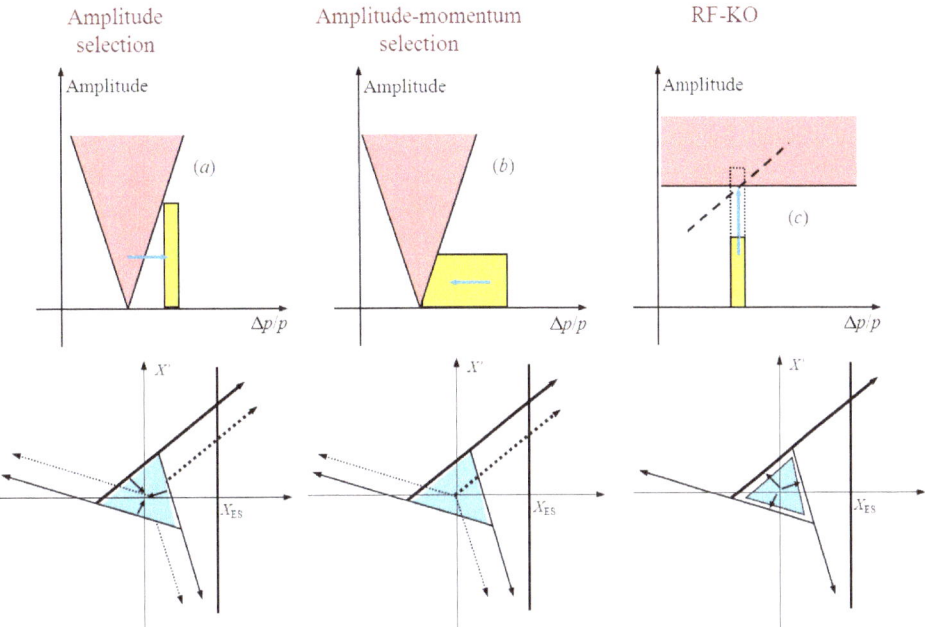

Fig. 23. Extraction methods. The pink areas represent the unstable region, while the yellow ones indicate the beam in the Steinbach diagrams above. In the lower part of the picture the action in phase space is represented with the separatrices for small and large amplitude particles.

In the *amplitude selection* method, the beam is narrow in momentum with respect to the stop band. The tune of the synchrotron is then varied using the quadrupoles, moving the unstable region onto the beam (whose momentum is not changed). During the extraction the separatrices move as shown in the lower part of the picture, which makes the extracted beam move at the entrance to the extraction line.

The extracted beam size changes as well during the spill and only one amplitude at a time is extracted. This last point will be explained later on.

In the *amplitude momentum selection* method, the beam is wide in momentum with respect to the stop band. The tune of the particles is then varied by accelerating them with a betatron core while leaving untouched the tune of the machine.

During extraction the separatrices do not move, and thus the extracted beam size and position do not change. Since ions with the same single-particle emittance always become unstable at the same momentum, the extracted beam energy does not vary during the spill.

Particles of all amplitudes participate in the extraction at the same time. The Hardt condition [22] for minimizing losses at the septum can be applied.

In the RFKO method, everything stays constant apart from the beam which is subject to a transverse RF perturbation (noise or single frequency swept through the betatron frequencies) that blows its emittance up. When the particle's amplitude reaches the separatrix, they become unstable and are extracted. The extracted beam size, position and momentum stay constant during the spill. Only large amplitude particles are extracted. The Hardt condition can be applied.

6. Tune Ripple

As said above, a variation in the tune of the synchrotron corresponds to a displacement of the unstable region in the Steinbach diagram. A tune ripple, created by a ripple of the power supplies, corresponds therefore to an oscillation back and forth of the resonance "V," as shown in Fig. 24.

If the *transit time*, the time needed by a particle to reach the septum after entering the unstable region, is neglected (instantaneous extraction of unstable particles), then the flux of extracted ions is given by

$$\frac{dN}{dt} = \frac{dN}{dQ}\frac{dQ}{dt} = \chi \cdot (\dot{Q}_0 + \dot{Q}_r)$$

$$= \chi \cdot (\dot{Q}_0 + \omega \delta Q_R \cos(\omega t)), \qquad (4)$$

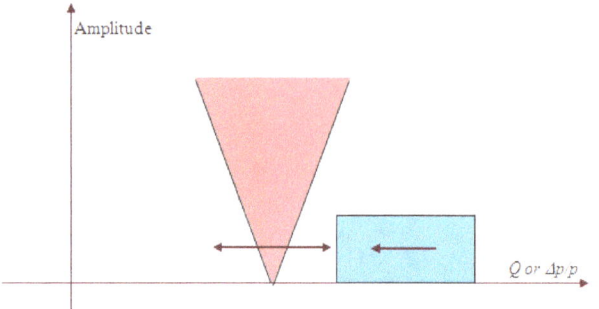

Fig. 24. Tune ripple representation in the Steinbach diagram.

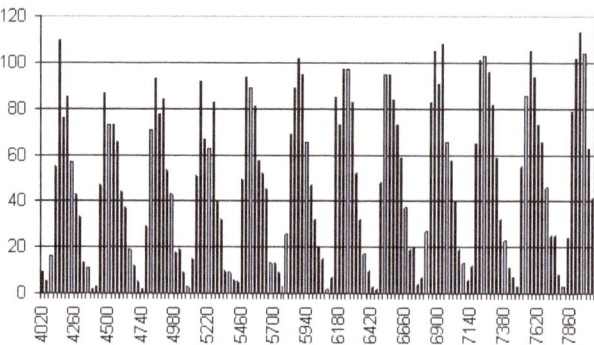

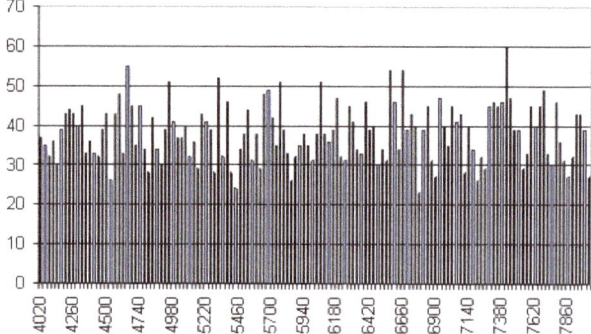

Fig. 25. Tune ripple effect on the extracted spill for amplitude selection (above) and for amplitude momentum selection (below). For the latter, a Gaussian amplitude distribution has been used. Ripple frequency-4 kHz; amplitude $(\Delta I/I)_{\text{quad}} = 1.3\text{E-}6$.

where $\dot{Q}_o$ is the average velocity with which the beam enters the resonance, δQ_R the amplitude of the tune ripple and ω the ripple frequency. When the resonance "escapes" from the beam, the flux is lower; if $\omega \delta Q_R$ exceeds $\dot{Q}_o$, there are instants during which no beam enters the resonance and is extracted.

If the transit time is considered, it can be shown [21] that the amplitude momentum selection method offers an intrinsic protection against ripple in the kHz region. Particles becoming unstable at large amplitudes reach the septum in a shorter time than those becoming unstable at small amplitudes. If particles get unstable at the same time at many amplitudes, a reservoir of particles is stored in the resonance that fills the "hole" created by the "escaping" resonance in the spill.

Numerical simulations of the slow extraction process for a single amplitude population of particles (like for amplitude selection or RFKO) and for amplitude momentum selection show the ripple on the extracted beam and the intrinsic smoothing mentioned above (Fig. 25).

To perform the amplitude momentum selection extraction, the beam has to be "pushed" into the resonance. To provide the corresponding acceleration, a Betatron core is used, like the one shown in Fig. 26. The B field of the betatron core links the beam path and therefore, when the betatron is ramped, it induces an accelerating voltage on the beam.

The drawback of the amplitude momentum selection scheme is that for the same reason for which the beam smooths the ripple effect, it cannot be stopped very quickly and a fast magnet (or, better, a beam chopper; see later) is required in the extraction line.

RFKO is the extraction method more frequently used in existing and planned synchrotron-based

Fig. 26. Betatron core at CNAO.

facilities because of its simplicity and rapid switch-off time.

7. Rapid-Cycling Synchrotrons

It has been said above that typical voxel durations are in the order of 5–10 ms, which corresponds to

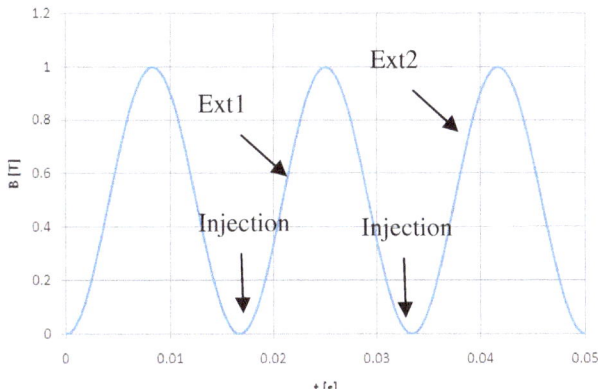

Fig. 27. Rapid-cycling synchrotrons can change the beam energy in a few ms.

100–200 voxel/s. When the duty cycle is considered, this number decreases to approximately 60 voxel/s.

Rapid-cycling synchrotrons might work at these frequencies, but since they would deliver the whole dose in a single bunch, to perform active scanning it would be necessary to be able to change the bunch charge cycle by cycle (a few ms), to adapt it to the dose required by the next voxel, and to be certain that the delivered amount of particles is the one desired to within 2%.

Moreover, the beam position cannot be steered to the wanted position and has to be correct at the first shot.

Rapid-cycling synchrotrons, on the other hand, offer the possibility of changing the energy in a few ms (as shown in Fig. 27), which can be useful for tracking a moving tumor in 3D.

8. Hadrontherapy Synchrotrons in the World

The number of centers dedicated to ion beam therapy is growing steadily and nowadays approximately 30 centers are in operation [23]. Most of the facilities are based on proton cyclotrons and provide treatment with passive beam spreading. Also, a few synchrotrons fully dedicated to ion beam therapy are in operation worldwide. Most of them are in Japan: HIMAC, Chiba; PATRO, Hyogo, Shizuoka, Tohoku; PMRC, Tsukuba; WERC, Fukui. Two are in the US: LLUMC, Loma Linda; M. D. Anderson Cancer Center, Houston.

A few more are under construction in Japan (Gunma, Fukui, Kagoshima), Russia (PMHPTC, Protvino) and Europe (HIT, Heidelberg; CNAO,

Pavia, Marburg; NRoCK, Kiel; MedAustron, Wr. Neustadt; Etoile, Lyon).

All the carbon ion accelerators (existing and under construction) are synchrotrons.

The dimensions of a synchrotron depend mainly on the maximum beam rigidity that the accelerator has to provide. Thus, machines for protons only ($B\rho = 2.4$ Tm) have a diameter of approximately 6–7 m, while synchrotrons for carbon ions ($B\rho = 6.6$ Tm), have a diameter of 20–25 m.

8.1. *LLUMC*

The first dedicated hadrontherapy machine was the Loma Linda synchrotron (shown in Fig. 28), which started operating in 1992.

It is the most compact machine, with a diameter of only 6 m and the injector sited on top of the main accelerator.

The injection is made at 2 MeV, with single turn injection. Extraction energy ranges from 70 to 250 MeV and is made with a half-integer resonance scheme.

The bending magnets are of the combined function type, providing both bending and focusing. The working point in the tune diagram is $(Q_x, Q_y) = (0.6, 1.29)$, near to the $Q_x = 0.5$ resonance line used for extraction.

The machine can deliver up to 2×10^{10} particles per spill.

8.2. *Hitachi synchrotron*

The Hitachi synchrotron, installed at PMRC (Tsukuba), WERC (Fukui) and M. D. Anderson (Houston), is 7 m in diameter; see Fig. 29.

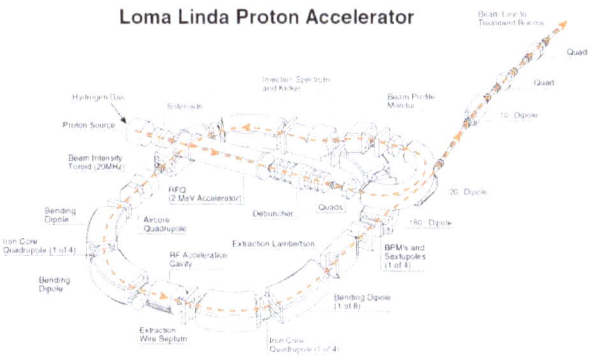

Fig. 28. Loma Linda synchrotron. It has a diameter of 6 m and can accelerate protons to 250 MeV [18].

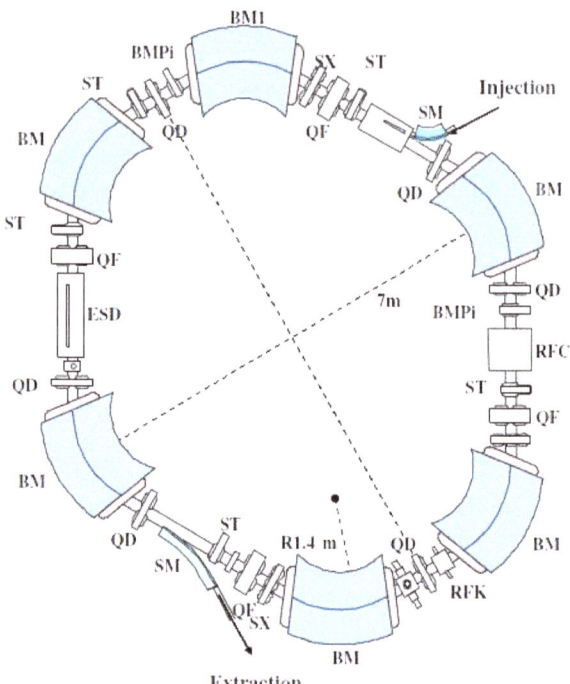

Fig. 29. Hitachi synchrotron. It has a diameter of 7 m and can accelerate protons to 250 MeV [24].

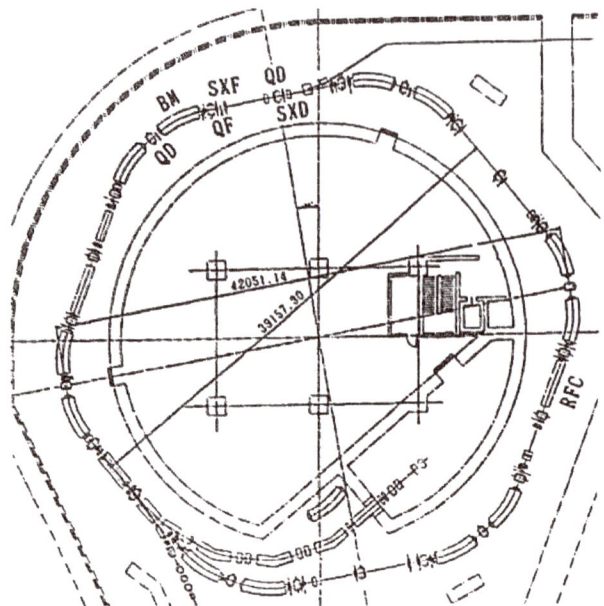

Fig. 30. The HIMAC synchrotrons are designed for almost 10 Tm and are consequently very large. The upper one is shown.

It has similar performances to the LLUMC one, but it provides a few improvements. First, it can give a higher dose rate, with 10^{11} particles per spill.

Second, it offers an improved extraction method, with the third integer RFKO, which can be used for both passive and active spreading, as it has happened at M. D. Anderson since May 2008 [25].

The main characteristics of the accelerator are summarized in Table 4.

8.3. *HIMAC*

The first treatments with high LET ions have been administered at Berkeley with the Bevalac using

Table 4. Main characteristics of the Hitachi proton synchrotron.

Injector	7 MeV linac: PL-7i
Injection scheme	Multiturn horizontal injection
Extraction beam energy	70–250 MeV
Extracted beam intensity	10^{11} ppp
Extraction type	RFKO
Circumference	23 m
Bending magnet	60° magnet with radius of 1.4 m
Revolution frequency	1.6–8 MHz
Betatron tune Q_x/Q_y	1.70/1.45 at injection
	1.68/1.45 at extraction
Flat top length	0.5–5 s

nuclei from He to Si, but the first dedicated machine is the HIMAC synchrotron at Chiba, (Fig. 30), which started operating in 1994. In the facility there are two synchrotrons, installed at the upper and lower floors (Fig. 31).

HIMAC [26] is much larger than the other machines, because it was designed to be able to deliver 800 MeV/u Si ions ($B\rho = 9.73$ Tm), which resulted in a 42 m diameter.

The irradiation field is specified to have a 220 mm diameter and the beam intensity should be sufficient to give a dose of 5 Gy into the irradiation field within a time duration of several to a few tens of seconds.

Both fast and slow extraction are possible from the upper ring, while only slow extraction is available from the lower one.

HIMAC started with ordinary third order extraction, driven by tune change, and then switched to the RFKO extraction with AM and FM [27].

The injector is made with two sources: one PIG for low Z and one ECR for higher Z, connected through an LEBT to a 7 m RFQ (8–800 keV/u) and a 24 m Alvarez LINAC (0.8–6 MeV/u), both working at 100 MHz. The length depends on the fact that the injector has to be able to accelerate ions with $Q/A > 1/7$.

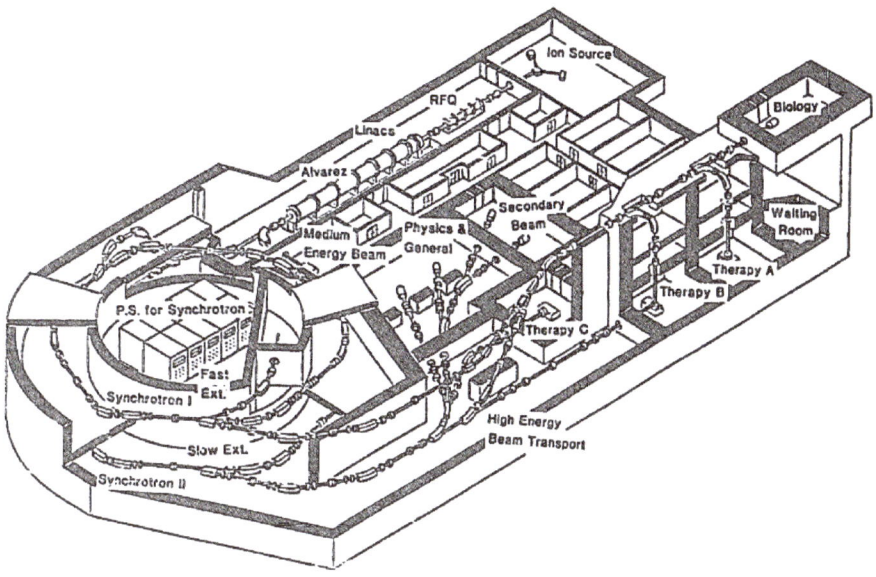

Fig. 31.   Two synchrotrons are available in the HIMAC facility.

The injection energy is 6 MeV/u and multiturn injection in the horizontal plane is used.

The lattice is made up of six FODO periods.

8.4. *PATRO*

The second accelerator dedicated to hadrontherapy is PATRO at Hyogo Ion Beam Medical Center, Fig. 32, where therapy has been delivered with protons since 2001, and carbon ions since 2002.

As in the case of HIMAC, the synchrotron is made up of 6 periods and is based on a FODO lattice. Also, PATRO utilizes the third order resonance extraction driven by RFKO.

Fig. 32.   PATRO accelerator at HIBMC [28].

The injector is composed of two ECR sources: a 1 MeV/u RFQ and a 5 MeV/u Alvarez LINAC. The LINAC and RFQ operation frequency is 200 MHz. Both single-turn and multiturn injection can be made.

Five treatment rooms are available at HIBMC: room A with an Oblique (45°) port, room B with horizontal and vertical ports, room C with a horizontal port for a sitting patient (A–C for carbon and proton) and G1, G2 with isocentric proton gantries.

The beam delivery system is passive, with wobbling to spread the beam.

The performance specifications are summarized in Table 5 [29].

8.5. *HIT*

In 1994, a pilot project was started by GSI Darmstadt, DKFZ Heidelberg, the University of Heidelberg and the Forschungszentrum Rossendorf, with the aim of installing an experimental therapy unit at the accelerator complex of GSI.

In about 10 years of operation this collaboration has led to valuable results, among which is the development of raster scanning.

Based on the results of this pilot project, a dedicated facility has been proposed and built in Heidelberg — the Heidelberg Ion Therapy (HIT) center [30].

Table 5.   Main characteristics of the Hyogo synchrotron.

Particles	Proton, helium and carbon
Energy range	70–230 MeV/u for p, He
	70–320 MeV/u for carbon
Beam intensity	7.3×10^{10} pps for p
	1.8×10^{10} pps for He
	1.2×10^{9} pps for carbon
Dose rate	5 GyE/min
Circumference	93.6 m
Beam range	40–300 mm for p and He
	13–200 mm for carbon
Field homogeneity	$\pm 2\%$ (over treatment field)
Field size	15 cm × 15 cm for ports A, B
	10 cm φ for port C
	15 cm φ for gantry ports G1, 2
Displacement of beam	± 2 mm (from isocenter)
Spill length	400 ms
Maximum repetition	0.5 Hz for He and carbon
	1 Hz for proton
Injection energy	5 MeV/u

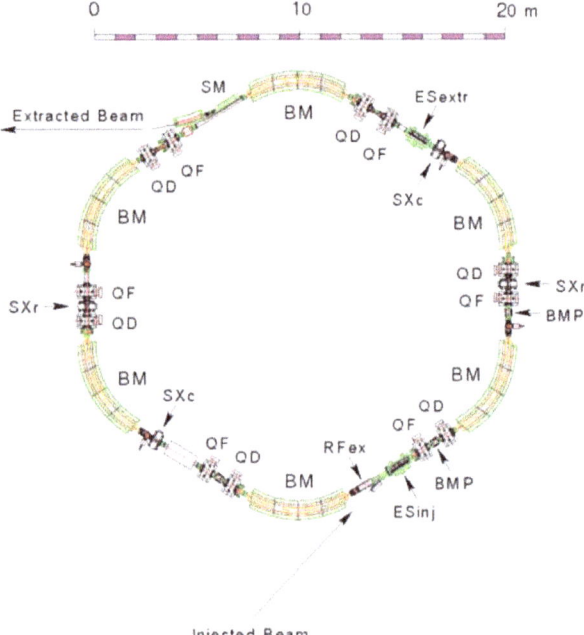

Fig. 33.   The synchrotron of the Heidelberg Ion Therapy center is approximately 20 m in diameter.

The HIT facility is expected to start treatments by the end of 2009.

The accelerator, a 20-m-diameter synchrotron with doublet focusing and only six dipoles, is shown in Fig. 33.

The beam is produced by two permanently running ECR ion sources with independent spectrometer lines that deliver the beam via a low energy beam transport (LEBT) system to the linear accelerator. The LEBT is equipped with a system to quickly change the intensity over three orders of magnitude.

The injector is a combination of RFQ and IH-DTL structures that accelerate the injected particles from 8 keV/u to 7 MeV/u within a length of only 6 m.

Multiturn injection is performed in the horizontal plane and slow extraction relies on the RFKO method.

Three treatment rooms are available at HIT, one of which contains the only carbon ion gantry in the world.

The main parameters are summarized in Table 6.

8.6. *CNAO*

At the beginning of 1996, the design of an optimized synchrotron for light ion (and proton) therapy was started at CERN under the leadership of P. Bryant with the acronym PIMMS (Proton-Ion Medical Machine Study) [31]. PIMMS was a collaboration of CERN, Med-AUSTRON (Austria), Oncology 2000 (Czech Republic) and TERA (Italy). GSI contributed and gave expert advice.

The PIMMS group had as mandate the design of the synchrotron and the beam lines for a light

Table 6.   Main characteristics of the HIT synchrotron.

Particle species	Protons, He, C, O
Beam energy	50–430 MeV/u
Beam intensity	p: 4×10^{10}
(particles per spill)	He: 1×10^{10}
	C: 1×10^{9}
	O: 5×10^{8}
Beam spot size	4–10 mm FWHM
	(2D-Gaussian)
Treatment rooms	2 fixed-horizontal-beam
	treatment rooms
	1 gantry room
Beam delivery technique	Intensity-controlled raster
	scan
Field homogeneity	$\pm 5\%$
Gantry type	360° rotating scanning
	gantry, isocentric
	geometry, normal
	conducting magnets
Treatment field	20×20 cm^2
PET	*In situ* verification of
	irradiation procedure
Number of patients per year	> 1000
Building area	= 70× ∼ 60 m^2

ion hadrontherapy center unconstrained by financial and/or space limitations. In fact, PIMMS was never intended to be built in its final layout. It was, rather, an open design study from which different modules could be taken for the design of various centers according to their requirements.

Based on that work, the TERA Foundation, with the collaboration of many institutions, including INFN, CERN, GSI and a few Italian universities, has engineered the PIMMS synchrotron and made a more compact design of the extraction and of the injection lines. The resulting project has been given, together with the design group, to the CNAO Foundation (Centro Nazionale di Adroterapia Oncologica) and is presently under construction in Pavia.

The CNAO synchrotron, shown in Fig. 34, has a diameter of approximately 25 m and accelerates protons and carbon ions respectively to 250 MeV and 400 MeV/u.

The injector is almost identical to that of Heidelberg, differing only in the geometry of the LEBT. Sources and LINAC are placed inside the main ring, making the accelerator very compact. The two ECR sources run continuously and can be individually monitored; the particle species to be accelerated is selected by just changing the LEBT magnets settings. At the end of the LEBT a fast electrostatic chopper (1 μs rise and fall time) permanently dumps the beam on the vacuum chamber and lets it reach the RFQ only when the system is ready for injection. The RFQ accelerates the beam from 8 keV/u to 400 keV/u and the IH-DTL LINAC accelerates it up to 7 MeV/u.

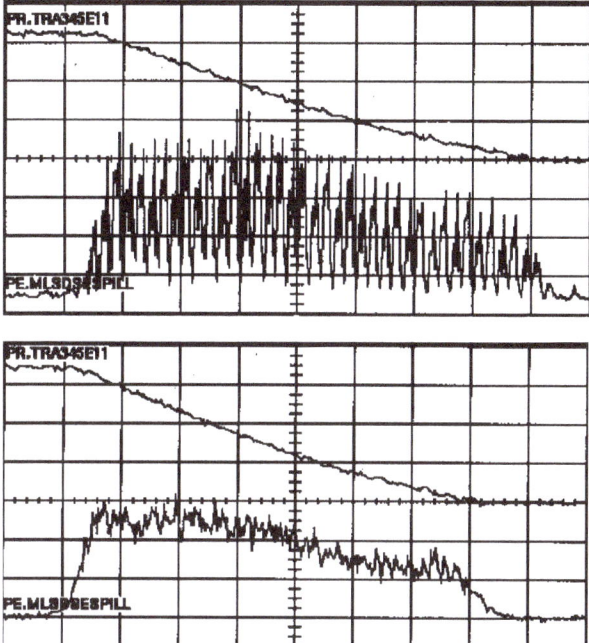

Fig. 35. Beam extracted from the CERN PS without (above) and with (below) air core quadrupole ripple compensation.

Injection follows a multiturn scheme in the horizontal plane.

Slow extraction can be made both with amplitude momentum selection (nominal method) and with RFKO; additional measures to reduce the spill modulation have been adopted, like the creation of an "empty bucket" in front of the circulating beam [32] and the installation of an air core quadrupole for feedforward compensation of the ripple; Fig. 35 shows the effect of feedforward compensation on the CERN PS [33].

Three treatment rooms are foreseen at CNAO in which the beam can be delivered with horizontal lines. In one of the rooms, a vertical line is also available for delivering the beam at the same isocenter as the horizontal one.

Active beam delivery is foreseen for all the treatment lines.

The extraction lines are designed following the "empty ellipse" approach, developed in PIMMS, to take into account the very asymmetric distribution of the slowly extracted beam.

While the vertical phase space shows a normal distribution, a slowly extracted beam has a "bar of charge" shape in the horizontal one. Its distribution, indeed, is the final part of the extraction separatrix

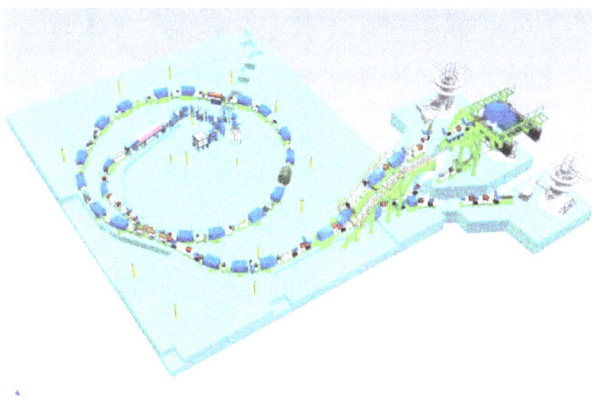

Fig. 34. CNAO accelerator and lines.

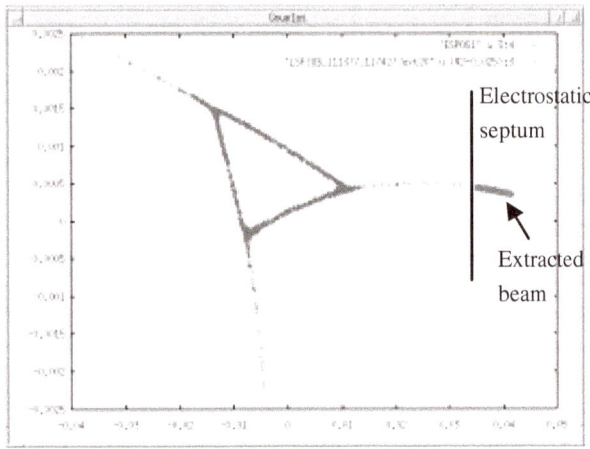

Fig. 36. The extracted beam is distributed along the separatrix and has the shape of a "bar of charge."

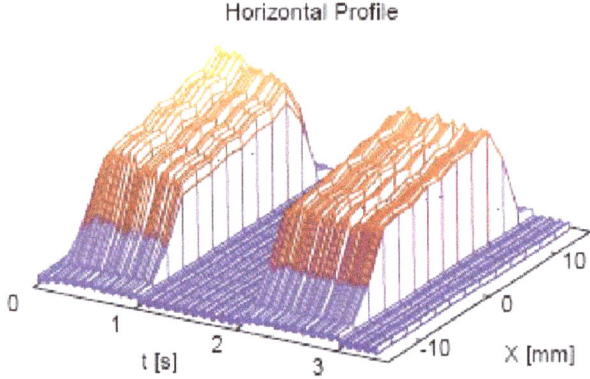

Fig. 37. Beam extracted at Heidelberg.

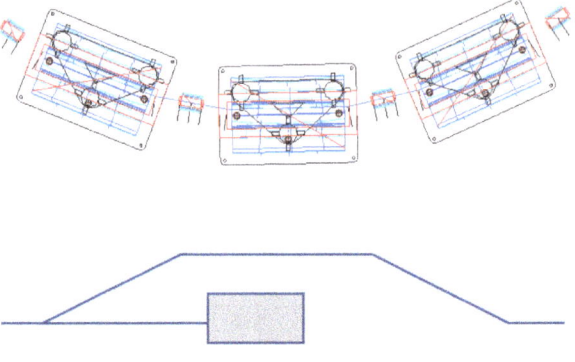

Fig. 38. Beam chopper; when the four small magnets are powered, the beam avoids the dump.

Table 7. Main characteristics of the CNAO synchrotron.

Particle species	Protons, He, C, O
Circumference	78 m
Beam energy	60–400 MeV/u
Beam intensity	p: 2×10^{10}
(particles per spill)	C: 4×10^{8}
Spill length	0.25–10 s
Beam spot size	4–10 mm FWHM
Treatment rooms	2 H
	1 H + V
Beam delivery technique	Intensity-controlled raster scan
Treatment field	$20 \times 20 \, cm^{2}$
Field homogeneity	±2.5%

and has therefore a small emittance and trapezoidal distribution; Fig. 36 illustrates the bar-of-charge distribution. The triangle has the emittance of the circulating beam; the emittance of the extracted beam is evidently much smaller, being the area of the bar of charge.

Figure 37 shows the measured beam at Heidelberg [34], indicating that the distribution is as expected.

Moreover, as required in Section 4.4, to start and stop the beam quickly, the CNAO extraction lines comprise a fast *chopper*, i.e. a set of four identical fast magnets powered in series according to a "+1, −1, −1, +1" scheme. The four magnets create an orbit bump that permits the beam to avoid a beam dump, as shown in Fig. 38. Also, during the

rise and fall of the field, the beam stays on target. The CNAO chopper has a rise and a fall time of 160 µs.

The main characteristics of CNAO are summarized in Table 7.

9. Conclusions

Proton therapy is nowadays accepted by the medical community and is the elective therapy for certain types of tumors. The number of centers where it can be administered is increasing steadily and proton therapy centers can be bought "turnkey." Worldwide there are approximately 30 centers in operation and almost 20 centers are planned or under construction.

Despite the presence on the market of synchrotrons for proton therapy only a few centers use this type of accelerator and the market is dominated by cyclotrons.

Carbon ion therapy is far less available, but interest has grown and six new facilities are under construction, which will give a boost to the clinical studies.

For carbon ions only synchrotrons have been built or are being built at present. The *de facto* standard ion therapy synchrotron is a carbon ion synchrotron delivering beams of the order of 400 MeV/u via a third order slow extraction.

Proton therapy and carbon ion therapy should not be considered to be in direct competition: the ways in which these particles kill cells differ strongly and this difference is a source of complementarity.

The more hadrontherapy demonstrates its efficacy, the more it gets accepted and the more it is considered for use in new locations in the body, which creates new challenges like moving targets.

Activity in the field is "sparkling" and offers many interesting subjects of development.

Acknowledgments

The author would like to thank V. Vitolo, M. Ciocca and S. Molinelli for their help and comments on the medical and medical physics apects, M. Popovic for the useful information and discussions, and M. Donetti for his help in finding references.

The author expresses his gratitude to T. Linnecar for his kind help with the English and for the useful discussions.

Many of the pictures used in this article, are familiar to the hadrontherapy community and appear in many talks/presentations. The author would like to thank the original creators of the images and also U. Amaldi, G. Coutrakon, H. Eickhoff, J. Flanz, T. Haberer, G. Kraft, K. Noda and E. Pedroni, whose presentations have been extensively ransacked.

References

[1] R. R. Wilson, *Radiobiology* **47**, 487 (1946).
[2] A. Mazal *et al.*, Some physical bases of particle therapy with protons and ions (PTCOG, Jacksonville, 2008).
[3] R. Slopsema, Proton beam delivery techniques and commissioning issues (PTCOG, Jacksonville, 2008).
[4] U. Amaldi, *Proc. NUPECC* (2001).
[5] M. Belli, D. Bettega, R. Cherubini, F. Ianzini, G. Kraft, G. Simone and M. Tabocchini, *Physical and Radiobiological Properties of Hadron Beams* (the

TERA project and the center for oncological hadron therapy), eds. U. Amaldi and M. Silari.
[6] A. Wambersie *et al.*, Dose reporting in ion beam therapy. IAEA-TECDOC-1560.
[7] R. Slopsema, PTCOG 47 (May 2008).
[8] PSI annual report (1989).
[9] W. T. Chu, B. A. Ludewigt, K. M. Marks, N. M. Nyman, T. R. Renner, R. P. Singh and R. Stradtner, *Proc. NIRS Int. Workshop on Heavy Charged Particle Therapy and Related Subjects* (4–5 July 1991, Chiba, 1991), pp. 110–123.
[10] T. Haberer, W. Becher, D. Schardt and G. Kraft, *Nucl. Instrum Methods Phys. Res. A* **330**, 296 (1993).
[11] U. Amaldi, lectures at EPFL.
[12] Courtesy of E. Pedroni, PSI website, http://p-therapie.web.psi.ch/spot-scanning.html.
[13] N. Saito, C. Bert, N. Chaudhri, A. Gemmel, D. Schardt, M. Durante and E. Rietzel, *Phys. Med. Biol.* **54**(16), 4849 (2009). Epub 2009, July 27.
[14] S. Yamada, *Proc. 1995 Particle Accelerator Conference* (1995), p. 5.
[15] T. R. Renner and W. T. Chu, *Med. Phys.* **14**, 825 (1987).
[16] SRIM 2008, http://www.srim.org/SRIM/SRIM 2008.htm.
[17] M. Kraemer, O. Jaekel, T. Haberer, G. Kraft, D. Schardt and U. Weber, *Phys. Med. Biol.* **45**, 3299 (2000).
[18] G. Coutrakon, PTCOG 47.
[19] J. Slater, D. Miller and J. Slater, *Particle Accelerators Conference* (1991), p. 532.
[20] Y. Kobayashi and H. Takahashi, Improvement of the emittance in the resonant ejection, in *Proc. VIth Int. Conf. High Energy Accelerators*, Massachusetts (1967), pp. 347–351.
[21] M. Pullia, Detailed dynamics of slow extraction and its influence on transfer line design. Ph.D. thesis (1999).
[22] W. Hardt, Ultraslow extraction out of LEAR (transverse aspects). CERN, PS/DL/LEAR Note 81-6 (1981).
[23] http://ptcog.web.psi.ch/ptcentres.html.
[24] K. Hiramoto *et al.*, *Nucl. Instrum. Methods Phys. Res. B* **261**, 786 (2007).
[25] http://www.mdanderson.org/patient-and-cancer-information/care-centers-and-clinics/specialty-and-treatment-centersv/proton-therapy/services/pencilbeam%20backgrounder.pdf.
[26] Y. Hirao *et al.*, *Nucl. Phys. A* **538**, 541c (1992).
[27] K. Noda *et al.*, *Nucl. Instrum Methods A* **374**, 269 (1996).
[28] http://www.hibmc.shingu.hyogo.jp.
[29] A. Itano *et al.*, *Workshop on Accelerator Operation* (2003).
[30] Th. Haberer, J. Debus, H. Eickhoff, O. Jäkel, D. Schulz-Ertner and U. Weber, *The Heidelberg Ion*

Therapy Center, Radiotherapy and Oncology, Vol. 73 (Suppl 2), pp. S186–S190 (2004).

[31] L. Badano, M. Benedikt, P. J. Bryant, M. Crescenti, P. Holy, P. Knaus, A. Meier, M. Pullia and S. Rossi, Proton–Ion Medical Machine Study (PIMMS). CERN/PS 2000- 007 DR (Geneva: CERN).

[32] M. Crescenti, CERN note, ps-97-068.

[33] R. Cappi, M. Gourber-Pace, S. Hancock, M. Pullia, J. P. Riunaud and Ch. Steinbach, PS/OP/Note 97-30 (MD).

[34] D. Ondreka, EPAC08.

Marco G. Pullia has been a physicist at the CNAO Foundation since 2003, where he is responsible for the accelerator department. Since 1993 he has been working on the design of an accelerator for hadron therapy with the TERA Foundation and he has participated in the Proton Ion Medical Machine Study (PIMMS) at CERN. In 2000 he was awarded the Christoph Schmelzer Prize for his Ph.D. thesis on resonant slow extraction.

Reviews of Accelerator Science and Technology
Vol. 2 (2009) 179–200
© World Scientific Publishing Company

Beam Delivery Systems for Particle Radiation Therapy:
Current Status and Recent Developments

J. M. Schippers

Paul Scherrer Institut, 5232 Villigen, Switzerland
marco.schippers@psi.ch

An overview is given of different techniques of dose delivery applied in currently operating and planned particle therapy systems. Their advantages and disadvantages will be compared and consequences of the methods for the rest of the instrumentation will be discussed. The interrelationship between beam delivery at the patient and the accelerator system is shown by means of several concrete examples. Apart from a description of several subsystems in a particle therapy facility, design rules for optimizing the reliability of an accelerator and beam delivery system will be discussed, as well as some remarks concerning how to deal with future developments.

Keywords: Accelerator; particle therapy; cyclotron; synchrotron; degrader; nozzle; gantry; scanning; scattering.

1. Introduction

Accelerator technology developed for nuclear physics or high energy physics is playing an increasingly important role in radiation therapy with hadrons (protons or ions) [1]. Originating in nuclear physics laboratories, the techniques developed by the particle therapy pioneers at the Harvard Cyclotron Laboratory [2] and at the Bevelac in Berkeley [3, 4] have been enhanced by several research institutes. Although the first hospital-based proton therapy facility came into operation at Loma Linda (California, USA) [5] in the 1990s, most of the proton therapy facilities were located in physics laboratories and operated in collaboration with nearby hospitals. At the end of the previous century the focus of nuclear physics experiments drifted away from these laboratories. The potential relevance of accelerator physics and nuclear physics to health care, especially proton therapy, has been an important factor in improving public relations for such laboratories, however. This enabled successful programs in several laboratories [6]. The number of patients treated worldwide until 2009 is now more than 70,000 [6]. The real breakthrough became visible at the beginning of this decade, when more and more commercially produced hospital-based facilities came into operation.

However, the innovative spirit, high tech environment and tradition of open communication have remained important advantages of the laboratory-based facilities, which therefore still play an important role in the stimulation of new developments. Thanks to the many efforts during the last 50 years in these laboratories and the initiatives of commercial companies since the 1990s, these techniques are reaching a mature status today.

The high risk decision by industry in the 1990s to step into this field and produce *off-the-shelf* proton therapy facilities [7], and later also heavy ion therapy facilities, has largely contributed to the acceptance of particle therapy. The availability of industrially produced irradiation facilities prompted the start of large clinical programs at the hospital-based facilities. Today, for several clinical indications, the outcomes are recognized as being very promising and the two fundamentals of particle therapy, namely avoiding a dose in healthy tissue and increasing the dose in the tumor, have resulted in an increase in cures for several types of cancer, without an increase in serious complications; see for example Refs. 8–10.

A particle therapy facility (see for example Fig. 1) usually has a layout clearly showing separate modules: accelerator, beam energy adjustment, beam transport system, gantries and/or fixed beam lines. However, the interplay between these modules is extremely important and the specifications of one subsystem are dependent on those of other

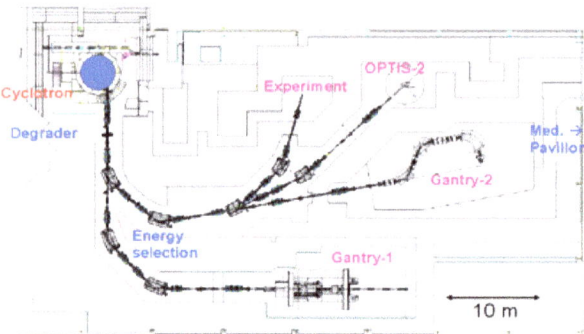

Fig. 1.   Floor plan of the proton therapy facility at the Center of Proton Therapy of the Paul Scherrer Institut (PSI) [11].

subsystems. The beam delivery method used in the treatment has direct consequences for the accelerator design. This interrelationship of functions imposes a large challenge on the system designers, and a strict separation of the system into modules with different functions has therefore not always been achieved. As a consequence, the design of the control system has become one of the most difficult and frequently underestimated tasks.

Excellent reviews of the equipment for particle therapy have been published in past years, such as Refs. 4, 12–14, and to acquire some basic knowledge of article the techniques used, these may be of help. In this article an overview will be given of different techniques of dose delivery recently applied in currently operating and planned systems. Their advantages and disadvantages will be compared and consequences of the methods for the rest of the instrumentation will be discussed. Since the accelerator system has a direct impact on the beam delivery, first, relevant aspects of the accelerator systems currently in use will be discussed. Then an overview is presented of the beam-shaping methods that are mostly (to be) used, namely *beam scattering* and *pencil beam scanning*, followed by a discussion on gantry systems and control systems. Finally, after discussion of several aspects related to reliability, conclusions and some remarks concerning how to deal with future developments will be given.

2. Accelerator Systems

The typical beam energies for particle therapy of deep-seated tumors are in the range 230–250 MeV for protons and 400–450 MeV/nucl. for carbon ions.

The two types of accelerators currently used in particle therapy to obtain these energies are the cyclotron and the synchrotron. Although other types of accelerators exist, they are not suitable for hadron therapy. A good textbook review of synchrotrons and cyclotrons is given, for example, in Ref. 15. The choice of the accelerator depends on many aspects. In this article the most relevant properties of these two machines for proton/ion therapy will be discussed. Since the degrader and the beam analysis system used in a cyclotron facility have drawn an increasing amount of attention in new proton therapy facilities, these important components will be discussed in a separate section.

2.1. *Cyclotron*

The most important advantage of a cyclotron is the continuous availability of the beam, and that its intensity can be adjusted very quickly to virtually any desired value. Although pulsed at the RF frequency (typically between 50 and 100 MHz) of the accelerating voltage, the beam intensity can be considered continuous for all applications in particle therapy. Another advantage of a cyclotron is the capability to change the beam energy very quickly, when using the appropriate energy degrader and beam line components, as discussed in Subsec. 2.2.

However, at energies relevant to therapy, the magnetic rigidity of carbon ions is approximately three times larger than that of protons. In the case where a conventional cyclotron would be used, this would need an excessively large machine. Therefore, until now the cyclotrons for particle therapy have only been used for protons. However, new designs using superconducting magnet technology have been presented recently. Cyclotron(s) (systems) of acceptable size, capable of accelerating carbon ions to clinically relevant energies, are being designed [16–18] and initiatives to build a cyclotron for carbon ions are being discussed at the time of writing.

Modern cyclotrons use magnetic fields that have been shaped to compensate for the relativistic effects of particles with very high velocity, whilst confining the beam in the vertical direction. This can be done very effectively for magnetic fields that do not saturate the iron completely. The magnetic field strength in the commercially available cyclotrons is between 2 T and 4 T. Conventional copper magnet coils are used to achieve the desired field, but since a few years also superconducting coils have been used. Until the late 1950s, the method for accelerating protons to energies higher than about 30 MeV in a cyclotron was to modulate the RF frequency to compensate for the relativistic effect of mass increase with energy as well as the decrease in the magnetic field strength toward the outer radius (conveniently providing vertical focusing). As a consequence the beam intensity extracted from these *synchrocyclotrons* (the RF has to modulate its frequency *synchronous* to the mass increase) is pulsed at a typical frequency of 1 kHz. In the development of proton therapy several of the 160–200 MeV synchrocyclotrons (for example the Harvard cyclotron [2]) have played an important role.

Some of the recent developments in cyclotron technology are aiming at very compact systems [19], which would enable single treatment room proton therapy facilities. The proposed cyclotrons have superconducting coils and are very small (< 2 m in diameter and about 20 tons), so that mounting on a rotating gantry is possible in principle. A first design of a 250 MeV SC cyclotron mounted on a gantry was already proposed by Blosser *et al.* in 1989 [87]. In the small 250 MeV cyclotrons the iron is highly saturated due to the strong magnetic fields (5–8 T). Therefore the very compact cyclotron systems are exploiting the synchrocyclotron concept again, of course using more modern RF and magnet technology.

Currently used cyclotrons, designed solely for proton acceleration to a fixed energy [20, 23], have a typical diameter between 3.5 m (superconducting coils) and 5 m (room temperature coils) and a magnet height of approximately 1.5 m, as can be seen in Fig. 2. Usually some extra space is needed above and below the cyclotron for the support devices of the ion source and equipment to open the machine. The total weight is typically 100–200 tons (for superconducting coils and room temperature coils, respectively).

The protons are created in an internal ion source, located at the center of the cyclotron; see Fig. 3. It consists of an upper and a lower cathode, and a hollow tube (*chimney*) in between. The source is flushed with a regulated flow of a few cm^3/min H_2 gas. Free electrons trapped in the chimney ionize the gas and a plasma is created. Protons can leave the chimney though a narrow slit and are pulled toward the first RF electrode in the cyclotron (*puller*). When enough space is available, the amount of electrons and thus

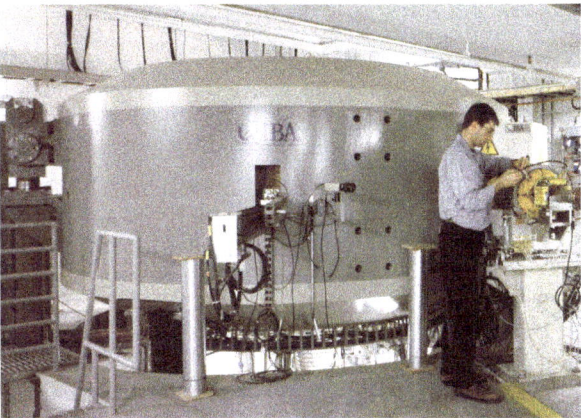

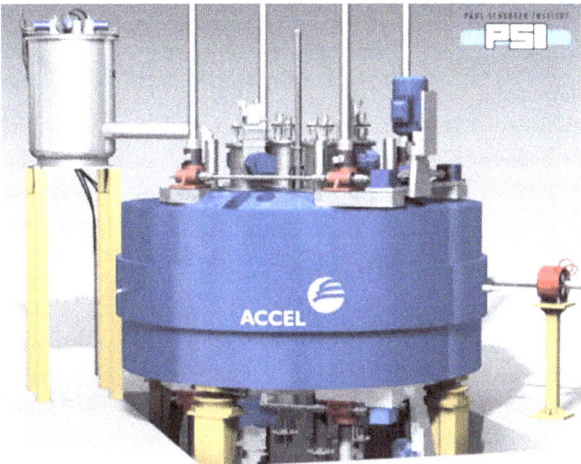

Fig. 2. Top: The 230 MeV proton cyclotron of IBA (Louvain la Neuve, Belgium); the first one has been installed in Boston, Massachusetts, USA [20]. Bottom: The 250 MeV superconducting cyclotron from Varian (previously ACCEL); the first one has been installed at PSI, Switzerland [23, 24].

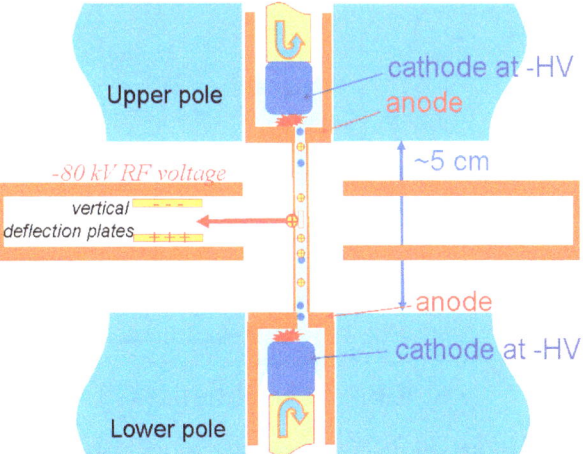

Fig. 3. Schematic view of an internal proton source in the center of a cyclotron and the vertical deflection plate.

the amount of protons can be increased by adding a filament close to one of the cathodes.

The beam intensity from an internal ion source suffers from noise, due to emittance fluctuations [21]. By selecting the central part of the emittance, the obtained beam intensity is less sensitive to emittance changes and a very stable beam intensity of better than 2% at kHz bandwidth can be obtained [22]. The intensity of the extracted proton source can be regulated by the arc current, the current through the filament (if present) and the gas flow. The response time depends on the source characteristics and can vary between milliseconds and seconds.

In the IBA cyclotron, therefore, a sophisticated control loop is used to set the intensity within milliseconds to the required value. At the PSI SC cyclotron, the ion source is operated at a constant arc current and intensity control is performed by means of a set of deflection plates, installed in the central region (see Figs. 3 and 4). By employing a vertical electric field between these plates, the beam is deflected in the vertical direction. A vertically limiting collimation system can intercept (part of) the deflected beam, so that this system can be used for very fast ($< 100\,\mu$s) beam intensity adjustment and/or fast ($50\,\mu$s) complete suppression of the beam intensity. Figure 5 shows a sequence of short (~ 3 ms)

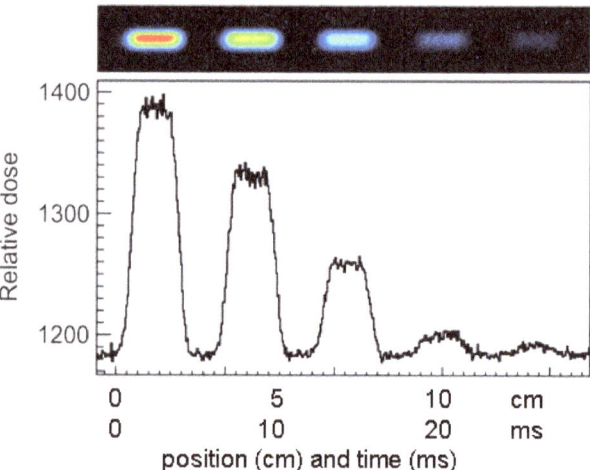

Fig. 5. The beam intensity as a function of position, of a beam swept over a scintillating screen (top), thus indicating the beam intensity modulation as a function of time and position (bottom).

beam pulses with different intensities, set by the deflector plates, while sweeping the beam along a line with a fast scanning magnet.

The finally obtained beam intensity is only a fraction of the proton current extracted from the source. Usually one or more apertures in the central region of the cyclotron select the central part of the emittance and determine the phase width of the beam pulse with respect to the RF frequency. This selection at low energy prevents particles that cannot be extracted anyway from being accelerated, thus optimizing the extraction efficiency (at PSI 80% is achieved routinely) and minimizing activation.

The emittance of the beam extracted from the cyclotron is in the order of a few π mm · mrad; usually the vertical emittance is larger than the horizontal emittance. Since the cyclotron is followed by a degrader system that increases the emittance so much that the initial emittance is insignificant, the asymmetry and exact beam shape of the extracted emittance play only a minor role.

2.2. *Energy degrader*

The proton beam extracted from a cyclotron is focused at a degrader consisting of one or more carbon blocks in the beam line, which slows the protons down (*degrades*) to the desired energy. To set the outgoing energy, the thickness of the total amount of degrader material must be adjusted. A carbon

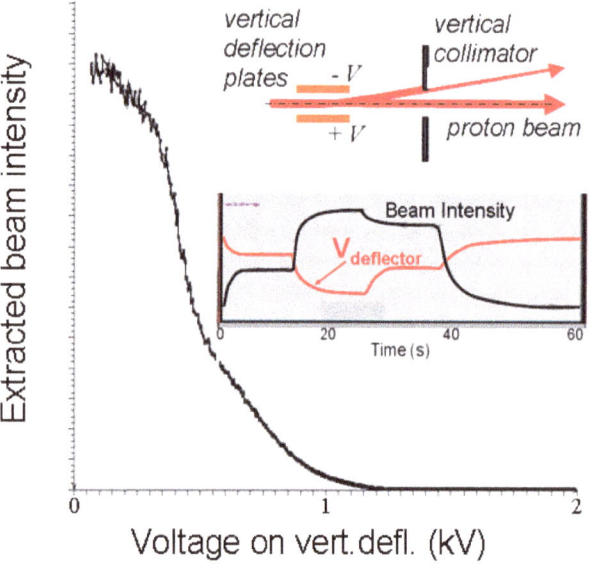

Fig. 4. The beam intensity from the PSI cyclotron as a function of the voltage on a vertically deflecting electrode pair. The inset shows the beam intensity as a function of time, with varying deflector voltage.

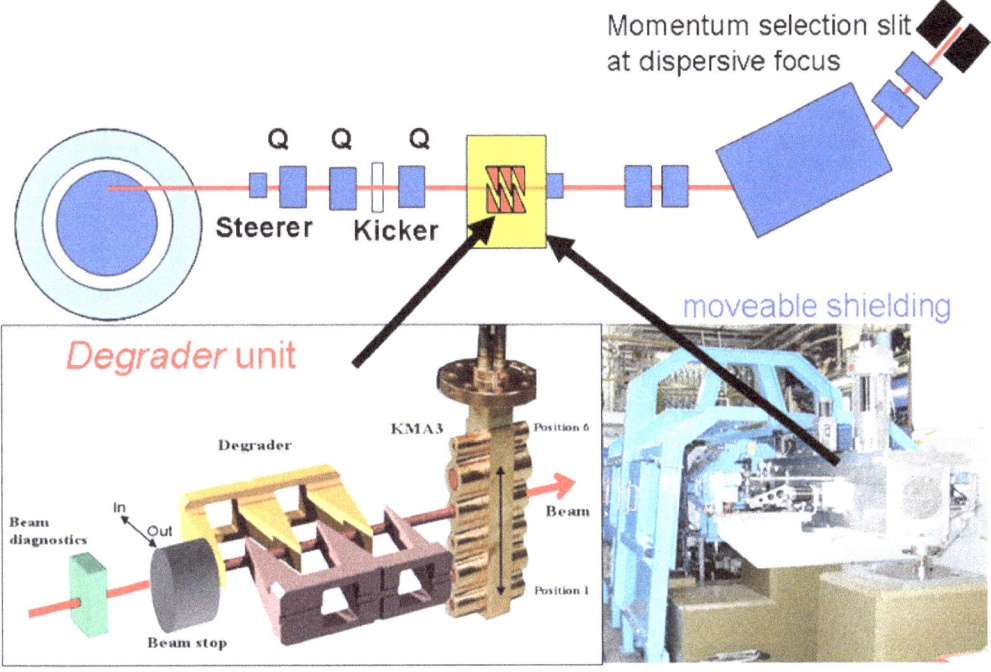

Fig. 6. Schematic drawing of a multiwedge carbon degrader layout and the beam transport system from the cyclotron until momentum analysis [25]. The red arrow indicates the proton beam direction. The picture at the bottom right shows the box in which the degrader, beam diagostics and collimators are mounted, with its dedicated moveable lead shield.

wedge, "rolled up" along the surface of a wheel, is very compact and is used in the IBA systems. A multiwedge degrader design optimized for very fast energy change has been designed at PSI and is shown in Fig. 6.

The degrader can have two functions. Firstly, it can be used as a range shifter to set the maximum needed energy in a certain treatment field, where the energy modulation is performed in the treatment nozzle just before the patient. Secondly, it can be used directly as an energy modulator, which then requires a fast response of the magnets [24]. Figure 7 shows the most important parameters that change synchronously to perform this quick energy change: the degrader setting and the current through a dipole magnet, as well as the error in the obtained beam energy behind the degrader.

The energy spectrum of the degraded protons depends on the amount of energy loss and, due to variation in the energy loss (*energy straggling*), it can obtain a substantial width; see Fig. 8 [25]. The width of the energy distribution contributes to the distal falloff of the Bragg peak. Figure 9 shows the effect on the distal edge of the Bragg peak for different selections of the accepted momentum width of a 75 MeV

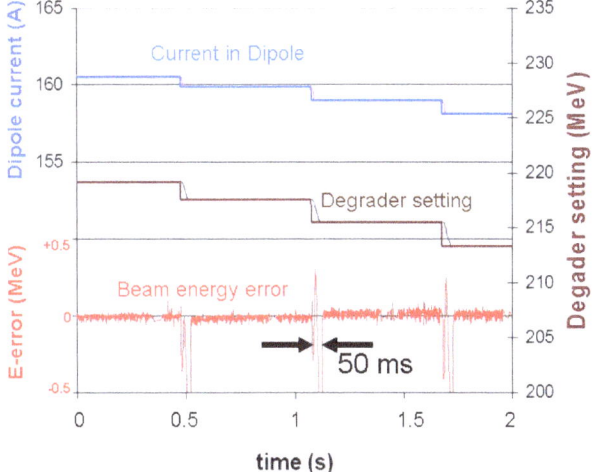

Fig. 7. Degrader setting and magnet current as a function of time. The lower line indicates the eventual discrepancy between magnet setting and degrader setting. The wiggles occur when the beam is suppressed during energy switching. This takes 50 ms.

beam. For eye treatments, the distal edge of the dose distribution is usually desired to be as sharp as possible, requiring a beam momentum spread down to $\pm 0.5\%$ or $\pm 0.25\%$. For deep-seated tumors the distal penumbra is usually dominated by the range

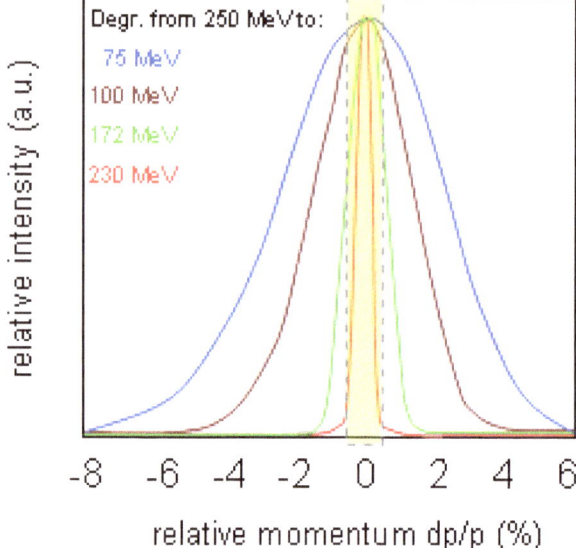

Fig. 8. The momentum spectrum of 250 MeV protons degraded to 230, 172, 100 and 75 MeV. The vertical dashed window indicates the typically selected momentum spread of ±0.5%.

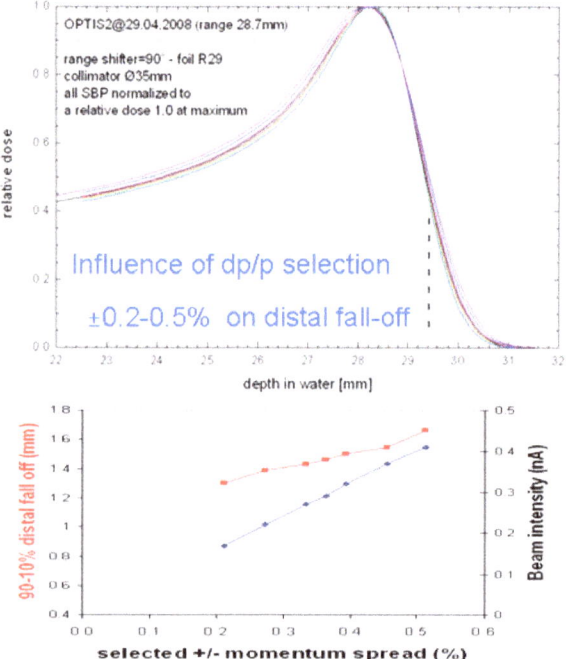

Fig. 9. The distal slope of 75 MeV Bragg peaks for different selected momentum spreads between 0.2% and 0.5%.

straggling of the protons in tissue, and therefore a larger momentum spread can be accepted.

The momentum selection is performed in a *beam analysis system* behind the degrader. This consists of two large dipole magnets, each bending the beam

over typically 45° or more. Using the quadrupoles in the beam line, the beam optics is designed to have a *dispersive focus* halfway between the two bending magnets. The ion optics also images the beam size defining aperture immediately behind the degrader, to this location. At this location the beam is spread out in the horizontal plane and a correlation exists between the energy of the protons and their distance to the optical axis (*dispersion*). Here a slit system selects the energy (magnet setting) and energy spread (slit aperture) that are accepted for the treatment. The second bending magnet compensates for the dispersion again and makes the beam transport almost independent of the beam energy (*achromatic*). When a small beam momentum spread is desired, one should be able to resolve a momentum spread of 0.2% at the slits. In order to prevent spoiling of the beam by scattering in an excessively narrow slit opening, and to obtain more spreading of the beam by dispersion than the beam size of the image (*resolving power*), a momentum dispersion of 30–40 mm per percent momentum spread is needed at the dispersive focus. This can be achieved with typically 45° of bending.

The beam losses at the slits performing the momentum selection increase with decreasing beam energy and/or decreasingly accepted energy spread, as can be seen in Figs. 8 and 9. However, activation at this location is hardly an issue, since the absolute beam intensity is already small (up to several nA).

The range in water must be measured frequently to verify the energy stability over time. The measured range stability depends mostly on the stability of the first bending magnet performing the energy selection. At higher energies the variations in beam energy from the cyclotron will become visible, since these beams do not fill the momentum slit completely (Fig. 8). It is therefore important to stabilize the vault temperature and/or monitor the magnetic field of the cyclotron (for example by means of a measurement of the beam phase with respect to the RF frequency) as well as of the analyzing magnet (for example by means of a Hall probe). Applying these methods, one can obtain a stability of the measured proton range in water of a few tenths of a millimeter during several months.

An undesired effect of the degrader is that it also causes a broadening of the beam due to multiple scattering, a process that has been investigated by many authors; see for example Refs. 39, 88–90. In addition, about 1% of the protons can be lost due to nuclear reactions per cm of traversed degrader material. Leaving the degrader, the scattered proton beam has an emittance that is too large to be accepted by the following part of the beam line. In order to define the beam emittance properly, and in order to confine the beam losses to well-known locations, the degrader is followed by an aperture system that can be adjusted to match the beam emittance to the desired beam transport mode. To understand the combination of a relatively large energy loss and sometimes a complex geometry, often the processes in the degrader need to be modeled by Monte Carlo methods [25].

The largest beam losses in the beam transport (up to 97%) occur at this stage. It is therefore advantageous to design the degrader and acceptance defining system such that the created activation consists of isotopes with a short half-lifetime. Appropriate shielding (see Fig. 6) is a necessity, as well as quick parts-exchange procedures. Figure 10 shows the transmission of 250 MeV protons from the entrance of the degrader until a location behind the energy selection. The beam intensity that must be transported by the beam lines to the treatment nozzle depends on the beam delivery method in the treatment area (scattering or scanning; see Sec. 3), as well as on the required dose rate at the patient.

It is worthwhile to optimize the design of the beam lines to obtain a high acceptance and to optimize the degrader layout and/or materials, with the goal of reducing the effect of multiple scattering [25]. In cases of limited available beam intensity, beryllium is often considered as a degrader material, since it has a lower scattering power. However, due to the larger length needed to get the same energy loss, the transmission is only a factor of 1.3 larger than with carbon. Since beryllium is rather brittle, toxic and creates more neutrons, the effort is probably worthwhile only when high intensities are needed at low energies (such as for eye treatments).

If a cyclotron is used to accelerate carbon ions, these also need to pass a degrader to obtain the appropriate energy. In this case the multiple scattering is much less because of the larger mass of the carbon ions. Nuclear reactions play a more important role due to larger cross sections and fragmentation of the carbon ions. The fragments, of which a substantial part travels typically with beam velocity, need to be separated from the ion beam. Due to the presence of fragments with magnetic rigidities similar to the carbon ions, slight contamination of the beam with helium ions cannot be avoided after beam analysis. However, the contribution of these fragments to the dose in the patient is small compared to that of the fragments that are created within the patient.

2.3. *Synchrotron*

Apart from the size of the machines, the most obvious difference between a cyclotron and a synchrotron is that an energy degrader is not needed when one is employing a synchrotron. But, especially, the

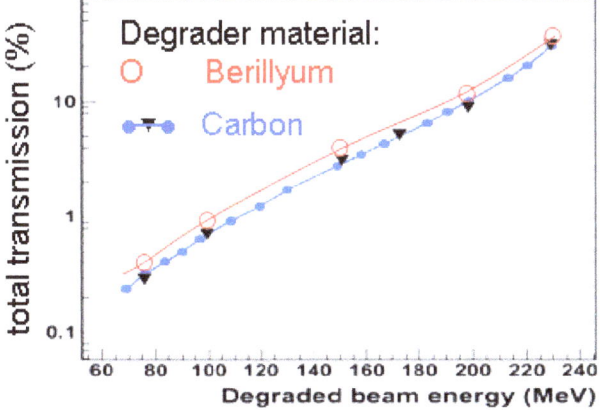

Fig. 10. Total transmission as a function of the degrader setting through the degrader, emittance-limiting apertures (44π mm · mrad in each direction) and momentum selection ($\pm0.6\%$) in the PSI beam transport lines. The transmission through the carbon degrader has been modeled (dots) and measured (triangles). The model has been used to calculate the transmission of a beryllium degrader [25].

Fig. 11. The 8-m-diameter proton synchrotron and part of the treatment rooms in the proton therapy facility at the MD Anderson Cancer Center in Houston [26].

capability to accelerate heavy ions has made the synchrotron the machine of choice for carbon ion acceleration to date.

Dedicated proton synchrotrons [5, 26, 27] with a typical diameter of 6–8 m, as well as synchrotrons for carbon ion therapy [27–35] with a typical diameter of 25 m, are offered by companies. The latter can also be used to accelerate other ion species, including protons. Acceleration by means of a synchrotron involves several steps. A high intensity particle beam (10–20 mA protons [26]; 0.25 mA carbon ions [27]) is obtained from an external ion source. If the synchrotron is designed for ions, several ion sources are switched on continuously, each optimized for one specific ion type. The ion sources can be commercially available proton sources, or ECR (electron cyclotron resonance) sources for carbon or other ions. By means of a switching magnet the source is selected. The particles need preacceleration before injection into the synchrotron ring. This is usually done by means of two subsequent linear accelerators; an RFQ (radiofrequency quadrupole) and a DTL (drift tube linac).

When one is filling the synchrotron, the particles must have a minimum energy of typically 4–7 MeV/nucl. The higher this energy is, the more particles can be stored in the ring, since the repelling effect of the space charge forces is less destructive at higher energy. When the filling process has been completed and the ring is filled with about 10^7–10^8 carbon ions or 10^8–10^{11} protons, the particles are accelerated to the desired energy by means of an RF cavity in the ring. This acceleration phase lasts about a second. The RF cavity needs to vary its frequency from typically 1 to 8 MHz, synchronous to the velocity of the particles in the beam. Therefore a "tune-free" cavity, thus with a low quality factor, is used. It is usually equiped with a magnetic core to obtain some power efficiency. Synchronous to the increasing momentum of the particles, the magnets of the ring are ramped up. When the desired energy (which can be anywhere between the injected energy and the maximum energy) has been reached, the acceleration is stopped and the particles are stored in the ring.

The stored beam circulates with stable betatron oscillations. Slow extraction from the ring is done by careful kicking of the beam with an RF kicker followed by an electrostatic extraction septum and a septum magnet [26, 29]. Only when particles need

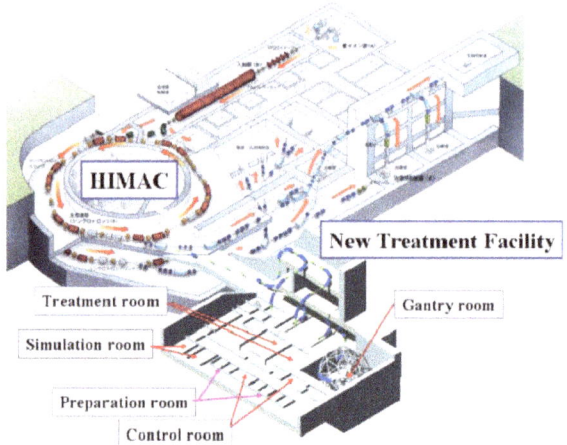

Fig. 12. The HIMAC double synchrotron and carbon ion therapy facility, together with the planned extension with new treatment rooms and a gantry for carbon ions [35].

to be extracted an RF perturbation, which includes frequencies of the betatron oscillations, is switched on to excite the betatron oscillation. The amplitude of the betatron oscillations increases, and when particles reach the other side of the septum, they are extracted. Thus, a sliver of the circulating stored beam is being "peeled off" and sent without further losses to the treatment area. The advantages of this method are a very fast (100–200 μs) response to switch extraction on and off, and a beam position stability of 0.5–1.0 mm.

When the necessary amount of particles is extracted, the remaining content of the ring is deposited on a beam dump. In several synchrotron systems, the particles are decelerated prior to dumping, to minimize the amount of radioactivity at the beam dump. One sequence of filling, accelerating and emptying the ring is referred to as a *spill* and typically takes 3–10 s.

The rather transparent structure of a synchrotron enables an accurate determination of the extracted beam energy. This can, for example, be used to compare this energy with results of accurate measurements of the particle range in phantom material [91].

The extracted beam is peeled off from the circulated beam, and is thus very sensitive to orbit fluctuations in the synchrotron. This causes fast intensity fluctuations of the extracted beam, which can be up to 50–100% (see, for example, Fig. 13) [26, 28, 33]. This is no problem for beam delivery systems based on scattering of spot scanning (see Subsec. 3.2), but

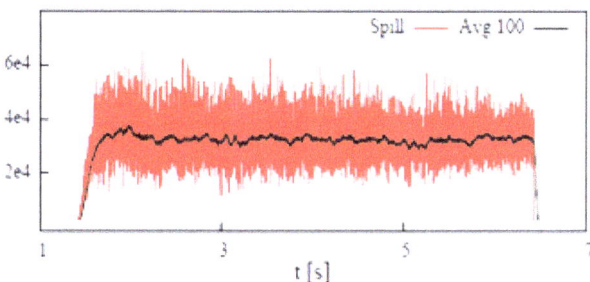

Fig. 13. The extracted beam intensity from the carbon synchrotron at HIT during one spill [33].

when continuous beam scanning is applied these fluctuations compromise this method.

Also, the spill structure of the synchrotron has implications for the types of beam delivery systems that can be used in the treatment areas. When *beam gating* (interruption of the beam extraction to cope with, for example, breathing motion; see Subsec. 3.3) or line scanning (see Subsec. 3.2) is applied, it is advantageous when the storing quality of the synchrotron as well as the stored intensity allows spill lengths of tens of seconds.

The horizontal emittance and momentum spread of a beam from a synchrotron can typically be up to a factor of 10 smaller than the emittance of a cyclotron beam. However, the emittance is not nicely Gaussian-shaped and shows large asymmetries. As a consequence of the horizontal extraction process at the HIT facility in Heidelberg, for example, the vertical profile has a Gaussian-shape, but the horizontal profile resembles a trapezoid with rounded corners [33]. For smaller beam widths and lower energies the effect is compensated for to some extent, due to scattering in the beam exit window and the treatment monitoring system. However, in cases where there is no smoothing component like a vacuum window or a degrader and the preservation of the small emittance is used advantageously for the beam transport, other methods are being studied to make the emittance more symmetric while keeping it small [28, 36]. This symmetry is essential for gantry systems, where the beam shape at the patient should not depend on the gantry angle. In cases where there is an extraction energy dependence of the beam emittance, the beam line settings must be given a beam-energy-dependent correction, in addition to the normal momentum scaling. Often the position as well as emittance of the extracted beam varies with energy.

The typical advantages of the synchrotron are the relatively small size of the components and the easy access to the machine parts for maintenance. Furthermore the activation of machine components is very low [37]. This can be attributed to very small beam losses during acceleration and beam transport in the synchrotron and beam transport lines, and to the fact that no degrading is performed.

Current developments for synchrotrons are concentrated on the improvement of the beam intensity. A proposal for a rapid cycling synchrotron [93] is aimed at a faster spill sequence. Other ongoing efforts at existing synchrotrons are aimed at an increase in the lifetime of the stored beam and at an increase in the stability of the extracted beam. At the HIMAC synchrotron, the intensity stability of the carbon beam has been improved to approximately 10% [28].

Recently a proposal from Balakin [92] has been pursued at MIT–Bates, USA, by building a small ring of 5 m in diameter for proton acceleration of up to 330 MeV. The protons are injected by a 1 MeV linear accelerator. Similar studies on a "tabletop" proton/carbon ion synchrotron have been reported by Endo *et al.*, of KEK, Tsukuba, Japan [95]. The relatively low cost of the facility is expected to be the major advantage of these small machines.

3. Shaping of the Dose Distribution

The final shape and quality of the dose distribution are made and determined in the last part of the beam line, which, for this purpose, is equipped with dedicated instruments to shape the beam, to verify beam parameters and to perform part of the verification of the patient alignment. Within this beam line section, usually referred to as the *nozzle*, many conflicting needs have to be satisfied and important choices, such as the beam spreading method but also the alignment verification method, must be made. Furthermore, space and weight impose very critical design conditions when the nozzle is mounted on a rotating gantry.

In order to shape the dose distribution to the tumor size, two different methods have been developed, briefly denoted as *scattering* and *scanning*. In the scattering technique, the particle beam is enlarged by scattering foil(s) placed in the beam path. In the scanning technique, a *pencil beam* < 10 mm in diameter is steered with fast magnets

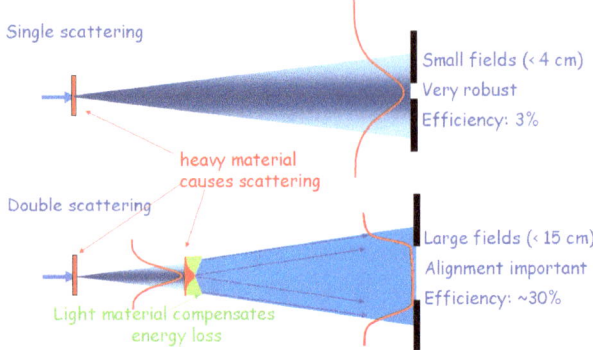

Fig. 14. The principle of single scattering and double scattering systems

and aimed sequentially at the different volume elements (*voxels*) inside the target volume.

3.1. *Scattered beam*

The most widely used and technologically simplest method (in principle) is to scatter the beam (Fig. 14). A good, detailed description of this technique is given by B. Gottschalk [39]. In this method basically the beam from the accelerator system, with a diameter of about 1 cm, is enlarged in the transversal direction by guiding it through one (*single scattering system*) or two (*double scattering system*) pieces of material that deflect the particle trajectories due to multiple Coulomb scattering. After passing through one or the first scatter foil, the beam broadens with an approximate Gaussian distribution. Depending on the required efficiency of the beam use, one can use only the central few percent of this distribution by employing a collimator. Inside the aperture the dose distribution is then typically flat within 3%. This is a very robust method and is often used at eye treatment facilities with a cyclotron directly providing proton beams of ~ 70 MeV with ample beam intensity [40].

In a double scattering system [39, 41] the beam is used more efficiently. Approximately 20 cm downstream of the first scatterer, a second scatter foil is mounted, with the purpose of suppressing or reducing the intensity in the center of the beam and scattering a part of these particles into the outer part of the beam. To achieve this, the central part of the foil is thicker. With an optimal shape of the scatter foil [39, 42], the sum of the scattering by the two foils is a flat dose distribution with a diameter that can be up to 15 cm (see Fig. 14). The first scatterer is usually

a flat foil of for example lead or combined with a range-shifting system. The second scatter foil can be designed in different ways. For low energies often a metal foil is used, with a disk or ring mounted at the center of the foil, which is thick enough to stop the particles that are close to the beam axis. To improve the efficiency of the system, more-complex-shaped ("contoured") foils are used. In these cases the foil has larger thicknesses in the center and sometimes [43] also at the edges. At the new eye treatment facility OPTIS2 at PSI application of a multi-ring foil is yielding sufficient intensity with 75 MeV protons (see Fig. 15).

Since protons traverse different amounts of material in a contoured scattering system, protons on the central axis will have a lower energy than protons at a larger distance from the beam axis. To avoid this, a sheet of light material can be added to the second scatter foil, also with a varying thickness, as shown in Figs. 14 and 16 [39]. The thickness is chosen such that the total amount of water-equivalent material is independent of the distance to the beam axis. Such compensated double scattering systems are the most widely used in systems designed for treatment of deep-seated tumors.

Range modulation is usually done with a low-Z material (for example lucite) wheel (see Fig. 16), which is placed in the beam with its rotation axis parallel to the beam direction. The beam traverses the wheel disk at some distance from the rotation axis. At the radius where the beam crosses the wheel, the thickness of the wheel varies along the circumference. A fast rotation of the wheel then modulates the range of the protons in tissue. A library of wheels with different modulation widths is used to adapt this width to the tumor thickness.

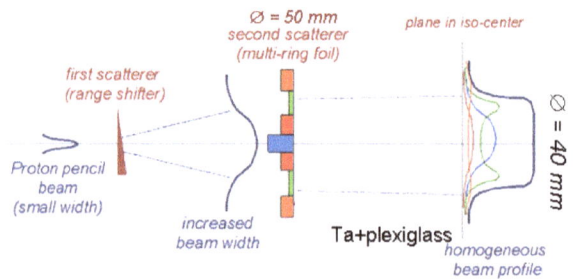

Fig. 15. The dual scattering system at the new eye treatment beam line OPTIS2 at PSI, employing a multi-ring foil. The thus obtained beam intensity is high enough to provide the required dose rate of ~ 15 Gy/min with 75 MeV protons.

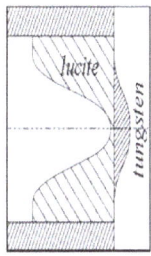

Fig. 16. The typical components of a dual scattering system for deep-lying tumors and large fields. From left to right: Compensated scatter foil, range modulator wheel, collimator and bolos [39].

The distal edge of the spread-out Bragg peak is shaped to the distal edge of the target volume by means of a *bolus* or range compensator, mounted as close as possible to the patient. A collimator defining the transversal limits of the dose distribution is mounted just before the bolus. The modulation wheel, collimator and eventual bolus are field-specific. So, if a gantry is used, these components have to be exchanged for every gantry angle. This has motivated current ongoing efforts to develop more remote control.

The energy modulation and range shifting act on the total beam cross section, which is designed to coincide with the tumor cross section, projected in the beam direction. Along particle tracks that cross the tumor where it is thinner than the maximum thickness, healthy tissue just in front of the tumor receives the same dose as the tumor. This is where pencil beam scanning is advantageous, as illustrated in Fig. 17. Another disadvantage of scattered beams is the production of neutrons in the elements interacting with the beam. Here also, pencil beam scanning systems are advantageous, since the neutron dose to the patient is typically a factor of 10 lower [44, 45].

3.2. *Pencil beam scanning*

Pencil beam scanning currently offers the best flexibility for shaping the dose distribution. Apart from a reduced integral dose with respect to scattered beams, robust field patching as well as intensity modulation is possible [46–48]. When "painting the dose" with a pencil beam, the lateral displacement of a pencil beam is performed by fast scanning magnets. The speed of the scanning magnets must yield a pencil beam displacement with a velocity of approximately 1 cm/ms at the isocentre, in order to minimize treatment time.

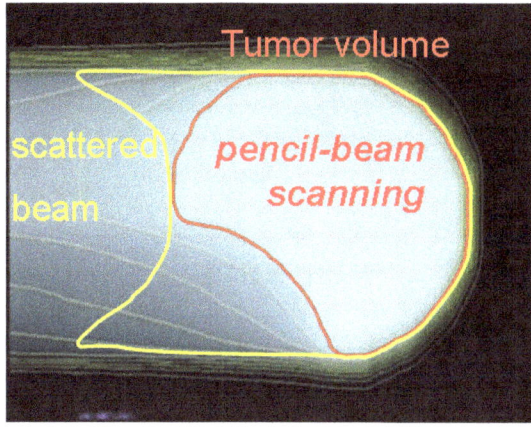

Fig. 17. Comparison of the dose distributions for a scattered beam (the light curve represents the 95% isodose line) and one obtained with pencil beam scanning (gray value; the dark curve is the 95% isodose line and coincides with the target volume).

Scanning magnets have also been applied in scattered beam systems. The beam can be *wobbled* or *scanned over a raster* with a periodic signal on two orthogonal scanning magnets, to obtain an extra beam spread, sometimes in addition to the one obtained by the scattering system. For proton beams this is applied to increase the field sizes up to 30–40 cm [4, 7, 12] and is sometimes referred to as *passive scanning*. Extra beam spreading by wobbling or raster scanning is especially important for carbon ions, which obtain much less angular deflection in a scattering system than protons. At HIMAC a wobbling system is used, followed by an additional scattering system. Within a useful area of $15 \times 15\,\mathrm{cm}^2$ a dose homogeneity of better than $\pm 2.5\%$ is obtained [97].

Figure 18 illustrates the two main methods for employing pencil beam scanning. The first method, discrete spot scanning, has been developed and

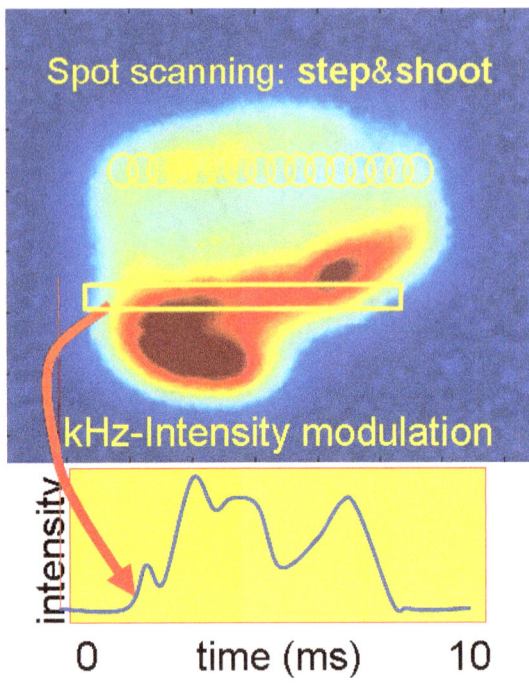

Fig. 18. Dose application with pencil beam scanning by means of spot scanning (top row of circles) and line scanning, where the beam intensity (curve at bottom) varies in time during the continuous sweep made by the scanning magnet.

implemented in Gantry 1 at PSI [49]. The pencil beam is aimed at a certain volume element in the target volume and the prescribed dose is applied. Then the beam is interrupted and the scanning magnet set for the next spot. This process is repeated until the appropriate dose has been applied to all voxels. In the system developed at GSI, Darmstadt (D) [50], the beam is not interrupted between the spots and the corresponding additional dose is taken into account.

For of these two schemes with discrete spot scanning, the specifications on the beam intensity and its fluctuations are not too stringent, so that both cyclotrons and synchrotrons are suitable to be used. Recently spot scanning has been introduced in the clinical operation at the proton therapy facilities at MD Anderson Cancer Center, Houston, USA and at RPTC in Munich (D), and recently the first patient has been treated with spot scanning at the Massachusetts General Hospital in Boston, USA.

Continuous scanning offers several possibilities, and is being explored at PSI and IBA at the moment.

In this advanced method the beam is swept over a line and the beam intensity is varied according to a time-dependent pattern prescribed for this line. A second scanning magnet then directs the beam one line further and the next sweep is done. In principle this is the fastest method, since it is limited only by the speed of the scanning magnets. However, complex control issues need to be dealt with to control the beam intensity with sufficient accuracy. In this scheme the parameter values (such as beam position and intensity) are *time-driven*. If the beam intensity cannot be controlled with sufficient accuracy, one can either rescan many times (then a very fast system is needed) or use an *event-driven system* in which the scan speed is corrected for the instantaneous beam intensity. Depending on the possibilities of the control system and the speed of the equipment, scanning patterns different from straight lines, such as contours following the tumor shape, can also be designed. The two scanning magnets must then be designed to be equally fast. If the beam transport system permits, the spot size can also be varied. When using a large spot size in the center of the target, an increased beam intensity can speed up the treatment a bit.

The scanning magnets perform a scan in the two transversal directions. The depth dimension must be covered by changing the beam energy. This can be done by inserting range shifter plates into the beam, just before the patient (Gantry 1 at PSI). Here the dose calculation program must take care of the increase in the transversal pencil beam size with the amount of range shifter material. This can be prevented when the energy of the beam is set before it enters the gantry, for example by choosing the appropriate energy at every spill from a synchrotron or with a degrader, as described in Subsec. 2.2. The beam line magnets and their power supplies must be suitable for following these fast changes. In all scanning systems, the time needed to change energy is the dominating factor that determines the total time needed for volume (re)scanning.

Pencil beam scanning gives rise to high local dose rates in the ionization chambers that are typically used for online dosimetry (beam monitoring) in the nozzle. Therefore thin air gaps or the application of shielding grids is necessary for obtaining fast-enough signals, not suffering from recombination. When employing pencil beam scanning, not only the integrated dose is of importance, but also the dose as a function of time and position. The ionization chambers need therefore to be equipped with segmented, stripped or pixelized electrodes. Also, scintillating gas detectors have been proposed for this purpose, because of their fast response and spatial resolution of 1 mm [96].

3.3. *Timing considerations*

The application of pencil beam scanning, being a dynamic treatment, has the risk of interplay effects with time-dependent processes in the beam delivery system, such as the spill sequence of a synchrotron, or the beam pulse frequency of a synchrocyclotron. This, however, can be dealt with by proper design and control.

A bigger problem exists for interplay effects due to motion of organs and tumors [51]. Because of range effects one can state in general that the motion problem is more problematic for particle therapy than for photon therapy. Currently several groups are working on strategies to deal with this problem; see for example Refs. 52–55, 73. Apart from developments in different new imaging techniques and control system developments, these studies concentrate mostly on the design of irradiation strategies, rather than rigorous adaptations of the beam delivery technology and equipment itself.

Assuming that one knows where and when to irradiate, one can pursue one or a combination of the three different strategies described below.

Beam *gating* (Fig. 19) has been applied for a long time by the Japanese groups [26, 56] and recently

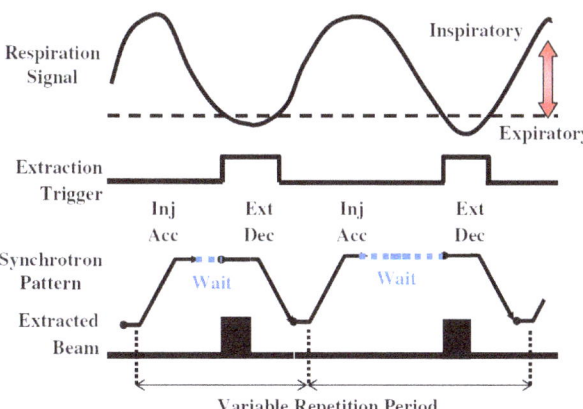

Fig. 19. An example of the operation pattern for respiration-synchronized (gated) irradiation [26].

also by the group in Houston [57]. A periodic motion (breathing) is monitored and only during one specific phase interval of this motion is the beam switched on. In this method one strongly relies on the periodicity and reproducibility of the motions. Methods requiring active cooperation of the patient (such as breath holding, or following a prescribed breathing rhythm) are also explored in combination with this technique.

The strategy of *fast rescanning* is explored at PSI [51, 58]. By using very fast 3D pencil beam scanning (for example 7 s for a 1 L volume), one can repeat this volume scan 15–30 times within 2 min. This may wash out dose errors, but still care must be taken for interplay effects. Also, the treated volume is larger than one may consider as optimal. A compromise if one cannot do fast energy scanning is fast rescanning per energy plane. Especially for synchrotron facilities this may be an option, since during the extraction, which lasts several seconds, the beam may be regarded as CW, as needed to perform fast 2D scanning. To date no clear data are available that demonstrate whether this simpler (and cheaper) method can deal with the organ motion problem as accurately as expected with volume scanning. Especially in cases where more *fields* (beam directions) are applied, this may be the case.

The third strategy is *adaptive pencil beam scanning*, in which the pencil beam location/direction or even energy [59, 73] is corrected by a signal from a motion detection system. This method is still in an early stage of development, however.

3.4. *Sharpness of the dose distribution*

The penumbra or sharpness of the dose distribution is of major concern when a tumor is located close to an organ needing to be protected from the irradiation. In pencil beam scanning the lateral penumbra is determined by the width of the pencil beam. It does not make sense to use a pencil beam with a width much narrower than the contribution of the multiple scattering in the patient. Therefore, typically, a pencil beam width of 7–10 mm FWHM is considered optimal, although for superficial tumors one might desire a smaller width. The pencil beam width directly determines the sharpest possible penumbra, when one is not using additional collimators.

In scattered beam systems a collimator is always used and mounted as close as possible to the patient.

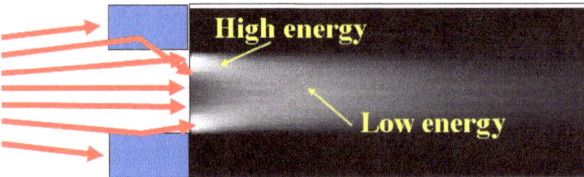

Fig. 20. Distortions of the transversal dose distributions, due to collimator scatter.

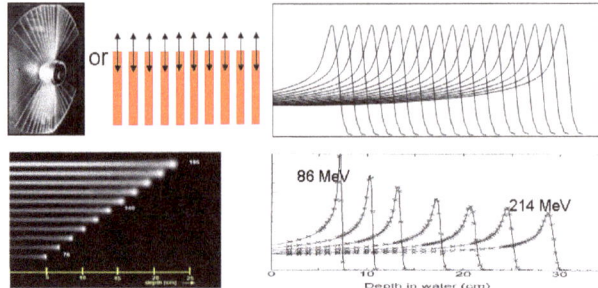

Fig. 21. The shape of the Bragg peaks for range modulation in the nozzle (top) and with an upstream degrader and beam analysis system (bottom).

The penumbra is determined by the initial beam size and scattering method (this determines the size of the virtual source), the distance from the scattering system to the collimator, the distance from the collimator to the patient and the material in the beam path (range shifter, energy modulator, bolus, monitors, etc.). A lot of work has been done to find rules for optimizing the beam characteristics in this respect [39, 60]. More recently, extensive Monte Carlo calculations have helped considerably to quantify the distorting effect of the different components [61–63]. The results of Monte Carlo calculations shown in Fig. 20 indicate that especially dose distributions in small fields can be distorted due to collimator scatter [64].

For pencil beam scanning, the amount of air between the vacuum exit window of the beam transport system and the isocenter is also of importance. Traversing 1 m of air already leads to a pencil beam increase of a few millimeters at the lower energies. However, often a vacuum window as close as possible to the patient is in conflict with the nozzle equipment needed to shape and monitor the beam. The general rule is that components causing scattering should be mounted either as close as possible to the patient (to minimize the lateral scatter distance) or as much upstream as possible to allow (re)shaping of the beam by collimators that intercept particles that are too far off in divergence or energy. Intermediate solutions give the worst results in this respect.

One of the advantages of pencil beam scanning is that "no" field-specific devices have to be mounted on the gantry nozzle. However, also with pencil beam scanning collimation systems may be advantageous for improving the lateral penumbra, such as for locations at shallow depth. One could think of remotely controllable systems to prevent the need to enter the room, such as an adjustable block or *multileaf collimators*, similar to those used in photon therapy.

As discussed in Subsec. 2.2, the energy spread in the beam contributes to the distal penumbra. Figure 21 illustrates that energy modulation in the nozzle creates a Bragg peak that does not change the width with the range. If the energy modulation is performed upstream of the nozzle, the maximum beam energy spread is always limited to 1% or 2% by the beam analysis system. As a consequence, a "new" beam with a relatively narrow energy spectrum is obtained and sent to the nozzle. The Bragg peaks of low energy beams will therefore be sharper than in the case of modulation inside the nozzle.

Sharp Bragg peaks can be advantageous when a sharp distal penumbra is needed. However, when they are stacked for range modulation, many more steps are needed in the modulation to obtain sufficient flatness in the longitudinal dose distribution, compared to the typical 5 mm steps used in the energy scanning at larger depth. As a remedy one could insert a range shifter into the beam nozzle, so that the minimum energy used from the degrader is about 100–120 MeV. The Bragg peaks in the patient will then be wide enough to perform upstream energy modulation with 5 mm steps. Another possibility (which can be combined with a range shifter) is to add a ripple filter [4, 65]. This increases the Bragg peaks of the low energies to the appropriate width. In both cases the distances of material to the patient must be as short as possible to avoid poor lateral penumbras.

Apart from the effects on the beam quality, attention should be paid to ergonomics and the time and/or effort it takes to replace or mount these devices. Therefore remote control to move components in or out or change their setting is of the utmost importance.

Last but not least, the struggle for available space around the nozzle should be given attention. Even in scanning systems, equipment must be mounted in the nozzle. The needed short distance to the patient may necessitate bulky mounting constructions, which are, however, in the way when one needs close proximity to the patient, such as in at treatments of the head and neck (shoulder) or when beams graze the skin.

4. Gantry

In order to direct the particle beam from different directions to the tumor, several methods are being developed. In ion therapy facilities, normally use is made of different fixed beams, using beam directions in the horizontal direction, the vertical direction and/or an inclined direction. A gantry — a mechanical system that rotates magnets of the last part of the beam line system around the patient [69] — offers the most flexibility.

The magnetic rigidity of the protons, however, implies the use of strong (> 1.5 T) and large magnets. A distance of at least 2 m is needed for the necessary distance to spread the beam behind a scattering system (*throw*). For pencil beam scanning a similar amount of space is needed to get sufficient lateral displacement of the scanned beam. Also, space is needed for dose monitors, range shifter, energy modulator and collimation systems between the exit of the last magnet and the patient. Therefore the typical diameter of these 100-ton gantries is 11 m, both for scattered beams and for scanned beams (see, for example, the Varian gantry — Fig. 23).

Due to the almost three times larger magnetic rigidity of carbon ions (approximately 6.8 Tm), the radius of curvature of particle trajectories is more than 3 m with conventional magnets. The huge difficulties and costs associated with a carbon gantry have resulted in only one existing gantry for carbon ions (570 tons), which has been installed in the HIT facility in Heidelberg; see Fig. 22 [66]. However, plans for a gantry for ion therapy also exist for the extension of the Japanese facility HIMAC [35]; see Fig. 12.

The first three gantries for protons originated from a design by the Harvard group [38] and were installed in the facility at Loma Linda [5]. The beam transport system layout is referred to as a *corkscrew* and consists of a double achromatic bending system, resulting in compact magnets and a rather short

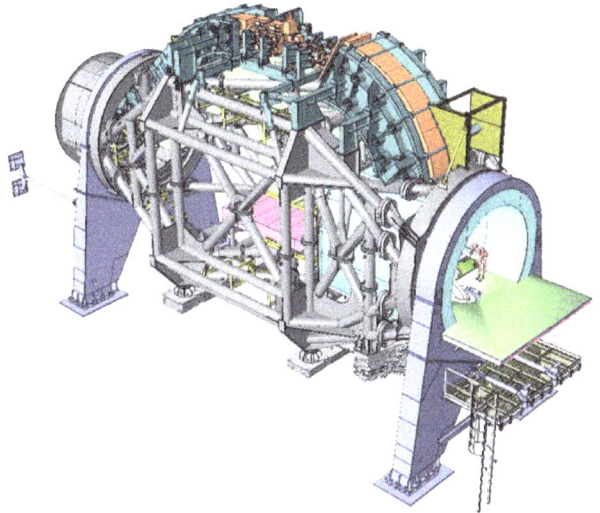

Fig. 22.   The gantry at HIT in Heidelberg for carbon therapy [32].

Fig. 23.   The construction phase of the Varian gantry designed for scanning beams (from Ref. 70).

gantry design. The nozzle has been designed for a scattered beam. Gantries developed later used larger magnets to comply with the large dispersion from the bending magnets [67–70], and designs have been made both for scattering and for scanning.

With the increasing demand for pencil beam scanning, dedicated nozzle designs have been made to allow (also or only) scanning. In the case of pencil beam scanning, the location of the scanning magnets, which are usually mounted at the exit of the last bending magnet, is the virtual source of the pencil beams. The large distance to the isocenter may require dedicated additional beam-focusing equipment, in order to the obtain a reasonably small pencil beam diameter at the isocenter.

At PSI, a compact proton gantry (Gantry 1) has been in use since 1996 [49]; it is optimized for pencil beam scanning. In this gantry a scanning magnet has been mounted before the 90° magnet that bends the beam toward the isocenter. The beam optics is designed such that the scanning magnet causes a parallel shift of the pencil beam at the isocenter. In this way the virtual source of the pencil beams is located at infinity and an orthogonal pencil beam arrangement is obtained. The other two orthogonal displacements are performed by inserting range shifter plates in the nozzle to shift the Bragg peak in depth and by shifting the table in the direction orthogonal to the magnetic scanning direction. The maximum speed of the dose application of 1 Gy/L/min is limited by the slow motion of the patient table. The thus-obtained orthogonal grid and parallel beam configuration have the advantage that there is virtually no field size limitation; by shifting the patient table, one can easily apply large fields.

The possibility to perform fast scanning in three dimensions has been a major design goal of Gantry 2, which has been built at PSI recently [58]; it allows for double magnetic scanning in a field of $12 \times 20 \, \text{cm}^2$, with parallel beam displacements in both directions; see Fig. 24. To permit scanning in the direction orthogonal to the bending plane, of course a relatively large gap of the last 90° bending magnet is needed. The magnets of Gantry 2 are laminated, so that a range change of 5 mm in water is achieved in $\sim$ 80 ms, complying with the fast energy change capability of the preceding beam line system (see Fig. 7).

If a collimator were used to sharpen the lateral penumbra, parallel scanning might not necessarily be optimal in the case of field sizes not much larger than the pencil beam size. In parallel scanning the effective source size is much larger than that of pencil beams coming from a virtual source [72], causing a less steep penumbra.

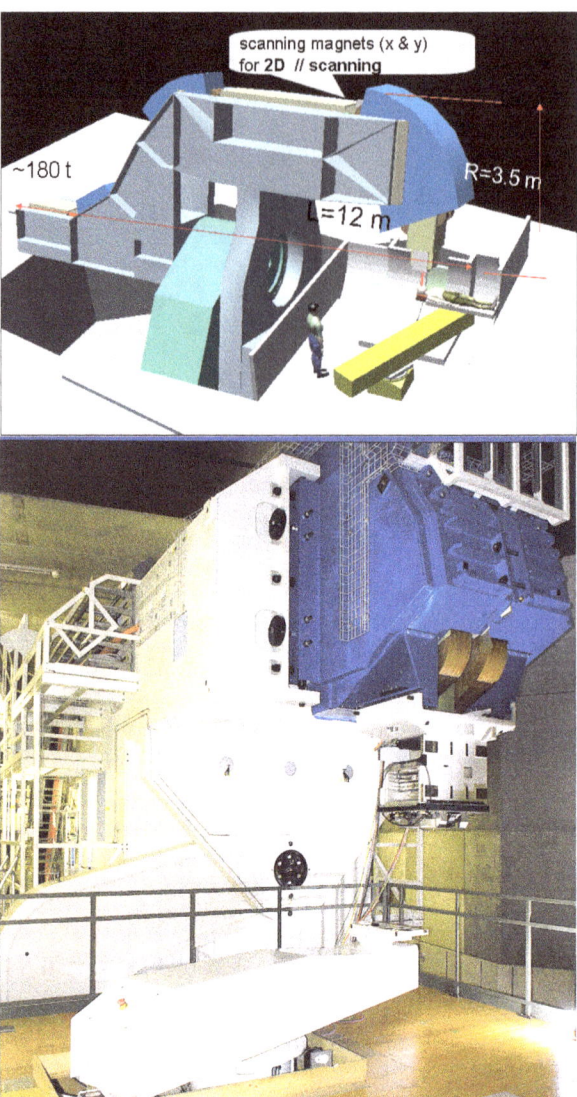

Fig. 24. Design and realization of Gantry 2 at PSI. This gantry has been designed for parallel beam scanning in two dimensions and fast energy scanning [58].

Figure 25 shows a scheme with one scanning magnet in front of the last bending magnet and a second one behind the magnet [71]. This provides a relatively compact system with a small pencil beam and moderate magnet gaps. Since it has no parallel pencil beam displacements, one should take care of the skin dose. Sophisticated treatment planning systems can design the dose weights of each pencil beam such that an orthogonal grid can also be obtained, offering the same advantages as parallel scanning. In addition, sharper lateral penumbras can be expected when collimation is added.

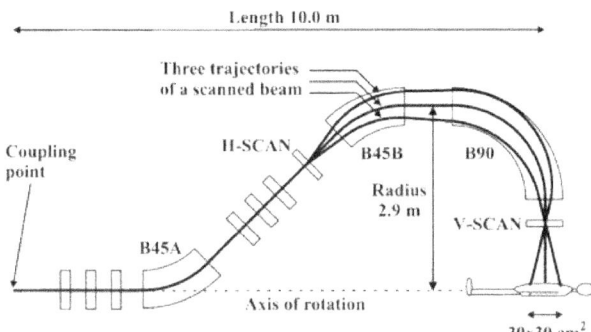

Fig. 25. Proposal for a gantry with double magnetic scanning, with a relatively small diameter [71].

A gantry design employing a beam focusing concept normally used in FFAG (fixed field alternating gradient) accelerators allows a very large momentum acceptance of ±15% [94]. The design of the magnet system shows very tightly packed focusing and defocusing magnets, with gradients up to 70 T/m, to be realized with superconducting magnets.

The required high precision and the need to deal with moving targets or organs necessitate advanced and fast methods for verifying the position of the tumor and critical organs with respect to the beam. Many methods exist or are under development for performing verifications immediately after the administration of the dose. In almost all gantry designs, provisions are made to x-ray the patient on the treatment couch for position verification. The position of the x-ray tubes in the gantry has important implications for the nozzle design as well as for the obtained beam quality. The x-ray tubes are typically mounted in or next to the nozzle, or opposite the nozzle. When they are mounted within the nozzle, a *beam's eye view* image can be made, which is very convenient for direct position comparisons with respect to the particle beam. However, the x-ray tube needs to be retracted before the proton beam is switched on and the presence of an x-ray tube in the nozzle may be in conflict with other components or with the beam line vacuum, which should end as close as possible to the patient. At PSI's Gantry 2 an x-ray tube is mounted at the outer curve of the 90° bending magnet, allowing a beam's eye view even during the proton beam irradiation, so that motion can be observed in directions orthogonal to the proton beam. Here one uses the advantage of a not-too-large distance to the isocenter as well as the absence of beam-shaping devices in the nozzle.

From the beam optics point of view there are several important constraints on the beam that is coupled to the gantry. At the rotation coupling point it is of the utmost importance that the beam is well aligned on the axis of the beam line mounted on the gantry, it must be symmetric in size as well as in divergence and it must be dispersion-free. Otherwise, the unfortunate situation would arise where the beam shape at the isocenter or even the transmission efficiency through the gantry would depend on the gantry angle. The alignment of the beam position and direction at the gantry entrance is especially important for gantries that use the pencil beam scanning technique, but small position errors are easy to compensate for, by small steering magnets in the gantry. This gantry-angle-dependent correction can vary from day to day or even between different treatments, when the beam has been directed to another area in the meantime. At PSI the beam optics of both gantries is designed such that the entrance of the gantry is imaged to the isocenter with a magnification factor close to unity. A collimator has been mounted at the entrance of the gantry, which thus defines the size of the pencil beam size at the isocenter. A misalignment of the fixed beam then results only in a loss of beam intensity at the collimator and has no effect on the position of the pencil beam at the isocenter.

5. Control and Safety Systems

The operation of the accelerator and beam lines (such as setting the current of a power supply, inserting a beam monitor or measuring the beam intensity) is performed by means of a *control system*. In addition to the control system, the *safety system* is the other major system controlling the facility. The purpose of the safety system is to protect patients from receiving an incorrect dose (distribution), to protect the personnel, patients and visitors from inadvertent exposure to radiation, and to protect the equipment and environment from heat, radiation damage or activation.

In the design of the safety system, a complex balance must be found between guaranteed safety for the patient and personnel on one hand and reliable, immediate and continuous availability of the beam on the other hand. Therefore it is important that safety-related functions are clearly identified as such and they must be performed by dedicated safety systems. These must work independently of the control system, and receiving and sending status information should be the only interactions between the safety systems and the control system. It is important that the concept of the control system architecture is related to the goals and the design of the safety systems. In their design, aspects need to be considered which concern questions like: which treatment room or which computer is controlling the beam (*mastership*), who can do what (*access control* to the control system) and when (*facility mode*, such as *therapy mode*, allowing only "start" and "stop" commands, and *maintenance mode*, allowing full access to all machine parameters), and how to guarantee a *separation of (safety) systems*. Often, an additional dedicated control system is controlling the therapy itself. It can be designed such that it also performs verifications during the irradiation process. Here it can be advantageous if the "set" and "verify" functions are performed by independent computer systems [49].

The implementation of the safety system is often highly site-specific. However, it is advantageous when the three protection fields (patients' treatment, protection of persons, and machine protection) are performed with different systems, since their performance and the rules for their design are not the same. For example, redundancy could be implemented for personnel and patient safety, but might not always be necessary for machine protection. Furthermore, the separation of functions reduces the risks and complexity that might occur in the case of a system in which the design is based on one combined operation and safety system in which "everything is connected to everything else." Of course, well-designed systems with a global function approach to the facility can be conceived without this separation, but the separated function approach leaves more freedom for further technical developments or equipment upgrades and allows easier testing.

In almost all recently built particle therapy facilities, the (complexity of the) control system and the safety systems have been a source of severe delay in the commissioning phase. The substantial effort to get the system working reliably and certified by authorities does not encourage the manufacturers to perform frequent upgrades, since these systems have become very inflexible to changes. This may hamper further developments, which are, however, inevitable during the ~ 30 years life time of the equipment.

6. Reliability

When new technology is developed for particle therapy, of course the focus is first to reach technical specifications that are of interest. However, nowadays proton and ion therapy have reached a stage that allows routine treatments of large patient numbers in a hospital environment, not much different from a standard large "normal" radiation therapy department. This routine, as well as financial constraints, implies the need for an efficient operation (i.e. an availability of the equipment of more than 95%) and logistics that assure a maximum patient throughput. Usually time for maintenance is limited and interruptions of more than three consecutive days would distort schemes of fractionated treatments. For accelerator and beam delivery devices this means that availability must be well integrated in the design concept. General design rules for optimizing the availability of a system encompass:

- Relaxed and well-known operating conditions of machine components, far away from breakdown limits;
- Quick access to components:
  - Reduce the number of actions to reach or exchange a component;
  - Reduce activation;
  - Design easy-to-use shielding measures;
  - Keep as much as possible equipment outside the vaults of the beam line and accelerator;
  - Optimize trajectories of cables and piping;
- Make replacement easy by using quick fits and dedicated tools, if necessary;
- Use modularity in all systems; make spares also as modules;
- Use uniformity in design to increase exchangeability and similarity of connections;
- Have redundancy, so that a repair can be done after the treatment hours;
- Ample diagnostics for all systems (beam diagnostics — e.g. Refs. 74, 75 — but also e.g. for RF, cooling and vacuum);
- A good user interface and easy tools for interlock analysis in the control system;

- A well-defined spare part policy and agreements with the vendor on delivery times;
- Specify the maximum collective dose annually received by service staff;
- Exercise and test actions with service staff;
- Specify reliability and perform acceptance tests.

With respect to relaxed operating conditions, measures like stabilization of the vault temperature and a relatively low temperature rise in cooling circuits help in decreasing wear as well as tuning time. Also, one can reduce downtime due to sparks in the RF cavity of the cyclotron by keeping the RF power on when the spark can be identified as a harmless *micro*-spark [74].

Aspects regarding service and availability need to be subjected to tests. One can specify the maximum time it may take to replace a component and/or to get a beam again within a specified time. Such specifications should be tested and demonstrations could be part of acceptance tests.

It should be noted that the definition of availability depends on who is using this number and that it does not always have the same meaning for all participants involved in a project. A definition related to the number of patients treated can give a very satisfying number for the financial administration. However, for the therapist who experiences the actual machine interruptions and how these are distributed over time, the same period could be experienced as a disaster. Therefore it is always important to give information on how the availability has been calculated.

7. Conclusions and Outlook

Proton and ion beam therapy have reached a mature state and are being applied routinely in an increasing number of hospital-based facilities [6]. When deciding for particle therapy, one of the first questions to be answered is: Which accelerator should be chosen? As has been made clear in the previous sections, this choice strongly depends on the type of beam delivery one wants to use or is planning for a future upgrade and relevant arguments have been presented.

Table 1 presents a comparison of the most relevant technical parameters of the (AVF) cyclotron and the synchrotron, the two accelerator types used today.

Table 1. Comparison of the most important parameters of the accelerators currently in use for proton or ion therapy.

	Cyclotron	Synchrotron
Protons	Established	Established
Carbon ions	In development	Established
Change particle	In development	Established
Machine size ($\varnothing$)	3.5 m (proton) 6 m (carbon)	6–8 m (proton) 20–25 m (Carbon) + injection system
Time structure beam intensity	Continuous	Spill structure, dead time > 10%
Fast energy scanning	Degrader	Wait for next spill
Activation	Degrader needs shielding	No problem
Beam intensity	"any," adjustable within ms	Limited in magnitude and range
Intensity stability	2–5%	10–20%
Scattering	Suitable	Suitable
Spot scanning	Suitable	Suitable
Beam gating	Suitable	Suitable
Fast continuous scanning	Suitable	Difficult due to pulse structure

At the time of writing, the majority of treatments are still performed using the scattering method. The advantages are that this method is well established and that a lot of experience exists. The more advanced pencil beam scanning is, however, coming out of the research laboratories and spot scanning is currently offered by several vendors. However, many developments are still needed for full exploitation of the possibilities that are available with fast pencil beam scanning.

The technological developments do not only include the hardware of accelerators, beam lines and gantries. A major and frequently underestimated part of the developments also encompasses control systems and systems taking care of the safety of the patients, staff and public.

With the increased commercial availability of particle therapy systems, rules for certification, safety and operation are playing an increasingly important role. On one hand, these help to create maintainable, standardized systems and clear protocols aimed at increased safety; on the other hand, the huge effort (and costs) needed to comply with these regulations hinders further development of systems implemented in the clinic. At the companies the high costs of such systems hardly allow big R&D departments equipped with prototype accelerator(s) and gantries. However, since technology will continue to develop, companies should make their products such that future upgrades are not hampered by rigid regulations or system complexity. Close research collaborations with existing research groups will therefore remain necessary, as well as the incorporation of end users early on in the definition process.

Presently the aims of new developments are focused on improving the current techniques (scanning, gating, tracking, image-guided, etc.), on increasing the efficiency of the patient treatment logistics and on reducing the costs of the equipment. This last goal is pursued through miniaturization and/or reduction of the versatility to attain a compact design. Interesting developments are going on among various groups specializing in accelerator physics, such as the LIBO concept (a linear accelerator using a cyclotron as an injector [77]), a linear accelerator using a dielectric wall that can withstand extremely high voltage pulses [78], particle acceleration driven by high power laser pulses [79–81] and FFAG accelerators, which can handle large momentum spread [82–85]. Some of the new ideas are being claimed to be ideal for a cost-efficient facility with one or two treatment rooms per accelerator, as in conventional radiation therapy. Although the costs per treatment room will not be much different from those of the 3–5-treatment-room facilities operating nowadays, the advantage is the lower cost to start at all.

Such new developments are highly interesting and important to support, but one should be aware of several major considerations [86]. First of all, it has to be ensured that the treatment quality of the new method is at least the same as what can be achieved with the current technology. If this is the case, one has to identify what the advantage of the new technology really is. Often one sees improvement on one parameter (such as the reduced size of the accelerator) but this can come with compromise(s) on other parameters (such as a pulsed beam intensity at a low repetition rate, complexity, or poor beam quality).

Especially when novel techniques are exploring the limits of current technology, one should question the reliability of the new technology, as well as how safe it is for the patient. In a laboratory one can allow for compromises on availability, but in the clinic one can only work with reliable, well-proven equipment. Last but not least, it is important not to underestimate the time it takes from a clever idea to certification of a product, and to apply it reliably and safely in clinical routine. Here sales offices of vendors are often more optimistic than the technicians and developers.

The biggest danger that particle therapy faces nowadays is that "cheap" systems will compromise the quality of the treatments. The standards of reliability and good practice developed during more than half a century of experience with particle therapy have been responsible for safe and high quality treatments to date. Compromises on quality would decrease the difference with photon treatments and could be very harmful to particle therapy in general, with consequences for the quality of patient treatment.

Acknowledgments

The author would like to thank F. Assenbacher, A. Coray, T. Böhringer, M. J. van Goethem, H. Reist, J. Verweij and S. Zenklusen of PSI, for (the input for) the figures with measurement results. In particular, he is grateful to G. Goitein, M. Goitein, T. Lomax and E. Pedroni for sharing their expertise in proton therapy, as well as the numerous colleagues at other particle therapy institutes for the countless discussions at conferences and workshops and the detailed information given during site visits.

References

[1] H. Eickhoff and U. Linz, *Rev. Accel. Sci. Technol.* **1**, 143 (2008).

[2] R. Wilson, *A Brief History of the Harvard University Cyclotron* (Harvard University Press, 2004).

[3] J. R. Alonso, *Nucl. Phys. A* **685**, 454 (2001).

[4] W. T. Chu, B. A. Ludewigt and T. R. Renner, *Rev. Sci. Instrum.* **64**, 2055 (1993).

[5] J. M. Slater *et al.*, *Int. J. Radiat. Oncol. Biol. Phys.* **22**, 383 (1992).

[6] Particle Therapy Coordinating Group (PTCOG), http://ptcog.web.psi.ch/patient_statistics.html

[7] A. Smith *et al.*, in *Hadrontherapy in Oncology*, eds. U. Amaldi and B. Larsson (Excerpta Medica, International Congress Series 1077, Elsevier, 1994), p. 138.

[8] H. D. Suit, *Clin. Oncol. (R. Coll. Radiol.)* **15**(1), S29 (2003).

[9] D. C. Weber *et al.*, *Bull. Cancer.* **94**, 807 (2007).

[10] M. R. Raju, in *Ion Beams in Tumor Therapy*, ed. U. Linz (Chapman & Hall, 1995), pp. 3–9.

[11] J. M. Schippers *et al.*, *Nucl. Instrum. Methods B* **261**, 773 (2007).

[12] Th. F. De Laney and H. M. Kooy (eds.), *Proton and Charged Particle Radiotherapy* (Philadephia, 2007).

[13] M. Goitein, *Radiation Oncology: A Physicist's-Eye View* (Springer, 2008).

[14] A. R. Smith, *Med. Phys.* **36**, 556 (2009).

[15] S. Humphries, *Principles of Charged Particle Acceleration* (John Wiley and Sons) (QC787.P3H86, 1986); http://www.fieldp.com/cpa.html

[16] Y. Jongen *et al.*, *Proc. Int. Conf. Cycl. Appl.* (Tokyo, 2004).

[17] L. Calabretta *et al.*, *Nucl. Instrum. Methods A* **557**, 414 (2006).

[18] J. M. Schippers *et al.*, *Proc. Conf. Heavy Ion Accelerator Technology 09* (Venice, 2009).

[19] Website of Still River Systems: http://www.stillriversystems.com

[20] Website of IBA: http://www.iba.be/health-care/radiotherapy/particle-therapy/particle-accelerators.php

[21] E. Forringer *et al.*, *Proc. 16th Int. Conf. Cycl. Appl.* (2001), pp. 277–279; and PhD thesis (NSCL, unpublished).

[22] J. M. Schippers *et al.*, *Proc. 18th Int. Conf. Cycl. Appl.* (Catania, Italy; 1–5 Oct. 2007), ed. D. Rifuggiato (L. A. Piazza, INFN-LNS Catania, Italy, 2008), pp. 300–302.

[23] M. Schillo *et al.*, *Proc. 16th Int. Conf. Cycl. Appl.* (2001), pp. 37–39.

[24] J. M. Schippers, *et al.*, *Proc. 18th Int. Conf. Cycl. Appl.* (Catania, Italy; 1–5 Oct. 2007), ed. D. Rifuggiato (L. A. Piazza, INFN-LNS Catania, Italy, 2008), pp. 15–17.

[25] M. J. van Goethem *et al.*, accepted for publication in *Phys. Med. Biol.* (2009).

[26] K. Hiramoto *et al.*, *Nucl. Instrum. Methods B* **261**, 786 (2007).

[27] Y. Hirao *et al.*, *Nucl. Phys. A* **538**, 541c (1992).

[28] T. Furukawa *et al.*, *Nucl. Instrum Methods A* **562**, 1050 (2006).

[29] L. Badano *et al.*, *Nucl. Instum Methods A* **430**, 512 (1999).

[30] Web site of Siemens http://www.medical.siemens.com

[31] A. Dolinskii, *Proc. EPAC 2000* (Vienna).

[32] T. Haberer *et al.*, *Radiother. Oncol.* **73**, S186.

[33] D. Ondreka *et al.*, *Proc. EPAC08* (Genoa, Italy, TUOCG01).

[34] P. Heeg *et al.*, *Z. Med. Phys.* **14**, 17.

[35] K. Koda *et al.*, *Proc. EPAC08* (Genova, Italy, TUPP125).

[36] J. Y. Tang *et al.*, *Phys. Rev. Spec. Top. Acccel. Beams* **12**, 050101 (2009).

[37] M. F. Moyers and D. A. Lesyna, *Radiat. Meas.* **44**, 176 (2009).

[38] A. M. Koehler, *Proc. 5th PTCOG Meeting: Int. Workshop on Biomedical Accelerators* (Lawrence Berkeley Lab., 1987), pp. 147–158.

[39] B. Gottschalk, unpublished book available at http://huhepl.harvard.edu/~gottschalk/

[40] E. Egger *et al.*, *Diagnostic Imaging and Radiation Oncology, Volume Radiotherapy of Intraocular and Orbital Tumors* (Springer-Verlag, 1993), pp. 57–72.

[41] A. M. Koehler *et al.*, *Med. Phys.* **4**, 297 (1977).

[42] E. Grusell *et al.*, *Phys. Med. Biol.* **39**, 2201 (1994).

[43] J. M. Schippers *et al.*, *PSI Scientific Report 2006*, pp. 62–63.

[44] X. G. Xu *et al.*, *Phys. Med. Biol.* **53**, R193 (2008).

[45] A. Pérez-Andújar *et al.*, *Phys. Med. Biol.* **54**, 993 (2009).

[46] A. J. Lomax, *Phys. Med. Biol.* **44**, 185 (1999).

[47] A. J. Lomax *et al.*, *Med. Phys.* **28**, 317 (2001).

[48] A. J. Lomax *et al.*, *Z. Med. Phys.* **14**, 147 (2004).

[49] E. Pedroni et al, *Med. Phys.* **22**, 37 (1995).

[50] T. Haberer *et al.*, *Nucl. Instrum. Methods A* **330**, 296 (1993).

[51] M. H. Phillips *et al.*, *Phys. Med. Biol.* **37**, 223 (1992).

[52] A. J. Lomax, *Phys. Med. Biol.* **53**, 1043 (2008).

[53] J. Seco *et al.*, *Phys. Med. Biol.* **54**, N283 (2009).

[54] J. Lambert *et al.*, *Phys. Med. Biol.* **50**, 4853 (2005).

[55] H. Paganetti *et al.*, *Phys. Med. Biol.* **50**, 983 (2005).

[56] S. Yamada *et al.*, *Proc., Asian Particle Accelerator Conference 1998, KEK Proc.*, pp. 98–100.

[57] Y. Tsunashima *et al.*, *Phys. Med. Biol.* **53**, 1947 (2008).

[58] E. Pedroni *et al.*, *Z. Med. Phys.* **14**, 25 (2004).

[59] N. Saito *et al.*, *Phys. Med. Biol.* **54**, 4849 (2009).

[60] M. Urie *et al.*, *Phys. Med. Biol.* **29**, 553 (1984).

[61] H. Paganetti, *Med. Phys.* **25**, 2370 (1998).

[62] H. Paganetti, *Phys. Med. Biol.* **49**, N75 (2004).

[63] S. W. Peterson *et al.*, *Phys. Med. Biol.* **54**, 3217 (2009).

[64] P. van Luijk *et al.*, *Phys. Med. Biol* **46**, 653 (2001).

[65] U. Weber *et al.*, *Phys. Med. Biol.* **44**, 2765 (1999).

[66] C. Kleffner and U. Weinrich, *Proc. EPAC08* (Genova, Italy, TUPP134).

[67] Y. Jongen, M. Abs, J. Bailey, *et al.*, *Proc. 6th European Particle Accelerator Conference: EPAC98*, (Stockholm, Sweden), p. 2354; http://accelconf.web.cern.ch/AccelConf/e98/PAPERS/WEP01C.PDF.

[68] J. B. Flanz, *Proc. 16th Int. Particle Accelerator Conference PAC95* (Dallas, Texas, USA; 1–5 May 1995) (2004).

[69] U. Weinrich, *Proc. EPAC 2006* (Edinburgh, Scotland, TUYFI01).

[70] http://medgadget.com/archives/2009/02/europe_approves_varians_proton_therapy_system_a_cancer_zipping_cyclotron.html

[71] H. Vrenken *et al.*, *Nucl. Instrun. Methods A* **426**, 618 (1999).

[72] A. M. Sabbas *et al.*, *Med. Phys.* **14**, 996 (1987).

[73] S. van de Water *et al.*, submitted for publication in *Phys. Med. Biol.*

[74] R. Dölling, *Beam Instr. Workshop 2004* (Knoxville, Tennessee, USA); *AIP Conf. Proc.* **732**, 244 (2004).

[75] P. A. Duperrex *et al.*, *Beam Instr. Workshop* 2004 (Knoxville, Tennessee, USA); *AIP Conf. Proc.* **732**, 268 (2004).

[76] P. Sigg *et al.*, *Proc. RF Systems Users Group Meeting APS* (Argonne National Laboratory, Mar. 2002).

[77] U. Amaldi *et al.*, *Nucl. Instrum. Methods A* **521**, 512 (2004).

[78] Caporaso *et al.*, *Nucl. Instrum. Methods B* **261**, 777 (2007).

[79] C. M. Ma, *Laser Phys.* **16**(4), 639 (2006).

[80] I. Veltchev *et al.*, *Med. Phys.* **36**, 2760 (2009).

[81] V. Malka *et al.*, *Med. Phys.* **31**, 1587 (2004).

[82] K. R. Symon *et al.*, *Phys. Rev.* **103**, 1837 (1956).

[83] E. Keil *et al.*, *PAC 2005* (Knoxville, Tennessee, USA), pp. 1667–1669.

[84] S. Antoine *et al.*, *Nucl. Instrum. Methods A* **602**, 293 (2009).

[85] Blumenfeld *et al.*, *Nature* **445**, 741 (2007).

[86] M. Goitein, *Int. J. Radiat. Oncol. Biol. Phys.* **70**(3), 654 (2008).

[87] H. Blosser *et al.*, *Proc. Particle Accelerator Conf., 1989. Accelerator Science and Technology; IEEE* **2**, 742 (1989).

[88] G. Molière, *Z. Naturforsch.* **3a**, 78 (1948).

[89] V. L. Highland, *Nucl. Instr. Meth.* **129**, 497 (1975); erratum, *Nucl. Instrum. Meth.* **161**, 171 (1979).

[90] B. Gottschalk *et al.*, *Nucl. Instrum. Methods B* **74**, 467 (1993).

[91] M. F. Moyers *et al.*, *Med. Phys.* **34**, 1952 (2007).

[92] V. E. Balakin *et al.*, abstract in *Proc. EPAC 1988*, p. 1505.

[93] S. Peggs *et al.*, *Proc. EPAC 2002*, pp. 2754–2756.

[94] D. Trbojevic *et al.*, *Proc. PAC07*, pp. 3199–3201.

[95] K. Endo *et al.*, *Proc. EPAC 2002*, pp. 2733–2735.

[96] E. Seravalli *et al.*, *Phys. Med. Biol.* **54**, 3755 (2009).

[97] M. Torikoshi *et al.*, *J. Radiat. Res. Suppl.* **48**, A15 (2007).

J. M. Schippers is Senior Physicist at the Accelerator Department of the Paul Scherrer Institut (PSI) in Switzerland and Professor in Particle Therapy at the University of Groningen, the Netherlands. In 1990 he initiated the plans for a proton therapy facility in Groningen and started several research projects and collaborations on proton therapy instrumentation (dosimetry and gantry design) and radiation biology (normal tissue damage). Since 2001 he is responsible for the accelerator and beam lines of the new proton therapy facility at PSI and working on the development of new techniques for particle therapy.

Reviews of Accelerator Science and Technology
Vol. 2 (2009) 201–228
© World Scientific Publishing Company

Laser Acceleration of Ions for Radiation Therapy

Toshiki Tajima

Fakultät für Physik, Ludwig-Maximilians-Universität München,
Am Coulombwall 1, 85748 Garching, Germany
and
Photo-Medical Research Center, JAEA, Kyoto, 619-0215, Japan
toshiki.tajima@physik.uni-muenchen.de

Dietrich Habs

Fakultät für Physik, Ludwig-Maximilians-Universität München,
Am Coulombwall 1, 85748 Garching, Germany
and
Max-Planck-Institut für Quantenoptik,
Hans-Kopfermann-Str. 1, 85748 Garching, Germany
Dietrich.Habs@physik.uni-muenchen.de

Xueqing Yan

Max-Planck-Institut für Quantenoptik,
Hans-Kopfermann-Str. 1, 85748 Garching, Germany
and
SKL of Nuclear Physics and Technology, Peking University,
100871, Beijing, China
xyan@mpq.mpg.de

Ion beam therapy for cancer has proven to be a successful clinical approach, affording as good a cure as surgery and a higher quality of life. However, the ion beam therapy installation is large and expensive, limiting its availability for public benefit. One of the hurdles is to make the accelerator more compact on the basis of conventional technology. Laser acceleration of ions represents a rapidly developing young field. The prevailing acceleration mechanism (known as target normal sheath acceleration, TNSA), however, shows severe limitations in some key elements. We now witness that a new regime of coherent acceleration of ions by laser (CAIL) has been studied to overcome many of these problems and accelerate protons and carbon ions to high energies with higher efficiencies. Emerging scaling laws indicate possible realization of an ion therapy facility with compact, cost-efficient lasers. Furthermore, dense particle bunches may allow the use of much higher collective fields, reducing the size of beam transport and dump systems. Though ultimate realization of a laser-driven medical facility may take many years, the field is developing fast with many conceptual innovations and technical progress.

Keywords: Coherent acceleration of ions by laser (CAIL); target normal sheath acceleration (TNSA); radiation pressure acceleration (RPA); adiabatic acceleration; phase stable acceleration; ion beams; cancer therapy.

1. Introduction

Advancement in the treatment of cancer is one of the major challenges in contemporary medicine and science at large. The number of deaths from cancer is increasing and this disease is the number-one cause of deaths in many countries. In Europe and America radiation treatment for cancer therapy is now applied in most cases (to more than 50% of patients) and in other countries it is increasing. For example, in Japan the percentage is currently around 20%. With the advent of superior diagnosis and early detection of cancer in the future, it is anticipated that this trend toward increase in radiation therapy will continue [1]. What is the drive behind this trend?

We would like to present here one possible school of thought for this based on an oncological consideration. This idea is based on the following observation [2–6]. Dr. Molls argues: "In chemotherapy the tumor cell kill depends on the transport of the substance to the clonogenic cells and molecular targets, DNA repair capacity, repopulation, pO_2, pH, MDR, etc. In macroscopic tumors not all the subvolumes of the tumor, clonogenic cells and relevant molecular targets are reached by those doses of the medical substance which are needed for cell kill. In other words," he goes on, "the chemotherapy dose distribution is intrinsically inhomogeneous." On the other hand, Prof. Molls points out: "In radiation therapy the tumor cell kill depends on intrinsic radiation sensitivity, DNA repair capacity, repopulation, oxygenation status, etc. However, the entire tumor can be irradiated matched with the dose, which is necessary to kill all clonogenic tumor cells, even the most resistant ones." In other words, "it delivers a matched dose distribution," citing the superior feature of radiation therapy.

Most (typically more than 90%) of current radiation therapy is based on photon therapy (using x-rays driven by accelerated electrons). This choice arose because of the technology available with the electron accelerator at an affordable cost and with a sufficiently compact setup. However, as is well known, most ionizing radiations, including x-rays, deposit a larger dose upon their entry or early propagation in the body than in their later propagation and at the all-important tumor site in the patient's body. Such a dose is exponentially diminishing with penetration depth (see Fig. 1). Because of this property, in most

applications of radiation therapy it is desirable to irradiate the tumor from many directions in such a way as to maximize the dose at the tumor and minimize the irradiation and deleterious toxic radiation damage. The best-known therapy for such a strategy is perhaps conformal therapy and it is done with intensity-modulated radiation therapy (IMRT).

A notable exception to the decaying dose as a function of the depth is ion beams. The dose curve as a function of the depth for ion beams has a peak near the end of their propagation. This is due to enhanced emission of secondary delta electrons when ions become sufficiently slowed down, these electrons being responsible for the cancer-killing dose. This peak is known as the Bragg peak (Fig. 1). Because of the Bragg peak, the irradiation effect can be peaked at the tumor and minimized in other parts of the body's healthy tissues, if the energy of ions beams is chosen such as to deposit most of the energy at the depth of the tumor. With or without IMRT, the ion beams may thus provide irradiation capable of tumor cell killing while healthy cells are kept intact. There is certainly less toxicity to the surrounding healthy issues, thus ensuring a higher quality of life for patients. This is the motivation for ion beam radiation therapy [8]. On the basis of this advantage, there are now more than two dozen ion therapy centers around the world [9] and there is a worldwide consortium that promotes this research and clinical experience, called PTCOG (Particle Therapy Co-Operative Group).[a] One such installation is that at Heidelberg, Germany, shown in Fig. 2. This occupies a large building for the synchrotron accelerator, beam transport, and gantry.

Clinical evidence of this therapy has accumulated to show [5, 6] that (1) the five-year survival rate of patients treated with particle therapy is as high as in the case of surgery, even though patients that receive particle beam therapy are often those declared not suitable for surgery and are thus often more severe cases; (2) the quality of life of patients in particle therapy is much higher than that of patients under surgery and the radiation side effects are much less than those from x-ray therapy, and often particle therapy can preserve vital organs that may not be spared by surgery. On the other hand, it is true that the accelerators (cyclotrons or synchrotrons)

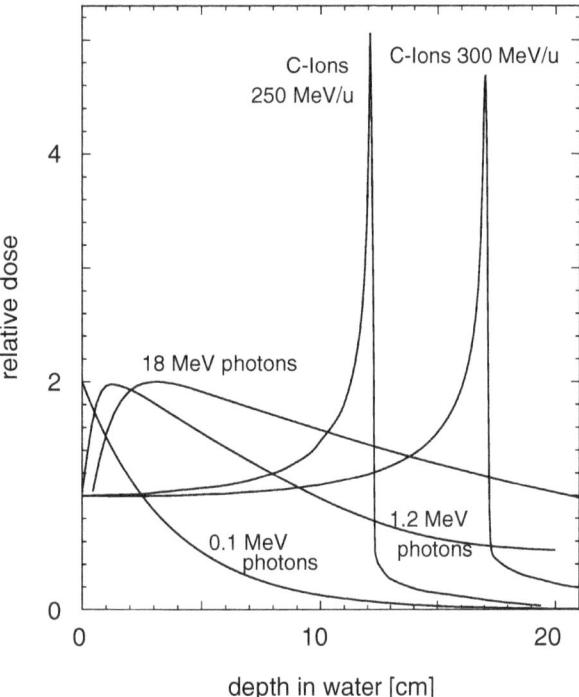

Fig. 1.   Comparison of depth–dose distributions in water as tissue equivalent. For large photon energies the dose decreases with depth, while ion (only carbon is shown, but it is similar to other ions) beams have a Bragg peak where a large fraction of energy is deposited [7].

[a]http://ptcog.web.psi.ch/index.html

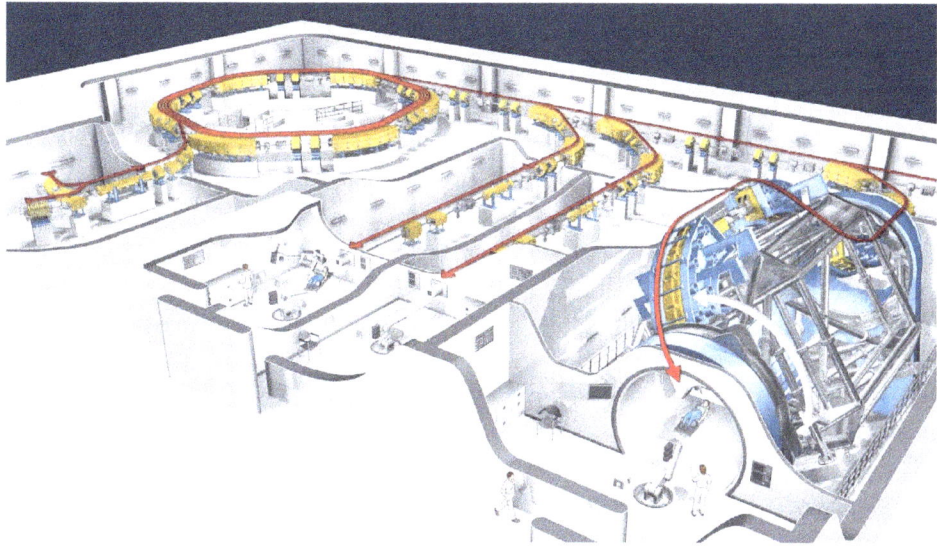

Fig. 2. Artist's view of the Heavy Ion Therapy Center (HIT) in Heidelberg [10], showing the large size of the installation. © Infografik: Martin Freiling/stern GESUND LEBEN

that drive ions (either protons or carbon ions) are larger than electron accelerators for x-ray therapy machines. Their gantry is also much larger than the counterpart to the x-ray machine. Thus the overall system of particle therapy is much larger than that for x-ray therapy. Moreover, the cost for the former is at least one order of magnitude higher than that for the latter. In many countries the cost to patients for the former is much higher than that for the latter, particularly because sometimes medical insurance does not cover "exotic and exorbitantly expensive therapies." Accordingly, a small fraction of the patient population receives treatment with particle therapy in spite of the above performance shown in clinical data. If the cost for particle therapy were to be reduced by an order of magnitude, it would become more affordable and more popular for many more patients and more different kinds of tumors.

In order to reduce the size and cost of the therapy installations now in use (shown in Fig. 2), many authors in recent years have ventured to state that the use of laser acceleration may introduce a novel way to accomplish this [11–16]. The size of cyclotrons and synchrotrons is primarily governed by the bending magnets and secondarily by the accelerating RF fields available. These, though varying in particular cases, tend to be of the order of a 10 m size machine for the synchrotron alone, not to mention the injector and beam transport, gantry, etc. The radiation shielding that has to be installed for the overall

accelerator adds to the size of the system. On the other hand, the present laser acceleration of ions provides a many-orders-of-magnitude greater acceleration gradient, of the order of 1 TeV/m. The first set of experiments on laser acceleration of ions [17–20] demonstrated this and initiated strong worldwide research efforts. Even though this feature of a very high acceleration gradient alone will not make an attractive and usable therapy accelerator, it nonetheless has attracted much research attention to the possibility of making a therapy system based on a laser. Progress in this field is so rapid that in a year or two old wisdom may already become obsolete in many respects. In order to assess the science and technology of laser ion acceleration accurately and examine its applicability to particle therapy for cancer (and some other diseases), we believe that it is imperative to review the latest progress comprehensively and consider its implications for the future. In addition, we want to make not just the accelerator more compact but also the rest of the system. This article deals with those problems and considers possible solutions.

We also note that in the future (and even currently) increasing early detection of tumors may become possible, so that the size of the tumor we need to cure by radiation should become smaller than what is typical at present. Of course, this should arrest the possibility of metastasis of the tumor as well as ensure better preservation of bodily functions.

It is thus entirely to the benefit of patients. The typical dose that is needed for therapy is 70 Gy, which is equivalent to 70 J of radiation energy deposited in a tumor weighing 1 kg. If the detected tumor is 1 g, then we need 70 mJ of particle energy. For a deep-seated tumor, protons with an energy of 200 MeV deposit about 10 MeV in it. (For a shallow tumor, the energy deposited in it per proton is about the same.) Thus the number of protons that need to be deposited amounts to approximately 10^{11}, allowing some room for a safety margin for covering the tumor. In other words, if we irradiate the tumor in 10 fractions and our laser-driven accelerator operates at 10 Hz over 2 min at a proton energy of 200 MeV, one shot should deliver 10^7 protons to the tumor. This means that even if the efficiency of laser-to-proton conversion is assumed to be as low as 1%, we need a laser energy of merely 1 J or so per pulse, if we can produce quasi-monoenergetic spectra. Such lasers satisfying many of these requirements are now commercially available, though certain elements of technology need further development. The question, rather, is whether we can deliver the necessary ions and how. We shall review the latest understanding in this respect.

We now identify the most crucial questions that need to be addressed in order to advance compact and affordable laser-driven ion accelerator systems for radiation therapy:

Issue #1: Can lasers accelerate ions (protons and carbon ions, for example) to 200 MeV per nucleon, and how?

Issue #2: Can this process be acceptable for use in radiation therapy? For example, are the dose, energy definition and discrimination and flexibility for uses such as with SOBP (spread-out Bragg peak) techniques. consistent with what we need? For this purpose, can we get a monoenergetic ion spectrum, or what kind of spectrum can we get? Is this suitable for the necessary use? How can we achieve this?

Issue #3: What are the other properties of laser ion acceleration, such as efficiency of laser-to-ion energy conversion, repeatability and fidelity, pulse shape and duration, and emittance?

Issue #4: How can we transport such beams and irradiate the patient? What is the property specific to laser-driven ion beams as to transport? Can we make the gantry for this system compact on the basis

of the specific property of this beam? Accordingly, what kind of system for the therapy machine may be envisioned?

Issue #5: What kind of radiation issue, such as shielding, has to be considered?

This review cannot cover all these issues with the full attention they may deserve. Rather, we try to address some of the main issues first, and we simply outline the current understanding of the others and refer to further literature.

2. Current Status of TNSA and Beyond

Energetic proton and ion beams with high beam quality, such as quasi-monoenergetic spectra, have been produced in the last few years from thick metallic foils (e.g. μm-thick aluminum) irradiated by ultraintense short laser pulses with appropriate target preparations [17, 21–23]. Most of the previous experiments fall under the regime known as target normal sheath acceleration (TNSA), with what we regard as relatively thick targets (Fig. 3). In this regime electrons are first accelerated by the impinging intense laser pulse and penetrate the target. Leaving the target on the rear side, the electrons set up an electrostatic field that is normal to the rear surface of the target. Ions accelerated from solids originate primarily from contaminant layers of water vapor and hydrocarbons on the target surface. As these targets are thick, the laser pulse is mostly

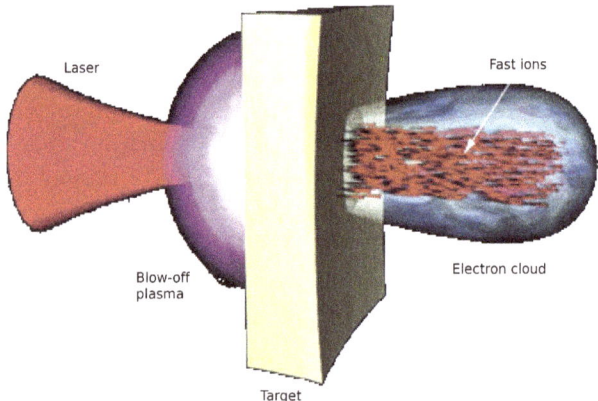

Fig. 3. Schematic picture of target normal sheath acceleration (TNSA). The high-intensity laser irradiates the front surface of a solid target, and electrons in the surface plasma are heated to high energies, which in turn propagate through the target and emerge from the rear surface. This induces large electrostatic fields that pull ions in the target or its rear surface [31].

reflected and the conversion efficiency of laser energy to ion energy is normally less than 1%; the maximum energy scales with less than a linear function of laser intensity. The maximum proton energy ε_p, based on the TNSA mechanism, has not improved since 2000 (~ 60 MeV). As reviewed by Robson *et al.* [24] [in particular their Fig. 1(a)], their fit of the data experimentally obtained (what they collected are all TNSA) shows that ion energies are proportional to the square root of the laser intensity I_L. The intensity I_L can be calculated from the electric field E_L using the relation

$$I_L = \frac{c \cdot E_L^2}{4\pi}. \tag{1}$$

In high field laser physics the amplitude of the laser pulse is usually not given in terms of the peak electric field E_L, but by the dimensionless normalized vector potential a_0,

$$a_0 = \frac{eE_L}{m_e c \omega_L}, \tag{2}$$

where m_e is the electron mass and ω_L the laser frequency. In more technical terms, a_0 is given by the laser intensity I_L and wavelength λ_L as follows:

$$a_0 = \sqrt{\frac{I_L[\mathrm{Wcm}^{-2}] \cdot \lambda_L^2[\mu\mathrm{m}^2]}{1.37 \cdot 10^{18}}}. \tag{3}$$

The proportionality $\varepsilon_p \propto \sqrt{I_L}$ is understandable (as we will examine more thoroughly in Sec. 5). In TNSA the laser energy is absorbed at the front surface of a solid target. Hot electrons are generated by a variety of mechanisms, among which is the Brunel mechanism [25]. Since electrons through interaction with the laser gain kinetic energy typically up to the ponderomotive potential

$$\Phi = \frac{m_e c^2}{e}\left(\sqrt{1+a_0^2} - 1\right), \tag{4}$$

the electron energy gain is approximately proportional to a_0 when a_0 is much greater than unity (which is the case for the data analyzed). Then through certain mechanisms the heated electrons transmit their energy to ions, which is thus again proportional to a_0.

In this review, however, we would like to report that there now emerges a new class of experiments and theory that supports such experiments, in which the regime of much more efficient and possibly higher-energy acceleration processes exists. In Sec. 5 we will scrutinize the process of this energy transfer

between electrons and ions, both for TNSA and for the new regime, which we now describe as coherent acceleration of ions by laser (CAIL), i.e. a regime of interaction less indirect than TNSA. However, it is worthwhile to look at one aspect of the laser interaction with electrons. At higher laser intensities above 10^{22} W/cm^2, numerical simulations seem to indicate that a laser could also accelerate protons to high energies [26–29]. With a compact high-repetition laser system, however, the highest proton energy is still lower than 20 MeV in all experiments [30]. In order to realize 200 MeV/u beams for proton/ion therapy, the laser intensity required should be as high as 10^{22} W/cm^2 [30], this not being available for therapy with state-of-the-art technology.

Experiments producing high-energy ions with submicrometer-to-nanometer-thick targets that are much thinner than those used so far have recently attracted strong interest. A typical physical situation is depicted in the sketch in Fig. 4. With the emerging nanometer target of diamond-like carbon (DLC), the conversion efficiencies are one to two orders of magnitude higher ($> 10\%$) even with modest-energy lasers (less than 1 J per pulse, and highly repetitive lasers) compared with those in the regime of TNSA with the thicker targets, and so the laser pulse can accelerate the ions to higher energies. The experiments show that the proton energy increases as the target thickness decreases for a given laser intensity, and that there is an optimum thickness of the target (several nm) at which the maximum proton energy peaks and below which the proton energy now decreases.

This optimum thickness for the peak proton energy is consistent with the thickness dictated by the relation [32–37]

$$a_0 \sim \sigma = \frac{n_0 d}{n_c \lambda_L}, \tag{5}$$

where σ is the (dimensionless) normalized electron areal density, n_0 is the electron density of the target, $n_c = m_e \omega_L^2 / 4\pi e^2$ and a_0 and d are the (dimensionless) normalized amplitude of the laser electric field and target thickness. Furthermore, we introduce the dimensionless parameter of the ratio of the normalized areal density to the normalized laser amplitude:

$$\xi = \frac{\sigma}{a_0}. \tag{6}$$

Henceforth, we simply call ξ the thickness parameter. The relation $a_0 \sim \sigma$ is understood as arising from the

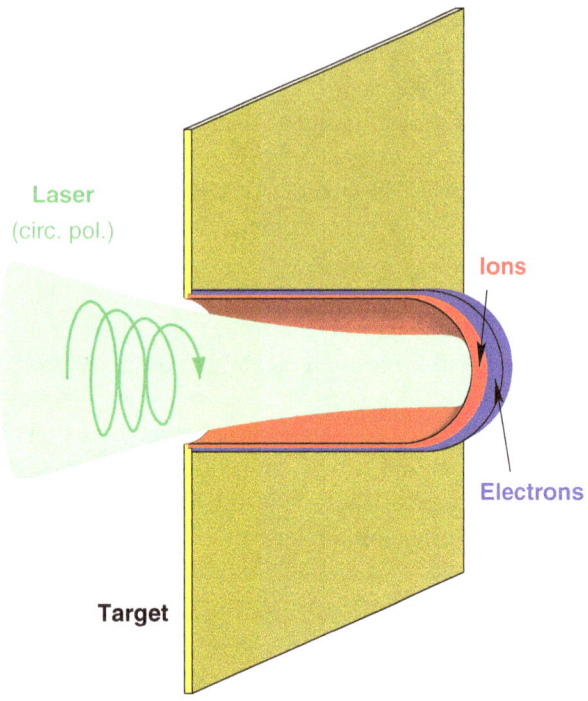

Laser
(circ. pol.)

Ions

Electrons

Target

Fig. 4. Schematic picture of coherent acceleration of ions by laser (CAIL), particularly radiation pressure acceleration (RPA). When the solid target is very thin, as characterized in the text, the laser can either substantially penetrate the target or directly move the target, while the laser radiation pressure pushes electrons in the target. The ponderomotive (light) pressure pushes electrons in the target with limited mass forward, while the induced electrostatic fields simultaneously pull ions in the target. The dynamics of electrons remains coherently slaving to the laser fields, in sharp contrast to TNSA.

condition that the radiation force pushes out electrons from the foil layer if $\sigma \leq a_0$ or $\xi \leq 1$, while with $\sigma \geq a_0$ or $\xi \geq 1$ the laser pulse does not have sufficient power to cause maximum polarization to all electrons. Note that this optimum thickness for typically available laser intensities is much smaller than the previously attempted target thicknesses (for ion acceleration). Thus we attribute the observed singularly large value of the maximum proton energy in the recent experiment [38] to the ability to identify and provide prepared thin targets of the order of nm to reach this optimum condition. In reality, at this target thickness the laser field comes to the point of partial penetration of the target, rendering the realization of the optimum rather sensitive. The experiments show that transparency plays an important role in energy enhancement. As we shall show, in this new regime (CAIL) with ξ of the order of unity (as opposed to $\xi \gg 1$ in TNSA) the electron dynamics

remains coherent, directly following the laser field. Thus we call this regime of acceleration coherent acceleration of ions by laser (CAIL). For details see Secs. 4 and 5.

3. Principal Direction

Having reviewed most of the previous laser ion acceleration experiments, we have reached the conclusion that: (i) laser ion acceleration has great potential, particularly in its accelerating gradient (of the order of TeV/m) and thus compactness of acceleration; (ii) however, its progress has slowed since the initial observation [17–19] in 2000 in enhancing its energy and other aspects with less energy laser drive, often limited to several-MeV energy gain [24, 30]; (iii) further, the energy spectra of ions remain broad [30] except in exceptional cases [21–23]; and (iv) the efficiency remains low [30]. On the other hand, although the numbers achieved are so far not overwhelming, some reports indicate one or two possible ways out, when, for example, the plasma density is near the critical value [33, 61].

The above situation may be broadly summarized as follows. The intense laser somehow heats electrons of the solid target to high energies, which contributes to large space charge separation on a very rapid time scale of electron runaway from the surface of the target (the rear surface) which pulls ions and makes them run after the escaped electrons. The electron heating involves complex processes — both coherent and individual particle processes (such as collisions) — and the original electron motion in the intense laser field is cascaded down to a thermal spectrum of electrons that drive ions, as described above. Firstly, this means that the electron spectrum is broadly spread (such as Maxwellian) and is certainly limited to or less than the ponderomotively driven electron energy of $m_e c^2 \sqrt{1 + a_0^2}$. The scaling to the intensity of the laser never greatly exceeds $\sqrt{I_L}$, as shown in the limiting energy of electrons [24, 39]. Secondly, hot electrons suddenly escape from the target, so that ions are unable to follow electrons, with the result that some fraction of these electrons run away from ions and the rest of them are pulled back toward ions. In any case, ions are unable to be smoothly accelerated; in other words, the adiabatic acceleration process is nonexistent. This nonadiabatic nature of the ion dynamics is the underlying reason for exhibiting properties (ii)–(iv). These features arise essentially from the mismatch between the group velocity of photons and the velocity of electrons subsequently energized and that of ions. Ions remain slow and nonrelativistic, while photons and electrons are relativistic.

Thus, our principal direction is firstly to utilize the photon energy more directly rather than cascading through multiples of collisional processes, and secondly to transfer laser energy to electron energy and to ions more adiabatically. When the solid target is too thick, most ions remain stationary and the momenta of photons are spread over broadly. Thus, we should limit the number of ions influenced by laser acceleration. Secondly, in order that the twofold interaction process of laser to electrons and electrons to ions can become more adiabatic, we need to slow down the photons and electrons and make them match the sluggish ions, at least initially until they reach a high speed [88]. Laser electron acceleration [40] may be easier in this sense, because light electrons at rest may be more easily trapped by the speeding photon-driven waves, whose velocity is near c, whereas the trapping velocity width [41, 42] is $\sqrt{e\Phi/m_e} \sim c$, where Φ is the ponderomotive potential. This ponderomotive potential is capable of inducing the wakefield with an amplitude of the order of $E_w \sim m_e \omega_p c/e$. This is why in the laser electron acceleration the fast wakefield $v_{gr} \sim c$ can still trap stationary electrons ("self-injection"). Meanwhile, for ions $\sqrt{e\Phi/m_i} \ll c$, and this value can only become $\sim c$, when $a_0 = \mathcal{O}(m_i/m_e)$ (ultrarelativistic), where m_i is the mass of ions. More importantly, in order to trap ions, the trapping velocity width of ions is much smaller than that of electrons:

$$v_{i,\text{tr}} \sim \sqrt{\frac{m_e}{m_i} a_0 Q} c, \qquad (7)$$

where Q is the ion charge in units of e (electron charge). Thus, in the regime of our interest, $1 \leq a_0 \ll m_i/m_e$, the accelerating structure has to move at a velocity within this $v_{i,\text{tr}} \ll c$. For ions to obtain net energy gain, the ion velocity needs to be within the trapping separatrix, which is situated over the velocity band that is centered at the phase velocity, v_{ph}, of the accelerating structure (we will call it a "bucket" sometimes) with a trapping width $v_{i,\text{tr}}$ [41]. Ions outside of this band (either above or below) simply oscillate in energy, but obtain no net energy gain. Even when the bucket velocity v_{ph} increases in time, ions that are trapped deeply enough may be kept trapped and, therefore, continue to gain energy from the bucket. This is the principle of adiabatic acceleration. Either when the velocity v_{ph} increases too suddenly or when ions are outside of the trapping separatrix, ions spill out or are left out of the accelerating structure. Although this is precisely the goal discussed in Subsec. 5.2, no published laser ion acceleration experiments have actually realized this condition.

In order to accomplish the first goal, one way is to adopt a very thin foil so that the mass contained in it is tiny. Alternatively, a diluter medium such as a dense gas or matter with clusters could be used. In order to accomplish the second goal, the most direct way to do so is to choose the density of the target material to result in a vanishing group velocity of photons such that ions can respond adiabatically. In this regard, it further helps if we can control the velocity of the accelerating structure to match the accelerated ion velocity. These considerations lead us to consider looking at very thin foil targets and, alternatively, at matter at or close to the critical density.

Another consequence of this general consideration entails a strategy to slow down the electrons motion after they are emitted from the target. This may be done by providing a concave geometry of the surface [43, 44] or other target preparations. One of the recent hopes is to employ circularly polarized laser pulses (CP). These, unlike linearly polarized laser pulses, do not have sudden high-frequency ($2\omega_L$) motions by linearly polarized photons, but result only in a smooth ponderomotive acceleration of the target [34, 37, 45–48]. If the CP ideally works, the ponderomotive force on electrons induces and matches the electrostatic force generated between the charge separated ions and electrons, and this keeps the overall dynamics smooth and adiabatic. A somewhat extreme and earlier rendition of this concept may be that of the radiation-dominant acceleration by Esirkepov *et al.* [49], where the laser photon pressure drives electrons to relativistic speed that drags ions also to relativistic speed closely following the electrons.

We now look at the accelerating energy gain as a function of the laser intensity or, equivalently, of a_0. The equation of motion of electrons is

$$\frac{d\boldsymbol{p}}{dt} = q\left(\boldsymbol{E}_L + \frac{\boldsymbol{v}}{c} \times \boldsymbol{B}_L\right). \qquad (8)$$

As is clear from this, the first term, due to the electric field, drives the electron primarily in the transverse direction, whose term is of the order of a_0. The second term, the Lorentz acceleration term, is proportional to a_0^2, if the electron motion is nonrelativistic, $a_0 \ll 1$. However, it is obvious that it is no longer proportional to a_0^2 when the velocity v becomes close to c, i.e. relativistic. In the relativistic case this term is also proportional to a_0. This is in agreement with the trend of many experimental scalings proportional to a_0 [24]. In other words, both terms scale as a_0 in the relativistic regime. There are, however, very important differences between these two terms. First, the directions of the forces are different. The electric term is transverse, while the magnetic force is in the direction of the propagation of the laser. Second, as a consequence of this, the transverse acceleration typically lasts while the laser accelerating phase passes by the individual electron, which is $\pi \cdot c / \omega_L$, while the longitudinal acceleration from the Lorentz term acts over the time that is potentially longer, particularly if the electron is now driven fast forward in the longitudinal direction. This is because the electron under the influence of this force may be accelerated to a relativistic energy in the direction of the laser propagation so that the time to be accelerated is longer. If this happens, the total momentum gain Δp can be proportional to a_0^2, even though dp/dt remains on the order of a_0. This means that it is crucial to have a long interaction time for electrons (and thus ions) with the laser pulse so that the time of interaction multiplied by the longitudinal force can give rise to the power of a_0 greater than a linear dependence on a_0. Then we can rapidly increase the ion energy by increasing a_0. This is unlike all or nearly all the experiments carried out so far under TNSA mechanisms. This is because all the past experiments have failed to make this adiabatic long time interaction. This is where we need innovation. For the CAIL mechanism, in particular for radiation pressure acceleration (RPA), one tries to make a break from this bind. This is a method for controling the group velocity of photons based on a circularly polarized pulse, by which one tries to keep electrons from gaining much energy.

4. Recent Experimental Progress

In this section we present recent experimental progress in laser ion acceleration, which shows

marked improvements over experiments in the regime of TNSA: in (i) the total conversion efficiency of laser energy into ion energy, (ii) the maximum observed ion energies, and (iii) the production of monoenergetic peaks in the ion energy spectra. We compare these results to those of previous experiments, which were based on the TNSA mechanism, established over the last 10 years in experimental and theoretical efforts. In TNSA one has observed for certain laser parameters, e.g. 1 PW lasers with 500 fs, maximum conversion efficiencies of 12%, or one has observed maximum proton energies of 58 MeV [17] and by filtering out small target regions could produce quasi-monoenergetic ion spectra [21–23]. At present, similar experimental values of efficiency, energy, etc. can be obtained with much smaller lasers. The formerly predicted laser parameters based on TNSA [24] to reach for medical therapy facilities ion energies of 240 MeV for protons or 450 MeV/u for carbon ions have been rather big and the laser would end up being costly. The new regime, where coherent dynamics of electrons in accelerating ions by laser (CAIL) plays a significant role, can yield scaling laws that lead to the prospect that short-pulse, high-intensity lasers with a high repetition rate may drive ion beams competitive with classical radio frequency accelerator systems.

4.1. Laser ion acceleration with ultrathin foils (CAIL)

In order to realize the CAIL regime, one wishes to employ nanometer-thick target foils together with high-intensity, short-pulse lasers. In the search for ultrathin freestanding foils that withstand strong ion and electron bombardment, DLC foils appear to be eminently suited. These have very high tensile strength, large hardness, good heat conduction, high heat resistance and, when used as stripper foils, show a large survival rate for ion bombardment. With a special production technique, freestanding foils with 75% sp^3 bonds — diamond-like bonds — can be produced.

The thickness of the DLC foils has been characterized by means of an atomic force microscope (AFM) with an accuracy close to 0.5 nm (see Fig. 5). Furthermore, the detailed depth composition — showing also front layer contaminations — was measured via elastic recoil detection analysis (ERDA; see Fig. 6).

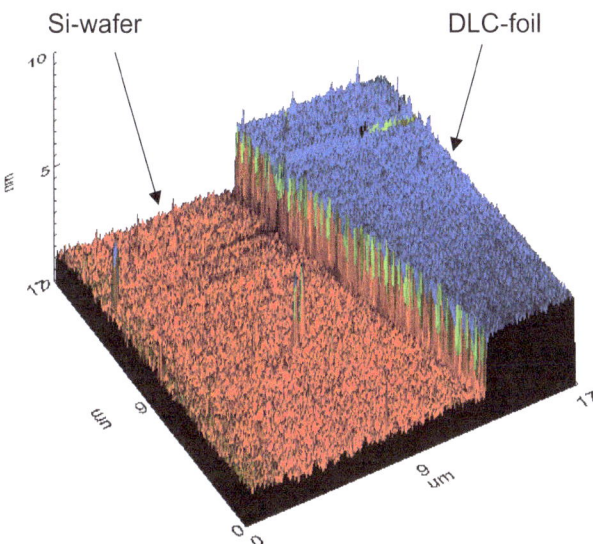

Fig. 5. The thickness of the deposited DLC foils can be measured with an AFM, by partially depositing on a small Si wafer.

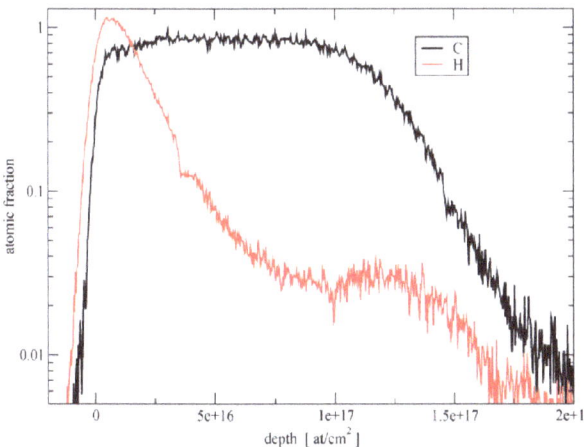

Fig. 6. Depth distribution of the carbon and hydrogen atoms of a DLC foil determined by ERDA measurements at a Tandem accelerator.

A second ingredient in these laser acceleration experiments [38, 53, 54] is an ultrahigh contrast of the laser pulses to avoid the preheating and expansion of the target before the interaction with the main laser pulse. The intensity of prepulses and the amplified spontaneous emission (ASE) pedestal are characterized with a third order autocorrelator, yielding typical values of 10^{-7} at 10 ps before the main pulse. This value was further improved by a recollimating double plasma mirror, which lets the low-intensity prepulse pass through, while it reflects the high-intensity part of the pulse. In this way an

estimated contrast of $\sim 10^{-11}$ was achieved. For the longer laser pulses the contrast was improved by self-pumped optical parametric amplification (SPOPA) [50] using nonlinear optical effects and thus avoiding the 50% energy loss of the double plasma mirror.

Ultrathin foils in two regimes have been investigated so far: (i) for laser pulses of 45 fs duration at the laser of the Max-Born Institute (MBI) in Berlin and (ii) for laser pulses with 700 fs duration at the Trident laser in Los Alamos. Laser-accelerated ions were measured with a Thomson parabola spectrometer.

4.1.1. *Using high-contrast lasers with 45 fs laser pulses*

The first class of experiments is as follows [38, 53]. At the MBI experiments with ultrathin DLC foils, irradiated by a 30 TW Ti:Sa laser have been carried out. After contrast improvement with a recollimating double plasma mirror (better than 10^{-11}), the laser had 0.7 J and a focused intensity of $5 \cdot 10^{19} \, \text{W/cm}^2$. The pulse duration was 45 fs and the wavelength 810 nm. With a mica crystal, used as a $\lambda_L/4$ plate, one can change the polarization from linear to circular. The normalized vector potential of the laser was $a_0 = 5$ for linear polarization and $a_0 = 3.5$ for circular polarization. For normal laser incidence we studied the ion spectra for various target thicknesses. In Fig. 7 the measured maximum proton and

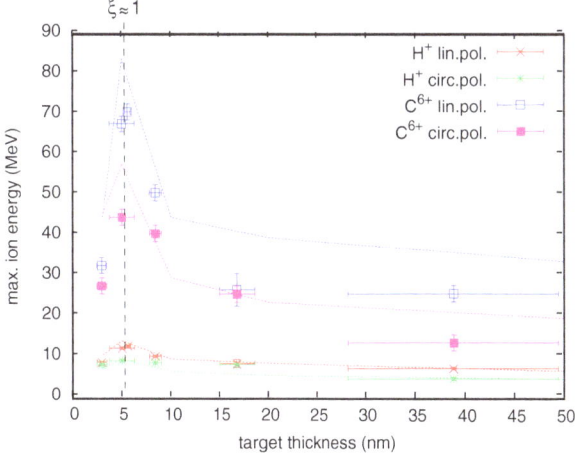

Fig. 7. Maximum cutoff ion energies as a function of target thickness in the regime of CAIL experiments [53]. Theoretical curves are from the CAIL theory in Subsec. 5.1.2. Observed values and theory (CAIL) are in good agreement over a broad parameter range. The optimal condition is realized at the thickness parameter $\xi \sim 1$.

carbon energies are shown as a function of target thickness. A pronounced maximum of ion energies at a particular target thickness clearly emerges at a target thickness of 5.3 nm. For linear polarization maximum energies of 13 MeV and 71 MeV for protons and carbon ions respectively were obtained; for circular polarization, 10 MeV and 45 MeV. This energy maximum in these experiments [38, 53] agrees with the condition of $a_0 \approx \sigma$ or $\xi \approx 1$ [32–34, 37].

We now compare the measured total conversion efficiency of laser energy into ion energy and the maximum ion cutoff energies with those of former TNSA measurements. In both cases more than an order of magnitude improvement is observed for these 5.3 nm targets.

A further advantage of ultrathin targets is that for circularly polarized light a quasi-monoenergetic peak is observed for the optimum target thickness for fully ionized carbon ions [53]. This is demonstrated for three consecutive shots in Fig. 8. In 2D particle-in-cell (PIC) simulations this monenergetic feature is nicely reproduced. We observe a more than 40-fold increase in conversion efficiency for this peak, when we compare it to an experiment [21], where a similar monoenergetic peak was obtained, by allowing carbon ions to be accelerated from a small area, which is in the regime of TNSA.

These three features, namely enhanced total conversion efficiency from laser energy to ion energy, increased ion energies, and the first indication for a monoenergetic peak in the carbon spectrum, are important improvements for reaching the required ion beams for cancer therapy with a high repetition rate (such as 10 Hz) and compact lasers. As long as we can keep the adiabatic laser pressure acceleration, a linear increase in ion energy with laser intensity I_L may be expected (see later), in contrast to the TNSA mechanism. The optimum laser pulse duration also needs to be explored.

4.1.2. Using high-contrast lasers with 700 fs laser pulses

Experiments have been carried out at the Trident laser (Los Alamos National Laboratory, LANL) using a short-pulse 200 TW laser of 700 fs at a central wavelength of 1053 nm [54]. Typical pulse energies were 80–100 J and repetition rates 1 shot per hour. In the experiments a double plasma mirror was used in front of the target to obtain a good laser contrast of about 10^{-12} at 50 ps before the main pulse. Due to the 50% losses at the double plasma mirror we had 40–50 J on target and a typical pulse duration of 700 fs. The peak intensities were $\sim 7 \cdot 10^{19} - 9 \cdot 10^{19}$ W/cm^2 and typical normalized vector potentials were $a_0 \approx 7$. In Fig. 9 three carbon spectra are shown, obtained for target thicknesses of 50, 30 and 10 nm for elliptic polarization $[I(p-\text{pol})/I(s-\text{pol}) = 0.5]$ and one measurement for linear polarization with a similar result.

In these experiments the maximum ion cutoff energy for DLC targets was obtained at a thickness

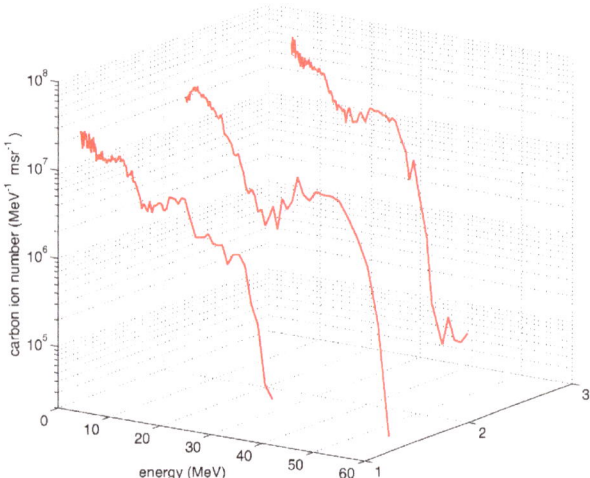

Fig. 8. Carbon spectrum for three consecutive shots using circularly polarized light at $5 \cdot 10^{19}$ W/cm^2 and a DLC foil target thickness of 5.9 nm in the regime of CAIL.

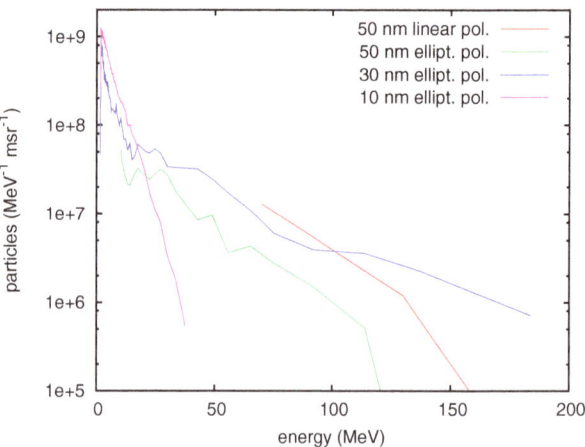

Fig. 9. Carbon spectra for DLC foils with thicknesses of 50, 30 and 10 nm in another experiment in the regime of CAIL, using the Trident laser with 40–50 J at $7 \cdot 10^{19}$ W/cm^2 and 700 fs.

of 30 nm, while the condition $\sigma \sim a_0$ would have predicted a value of about 7 nm. Thus, an approximately five times larger target thickness is required for these longer laser pulses, which are explained in Refs. 54 and 79 by an enhanced acceleration which occurs after a self-induced transparency of the target (see Subsec. 5.1.3). For the optimum target thickness 15 MeV/u and an estimated total conversion efficiency of laser energy to ion energy of 2% have been observed. In comparison, at the VULCAN laser McKenna *et al.* [56], measured with a 700 fs laser pulse, $\sim 2 \cdot 10^{20}$ W/cm^2, 400 J and iron targets of 100 μm thickness. The targets were heated to evaporate the hydrocarbon surface contamination. Then they accelerated predominantly iron ions and observed a cutoff energy of 12 MeV/u and a total conversion efficiency of 4%. These values are similar. However, in the TNSA regime 8–10 times larger laser pulse energies have been used.

4.1.3. *Common results for ultrathin targets*

Let us now discuss jointly all results for ultrathin targets in comparison with the thicker target results, where TNSA is the dominant acceleration mechanism.

In Fig. 10 we compare the conversion efficiency from laser energy to ion energy. At the optimum target thickness and 45 fs laser pulses an optimum conversion efficiency of 10% for laser energy into

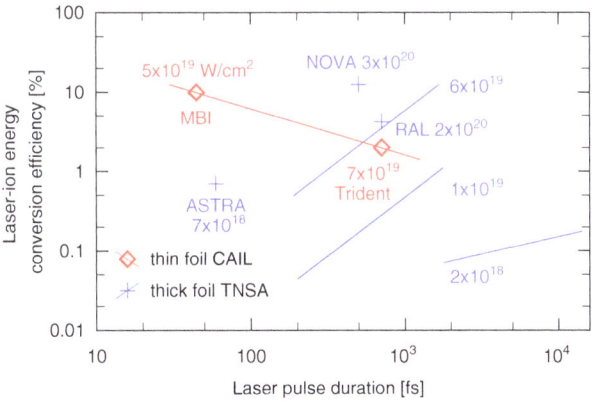

Fig. 10. Conversion efficiency of laser energy to ion energy, comparing results from thick targets and the TNSA mechanism to measurements with ultrathin targets in the regime of CAIL (red diamonds and line). For the TNSA mechanism smooth curves from the fluid model by J. Fuchs [30] are shown together with some experimental points: ASTRA [57], NOVA [17], RAL [56].

ion energy was obtained by integrating all protons above 2 MeV and all carbon ions above 5 MeV for the linearly polarized laser. Correspondingly, $\sim 9\%$ was obtained for circular polarization. For the 700 fs experiment a lower efficiency of about 2% was observed. The values are shown in Fig. 10 in comparison of the CAIL results with efficiencies for thicker foil targets (TNSA). We show the general trends of the TNSA mechanism by theoretical results from the fluid model [30], which describes the experimental data quite well. In addition, we show specific experimental results for the ASTRA laser [57], the RAL PW laser [56] and the NOVA PW laser [17]. We observe approximately a 50-fold increase in conversion efficiency for thin targets with the 45 fs pulses compared to TNSA at the same laser intensity, also taking into account our own measurements at larger target thicknesses. For 700 fs the efficiencies of thick target TNSA results and thin target results are comparable. If one increases the pulse duration, for short pulses the optimal target areal electron density $\sigma \approx a_0$, while for longer pulses σ has to be significantly larger to reach the maximum cutoff energy. Here one first has to reach the relativistic transparency of the target by expanding the target (see Subsec. 5.1.3), explaining in part the reduced conversion efficiency. For somewhat shorter laser pulses and cold adiabatic RPA, a 60% conversion efficiency has been predicted theoretically for higher laser intensities in an idealized 1D PIC simulation [47].

Experimentally the optimum conditions depend on many parameters, such as optimal laser focusing, to prevent heating of the walls of the bulged-out target [46]. In future experiments these optimum conditions need to be explored.

In Fig. 11 we compare the maximum cutoff ion energy for protons and carbon ions between measurements with ultrathin targets and μm-thick targets, where the TNSA mechanism dominates. In the figure, for TNSA only proton energies are shown for model calculations, which reproduce experiments quite well [30]. Approximately an increase by a factor of 10 is observed for the short laser pulses of 45 fs in the cutoff energies between TNSA and CAIL. An overall increase in energy occurs for both process for longer laser pulses. At the PW level the proton energies for TNSA vary from 58 MeV [17] to 13 MeV [56] for similar pulse energies of 500 J and 400 J. Thus, it is difficult to obtain a good comparison with the

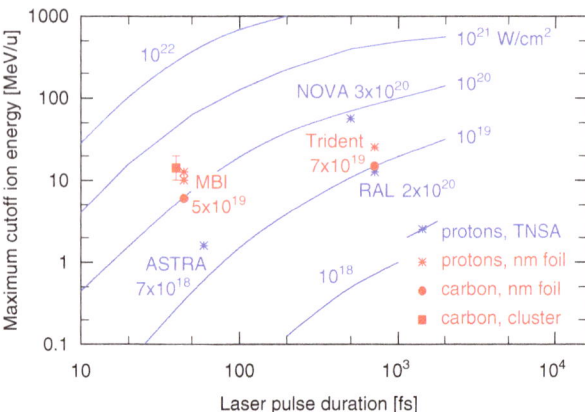

Fig. 11. Maximum cutoff energies of ions given in MeV/u as a function of laser pulse duration. The energy gain by CAIL experiments is embedded with red dots in the predicted curves of TNSA. Note that in shorter pulses, energies by CAIL are more than an-order-of-magnitude higher than TNSA. Here also, the results from the cluster target of Subsec. 4.2 are shown.

results of ultrathin targets. For the carbon ions and the longer pulse of 700 fs, a-factor-of-4 larger energies were observed for a factor of 4 smaller pulse energies, pointing to a clear advantage of the ultrathin targets in CAIL. For the short pulses of 45 fs, again an-order-of-magnitude improvement is seen for the ultrathin targets (CAIL).

4.2. *Laser ion acceleration with a gas target mixed with sub-μm clusters*

Fukuda *et al.* [58] have explored a different path of laser ion acceleration for hadron therapy: they use a gas jet target mixed with submicron clusters. They employ a rather small 4 TW TiSa laser with 40 fs (FWHM), 150 mJ and a 1 Hz repetition rate. The laser has a contrast of 10^{-6} and a focused intensity of $7 \cdot 10^{17}$ W/cm^2. By self-focusing in the gas jet the normalized vector potential of the laser probably reaches $a_0 \approx 2$ or a self-focused intensity of $1 \cdot 10^{19}$ W/cm^2. The target consists of a He gas jet with a density of $2 \cdot 10^{19}$ cm^{-3}, into which solid-density CO_2 clusters with an average diameter of 400 nm are dispersed at a cluster density of $3 \cdot 10^9$ cm^{-3}. This constitutes an average density at or near the critical density. A well-formed self-channeling phenomenon coincides with detection of high-energy ions. (Self-channeling coincided with the x-ray laser for the group of Rhodes [59, 60]). They observe in their ion spectrometers rather high ion energies in the range of 10–20 MeV/u for

carbon, oxygen or helium ions with a small divergence angle of $3, 4°$. This maximum energy value is much higher than expected from TNSA. However, an experimental parameter search has yet to be carried out to gain more knowledge, such as on efficiency and scaling.

It is noted that the average density near the critical density may have played an important role, perhaps as in the expriments of Matsukado *et al.* [33] and Yogo *et al.* [61]. In these experiments the enhancement of ion energies was noted when the density is in the neighborhood of critical. Thus the acceleration with clusters has commonality with the dynamics observed in the long-pulse thin-target experiments (Subsec. 4.1.2) after the laser burns through the target (see also Subsec. 5.1.3), when the density becomes critical through relativistic transparency. Near the critical density, as we noted, the group velocity of photons is small. In recent simulations [63] the maximum ion energy is observed to scale with the pulse length, with the intensity fixed and inversely proportional to the size of clusters. Thus, the nanostructured targets may provide an enhanced coupling of laser and ions [62]. Larger efficiencies may be obtained by increasing the pulse energy, the contrast of the laser, and using much smaller clusters with higher density.

4.3. *Dense relativistic electron bunches from ultrathin DLC foils used as relativistic mirrors to produce intense x-rays for medical diagnostics*

For modern ion cancer therapy a very good medical diagnostics is required in the same setup to have good feedback control of irradiation of a tumor at the requested position. For this the use of ultrathin diamond foils in combination with electron acceleration to produce brilliant x-ray sources is underway, leading to a combined setup of x-ray diagnostics and ion therapy.

The idea is to produce, first, very dense electron bunches by inducing their breakout from ultrathin foils with an intense laser. Then the electron sheet is accelerated in a laser half-cycle to electron energies of 15–100 MeV ($\varepsilon_e = \gamma \cdot m_e c^2$; $\gamma = 20$–300). A second laser beam is then injected opposite to the dense electron sheet with photon energies of $\hbar\omega_L$. After the coherent reflection one obtains x-rays with energy

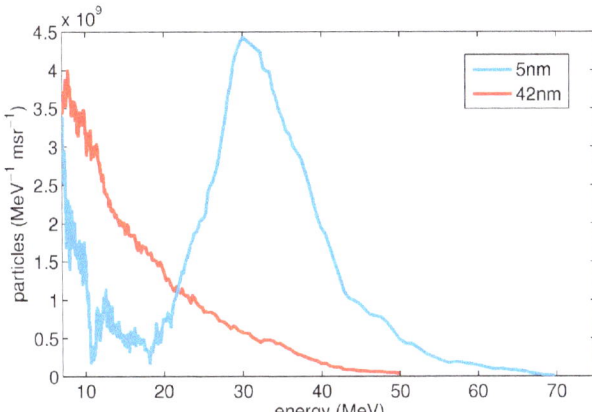

Fig. 12. Electron spectra, when irradiating 5 nm and 42 nm foils with $2 \cdot 10^{20}$ W/cm^2 and 500 fs pulse length at the Trident laser [51].

$\varepsilon = 4\gamma^2 \cdot \hbar\omega_L$. By shaping the DLC foil properly the focus of the x-ray beam can be chosen appropriately.

In Ref. 51 the breakout of electron bunches from ultrathin DLC foils has been observed for the first time. Figure 12 shows the electron spectra for a 42 nm foil without electron breakout and a 5 nm foil with breakout at the Trident laser with 80–100 J. Electron bunches with 30 MeV energy and a typical charge of a few nC have been obtained [64].

In the near future the study of the electron breakout at a 100 TW 25 fs TiSa laser will be performed. From the primary laser beam a secondary beam may be split off, which is then reflected from the laser-driven electron sheet to obtain x-rays. In this way the properties of the electron bunch during acceleration can be studied via x-rays, while the present electron spectra are measured only after a flight length of 1 m. Theoretical 2D simulations show that the dense electron sheet preserves its high density during the first part of its acceleration with good energy resolution [52].

If successful, the use of the intense x-ray beam in medicine with phase contrast imaging is desired, with a grid-based setup [65, 66], which leads to high resolution at a much smaller dose compared to present absorption techniques. This combined usage of laser acceleration and ultrathin foils provides a compact simultaneous setup for diagnostics and therapy.

4.4. *Summary of experimental status*

If one compares recent results of the ultrathin foil experiments (CAIL) with earlier thick target results

(TNSA), about a factor of more than 10 in improved conversion efficiency and a factor of more than 10 in improved maximum ion cutoff energies (and also the jet microclusters) have been observed. Important is the fact that theory (and simulation, discussed in Sec. 5) now has good agreement with this new class of experiments and thus has a predictive capability. It may be expected that future experiments in higher a_0 and other parameters in the CAIL regime will experimentally confirm the scaling laws with a_0 and laser pulse duration.

It has been demonstrated recently that nanostructured DLC foils may be produced. An example is shown in Fig. 13 for a DLC foil consisting of half-spheres with a 15 μm diameter. 2D PIC simulations for these structured surfaces predict a further increase in ion energy by a factor of 2–3.

If the scaling of the ion energy with laser intensity for circularly polarized laser light by adiabatic laser pressure acceleration is confirmed and can be maintained for longer laser pulses like 100 fs, the required ion energies for medical cancer therapy with a commercially available 200 TW TiSa 10 Hz laser may be within reach.

5. Understanding of These Experiments (Towards More Adiabatic Acceleration)

It is crucial to develop a theoretical framework that links the class of new experiments we have introduced in Sec. 4 (CAIL) with those that have been based on TNSA. Little attention has been paid so far to controlling the gradual increase of the accelerating

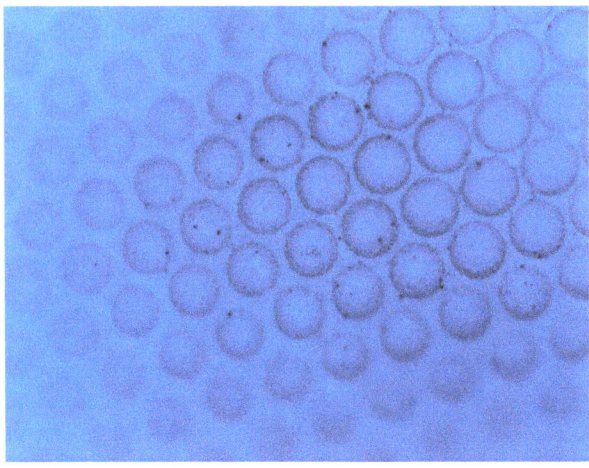

Fig. 13. Nanostructured DLC foil.

structure of ions. Because ions are nonrelativistic, unlike electrons in laser acceleration, it leads to a huge jump in energy gain, spectral narrowing, and gain efficiency if we can accomplish this controlled adiabatic acceleration. Toward this goal, we look at a theoretical understanding of this process.

5.1. *Theory of CAIL*

Most of the theories for TNSA [68, 69] have been motivated by much thicker and massive targets as compared to mass-limited targets such as those mentioned in Sec. 4. In the TNSA regime electrons are first accelerated by the impinging relativistic laser pulse and they penetrate the target driven by the ponderomotive force. Leaving the target on the rear side, electrons set up an electrostatic field that points normal to the target rear surface. That was the origin of the terminology of the TNSA mechanism. Most electrons are forced to turn around and build up a quasi-stationary electron layer. These fast electrons are assumed to follow thermal distribution in theoretical studies of the conventional TNSA mechanism for thicker targets [68–74], where the acceleration field is estimated by the exponential potential dependency in the Poisson equation. Though this mechanism is widely used in the interpretation of the experimental results, it does not apply to the ultrathin nanometer scale targets, because the direct laser field and partially transmitted laser pulse play an important role in electron dynamics and the energetic electrons oscillate coherently, instead of chaotic thermal motions. Based on a self-consistent solution to the Poisson equation and TNSA model, Andreev *et al.* [73] had proposed an analytical model for thin foils and predicted the optimum target thickness to be about 100 nm. It obviously does not explain the recent experimental results [38, 53].

5.1.1. *Electrostatic potential in coherent dynamics*

In the case of a thin target, electron motions maintain primarily those organized characteristics directly influenced by the laser field in the CAIL regime, rather than chaotic and thermal motions of electrons resulting from laser heating on the front surface where the laser is either absorbed or reflected for the TNSA regime. In PIC simulation (Fig. 14) it is observed that momenta of electrons show in fact

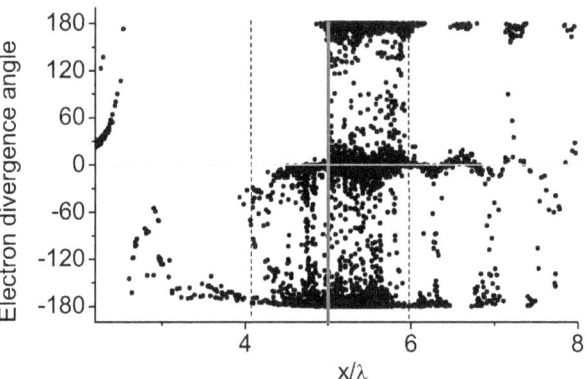

Fig. 14. Coherent electron motions in the laser irradiation on a thin target in 1D 3V PIC simulations. The electron divergence angle $\tan^{-1}(p_y/p_x)$ versus position x at $t = 22$, where space x is measured in wavelength λ_L and is in the direction of laser propagation and y is the polarization direction, while time t is normalized by the laser cycle. On the left of the target we see electrons backwardly spewed out near an angle of $-180°$. In the forward direction we see forward electrons at $0°$ due to the ponderomotive force, and electrons reflexing by the electrostatic fields ($180°$ or $-180°$). We further see electrons trapped in some wavelike structure, changing swiftly their directions. All these are indicative of the direct imprint of the electron motion in the laser fields. Note also that even within the target we discern structured electron loci, showing electrons driven by some minute structured (perhaps the wavelength of $2\pi c/\omega_p$) fields in the target. (Laser amplitude $a_0 = 3.6$, $\xi \sim 1$ and normal incidence. The vertical bold line represents the initial target located at $x = 5\lambda_L$, and the two dotted lines show the boundaries of the expanded target.)

coherent patterns pointing in either the ponderomotive potential direction, the backward electrostatic pull direction, or the wave trapping motion direction, in stark contrast to broad momenta of thermal electrons. In other words, through a very thin target the partially penetrated laser fields enable the electrons to execute dynamic motions still directly tied with the laser rather than the incurred thermal motions from the front surface.

It may be instructive to compare conceptual differences between the regime of TNSA and that of CAIL. In order to do so, a schematic comparison of the phase space dynamics of several useful cases is shown in Fig. 15. Here we introduce not only the case of TNSA (a) and that of CAIL (c)–(e), but also a case (b) of electron beam injection through a metallic boundary (the Mako–Tajima case, [75]). In the TNSA case (a) the laser interacts primarily at the front surface of the target, while in all the cases of CAIL (c)–(e) interaction takes place at the rear surface. In the regime of TNSA, electrons that

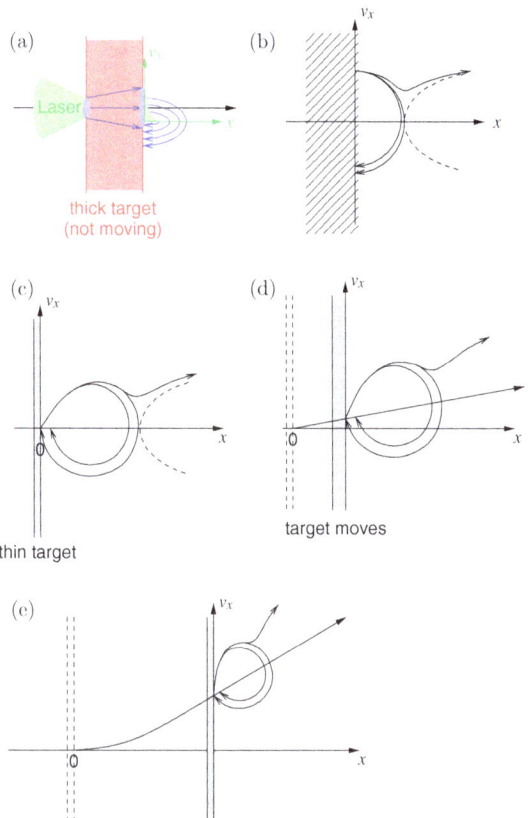

Fig. 15. Conceptual comparison of the phase space dynamics in various acceleration mechanisms. (a) TNSA. The laser generates energetic electrons on the front surface of the thick target. Electrons travel through the target to emerge from the rear side with a broad energy spread. They exit into the vacuum to pull ions. However, most electrons are pulled back to the immobile target before ions gain much energy. Electrons at the margin of the electron cloud are ejected from the electron space charge. (b) The Mako–Tajima scenario. Electrons with the delta function energy spectrum enter from the metallic immobile (real) surface. They rush out in vacuum to pull ions. However, most electrons are pulled back to the immobile boundary before ions gain large energy. Some electrons are ejected forward. The electron dynamics is much in common with case (a), although the electron spectrum is broad and has a tail in (a). (c) A case study with an ultrathin target that is immobile. One significant difference of (c) from (a) is that the electron energy is directly determined by the laser and its ponderomotive potential beyond the rear surface of that target. Thus, the energy of ions is expected to be narrow in its width and to have a higher maximum than (a). (d) When the target is sufficiently thin, the rear surface of the target (and sometimes the entire target) begins to move, while the laser interacts with the target. (e) When the target is pushed with the laser ponderomotive force (such as the circularly polarized laser pulse) without too much heating of electrons, ions in the target as a whole are trapped in an accelerating bucket with tight phase space circles. If and when the laser leaks through and electrons are ejected forward, the bucket may begin to collapse. Cases (c)–(e) belong to the regime of CAIL, while (e) is in particular under the RPA conditions.

have gained energy at or near the front surface propagate through the target and escape from the rear surface with a broad energy spread [see Fig. 15(a)]. In the Mako–Tajima problem, the electron beam (with a delta-function energy spectrum) enters from the metallic surface which may be regarded as the rear surface. Thus, the Mako–Tajima problem may be considered to serve as a stepping stone for quantitative analysis of the collective acceleration process we have in hand. In the cases (c)–(e), once the laser penetrates the target and electrons gain energy from the laser, the electron dynamics in the presence of the rear surface is once again similar to the Mako–Tajima problem. In the Mako–Tajima work the technical word "reflexing" was introduced to analyze why the injection of an electron beam was not sufficient to cause adiabatic acceleration of ions; electrons are "reflexed" and stay around the rear surface, instead of moving gradually together with ions. In this sense all cases, except possibly (e), show this lack or destruction of the adiabatic acceleration process, which was endemic to the Mako–Tajima problem. Unlike (a)–(d), (e) may have a chance to cause adiabatic ion acceleration, as the target is moving together with the laser and forming a sufficiently compact bucket. In the case (d) a similar phenomenon is induced, but the laser ceases to move together with the target and eventually runs away from the rear surface, leaving the bucket starting to collapse by ejecting electrons forward.

So what we have learned is that in a typical sheath acceleration scheme the termination of ion acceleration commences due to this electron reflexing by the charge separation of the electrostatic field. In what follows we first study the experimental situations corresponding to Figs. 15(c)–15(d) in Subsec. 5.1, and then go on to the situation created by the irradiation of circularly polarized laser pulses to provide the more adiabatic situation depicted in Fig. 15(e) in Subsec. 5.2.

In an ultrathin target, the laser electromagnetic fields largely sustain coherent motions of electrons. As partially or fully penetrated laser fields in addition to the laser fields in the target, the electron motion under laser fields is intact and is characterized by the transverse field. The electron energy consists of two contributions: the kinetic energy of (organized) electrons under the laser and the ponderomotive potential of the partially penetrated laser fields that help sustain the electron forward momentum. Following the analysis of Mako and Tajima [75], the plasma density can be determined by

$$n_e = 2\int_0^{V_{\max}} g(V_x)\mathrm{d}V_x, \tag{9}$$

$$V_{\max} = c\sqrt{1 - m_e^2 c^4/(\varepsilon_0 + m_e c^2)^2}, \tag{10}$$

where g is the electron distribution function and ε_0 is the maximum electron energy in this theoretical distribution and we call the latter characteristic electron energy henceforth.

The forward current density of electrons J and electron density n_e are related through

$$J(v) = -e\int_v^{V_{\max}} V_x g \mathrm{d}V_x, \tag{11}$$

$$n_e = \frac{2}{e}\int_0^{V_{\max}} \frac{\mathrm{d}J/\mathrm{d}v}{v}\mathrm{d}v. \tag{12}$$

At a given position in the reflexing electron cloud where the electrostatic potential is ϕ, the particle kinetic energy (disregarding the rest mass energy) is given by

$$\varepsilon = (\gamma - 1)m_e c^2 - e\phi. \tag{13}$$

Current density can be determined from the 1D simulation results. We find that the current density dependence on ε is not exponential, but is rather well fitted by a power law. The power law dependence may be characterized by two parameters: the characteristic electron energy ε_0 and the exponent α of the power law dependence on energy ε.

$$J(\varepsilon) = -J_0\left(1 - \frac{\varepsilon}{\varepsilon_0}\right)^\alpha, \tag{14}$$

where $J_0 = en_0 c/2$. The index α designates the electron energy dependence and is a measure of the coherence of the electron motion. In other words, the greater α is, the more electrons in coherent motion are contributing to the overall current of electrons. Thus, α is called the coherence parameter of electrons. Usually the most energetic electrons are lost from the system and make a minor contribution to the ion acceleration [76–78]. We find [79] that α as a function of the target thickness d (or normalized thickness σ) peaks at about 3 around where the optimal thickness condition for acceleration of $\sigma = a_0$ is fulfilled. The maximum of the coherence parameter α is realized and the coherence of electron dynamics

is best preserved at (or near) the thickness parameter $\xi = 1$. The coherence parameter decreases from the peak value away from the optimal value toward zero. When ξ becomes too large ($\gg 100$), the coherent electron dynamics is lost and approaches the typical TNSA regime.

5.1.2. *Ion dynamics and maximum ion energy*

The system's evolution needs to be tracked self-consistently with electrons, ions and the interacting electrostatic potential in time. These consist of a highly nonlinear coupled system of equations. Electrons are treated as discussed in Ref. 79 and Subsec. 5.1.1, while ions are described in nonrelativistic nonlinear equations. The fully nonlinear nonrelativistic fluid equations are used to describe the response of the ions to the electrostatic field.

In order to solve the equations self-consistently, the self-similar condition may be invoked with the self-similar parameter $\zeta = x/v_0 t$ [79]. An exact solution for the ion density, velocity, and potential is obtained (shown in Ref. 79) when the self-similar evolution is justified as a function of space x and time t through the self-similar parameter ζ. Based on this solution, the maximum ion energy is assessed by applying the Mako–Tajima boundary condition [75]. This yields

$$\varepsilon_{\max,i} = (2\alpha + 1)Q\varepsilon_0, \tag{15}$$

where the characteristic energy is equal to the ponderomotive potential

$$\varepsilon_0 = mc^2\left(\sqrt{1 + a_0^2} - 1\right). \tag{16}$$

In Eq. (15) we see that the ion energy is greater if the coherence parameter of electrons α is greater. It also shows that the ion energy is directly tied to the electron energy in the laser field and thus is a direct function of a_0. A common feature of this with TNSA is the proportionality with a_0. However, the similarity ends here and the differences are twofold: (a) the energy gain of the present case is several times higher than that of TNSA; (b) the energy gain maximizes at the optimal thickness of $\sigma \approx a_0$ mentioned earlier in Sec. 2 for CAIL, as opposed to a much thicker target for TNSA. These features are also seen in Fig. 16 later.

A more general expression for the time-dependent maximum ion kinetic energy at the ion front is [79]:

$$\varepsilon_{\max,i}(t) = (2\alpha + 1)Q\varepsilon_0[(1 + \omega_L t)^{1/2\alpha+1} - 1]$$
$$(t \leq 2\tau). \quad (17)$$

Here τ is the laser pulse duration and ω_L is the laser frequency. At the beginning the ion energy $\varepsilon_{\max,i}(0) = 0$ and the ion energy approaches infinity as long as the time $t \to \infty$. Normally, as the maximum pulse duration of a CPA (chirped pulse amplification) laser is less than picoseconds, the final ion energy from Eq. (17) is only about

$$\varepsilon_{\max,i}(t = 1\text{ps}) = 2(2\alpha + 1)Q\varepsilon_0.$$

5.1.3. *Relativistic transparency*

To understand the dynamics of the laser pulse and the target evolution is important when the pulse is longer ($\sim$ hundreds of fs) than tens of fs and/or the target is thicker than a few nm. We now consider the cases where the target is thicker ($\xi \gg 1$) than when it is immediately influenced by the laser fields but still ξ is less than for TNSA. In this case the laser does not immediately penetrate through the target. We can delineate at least three stages. The first stage is similar to the situation we described above for $\xi \approx 1$. The laser just impinges on the thin surface layer of the dense target. The second stage is after the target begins to expand by the laser interaction primarily in the direction of laser propagation until the plasma becomes relativistically transparent at time t_1. After this relativistic transparency (the relativistic transparency time t_1), the plasma expands in all three dimensions. The third stage begins when the plasma becomes underdense at time t_2 (the classical transparency time) until the pulse is over. (Here we have assumed a case where the pulse length is greater than both t_1 and t_2 for the sake of concreteness.)

Now we want to evaluate the plasma expansion in terms of the two characteristic times t_1 and t_2. In the solid-density plasma, the skin depth is so short that the ponderomotive force is opposed by the charge-separation force beyond the skin depth. Therefore, the foil expansion in the longitudinal direction may be written as

$$\frac{\mathrm{d}p}{\mathrm{d}t} = -Qe\nabla\Phi. \quad (18)$$

We integrate Eq. (18) over $\mathrm{d}t$ to obtain [79]

$$x^2 - d^2 = \frac{m_e}{m_i}\frac{Qc^2a_0}{3}\Xi^2 t^4, \quad (19)$$

where a laser pulse with the profile $a = a_0 \sin^2(\Xi t)$ is assumed ($\Xi = \pi/2\tau$).

Assuming expansion only in the x direction at the relativistic transparency, the expanded distance x_1 may be evaluated by

$$x_1 = \frac{Nd}{\gamma} = \frac{Nd}{\sqrt{a_0^2 + 1}},$$

where $N = n_0/n_c$ and $x_1 \gg d$, with n_c being the critical density.

One of the distinguishing features of the thin target CAIL regime, as compared with TNSA, is the presence of the relativistic transparency time t_1 before the pulse length τ, so that the laser pulse emerges or interacts with the entire target before the pulse is gone. We find this time to be [79]

$$t_1 = \left(\frac{m_i}{m_e}\frac{3N^2d^2}{Q\Xi^2c^2a_0^3}\right)^{1/4}$$
$$\cong \left(\frac{12}{\pi^2}\right)^{1/4}\frac{N^{1/2}}{a_0^{1/2}}\left(\frac{\tau d}{C_s}\right)^{1/2}. \quad (20)$$

Here the sound speed $C_s \cong (Qm_ec^2a_0/m_i)^{1/2}$. The relativistic transparency time t_1 in Eq. (20) is approximately the geometrical mean of the laser pulse length τ and the traverse time over the target by the sound speed. During this period, the laser pulse penetration is limited as expressed by the transmission coefficient [80]

$$T = \frac{1}{1 + (\pi\xi)^2}. \quad (21)$$

Thus, when we integrate the impact on the electron energy at the rear surface of the target to evaluate ε_0, we need to incorporate this effect to obtain the maximum ion energy as

$$\varepsilon_{\max,i} = (2\alpha + 1)Q\bar{\varepsilon}_0(t_1)((1 + \omega_L t_1)^{1/2\alpha+1} - 1), \quad (22)$$

where $\bar{\varepsilon}_0(t_1)$ is the time average of $\varepsilon_0(t_1)$ over the period of $(0, t_1)$.

Yin *et al.* [81] have found in their 3D simulation that for long pulse irradiation the pulse exhibits an epoch of burnthrough (and relativistic transparency). This phenomenon is when the laser goes through the target and eventually emerges from the rear end of the target. This corresponds precisely to the second period between t_1 and t_2 and in fact most of the acceleration takes place shortly after t_1.

We now characterize the physical processes including these phenomena. Beyond time t_1 the plasma is relativistically transparent so that the laser can now interact with the (expanded) target plasma in its entirety. It can also now expand in three dimensions. For 3D isotropic expansion, it takes time Δt, during which the normalized density reduces from γ to 1:

$$x_2^3 = x_1^3 \gamma. \tag{23}$$

We obtain

$$\Delta t = \frac{Nd(\gamma^{1/3} - 1)}{\gamma C_s} \frac{1}{\sin(\Xi t_1)}. \tag{24}$$

Now time t_2 when the plasma becomes underdense is given as

$$t_2 = \Delta t + t_1. \tag{25}$$

Now, as we examine the physical situation, at time t_1 the laser pulse has penetrated the entire target with the relativistic transparency and we may say that the laser begins to drive the entire plasma electrons from this already-expanded target. An expression in a closed form for the ion energy gain between time t_1 and time t_2 in the case of a laser pulse with the duration longer than the relativistic transparency (rt) time t_1 has been obtained:

$$\varepsilon_{\max,i,\mathrm{rt}} = (2\alpha + 1)Q\bar{\varepsilon}_0((1 + \omega_L(t_2 - t_1))^{1/2\alpha+1} - 1). \tag{26}$$

Here $\bar{\varepsilon}_0$ is evaluated over time interval (t_1, t_2) and also note that after t_1 transmission T is very close to 1. We have assumed that $t_1 < 2\tau$ and $t_2 < 2\tau$. As we remarked, Eq. (26) has been derived for one-dimensional self-similarity. It is thus considered that this would yield an overestimate of energy rather than a fully three-dimensional solution.

Combining these expressions in Eqs. (22) and (26), when $t_1, t_2 < 2\tau$, the total ion energy gain can be obtained. In Fig. 16 we plot the total energy gain in the case of carbon ions from this formula as a function of the target thickness. Once again the optimal thickness, for which the ion gain is maximum, is sharply realized in the CAIL regime. Toward the TNSA regime the ion energy decreases substantially.

It is noteworthy to consider how the photon pulse behaves right after the relativistic transparency time t_1. The group velocity of the laser pulse,

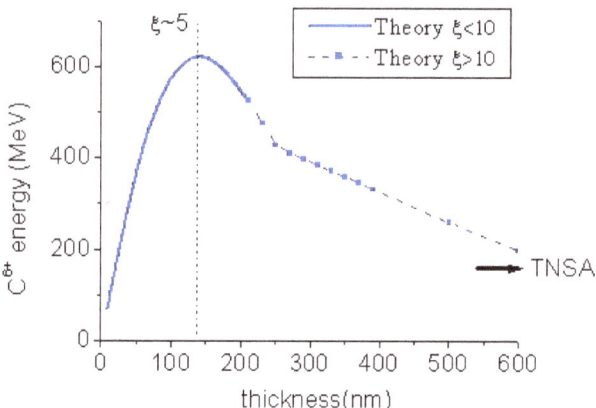

Fig. 16. The maximum ion energy driven by the laser pulse as a function of target thickness in the regime of CAIL. The optimum is reached at $\xi \sim 5$. Energies for thicker targets decrease from this value by a factor of a few, leading to the value often found in the TNSA regime (μm or more in this graph). The C^{6+} energy gain is estimated from Eqs. (22) and (26) as a function of target thickness and with $\alpha \cong 3$. Beyond $\xi > 10$, where α is supposed to quickly decrease and the model's predictiveness decreases (for a given laser pulse length at 700 fs and laser amplitude $a_0 = 20$). The contribution of the electron energy gain after the relativistic transparency has been reached is dominant.

$v_{\mathrm{gr}} = c\sqrt{1 - \omega_p^2/\omega_L^2}$, vanishes at $t = t_1$. At this moment the ponderomotive structure of photons is stationary, which pushes electrons forward as well as ions effectively. This is because the heavy and sluggish ions can respond easily to this stationary potential. As the laser penetrates further and the plasma density decreases below $n_{\mathrm{cr}} \cdot \gamma$, the group velocity begins to increase. In our model case of the laser temporal structure of $a_0 \propto \sin^2(\Xi t)$, the group velocity increases as

$$v_{\mathrm{gr}}(t) \sim 2c \cot(\Xi t_1)\sqrt{\Xi(t - t_1)}. \tag{27}$$

This means that the speed at which the accelerating structure — made up of the electron layer driven by the laser ponderomotive force and the ion layer that is attached to the former by the electrostatic force — moves is picking up quickly from zero. This suggests that if we can slow down the photon group velocity, the rate of increase of the photon group velocity and thus the accelerating structure is reduced and, therefore, the adiabatic nature of acceleration becomes more pronounced. This may be accomplished by increasing the density of the plasma behind the solid target by a further material.

5.2. *Synchrotron oscillations in RPA*

Monoenergetic ion beams are one of the important requirements of ion beam therapy. CAIL by linearly polarized pulses can efficiently accelerate ions to higher energy by using nanometer targets, as we have seen in Subsec. 4.1. However, the monoenergetic spectrum has not yet been obtained. Even under TNSA interaction, if the target is prepared to have certain characteristics, monoenergetic characteristics may be obtained [49], which was experimentally demonstrated [21–23]. Recently, theoretical attention has focused on the use of circularly polarized (CP) laser pulses in the CAIL regime to accelerate high-density ion bunches at the front surface of thin foils. For CP pulses, the ponderomotive force has no oscillating component, as discussed; hence, electrons are steadily pushed forward, inducing a charge separation field which can accelerate ions. There is expected to be more adiabatic interaction so that monoenergetic ion beams may be realized, in this case in the regime called radiation pressure acceleration (RPA). There is a regime of phase-stable acceleration in the interaction of a CP laser with a thin foil in a certain parameter range, where the proton beam is synchronously accelerated and bunched like in a conventional radio frequency (RF) accelerator. This synchronous acceleration leads to the acceleration regime in which the position of ions is well tied to the accelerating structure made up of the laser ponderomotive potential and electron layer. Therefore, ions may be trapped in this accelerating bucket, in which they may show a phase-stable behavior. That is to say, ions exhibit phase stable oscillations (synchrotron oscillations).

A simple model can be used to elucidate the bunch formation in the phase-stable acceleration [34], as shown in Fig. 17. A linear profile in both the electron depletion region ($E_{x1} = E_m x/D$ for $0 < x < D$) and in the compressed electron layer ($E_{x2} = E_m[1 - (x - D)/l_s]$ for $D < x < D + l_s$). The parameters E_m, n_{p0}, and l_s are related by the equations $E_m = 4\pi end$ and $n_{p0}l_s = nD \approx n_0 d$. As E_{x1} increases with x, the protons starting at initial positions $x < D$ are debunched (longitudinally defocused) and their density will decrease in the electron depletion region. On the contrary, because E_{x2} decreases with x, the protons inside the compression layer ($D < x < D + l_s$) can be bunched by the electrostatic field E_{x2}. The equilibrium between the

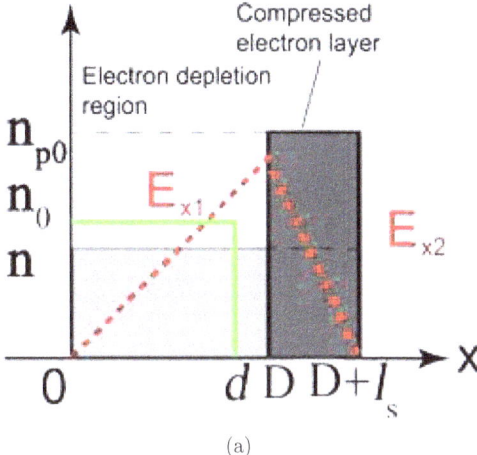

(a)

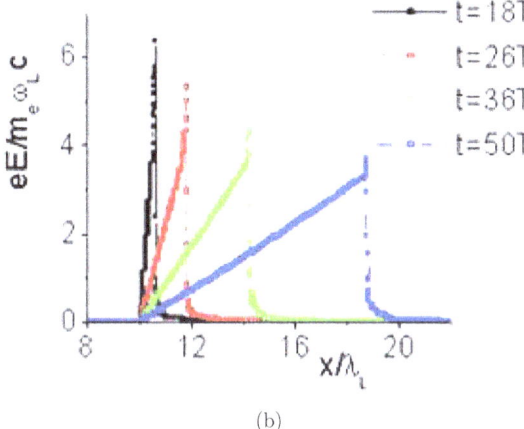

(b)

Fig. 17. Trapping of electrons in an accelerating bucket. (a) Schematic picture of the equilibrium density profiles for protons (n) and electrons (n_{p0}). The x position at $x = D$ indicates the electron front, where the laser evanescence starts, and it vanishes at $x = D + l_s$, where l_s is the plasma skin depth. The initial plasma density n_0 and target thickness d are also plotted. (b) Snapshots of the longitudinal electrostatic field obtained from PIC simulations.

electrostatic and ponderomotive forces on electrons is only temporarily lost and the electrons rearrange themselves quickly to provide a new equilibrium if the laser pulse is not over. Thus, the light pressure exerted on the electrons $(1+\eta)I_L/c$ is assumed to be balanced by the electrostatic pressure $E_m e n_{p0} l_s/2$. Here η is the reflecting efficiency.

As the purely hydrodynamic description is not adequate for describing the interaction between the protons and electrons, dynamic equations are derived based on this model. We introduce $\zeta = (x_i - x_r)$ with $-l_s/2 \leq \zeta \leq l_s/2$, where $x_r = D + l_s/2$ represents the position for the reference particle. The force acting on a test ion is given by $F_i = q_i E_m (1 - (x_i - D)/l_s)$. Thus, the equation of motion for the proton is

$$\frac{\mathrm{d}^2 x_i}{\mathrm{d}t^2} = \frac{Q e E_m}{m_i} \left(1 - \frac{x_i - D}{l_s} \right).$$ (28)

The phase motion (ζ, t) can be written as

$$\ddot{\zeta} = -\Omega^2 \zeta, \quad \Omega^2 = \frac{Q e E_m}{m_i l_s},$$ (29)

where Ω is the frequency of the synchrotron oscillation motion in the longitudinal direction [86]. For the reference ion E_m is assumed to be quasi-constant, so the longitudinal phase motion (ζ, t) is harmonic oscillations, called synchrotron oscillations. We obtain

$$\zeta = \zeta_0 \sin(\Omega t),$$ (30)

$$\dot{\zeta} = -\zeta_0 \Omega \cos(\Omega t).$$ (31)

Simulations show this dynamics and limitations of the process with a fully relativistic 1D PIC code with 100 particles per cell per species, with cell sizes of $\lambda_L/100$. In PIC simulations a laser pulse with $a_0 = 5$ and duration $100 T_L$ is incident on a purely hydrogen plasma (cold, step boundary, overdense plasma slab with $n_0/n_c = \omega_p^2/\omega_L^2$ and $d = 0.2\lambda_L$). Figure 17(b) shows snapshots of the electrostatic field profile. As the depletion region expands with time and the proton density in this region decreases, the slope of the field in the depletion region reduces gradually. In the compressed electron layer, it is found that the width of the compression layer remains to be equal to the skin depth ($l_s \cong \lambda_L/20$). Therefore, the charge separation field in this layer keeps approximately the same steep linear profile, even though the maximum separation field is decreased slightly.

The synchrotron oscillations appearing in the phase space (see Fig. 18) imply that the protons in the compressed electron layer have been trapped in the bucket, which is quite similar to the RF accelerator. It shows phase-stable synchrotron oscillations. The period of the phase oscillation can be estimated through Eq. (31) and it is about $8T_L$, which is consistent with simulation results as shown in Fig. 18. Because these protons can be accelerated and bunched over much of the laser interaction time, a monoenergetic proton beam of a few hundred MeV is observed in the 1D simulations [see Fig. 18(d)].

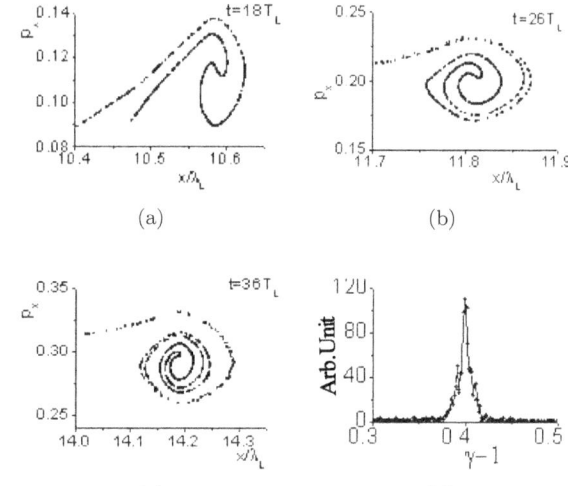

Fig. 18. Ions trapped in the stable accelerating bucket showing synchrotron motions obtained from PIC simulations. (a,b,c) Evolution of phase space distribution for protons; the first, second, and third oscillation periods are 8, 8, and 10 T_L, respectively. (d) Monoenergetic spectrum of protons trapped in the bucket.

The bucket velocity width may be determined from Eq. (31):

$$v_{i,\mathrm{buc}} = \zeta \Omega = \sqrt{\frac{m_e}{m_i} Q a_0} c N^{-1/4}.$$ (32)

This bucket size is close to (and slightly less than) the trapping velocity width from Eq. (7). As long as $N > \gamma$ (or $t \leq t_1$), $v_{i,\mathrm{buc}} \leq v_{i,\mathrm{tr}}$ [Fig. 18(a)], while for $N < \gamma$ ($t > t_1$) $v_{i,\mathrm{buc}} \geq v_{i,\mathrm{tr}}$ and some of the ions in the bucket begin to spill over, as seen in Fig. 19(b).

The energy spread of trapped ions in the bucket of the accelerating structure with respect to the energy ε_r of the reference particle is

$$\frac{\Delta \varepsilon}{\varepsilon_r} = \frac{2 \zeta_0 \Omega}{\sqrt{2 m_i \varepsilon_r}}.$$ (33)

If we take $\zeta_0 = l_s/2$ and $\varepsilon_r = 400\,\mathrm{MeV}$, the energy spread will be less than 4%, which agrees well with the simulation results.

In 1D simulations the plasma is kept cold and the target is pushed forward as a whole, so that an ideal monoenergetic ion beam can be generated. A quasi-monoenergetic carbon ion beam with a 17% energy spread has been observed in recent experiments [53]: $\Delta \varepsilon / \varepsilon_r \sim 17\%$ is about three times higher than the estimation by Eq. (33) because of

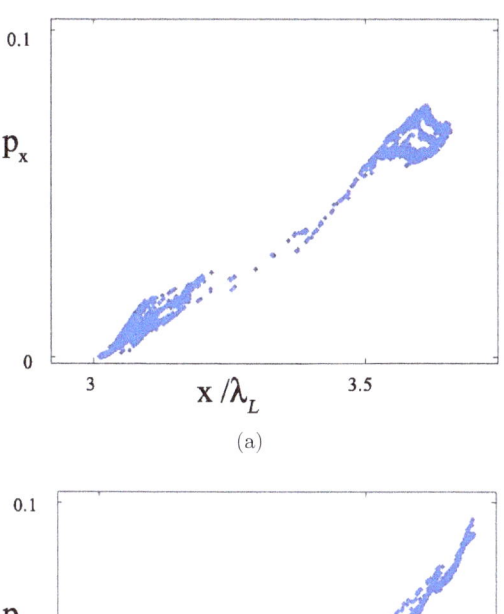

(a)

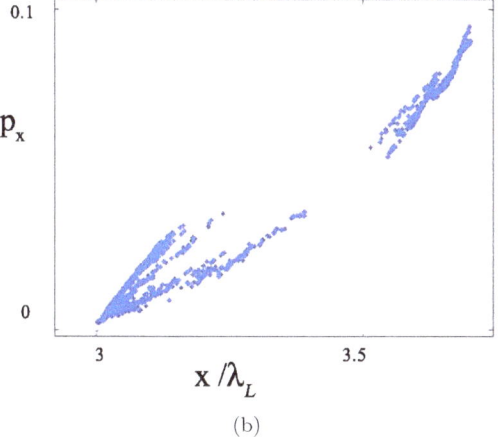

(b)

Fig. 19. Longitudinal phase space of carbon ions in 2D simulations: (a) stable bucket structure of synchrotron oscillation ($t = 28T_L$); (b) collapse of the accelerating bucket when the plasma becomes hot ($t = 40T_L$).

the multidimensional effects. In real situations typically the laser intensity is not uniform, the transverse profile tends to bend the flat target, and foil electrons are heavily heated by the oblique incident laser, in spite of the CP pulse. If and when electrons become hot or the laser leaks through, the bucket begins to collapse (see Fig. 19) and the energy spread drastically increases. We see not only an obscure quasi-monoenergetic peak in Fig. 8, but also the monotonically decreasing ion spectrum at its low energy domain.

Except for bending effects, Rayleigh–Taylor and Weibel-like instabilities also play an important role in laser plasma interactions [82, 83]. At a laser intensity of 10^{22} W/cm^2, self-focusing effects from these instabilites may help to improve the ion beam quality [84, 85].

5.3. Suggested improvements from theory

When the electron dynamics is slow enough that ions evolve less suddenly, i.e. adiabatically [86–88], the final energy gain of electrons (and thus that of ions) may not be that of the instantaneous energy dictated by the expression $\varepsilon_0 = m_e c^2(\sqrt{1 + a_0^2} - 1)$. For example, we have remarked on a case with a circularly polarized pulse. In the latter, for example, the pulse should cause less electron energy gain than the linearly polarized case. Therefore, the cloud of electrons cannot instantaneously shoot out of the foil, but more gradually leaves the target. This gives the possibility that the electron energy is not only proportional to the field strength (as proportional to a_0), but also to the time over which electrons are accelerated by $v \times B$ if this is much longer than the laser period. When electrons substantially comove with the laser pulse, this time can be proportional to a_0 or some fraction of it, leading to a proportionality greater than a_0 such as a_0^2. We expect more results to come in advancing the ion energy by laser acceleration spurred by the current theoretical understanding of the physics.

In PIC simulation, an a_0^2 scaling is observed in Ref. 37 using CP laser pulses, whereas it is only 1D simulation and the bending and boring effects are not considered.

With 2D simulations, the scaling with normalized vector potential a_0 is only $a_0^{1.1}$ in the linearly polarized case (and the fit looks much better than $\sqrt{a_0^2 + 1} - 1$), while for the circular case the maximum proton energy scales with $a_0^{1.6}$ at the lower intensity for $a_0 \leq 30$. It tends to be closer to a_0^2 scaling for larger a_0 values (see Fig. 20), including the highly relativistic regime [49]. This tendency of having less energies (or a smaller exponent to a_0) in the 2D simulation than the one shown in idealized 1D model may be due to a couple of factors. One is the bending or bulging (the convex shaping as viewed from the rear surface, where ions are emitted) of the thin target by the impinging laser pulse. This makes the excess plasma electron heating by the obliquely incident laser electric fields. This contributes to hot electrons that run away from the target leaving ions far behind, yielding nonadiabatic electrons and thus ion dynamics. Secondly, once the laser ponderomotive potential penetrates through the thin target, the slow motion of the ponderomotive potential till now

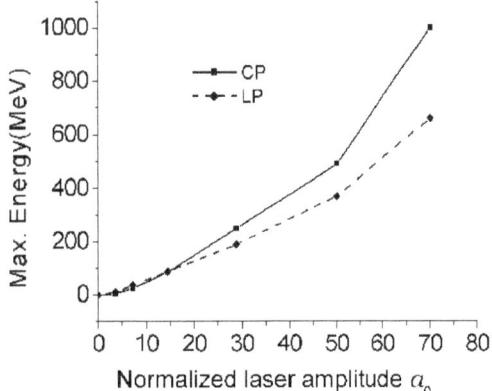

Fig. 20. Maximum proton energy versus a_0 obtained in 2D PIC simulations. In the modest a_0 regime ($15 \geq a_0 > 1$) the energy gains by LP and CP are not much different, while that by CP is much greater than by LP for large a_0 ($a_0 \geq 15$). The exponent to a_0 for CP seems to increase from less than 2 in $a_0 \leq 15$ toward 2 in $a_0 \geq 15$.

begins to pick up, as the density of the plasma that the laser sees is less than the relativistic transparent density (in terms of the timing of the interaction corresponding to time t_1 in Subsec. 5.1.3). Once the laser increases its group velocity in this less dense region, only electrons can catch up with photons, while ions are left behind. When this develops, we see that the nice closed phase space circle in Fig. 19(a) in Subsec. 5.2 is now skewed to form a squeezed parallelogram with the upper-right corner stretching further to the right-upward direction, eventually leading to a collapse of the parallelogram into a sharp tongue shooting up, ejecting ions from the bucket [Fig. 19(b)] and the drawing in Fig. 15(e).

In order to further improve these situations, we envision counteracting the convex bulging by giving a concave shape to the target. This little manipulation may improve the energy enhancement by a factor of a few [43, 44]. To further arrest the collapse of the trapping bucket by the accelerating photons after they transmit through the thin target, we could let them slow down again by adding supplementary target material, such as dense gas/clusters, a foam target, or a mesh of carbon nanotubes behind the rear surface. This would decrease the group velocity of the laser after it passes through the thin solid foil, and increase the interaction time between the laser pulse and the plasma and therefore the accelerating time for ions. With special target manufacturing techniques (for example, concave as seen from the rear surface) or hemispheric targets, it may be

possible to increase the power of a_0 in ion energy, which is important as an outlook, because in the one case we need a much bigger laser when we want to reach the medical energy of 200 MeV/u. If these additional measures can enhance the ion acceleration time, perhaps it is possible to reduce the necessary laser intensity to reach the same energies, contributing less demanding laser power for the necessary energy regime.

Even if the necessary energies are reached, there are many elements which the therapy system needs to satisfy, as was discussed in Sec. 1. These are by no means an easy task and pose challenges for us researchers. On the other hand, we do have a host of new ways to manipulate intense ion and laser beams by harnessing the relativistic dynamics of the laser and its plasma interaction, which is sometimes called relativistic engineering [89]. For example, our ability to simultaneously generate ion beams which we are discussing with the coherent x-rays from the laser–thin-target interaction [64] can allow us to make an unprecedented accurate x-ray diagnosis such as phase contrast imaging [65, 66]. This will give us the additional ability to make accurate imaging simultaneous with the therapy. Undoubtedly, we will marshal all these abilities to cope with the challenges we face.

6. Technical Developments and Perspectives

6.1. *Goals of therapy machines*

The science of laser acceleration of ions is important in realizing our goal. Even though it is still in an early phase, it is encouraging to see biomedical experiments using human cancer cells irradiated by laser-accelerated proton beams [90]. The compactification of the accelerator by itself, however, is not sufficient to realize a clinically acceptable solution. Here, we consider such issues; some of them concern laser technology and accelerator technology, while others are medical physics issues and clinical considerations. For example, it is not sufficient to render the accelerator more compact. A fairly large portion of the cost of radiation therapy comes from the beam transport and irradiation system as well as the infrastructure. It is true that so far most of the research has been on how to realize the laser acceleration of ions, and not on the latter issues we have raised in this section. Thus, the work on the latter issues remains relatively scarce and perhaps of a preliminary nature. This should not be, however, a reason not to consider these issues, as the sum of these considerations is necessary for success in creating such a device and practice, as well as their overall cost.

Compared with a conventional synchrotron, the beam transport system associated with laser acceleration is believed to be much more compact. This is because the ion acceleration takes place in a tiny space and can be accommodated easily in the requirement concerning how one wants to direct the beam that has been generated. This, for example, reduces the number of shields around the transport section of the laser-driven system. The issue of the gantry may be an entirely different one. The gantry plays a crucial role in the final beam transport to the area of the patient and in directing it appropriately toward the patient. It is also essential to avoid unintended radiation from the beam or associated systems. Finally, the beam delivered to the patient needs careful monitoring and control, equipped with a cleanup optics that separates needed ions from the rest of the ions, electrons, photons, etc. The first glimpse of this issue, thus, casts doubt that a gantry for the laser accelerator can be much more compactified than that for a conventional accelerator. If this were the case, there would be little hope of compactifying the overall system drastically, as the system size is ultimately determined by the largest of the components and the gantry remains among the largest, even if the laser accelerator is substantially compactified. However, upon further scrutiny, one finds that the exploitation of the laser pointing in the general direction of the patient can substantially make the system compact and flexible [12–15]. This is due to the point that once we can direct the laser (and thus the general direction of produced ion beams), the energy segregation and other optics based on magnets can be much more compact, because the necessary magnets need to bend ions by relatively small angles such as $10°$, instead of $90°$. This is because in the former the tiny laser accelerator can move together with the gantry magnets, while for a conventional installation the accelerator is on the floor and the beam has to be bent to the direction of the patient. The magnetic chicane [12, 13], for example, may be able to produce energy-specific therapy beams, e.g. appropriate for SOBP (spread-out Bragg peak) treatment.

The tiny size of the laser-driven bunches also allows a much tinier magnetic aperture, possibly leading to a reduced size of magnets.

However, upon further scrutiny such a system poses a problem. If the laser-driven ion beam has a broad energy spectrum and we have to dispense a substantial portion of its spectrum, the dispensed particles will hit the surrounding material, turning out nuclear collisions and subsequent neutrons and making this material radioactive over time. Thus, ideally it is hopped that the dispensed portion is only a very minor part of the beam that is generated by laser acceleration. For this, one needs substantially monoenergetic and less divergent ion spectra to begin with in order to reduce the amount of filtered-out ions and thus radioactivation.

One of the interesting properties of laser-accelerated ions is their potentially extraordinarily short bunch structure. We need to be cautious about the fact that even when the original bunch is short, a substantial energy spread and/or space charge effect cause the bunch to broaden upon its arrival at some distance. If the ion pulse length may be chosen to be smaller than the plasma wavelength in the medium (such as a gas) that is injected with the short bunch of ions, such an ion bunch may be subject to an enhanced collective decelerating field [67, 91]. The stopping power of this collective force is

$$-\left(\frac{\mathrm{d}\varepsilon}{\mathrm{d}x}\right)_C = m_e c \,\omega_{pe}\left(\frac{n_b}{n}\right), \qquad (34)$$

where n_b is the electron density of the beam that is to be decelerated in a plasma with electron density n.

We may be able to take advantage of such deceleration in manipulating the bunch. For example, we may be able to slow down the low-energy component of the laser-accelerated ion spectrum. (See Fig. 8 for circularly polarized laser irradiation on a thin target. The spectrum shows a monoenergetic peak as well as the low-energy component.) For such a collective deceleration to work, we need to satisfy, among other things, the condition $\sigma_L/\lambda_p < 1$. However, how effectively this can be done, we need to study.

More generally, we pursue the following goal. In order to control the beam over a short distance, it should be deflected like an optical beam or a beam of charged particles in a magnetic or RF field. We introduce the idea of collective optics which covers the entire field of acceleration, deceleration, and manipulation of these bunches, using the collective intense fields. Beside the above-mentioned collective deceleration [67], we envisage collective focusing and defocusing (see e.g. Ref. 92 and 93). The plasma lens concept of Chen may be realized by the charged particle bunch, or by injection of a short laser pulse (creating such a structure as a laser wakefield bubble [91, 94]). Combinations of such lenses, such as a combination of convex and concave lenses, would lead to a highly collimated and energetically discriminated beam. A convex lens for positively charged ions may consist of a positively dominant charged clump, while for a concave lens it is a negatively dominant charged clump. These clumps may be generated either by charged bunches (driven by laser) or by laser pulses of an appropriate design. These collective optics examples deserve much more vigorous study in the future, so that the laser-driven accelerating system becomes equipped with not just an accelerator, but an appropriate beam transport and gantry system made up of a variety of collective optics to deliver the needed beam specifications over a compact dimension.

There are, furthermore, issues of stable operation and the delivery of repeatable and prescribed amounts of beam energies. A related matter is determiniation of the deposited dose and possible feedback control in ascertaining the desired dose distribution. In order to realize these qualities in the system, we believe that laser stability and fidelity, stability and repeatability of the laser–target interaction, and a suite of diagnoses of the laser and beams as well as the deposited dose which ascertain the above are necessary.

In fact, the following has emerged as a consensus at the 440th WE Heraeus Seminar on laser ion acceleration for medicine:

(1) a small tumor may be identified and guided by advanced low-dose x-ray detection via phase-contrast imaging [65, 66, 95] (image-guided therapy).

(2) With the small emittance of laser-accelerated ion beams we are better equipped to handle such tumor therapy provided that we can verify the dose by a method that monitors the autoactivated radiation level, such as γ-rays induced by the irradiating ion beam in the tumor. This

is due in part to the pulsed nature of laser-accelerated ions and in part to their origin from an ultrafast point source.

(3) Such an approach of integrating the functions of diagnosis, therapy, verification, and repeating the cycle is expected to reduce the risk to the patient and to increase the efficiency of radiotherapy. This also reduces the risk of launching the new technology of laser acceleraton in radiotherapeutic clinics and eventually in the marketplace. These points respond in part to the central questions of laser-driven ion accelerator systems raised in Sec. 1 (items #1–5).

In total, these issues amount to consideration of the system design, and not just of a component such as an accelerator. For this a close collaboration among physicists, engineers, and physicians is called for.

6.2. A vision of a future ion therapy facility driven by laser

Let us assume that the theoretically predicted scaling laws for radiation pressure acceleration and the collective deceleration of electrons and ions can be realized. Then we can imagine a laser-driven ion therapy facility which is reduced in all three dimensions by about a factor of 5 compared to the classical facility shown in Fig. 2, covering an area of $\sim 70 \times 60\,\mathrm{m}^2$ for the accelerator and gantry. It may be equipped with many small stationary ion beam units and x-ray beam units, which are supplied with the laser beam via a gantry for the laser beam. Thus, the classical synchrotron with a 20 m diameter may be replaced by several laser accelerators with a few-cm diameter. The classical 600 t, 5-story-high gantry, consisting of large dipole magnets for the stiff carbon/proton beams, we may see replaced by a laser gantry system of 20-cm-diameter mirrors to guide the laser beam to the different accelerator units. Some of the mirrors rotate. Since it is likely to irradiate early detected small tumors, a very accurate shot-to-shot localization of the tumor is essential. This may be achieved by several laser-driven, brilliant, intense x-ray sources, where high-resolution images ($\sim 10\,\mu\mathrm{m}$) at a low dose are taken with phase contrast imaging detectors (see Subsec. 4.2). Fast neutrons, which are produced when the high-energy ion beams are slowed down and stopped inside the patient, require an ∼ 1-m-thick iron wall at the sides, followed by an ∼ 2-m-thick concrete wall. The fast neutron beams have a small opening angle of about 20° with respect to the primary ion beams. For beams coming — with respect to the patient — from above, a sufficiently thick floor can be used to stop fast neutrons. For beams coming from the upper hemisphere, the thick iron and concrete walls are needed. The very small average ion current of a few pico-A results in a negligible activation of the shielding structure or the groundwater. This shows that the laser-driven facility needs a basement, able to support the large weight of the required shielding. While the rooms below the classical accelerator and classical beam guidance system of Fig. 2 are usually filled by arrays of power supplies and amplifiers with a maximum power consumption of 8 MW [10], the laser of the new facility can be located in a separate room of the hospital, from which the laser beam is transported in an evacuated tube with mirrors to the facility.

The ion acceleration unit will use ultrathin DLC targets, where the laser with appropriately selected intensity produces by radiation pressure acceleration a neutral electron-ion bunch with the required energy. We expect that the intensity of the ion bunch will be too large to be used directly for irradiation of patients. Here we are exploring the collective deceleration of ions in thin foils with lower than usual solid-state density. A very small hole of a suitable diameter in the deceleration foil would let part of the beam go through undisturbed, allowing reduction of the usually high intensity of the laser-accelerated ion beam to the required level. It is important that this collective deceleration of the ion beam does not produce neutrons, but instead converts the ion beam energy into collective plasma oscillations. The possibility of operating with the same laser brilliant ion and x-ray units is important, in order to obtain the accuracy to verify the irradiation and localization for smaller tumors.

In the x-ray units, the laser beam should drive a dense relativistic electron sheet out of the ultrathin DLC foil. A fraction of the original laser beam would be reflected coherently from this electron sheet, boosting the photon energy via the Doppler effect to quasi-monoenergetic x-ray energies in the range of 50 keV [64]. The dense electron sheet has typical electron energies of 50 MeV. These dense bunches may be stopped collectively [67] in a buffer gas cell, while produced x-rays traverse the exit and entrance windows and the gas unperturbed. 10 μm silicon wafers may be used as entrance and exit windows, where the first wafer is also used to reflect the split-off part of the main laser beam onto the electron sheet. The low pressure in the gas cell is matched to stop electron bunches.

Certainly, many properties of the laser acceleration and deceleration have to be ascertained and optimized in the years to come. The distributed x-ray units are equivalent to a classical CT scanner. However, this has a much higher resolution and a significantly reduced dose. Thus, a facility without ion beam units and heavy shielding may be employed as a standalone diagnostic unit.

7. Conclusion

In conclusion, the collective acceleration of ions first envisioned in 1956 by Veksler [96] has been reincarnated in the new research field of laser acceleration of ions since 2000. Most experiments since then have belonged to the TNSA regime until recently. While rapid progress has been achieved in these efforts, problems in the TNSA regime have been identified. Thus, there has emerged a new set of experiments in the regime where more coherent electron dynamics driven by laser is utilized (the CAIL regime). Our main purpose in this review is to report to the reader this latest new development. These experimental features have been well characterized by theory, giving a predictive capability to scale parameters to the future use of laser-accelerated ion beams. For the purpose of ion beam radiation therapy for cancer, in accordance with this new understanding we now envisage that this progress with the further maturation of the science and technology of laser acceleration yet to come and the developments of laser target and system design will substantially compactify the radiation therapy facility. This includes not only the accelerator, but also the delivery part and associated systems. At the same time we remain humble, in that because of the extremely fast pace of laser acceleration research, we admit that some of what we have presented here will someday become obsolete. Even though the tasks at hand are difficult, the rapid progress in recent years has given us renewed hope of making even more rapid progress.

Acknowledgments

We thank M. Gross, A. Henig, D. Kiefer, D. Jung, R. Hörlein, J. Schreiber, M. Hegelich, S. Steinke, M. Schnürer, T. Sokollik, P. V. Nickles, W. Sandner, C.-H. Wu, J. Meyer-ter-Vehn, Y. Kishimoto, Y. Fukuda, S. Kawanishi, S. Bulanov, T. Esirkepov, K. Kondo, M. Molls, F. Nüsslin, C. Ma, P. Bolton, F. Pfeiffer, M. Murakami, H. Tsuji, K. Noda, M. Abe, J. E. Chen, Y. R. Lu, Z. Y. Guo, and Z. M. Sheng for fruitful discussion and help.

This work was partly supported by Deutsche Forschungsgemeinschaft (DFG) through Transregio SFB TR18 and the DFG Cluster of Excellence Munich–Center for Advanced Photonics (MAP). T. Tajima is also supported by the Special Coordination Fund (SCF) for Promoting Science and Technology, commissioned by the Ministry of Education, Culture, Sports, Science and Technology (MEXT) of Japan. X. Q. Yan acknowledges support from the Humboldt Foundation and NSFC (10855001, 10935002).

References

[1] *Global Action Against Cancer* (World Health Organization and International Union Against Cancer, 2005).

[2] M. Bamberg, M. Molls and H. Sack, *Radioonkologie — Grundlagen* (W. Zuckschwerdt Verlag, Munich, 2009).

[3] M. Molls, C. Nieder, C. Belka and J. Norum, Quantitative cell kill of radio- and chemotherapy, in *The Impact of Tumor Biology on Cancer Treatment and Multidisciplinary Strategies* (eds. M. Molls, P. Vaupel, C. Nieder and M. S. Anscher) (Springer-Verlag, Berlin, Heidelberg, 2009), pp. 169–190.

[4] O. Zlobinskaya, T. E. Schmid, G. Dollinger, V. Hable, V. Greubel, D. Michalski, G. Du, M. Molls and B. Röper, Differences in gamma-H2AX foci formation after irradiation with continuous and pulsed proton beams, in *Proc. World Congress of Medical Physics and Biomedical Engineering (2009)*, CD.

[5] M. Abe, *Proc. Jpn. Acad., Ser. B* **83**(6), 151 (2007).

[6] M. Abe, Report on particle beam therapy in Japan (MEXT, Tokyo, 2004).

[7] G. Kraft, *Nucl. Phys. News* **17**, 24 (2007).

[8] D. Schardt *et al.*, Heavy-Ion Tumor Therapy: Physical and Radiobiological Benefits, *Rev. Mod. Phys.* (2009), accepted.

[9] H. Tsuji, S. Minohara and K. Noda, this issue.

[10] Th. Haberer *et al.*, *Radiother. Oncol.* **73**(2), 186 (2004).

[11] T. Tajima, *J. Jpn. Soc. Ther. Radiol. Oncol.* **9**, suppl. 2, 83 (1997).

[12] C. M. Ma *et al.*, *Med. Phys.* **28**, 1236 (2001).

[13] E. Fourkal *et al.*, *Med. Phys.* **30**, 1660 (2003).

[14] S. V. Bulanov and V. S. Khoroskhov, *Plasma Phys. Rep.* **28**, 453 (2002).

[15] M. Murakami *et al.*, Radiotherapy using a laser proton accelerator, in *First Int. Symp. Laser-Driven Relativistic Plasmas Applied to Science, Industry and Medicine*, eds. S. V. Bulanov and H. Daido, AIP Conf. Proc. (AIP, 2008), p. 275.

[16] *First Int. Symp. Laser-Driven Relativistic Plasmas Applied to Science, Industry and Medicine*, eds. S. V. Bulanov and H. Daido, *AIP Conf. Proc.* (AIP, 2008) and references therein.

[17] R. A. Snavely *et al.*, *Phys. Rev. Lett.* **85**, 2945 (2000).

[18] E. Clarke *et al.*, *Phys. Rev.Lett.* **84**, 670 (2000).

[19] Maksimchuck *et al.*, *Phys. Rev. Lett.* **84**,4108 (2000).

[20] S. P. Hatchett *et al.*, *Phys. Plasmas* **7**, 2076 (2000).

[21] B. M. Hegelich *et al.*, *Nature* **339**, 441 (2006).

[22] H. Schwörer *et al.*, *Nature* **439**, 445 (2006).

[23] S. Ter-Avetisyan *et al.*, *Phys. Rev. Lett.* **96**, 145006 (2006).

[24] L. Robson *et al.*, *Nat. Phys.* **3**, 58 (2007) and references therein.

[25] F. Brunel, *Phys. Rev. Lett.* **59**(1), 52 (1987).

[26] S. V. Bulanov *et al.*, *C. R. Physique* **10**, 216 (2009).

[27] M. Chen *et al.*, *Phys. Rev. Lett.* **103**, 024801 (2009).

[28] M. Grech *et al.*, *New J. Phys.* **11**, 093035 (2009).

[29] X. Zhang *et al.*, *Phys. Plasmas* **14**, 123108 (2007).

[30] J. Fuchs *et al.*, *Nat. Phys.* **2**, 48 (2006), and references therein.

[31] S. C. Wilks *et al.*, *Phys. Plasmas* **8**, 542 (2001).

[32] T. Esirkepov *et al.*, *Phys. Rev. Lett.* **96**, 105001 (2006).

[33] K. Matsukado *et al.*, *Phys. Rev. Lett.* **91**, 215001 (2003).

[34] X. Q. Yan *et al.*, *Phys. Rev. Lett.* **100**, 135003 (2008).

[35] B. C. Liu *et al.*, *IEEE Trans. Plasma Sci.* **36**(4), 1854 (2008).

[36] X. Q. Yan *et al.*, *Chin. Phys. Lett.* **25**(9), 3330 (2008).

[37] S. G. Rykovanov *et al.*, *New J. Phys.* **10**, 113005 (2008).

[38] S. Steinke *et al.*, Efficient ion acceleration by collective laser-driven electron dynamics with ultrathin foils. submitted to *Phys. Rev. Lett.* (2009) [arXiv:0909.2334v1 (physics.plasm-ph)].

[39] P. McKenna *et al.*, *Plasma Phys. Control. Fusion* **49**, B223 (2007).

[40] T. Tajima and J. Dawson, *Phys. Rev. Lett.* **43,** 267, (1979).

[41] T. O'Neil, *Phys. Fluids* **8**, 2255 (1965).

[42] E. Esarey and M. Piloff, *Phys. Plasmas* **2**, 1432 (1995).

[43] T. Tajima, Laser-driven compact ion accelerator. US Patent 6,867,419B2 (filed Mar. 29, 2002; date of patent Mar. 15, 2005).

[44] T. Tajima, US Patent 6,906,338B2 (filed Jan. 8, 2001; date of patent Jun. 14, 2005).

[45] A. Macchi *et al.*, *Phys. Rev. Lett.* **94**, 165003 (2005).

[46] O. Klimo *et al.*, *Phys. Rev. ST Accel. Beams* **11**, 031301 (2008).

[47] A. P. L. Robinson *et al.*, *New J. Phys.* **10**, 013021 (2008).

[48] B. Qiao *et al.*, *Phys. Rev. Lett.* **102**, 145002 (2009).

[49] T. Esirkepov *et al.*, *Phys. Rev. Lett.* **92**(17), 175003 (2004).

[50] R. C. Shah *et al.*, *Eur. Phys. J. D*, online first (2009) [DOI:10.1140/ejpd/e2009-00152-3].

[51] D. Kiefer *et al.*, *Eur. Phys. J. D*, online first (2009) [DOI:10.1140/ejpd/e2009-00199-0].

[52] V. V. Kulagin *et al.*, *Phys. Rev. E* **80**, 016404 (2009).

[53] A. Henig *et al.*, Radiation pressure acceleration of ion beams driven by circularly polarized laser pulses. Submitted to *Phys. Rev. Lett.* (2009) [arXiv:0908.4057v1 (physics.plasm-ph)].

[54] A. Henig *et al.*, *Phys. Rev. Lett.* **103**, 045002 (2009).

[55] R. C. Shah *et al.*, *Opt. Lett.* **34**, 2273 (2009).

[56] P. McKenna *et al.*, *Phys. Rev. E* **70**, 034405 (2004).

[57] L. Spencer *et al.*, *Phys. Rev. E* **67**, 046402 (2003).

[58] Y. Fukuda *et al.*, *Phys. Rev. Lett.* **103**; 165002 (2009).

[59] A. B. Borisov *et al.*, *J. Phys. B* **40**, F307 (2007).

[60] A. B. Borisov *et al.*, *J. Phys. B* **40**, F131 (2007).

[61] A. Yogo *et al.*, *Phys. Rev. E* **77**, 016401 (2007).

[62] Y. Kishimoto and T. Tajima, *Strong Coupling Between Clusters and Radiation*, High Field Science, eds. T. Tajima, K. Mima and H. Baldis (Kluwer, NewYork, 2000), pp. 83–96.

[63] Y. Kishimoto, Laser cluster interaction and ion acceleration aiming at the understanding of Kansai experiments (2009). http://wwwapr.kansai.jaea.go.jp/pmrc_en/org/colloquium

[64] D. Habs *et al.*, *Appl. Phys. B* **93**, 349 (2008).

[65] F. Pfeiffer *et al.*, *Nat. Phys.* **2**, 258 (2006).

[66] F. Pfeiffer *et al.*, *Phys. Rev. Lett.* **98**, 108105 (2007).

[67] H.-C. Wu *et al.*, Collective deceleration. [arXiv:0909.1530v1 (physics.plasm-ph)].

[68] P. Mora, *Phys. Rev. Lett.* **90**, 185002 (2003).

[69] P. Mora, *Phys. Rev. E* **72**, 056401 (2005).

[70] M. Passoni *et al.*, *Phys. Rev. E* **69**, 026411 (2004).

[71] M. Passoni and M. Lontano, *Phys. Rev. Lett.* **101**, 115001 (2008).

[72] J. Schreiber *et al.*, *Phys. Rev. Lett.* **97**, 045005 (2006).

[73] A. Andreev *et al.*, *Phys. Rev. Lett.* **101**, 155002 (2008).

[74] T. Ceccotti *et al.*, *Phys. Rev. Lett.* **99**, 185002 (2007).

[75] F. Mako and T. Tajima, *Phys. Fluids* **27**, 1815 (1984).

[76] T. E. Cowan *et al.*, *Nucl. Instrum. Methods Phys. Res. A* **455**, 130 (2000).

[77] M. Allen *et al.*, *Phys. Plasmas* **10**, 3283 (2003).

[78] S. Kar *et al.*, *Phys. Rev. Lett.* **100**, 105004 (2008).

[79] X. Q. Yan *et al.*, *Appl. Phys. B*, DOI:10.1007/s00340-009-3707-5 (2009).

[80] V. A. Vshivkov *et al.*, *Phys. Plasmas* **5**, 2727 (1998).

[81] L. Yin *et al.*, *Phys. Plasmas* **14**, 056706 (2007).

[82] F. Pegoraro and S. V. Bulanov, *Phys. Rev. Lett.* **99**, 065002 (2007).

[83] M. Chen *et al.*, *Phys. Plasmas* **15**, 113103 (2008).

[84] X. Q. Yan *et al.*, *Phys. Rev. Lett.* **103**, 135001 (2009).

[85] X. Q. Yan *et al.*, *Phys. Plasmas* **16**, 1 (2009).

[86] A. Chao and M. Tigner, *Handbook of Accelerator Physics and Engineering* (World Scientific, Singapore, 1999).

[87] S. Pastuszka *et al.*, *Nucl. Instrum. Methods Phys. Res. A* **369**, 11 (1996).

[88] B. Rau and T. Tajima, *Phys. Plasmas* **5**, 3575 (1998).

[89] G. A. Mourou *et al.*, *Rev. Mod. Phys.* **78**, 591 (2006).

[90] A. Yogo *et al.*, *Appl. Phys. Lett.* **94**, 181502 (2009).

[91] A. Pukhov and J. Meyer-ter-Vehn, *Appl. Phys. B* **74**, 355 (2002).

[92] P. Chen, *Part. Accel.* **20**, 171 (1987).

[93] T. Toncian *et al.*, *Science* **312**, 410 (2006).

[94] A. Pukhov, *Phys. Rev. Lett.* **86**, 3561 (2001).

[95] D. Habs, in 440th WE Heraeus Seminar on Laser-Driven Particle and X-Ray Sources for Medical Applications (Frauenwörth; Sep. 13–17, 2009). http://www.ha.physik.uni-muenchen.de/veranstaltungen/workshops

[96] V. I. Veksler, in *CERN Symposium on High Energy Accelerators and Pion Physics* (CERN, Geneva, 1956), p. 80.

Toshiki Tajima served as Jane and Roland Blumberg Professor at the University of Texas at Austin, later as Director General of the Kansai Photon Science Institute of JAEA, and is currently Guest Professor of the University of Munich. He, with John Dawson, suggested laser acceleration in 1979. He is considered a founder of High Field Science and is currently Chair of the International Committee for Ultra-Intense Lasers. He is a recipient of the 2009 Blaise Pascal Chair, Nishina Memorial Prize, Suwa Prize of Accelerator Physics, and Farrington Daniels Award (for beam therapy).

Dietrich Habs has been an experimental physicist at the Ludwig Maximilian University of Munich since 1996, working on accelerator, laser and nuclear physics. He is director of the DFG cluster of excellence Munich-Centre for Advanced Photonics, where the development of high-power short-pulse lasers, laser driven secondary particle and x-ray beams, and their application in medical diagnostics and therapy are some of the grand goals. While he is now focusing on laser acceleration, he has been strongly involved in classical accelerators, setting up the radioactive ion beam accelerator REX-ISOLDE at CERN and the Heidelberg Test Storage Ring (TSR) at the Max Planck Institute for Nuclear Physics in Heidelberg.

Xueqing Yan is an associate professor at Peking University, working on conventional linear accelerators and novel concept accelerators. Since 2008 he has been Alexander von Humboldt Research Fellow, focusing on laser ion acceleration and high-harmonics generation.

Reviews of Accelerator Science and Technology
Vol. 2 (2009) 229–251
© World Scientific Publishing Company

FFAGs as Accelerators and Beam Delivery
Devices for Ion Cancer Therapy

Dejan Trbojevic

Brookhaven National Laboratory,
Upton, NY 11973, USA
trbojevic@bnl.gov

First, a review is given of fixed field alternating gradient (FFAG) accelerators, presenting a bit of their history and basic concepts. Special attention is paid to the concept of scaling (S-FFAG) and nonscaling (NS-FFAG) FFAGs. This notation is used only in the NS-FFAG part of the article. A discussion is then provided on operating FFAGs. A presentation is made of the designs being considered for S-FFAGs. A bit more is said about the concept of the NS-FFAG and a resonance crossing problem resulting from designs of the NS-FFAGs. A beam delivery system (gantry) employing the NS-FFAG concept is presented after that and, finally, future plans and R&D requirements are put forward.

Keywords: Fixed Field Alternating Gradient Accelerators (FFAG); Non-scaling FFAG; scaling FFAG; gantries; cancer therapy; isocentric gantries; gantry beam delivery devices; Bragg peak.

1. Introduction

As is well known, proton and carbon cancer therapy is expanding fast, and in this article we will consider the possible role of FFAGs in cancer therapy. We will, in accordance with the medical requirement of penetrating 30 cm of water, consider accelerators of protons with a range of 70–250 MeV, corresponding to a range in depth of $\sim$ 4–35 cm, and carbon ions up to 400 MeV/u, and light ions between protons and carbon that may be especially beneficial for treating particular tumors. In general, an ion cancer treatment facility includes an accelerator, which starts with an ion source with an extraction system and a preaccelerator, and beam delivery systems, such as possibly direct lines and rotating isocentric gantries. High cost and large size constitute a major concern for ion facilities, but despite this there are an impressive number of facilities already built or under construction. Capital and operating costs are an important consideration.

An ion therapy center must be capable of changing the energy of the beam (from patient to patient), so as to locate the Bragg peak exactly at the tumor depth with ability to perform 3D conformal treatment. There needs to be a capability of transverse scanning — preferably spot scanning with respiration adjustments with 2D conformal treatment. The scanning range should cover an area of at least $\pm$ 10 cm and energy scanning $\pm$ 20%. The intensity modulation of up to $\sim 10^3$ in current is preferable. The spot scanning could be established if the repetition rate is $\sim$ 100 Hz–1 kHz, to allow a few minutes for the whole treatment. A treatment dose is of the order of more than 5 Gy/min, requiring ion currents $>$ 100 nA. The treatment has to be efficient, easy to perform, and possess high maintenance ability, with small residual radioactivities. A respiration mode in the facility operation correcting for the organ movement is already present in many beam cancer treatment facilities.

Most of the proton cancer treatment facilities today use cyclotrons. The extraction efficiency of cyclotrons is usually of the order of 70%. This beam loss induces radioactivity and, in the case of access, the required "cooling" time is up to 10 days. Protection from continuous unavoidable small beam loss requires additional shielding. The fixed extraction energy in cyclotrons entails use of the degrader system to reduce the energy to the required value. This not only induces radioactivity around it but also reduces by even 100 times the beam intensity and blows up the emittance of the beam.

FFAGs address these requirements very favorably: the beam current of 100 nA is easily achievable, the variable energy is obtained by timing correctly the extraction kickers, the respiration mode does not represent a problem, and the multiextraction can be provided by many extraction ports. The cyclotrons can produce enough current as they can operate in the CW mode, the variable energy requires degraders, and the respiration mode and multiextraction are not easy. The synchrotrons might be able to produce the required beam current if they are rapid cycling > 20 Hz. The variable energy and the respiration mode in synchrotrons are not a problem. The multiextraction mode is not trivial.

The NS-FFAG ion delivery systems and gantries should provide important advantages:

- They should be easy to operate, due to the fixed magnetic field.
- They would reduce dramatically the weight of the transport elements as well as of the support structure with respect to the warm magnet designs.
- They would reduce the operating cost, due to either permanent magnets or cryogenic small magnets.
- They should reduce the overall cost. Usually the isocentric gantries and ion delivery systems represent a dominant cost of the whole ion therapy facility.

Ions deposit energy in the human tissue by Coulomb interaction with the atomic electrons and nuclei, by bremsstrahlung energy loss, and by nuclear interactions. As ions penetrate into the body the absorbed dose increases gradually, and near the end of the range the dose increases rapidly — this is the Bragg peak. The beam straggling and Coulomb scattering set the limit on the treatment accuracy. The effective beam size — penumbra in the tissue depends on these two effects which scale with energy. For example, the effective beam size due to Coulomb scattering (C) for a 200 MeV proton beam is $\sigma_C = 6.5$ mm [1], while the straggling effect (S) makes the beam size $\sigma_S = 6.5$ mm (even though the optical beam size at the patient's skin could be $\sigma = 0.23$ mm or $\sigma = 0.35$ mm with an optical amplitude function of $\beta = 0.45$ m and $\beta = 1$ m, respectively). Thus it is not important to have a

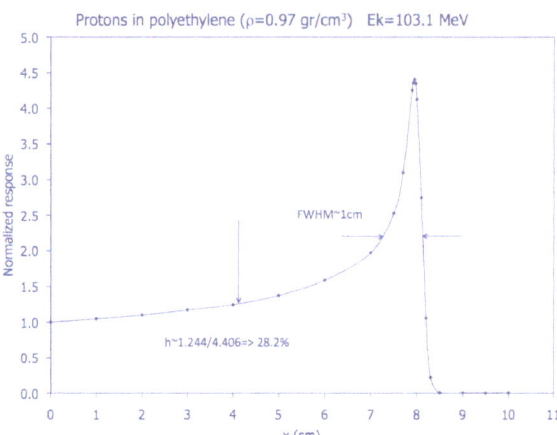

Fig. 1. Normalized exposure for the 103.1 MeV protons. Experimental results from the BNL–NSRL project.

very small beam (transverse emittance and energy spread) beyond a certain point given by the above phenomena. Small emittance and well-defined energy are important for beam acceleration and transport and will be discussed at some length below.

The experimental data of 103.1 MeV proton and 200.2 MeV/u carbon ion propagation in polyethylene ($\rho = 0.97$ gr/cm^3) produced very similar positions of the Bragg peak. Proton and ^{12}C^{6+} ion energy deposition data are shown in Fig. 1 and Fig. 2, respectively. The polyethylene resembles the human tissue very well. This is a part of the NASA Space Research Radiation Laboratory (NSRL) at Brookhaven National Laboratory, Upton, New York, and the data became available courtesy of A. Rusek. There are two additional very important effects if different ions are being considered in planning the

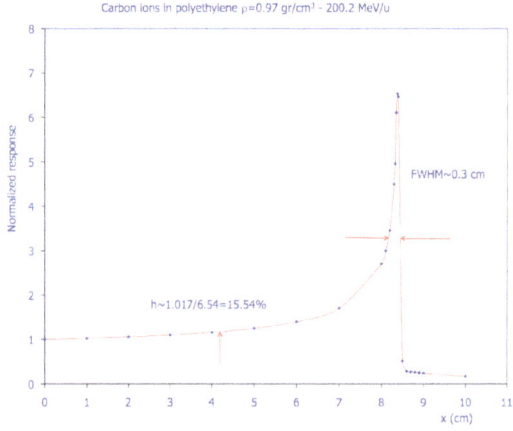

Fig. 2. Normalized exposure for the 200.2 MeV/u carbon ^{12}C^{6+}. Experimental results from the BNL–NSRL project.

patient treatment: relative biological effect (RBE) and linear energy transfer (LET). For protons, RBE ~ 1.1 compared to the radioactive cobalt ^{60}Co as a reference. As the ion loses energy while it penetrates the volume, the LET increases as well as the RBE if it is in the Bragg peak region. The RBE is still fairly low in the entrance region. The carbon ions have a higher RBE in both the Bragg peak and the entrance region. A comparison of the experimental results, with a very similar depth of penetration, shows very clear differences between the proton and carbon ion propagation:

- The straggling and multiple Coulomb scattering is much smaller in the case of carbon ions. This reduces the amount of radiation received by healthy human tissue.
- The Bragg peak, located at the tumor where most of the energy is deposited, is sharper for carbon ions. The carbon ion treatment provides better control of the longitudinal beam profile in the tumor. This is performed by successive ion depositions at different energies.
- The "leakage" of the carbon ions in the area behind the Bragg peak is larger than for the proton case. This is mostly due to breakup of carbon ions. The "leakage" is important if a sensitive organ is located behind the cancerous tumor.

In Sec. 2 we will discuss the FFAG history and the FFAG principles. In Sec. 3 we will discuss operating FFAGs, and in Sec. 3.4 the designs for S-FFAGs. In Sec. 4 we will describe a principle of the NS-FFAG and make a comparison with the S-FFAG. Three different approaches to applying the NS-FFAG in cancer therapy are described in Sec. 5. The NS-FFAG tunes change during acceleration, and multiple integer resonances are crossed. The acceleration tracking results of the few NS-FFAG designs are shown in Sec. 6. Section 7 deals with the subject of NS-FFAG application in the ion beam delivery system. Finally, in Sec. 8, we discuss advantages and disadvantages of different designs, and at the end, in Sec. 9, the future plans and R&D requirements are presented.

2. FFAG Accelerators

Most proton therapy facilities today use fixed magnetic field cyclotrons. Synchrotrons are present in both proton and carbon cancer therapy but are presently dominant in carbon cancer treatment facilities [Chiba (Japan), Hyogo, Gunma, Heidelberg (Germany), CNAO (Italy)]. Cyclotrons use either fixed frequency beam acceleration if they are classical CW cyclotrons or frequency modulation if they are pulsed beam synchrocyclotrons. The magnetic fields may alternate in sign around the circumference. If the acceleration frequency is fixed, then these are isochronous cyclotrons, and they become the S-FFAG if the frequency is modulated [2]. The S-FFAG can be considered as an extension of the sector-focusing cyclotron made into a ring of many sectors. Okhawa, Kolomensky, and Symon (1953–55) proposed [3–5] independently the FFAG concept.

The basic characteristics of the S-FFAG are a fixed magnetic field, large acceptances, high beam currents, and fast repetition rates limited by the RF capabilities. The S-FFAG has the betatron tunes fixed, with zero chromaticity for all particle energies as the orbit radii scale for different energies in such a way that the integral of $1/\beta$ around the orbit, i.e. the tune, is the same,

$$\nu = \frac{1}{2\pi} \oint \frac{ds}{\beta}, \tag{1}$$

and for β,

$$2\beta\beta'' - (\beta')^2 + 4\beta^2 K(s) = 4, \tag{2}$$

where β is the beam betatron amplitude function, s is the azimuthal coordinate, β'' and β' are the first and second derivatives with respect to s, and $K(s)$ is the normalized focusing parameter $B'/B\rho$ (ρ is the bend radius). If β scales with the radius, then Eq. (2) is satisfied for all radii. The tune remains the same for all radii and the orbit scales exactly. From Eq. (1) the tune is the same at all radii as β increases and ds increases for the same amount.

Resonances in rings are of concern if the acceleration is slow and occurs when

$$l\nu_x \pm m\nu_y = n, \tag{3}$$

where l, m, n are integers. The S-FFAG orbits of all energies are scaled replicas of one another, with the same tunes for all energies (avoiding a problem of crossing resonances), and chromaticity is fixed and close to zero. Variation of the magnetic field flutter is defined as

$$F^2 = \left\langle \left(\frac{B(\theta) - \bar{B}}{\bar{B}} \right)^2 \right\rangle, \qquad (4)$$

where $B(\theta)$ is the magnetic field on the trajectory, while $\bar{B}$ is the average magnetic field.

The equations of motion of particles around the linear orbit in the combined function magnet are

$$\frac{\partial^2 x}{\partial s^2} + \frac{1-n}{\rho^2} x = 0, \qquad (5)$$

$$\frac{\partial^2 y}{\partial s^2} + \frac{n}{\rho^2} x = 0, \qquad (6)$$

where x and y are the radial and vertical offsets, ρ is the radius of curvature, and n is the field index:

$$n = -\frac{\rho}{B} \frac{\partial B}{\partial x} \quad \text{or} \quad k \equiv \frac{r}{B_m} \frac{\partial B_m}{\partial r}, \qquad (7)$$

where B_m is the mean magnetic field along the orbit. The tunes in the radial and vertical planes are derived from Eqs. (5) and (6) (details can be found in Ref. 6),

$$\nu_r \sim \sqrt{n-1} = \sqrt{1+k}, \qquad (8)$$

and for the vertical tune:

$$\nu_y = \sqrt{n} \quad \text{or} \quad \nu_y = \sqrt{-k + F^2(1 + 2\tan^2 \varepsilon)}. \qquad (9)$$

The radial tune ν_r is fixed during acceleration if the parameter k is a constant, as seen in Eq. (8). The term "scaling FFAG" comes from the conditions for the fixed tunes accomplished for the radial sector FFAG by the magnetic field dependence on k as

$$B_R \sim B_o \left(\frac{r}{r_o} \right)^k \mu(r, \Theta), \qquad (10)$$

where r_o and k are constant numbers and $\mu(r, \Theta)$ is a periodic function with a number of periods N_a [5]. For the spiral sector S-FFAG is

$$B_{\mathrm{S}} = B_o \left(\frac{r}{r_o} \right)^k f(\psi), \qquad (11)$$

where $f(\psi)$ is a periodic function. The vertical tune ν_y is fixed if $F^2(1 + 2\tan^2 \varepsilon) = \text{const}$. In the case of the radial sector FFAG this is accomplished by introduction of opposite bends, while for the spiral sector FFAG the sector axis dependence is $R = R_o e^{\theta \cot \varepsilon}$. The relationship between the momentum and the

rigidity is $p = zeB\rho$, where z is the number of charges.

$$p_o = eB_o r_o, \quad p = eB_R r, \quad p = p_o \left(\frac{r}{r_o} \right)^{k+1}. \qquad (12)$$

There are radial and spiral sector S-FFAGs:

2.1. *Radial sector FFAG*

The first radial sector designs used two magents, both with magnetic fields of the form $B \approx B_o(r/r_o)^k$, where r is the distance from the machine center, and k is usually is of the order of 10. In a few S-FFAG lattice designs for maximum proton energy of 10 and 20 GeV, k reached the values of 192.5 and 150, respectively [5]. Both examples had a large number of periods, N_p. The orbits are symmetric and periodic in azimuthal angle Θ with period $2\pi/N_p$, as shown in Fig. 4. Due to restrictions on the particle motion stability in the vertical plane, the length of the opposite bending magnet cannot be shorter than 2/3 of the positive field bend. This increases the circumference.

The radial sector S-FFAG is madeup of N identical sectors [5]. One magnet has positive bending and radial focusing, while the other is defocusing and opposite-bending, as shown in Fig. 3. In Eq. (13) the periodic function $\mu(\Theta)$, dependent on the azimuthal coordinate $s = \Theta r$, is defined as

$$\mu(\Theta, r) = \frac{r}{\rho(\Theta, r)}, \quad dx = \eta(\Theta, r)dr \qquad (13)$$

and

$$\left[\frac{\varepsilon}{\eta} \right]_o = \tan \xi, \qquad (14)$$

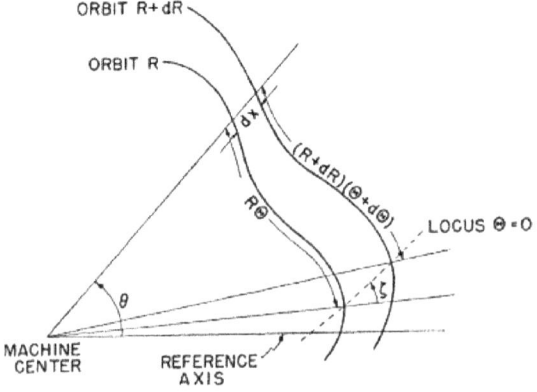

Fig. 3. Equilibrium orbit notation and definition of parameter premagnets on the right, from the original article by K. R. Symon *et al.*

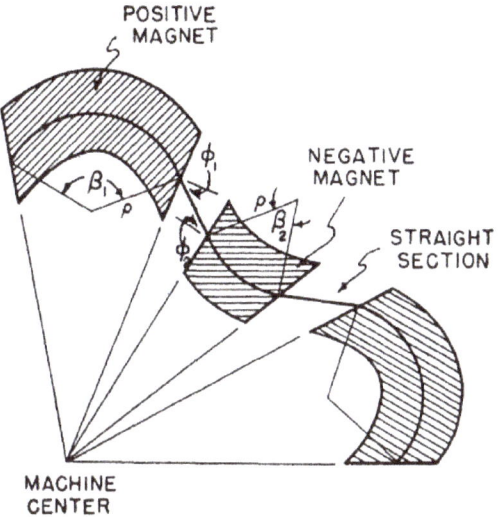

Fig. 4. The radial sector magnets from the original article by R. K. Symon [5] show essential definitions and the referenced orbit.

where ξ is defined as the angle between the reference orbit and the reference curve, for $\Theta = 0$, where the orbit crosses r as in Fig. 3: $d\Theta = \epsilon(\Theta, r)dr/r$. The radial sector FFAG requires opposite bending, as shown in Fig. 4.

The S-FFAGs were studied and electron models were built at MURA (Midwest University Research Association). There were many designs of the proton FFAGs by the MURA group, but the first one was built in the year 2000 in Japan, at KEK. The proof-of-principle, 1 MeV proton machine built and commissioned at KEK is shown in Fig. 5.

Fig. 5. The proof-of-principle, 1 MeV proton S-FFAG accelerator built and commissioned in 2000 at KEK, by the Y. Mori group.

2.2. *Spiral sector FFAG*

The spiral sector FFAG was developed by Kerst in 1954 [7]. He realized that the sectors do not need to be symmetric and by tilting the edges he achieved vertical focusing without the need for opposite bending. The edge focusing is used in both radial and spiral sector FFAGs; the vertical tune depends on $\nu_y^2 \approx -k + F^2(1 + 2\tan^2 \epsilon)$, where k is the average field index and ϵ is the spiral edge angle. For the spiral sector FFAG, ϵ is constant. The sector axis dependence is $R = R_o e^{\theta \cot \epsilon}$. One of the original "MURA" notes on the spiral sector is shown in Fig. 6.

$$B = B_o \left(\frac{r}{r_o}\right)^k \left\{1 + f \sin\left[N\theta - \left(\frac{1}{\omega}\right)\ln\left(\frac{r}{r_o}\right)\right]\right\}, \tag{15}$$

where it is required that $1/\omega = N\tan\xi = 2\pi/\lambda$, where λ is the ridge separation, and f is the flutter factor defined in Eq. (4).

It is interesting to note that by abandoning the scaling law for the magnetic field and using a spiral sector with a combined function magnet of the straight line edges and linear radial field dependence, stable orbits in the large momentum range $-50\% \leq \delta p/p \leq 50\%$ could be found [8] with orbit offsets $|x_{max}| \leq 25$ cm.

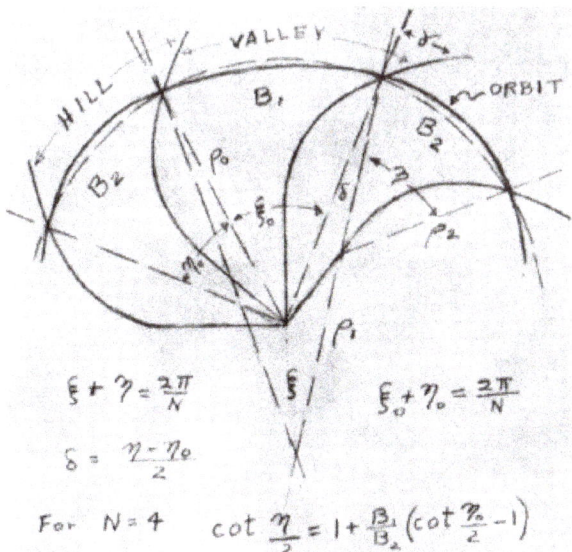

Fig. 6. The spiral sector principle from the original "MURA" notes. Periodicity is shown from $\xi + \eta = 2\pi/N$. The note corresponds to a case of $N = 4$.

3. Operating S-FFAGs in Medical Applications

There are multiple reasons for today's revival and success of the S-FFAG developments: one is excellent new tools (TOSCA-OPERA, POISSON, etc.), and others are technological improvements and better accuracy of designing and making complicated magnets, as well as accurate new computational tools like Polymorphic Tracking Code (PTC) written by E. Forest [9]; ZGOUBI, by F. Meot [10]; MADX-PTC [11]. The new tools compute particle motion without any approximation and the Hamiltonian is calculated without any error.

The technological innovations of the group led by Y. Mori [12] include RF acceleration using Finemet alloy water-cooled cavities allowing broadband operation with variation in frequencies between 1.5 and 4.6 MHz, with the RF output power of 55 kW. The other innovation is a development of the scaling triplet DFD (defocusing–focusing–defocusing) magnets as one unit, fulfilling high accuracy for very tight magnetic field requirements. Both were applied in the 150 MeV S-FFAG ring. There are already two commissioned S-FFAGs for medical applications: one is the radial sector S-FFAG developed and commissioned at KEK, Tsukuba, Japan (shown in Fig. 7), and the other is the storage ring ERIT (Energy/emittance Recovery Internal Target) at Kyoto University. The 150 MeV ring, built at KEK, was designed for proton therapy, but it was moved from KEK to the Ito campus at Kyushu University, Kyushu Island, Japan, in March 2008. It is not clear yet if the future research will include radiation

therapy. Both S-FFAGs had already been commissioned under the leadership of Y. Mori [12].

3.1. *150 MeV proton S-FFAG accelerator*

The 150 MeV ring is 12 the radial sector scaling FFAG with the field index $k = 7.5$ and an energy range between 12 and 150 MeV. A designed repetition rate is 250 Hz, with the closed orbit radii between 4.47 and 5.3 m. During the commissioning the ring operated at a 100 Hz repetition rate with one operating cavity. The injection cyclotron (shown in the middle of Fig. 6) provided during the commissioning protons with an energy of 10 MeV with a 1 μA current with the energy range between. The radial sector S-FFAG accelerator is madeup of 12 DFD cells. The magnet design, done by using the program OPERA 3D, is shown in Fig. 8. The maximum field in the focusing magnet is 1.63 T, while the field in the opposite bend defocusing magnet is 0.13 T. The horizontal and vertical betatron tunes are $\nu_x = 3.8$ and $\nu_y = 2.2$, respectively.

3.2. *ERIT*: *intense neutron source with Energy Recovery Internal Target*

A radial sector S-FFAG for boron neutron capture therapy has already been commissioned at Kyoto University. The required low energy spectrum neutron flux has to be $\Phi > 1 \cdot 10^9 \, \mathrm{n/cm^2/s}$. The neutron flux can be produced, for the first time by an accelerator, due to the multipass of the proton beam through a thin internal target. A large momentum acceptance is necessary in order to recover the lost proton energy by an RF cavity from interaction with the internal target. The RF reacceleration and ionization cooling should reduce the emittance growth

Fig. 7. The first proton therapy radial sector 150 MeV S-FFAG, built and commissioned at KEK in 2003.

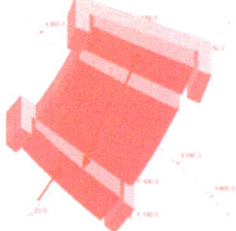 

Fig. 8. The radial sector 150 MeV S-FFAG magnets built and commissioned at KEK, Japan by the Y. Mori group. The magnet design is on the left and the actual radial sector triplet magnet on the right.

Fig. 9. ERIT (Energy Recovery Internal Target) at Kyoto University, Japan: a neutron source for boron neutron capture therapy. The 20 MHz cavity is the large yellow circular disk. The magnets are readily seen.

and recover the energy loss. The large momentum and transverse acceptance and zero chromaticity were essential in choosing the S-FFAG ring. The S-FFAG ring, presented in Fig. 9, stores protons with a 20 MHz RF frequency with an about 1000-turn lifetime, with a current of 70 mA at an energy of 11 MeV. A negative hydrogen beam H^- is accelerated up to 11 MeV, by a linac with a repetition rate of 20–200 Hz. Acceleration is created by a combination of RFQ (radiofrequency quadrupole) and DTL (drift tube linac). The production target is made of beryllium $10\,\mu$ foils placed in a rotating carousel system. The internal target is placed in a three-layer moderator to shield it from neutrons and γ-rays. Very soon the foils are to be replaced with the wedge type

to make cooling in momentum space possible. This is due to the scaling of the orbits with energy. The greater target thickness will produce a larger energy loss. After protons interact with the wedge foil target, they arrive after a half-turn at the the cavity, where they gain an energy kick. ERIT will be used for patient treatment.

3.3. *RACCAM spiral sector S-FFAG for proton therapy*

The project RACCAM (Research on ACCelerators and Applications in Medicine) has the goal of providing a proton cancer therapy accelerator with variable extraction energy covering the energy range of 70–180 MeV [13]. It represents the collaboration of a few laboratories and hospitals in France. The project design is a combination of a small energy cyclotron and a spiral sector FFAG with a single turn extraction. The injection H^- cyclotron has a span between 8 and 14 MeV of extraction energies by using a variable stripper position. The injection cyclotron with extraction orbits is shown in Fig. 10. The spiral sector S-FFAG was selected, with 10 cells and a spiral angle of $53.7°$, and radii of 2.78 m at injection and 3.46 m at extraction. The maximum of the orbit offsets is $\Delta r_{max} \sim 0.67$ m. The variable extraction energy from the 70–180 MeV is planned by adjustment of the magnetic field of the spiral magnets and synchronizing the extraction kicker time. At the time of this report a time scale for the project has not yet been determined. The first measurements of the magnetic field are very encouraging (the magnet is shown in Fig. 12).

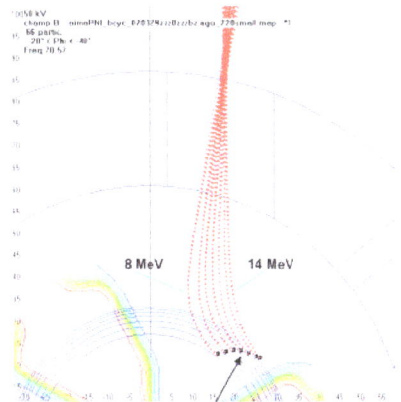

Fig. 10. The injection H^- cyclotron with orbits for energies of 8–14 MeV and the extraction stripper at variable positions for the French RACCAM project.

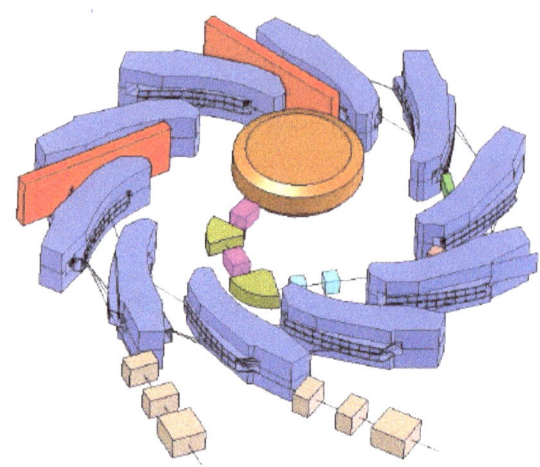

Fig. 11. The RACCAM proton therapy accelerator spiral sector FFAG, with the injection cyclotron in the middle.

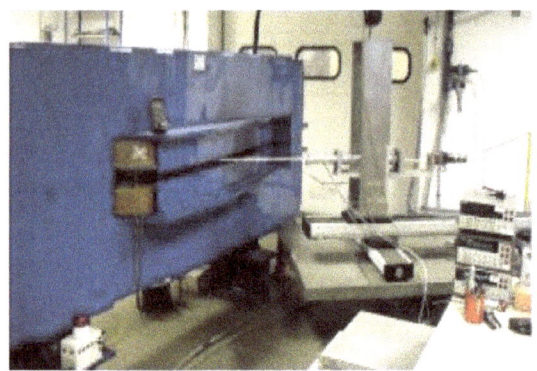

Fig. 12. The RACCAM spiral magnet with a probe for precise field measurements.

3.4. *Other designs of S-FFAG projects planned for ion therapy*

3.4.1. *Eight spiral-sector S-FFAG 230 MeV design for the Ibaraki Medical Facility*

The design of the S-FFAG project for the Ibaraki Medical facility is made with eight spiral sectors for 230 MeV proton therapy, with the cyclotron as an injector. Orbits at injection are at a radius of 2.2 m; at extraction, 4.1 m.

3.4.2. *MEICo C^{6+} ion therapy by the Mitsubishi Electric hybrid accelerator*

There is a report on design studies of the hybrid S-FFAG/synchrotron MEICo ion therapy accelerator by Mitsubishi Electric. The S-FFAG is madeup of 16 spiral sectors with a spiral angle of 64°. The maximum magnetic field is $B_{\max} \leq 1.9$ T. Maximum

orbit offsets are between radii of 7 and 7.5 m, with a machine diameter of 16 m.

3.4.3. *Chiba 12-radial-sector 3-concentric S-FFAG design*

A design study on three-concentric-radial-sector S-FFAGs for carbon ion cancer therapy, as a possibility for the Chiba Medical facility, was presented in 2004 [14]. The low energy five-radial-sector FFAG ring would start with a 40 keV/u ECR (electron cyclotron resonance) ion source and accelerate to 6 MeV/u. The medium energy, 12-cell FFAG ring has an energy range of 6–100 MeV/u, and the high energy, 12-sector ring a range of 100–400 MeV. A design of the three S-FFAG rings is presented in Fig. 13.

The repetition rate was assumed to be 200 Hz. The low energy FFAG has an RF frequency between 0.215 and 1.878 MHz and the accelerating voltage of 3.3 kV is provided by two cavities. The medium energy FFAG has an RF frequency range of 0.894–3.002 MHz and the required 18.4 kV is supplied by two cavities, while the high energy ring has a 1.991–3.098 MHz frequency range and four cavities supply the 45.5 keV needed. This design is not going to be implemented. A synchrotron was selected instead.

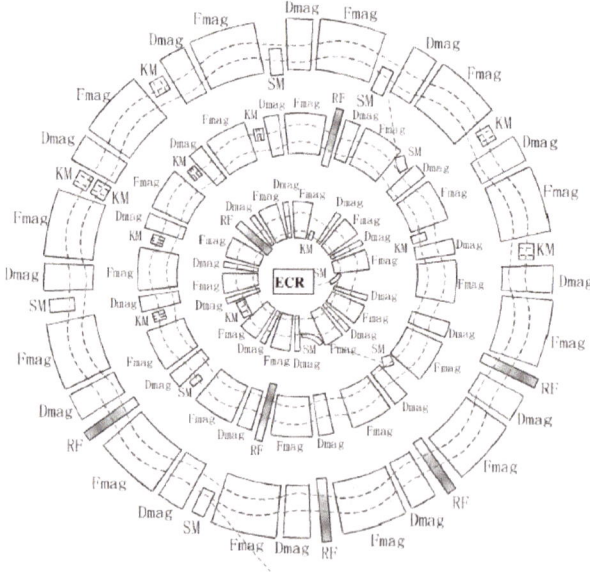

Fig. 13. A new design for $^{12}C^{6+}$ and proton, new radial sector S-FFAG accelerators for the Chiba facility in Japan. Cavities (RF), septum (SM), and kicker magnets (KM) are shown in each ring.

4. Nonscaling vs. Scaling FFAG

The aperture size of the S-FFAGs depends on the accelerator energy range. In the proposal for the radial sector scaling for the 10 GeV protons by K. R. Symon *et al.* [5] for a radial sector scaling 10 GeV proton ring, the value of $r_\mathrm{max} - r_\mathrm{min}$ was 2.3 m. The width of the radial sector magnet was 3.25 m and the height 1.6 m, with a magnet weight of 9650 tons and excitation power of 5.5 MW. Their spiral sector 20 GeV proton FFAG example had an aperture of 3.12 m with 5 MeV injection orbits and 20 GeV extraction orbits. Today the smaller 150 MeV proton radial sector S-FFAG built at KEK has orbits offsets of $\Delta r \sim 0.9$ m varying as $\Delta r = 4.4 - 5.3$ m. The NS-FFAG concept arose during a study of future muon colliders or neutrino factories [15]. The NS-FFAG reduced the cost by using more circulating turns than in the recirculating linacs; they replaced the high linac cost by using the same linac with multiple turns. The first suggestion for an NS-FFAG was made by L. Teng in 1956 [16], while the first publication of the new FODO NS-FFAG design came from C. Johnstone *et al.* in 1999 [17]. Crossing integer resonances should not be important due to the very fast acceleration.

The benefit of the NS-FFAGs is due to the relationship

$$\Delta x = D_x \frac{\Delta p}{p}, \qquad (16)$$

where Δx is the radial beam offset, D_x the lattice dispersion function and $\Delta p/p$ the fractional momentum deviation. The value for Δx may be kept at less than ± 50 mm for $\Delta p/p = \pm 50\%$, if D_x is < 0.1 m. The dispersion function or the dispersion action H is well controlled in minimum emittance lattices for synchrotron light sources where it must be minimized:

$$\langle \mathcal{H} \rangle = \frac{1}{L} \int_0^L (\gamma D^2 + 2\alpha D D' + \beta D'^2)ds, \qquad (17)$$

where α, β, and γ are the lattice betatron functions, D' is the derivative of the D with respect to s, and L is the lattice superperiod length. The NS-FFAG has a very strong focusing structure in order to obtain small values of D and β, and hence small magnet sizes. Linear magnetic field dependence $B \propto r$ is an additional simplification. The strongest focusing, smallest dispersion, and best circumference are achieved by the combined function magnets in the triplet FDF, in the middle of which is a larger

defocusing main bending element surrounded by two smaller focusing combined function magnets with opposite bends [18]. To reduce the number of magnets a doublet, instead of a triplet, is used in the basic cell, slightly increasing the orbit offset size. The orbit circumference or time of flight of the particles in the NS-FFAG varies quadratically with energy:

$$C(p) = C(p_c) + \propto (p - p_c)^2. \qquad (18)$$

A small dependence on momentum enables acceleration with fixed frequency in the case of high energy particles or when $v \sim c$. The tune dependence on momentum in the NS-FFAG basic cell varies in both planes in the range $\nu_{x,y} \sim 0.4$–0.1. The first NS-FFAG, the Electron Model for Many Applications (EMMA) [19], has been built and will soon be commissioned at the Daresbury Laboratory, England. This is a proof of principle of the NS-FFAG concept for the fixed frequency acceleration of relativistic electrons, muons, and other ions. This initial design aim may be extended to tests relevant for ion cancer therapy rings, by including a new RF system for frequency modulation acceleration of the nonrelativistic electrons, though this possibility requires further evaluation. These additional tests are very important for the FFAG proton driver to be used in muon production or high intensity neutrino beams, FFAG heavy ion acceleration, and the FFAG ADS (accelerator driven system) to drive the thorium nuclear reactors, etc. A very valuable review of recent developments in the S-FFAG and the NS-FFAG can be found in Ref. 20.

4.1. *Integer resonance crossing during acceleration*

The nonrelativistic acceleration of the proton or light ions with the NS-FFAG has to be fast enough to avoid emittance or beam amplitude blowup due to crossing integer resonances. The amplitude growth during the resonance crossings depends on errors in the magnetic field and on the speed of the resonance crossing. Results of many simulation studies [21–23] have shown very good agreement with analytical predictions by R. Bartmaan [24]. Because of the high importance of this problem in using the NS-FFAG for ion acceleration in cancer therapy treatment, more details of the analytical predictions and simulation results are presented. The resonance occurs when $m\nu = n = fN_p$, where ν is the tune of the ring,

m is the order of the resonance driving term, n corresponds to the n_{th} harmonic Fourier component of the $(m-1)_{th}$ derivative of the magnetic field, f is an integer, and N_p is the number of lattice periods. The intrinsic or systematic resonances of an accelerator depend on the number of lattice periods N_p through the resonance condition: $m\nu_x = N_p$. The betatron amplitude is defined as $A = \sqrt{2J/\nu}$ with a resonance condition $m\nu = n$, while the resonance strength parameter $b_{n,m}$ is

$$nb_{n,m+1} = \frac{\bar{R}}{\bar{B}} \frac{1}{m!} \frac{\partial^m B_n}{\partial x^m}, \qquad (19)$$

where the average magnetic field is $\bar{B} = B\rho/R$. The resonance crossing depends on how fast the tunes are changing: $\nu_\tau = \Delta\nu/N_o$, where $\Delta\nu$ is the tune change and N_o is the number of ring turns during acceleration. The action angle variables used are J and ψ, with $x = A\cos(n\theta + \phi) = A\cos\psi$, where θ is the azimuth around the ring. For $m = 1$, $\nu = n$ the amplitude growth for integer resonance crossing is equal to

$$\Delta A = \pi \frac{b_{n,1}}{\sqrt{\nu_\tau}} = \frac{\pi}{\sqrt{\nu_\tau}} \frac{\bar{R}}{\bar{B}} \frac{B_n}{\nu}. \qquad (20)$$

For the half-integer resonance crossing $m = 2$, and $n = 2\nu$, the resonance coefficient $b_{n,2}$ is

$$b_{n,2} = \frac{\bar{R}}{n\bar{B}} \frac{\partial B_n}{\partial x} \qquad (21)$$

and the ratio of the final to the initial amplitude is given by

$$\log \frac{A_f}{A_i} = \frac{\pi}{\sqrt{2}} \frac{b_{n,2}}{\sqrt{\nu_\tau}}. \qquad (22)$$

Although the higher order resonances are less of a problem, the third integer crossing could be a problem if the cell tunes cross values of 1/3, as any systematic sextupole fields in the ring also excite the intrinsic resonance [24], $3\nu = N_p$, for which the emittance growth is

$$\Delta\varepsilon^{-1/2} = \frac{\pi}{4\sqrt{3}} \frac{1}{\sqrt{\nu_\tau}} \frac{R}{B\rho} \left| \beta^{3/2} \frac{\partial^2 B_z}{\partial x^2} \right|_{n=3\nu}, \qquad (23)$$

where R is the radius of the ring, $\partial^2 B_z/\partial x^2$ the second derivative of the magnetic field, and β_x the transverse betatron amplitude. Faster acceleration in the NS-FFAG is preferable not only due to better resonance crossings without significant amplitude growth, but also for the higher ion delivery rate

reducing the time of patient treatment. To accelerate nonrelativistic ions with the NS-FFAG implies a large range of speeds, and some small orbit length change, and thus requires fast and large frequency variation or fast RF phase jump variation after every turn. In crossing the one-fourth integer, the cell tune may be of concern if transverse space charge forces are at a significant level.

5. Designs of the NS-FFAGs

There are three different approaches to using the NS-FFAG for the ion therapy and ion delivery systems. Two of them employ tune stabilization to avoid resonance crossing and can have slow variation in energy, but have large aperture requirements. The third approach is an NS-FFAG madeup of small magnets with linear magnetic field dependence, and a small aperture, but requires fast acceleration for resonance crossings.

5.1. *PAMELA–CONFORM project*

The first approach is the the proton and carbon ion accelerator design for the PAMELA–CONFORM project (PAMELA — Particle Accelerator for medical applications; CONFORM — Construction of a Nonscaling FFAG for Oncology Research and Medicine) in the John Adams Institute at Oxford University, UK. The lattice design uses the superconducting magnets with the maximum field of 5 T. Small variations of tune, with orbit offsets of < 0.2 m, are obtained in an NS-FDF (focusing–defocusing–focusing) triplet configuration by introducing nonlinear magnetic fields of the form $B \sim B_o(r/r_o)^k$, where $k = 38$. The lattice is madeup of 12 cells. The accelerator has a kinetic energy range of 30–250 MeV and of 8–70 MeV/u for protons and carbon ions, respectively. The radius is 6.25 m, with 1.7-m-long straight sections. The lattice design is presented in Fig. 14. The small variation of tune versus momentum in PAMELA is presented in Fig. 15. Acceleration in the PAMELA design is assumed to be with a frequency variation from 1.94 to 4.62 MHz with ~ 15 kV per cavity and 100 kV/turn making at a 1 kHz repetition rate. A crossing of the half-integer resonance at 50 kV/turn induced by emittance blowup, $\varepsilon_1/\varepsilon_o$, of more than 5, for a field error, $\delta B_1/B_1 = 2 \cdot 10^{-3}$, as shown in Fig. 16. This was confirmed by tracking particles with the program ZGOUBI [10].

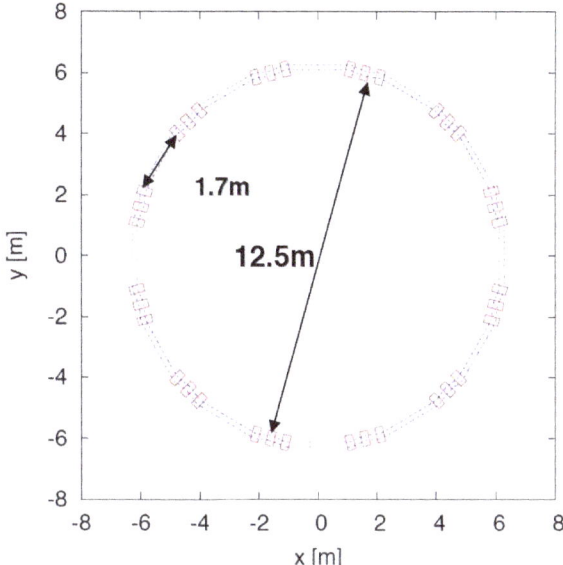

Fig. 14. Layout and orbits of PAMELA, a design from the John Adams Institute at Oxford University, UK.

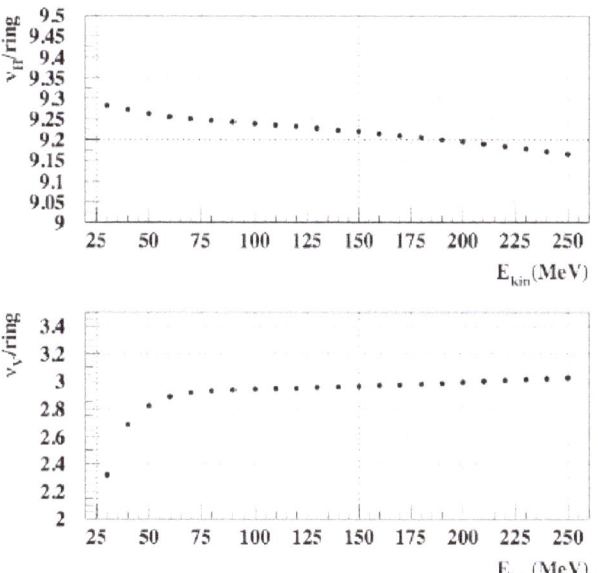

Fig. 15. Tune variation with energy in the PAMELA lattice design.

5.2. *Second nonlinear NS-FFAG for medical applications*

Another example of a nonlinear NS-FFAG with similar tune stabilization over an extended acceleration range and with a factor of 6 or more in momentum is designed by using the wedge effect with a nonlinear magnetic field dependence along the radial aperture [25], as in the FDF triplet cells of

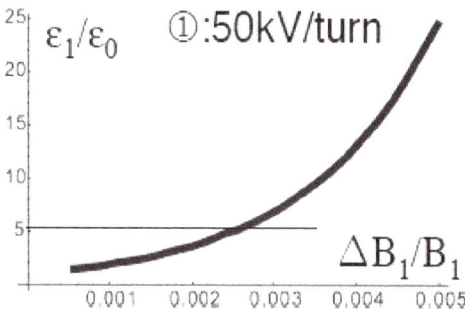

Fig. 16. Emittance blowup dependence on magnetic field error for the acceleration rate of 50 kV/turn. The acceleration of 100 kV/turn was selected instead.

PAMELA. Fast acceleration of the nonrelativistic particles is limited by the ferrite material response function for the RF frequency variation method. To reduce the RF power and slowdown acceleration with hundreds of thousands of turns, the tune variation needs to avoid crossing integers. The 250 MeV proton cancer therapy ring is madeup of eight triplet superconducting magnet cells. The maximum of the magnetic field is 3.41 T. Maximum orbit offsets are $x_{max} = 0.65$ m, while the maximum orbit radius is $r_{max} = 6.9$ m. Both rings are presented in Fig. 17.

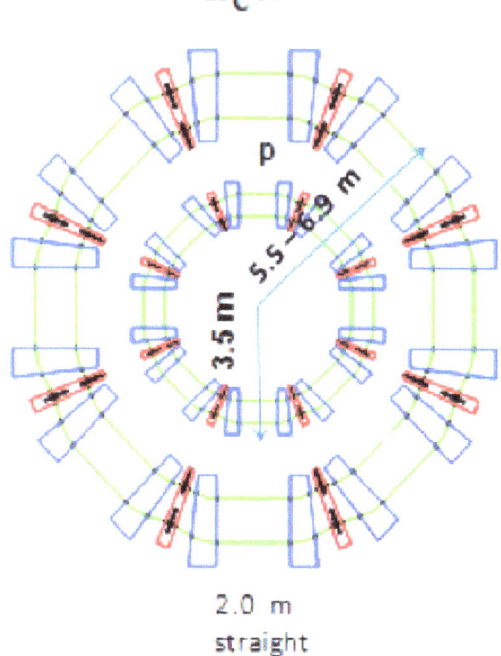

Fig. 17. Proton and carbon rings of the second nonlinear NS-FFAG design. The triplet magnet structure is FDF, as in the S-FFAG.

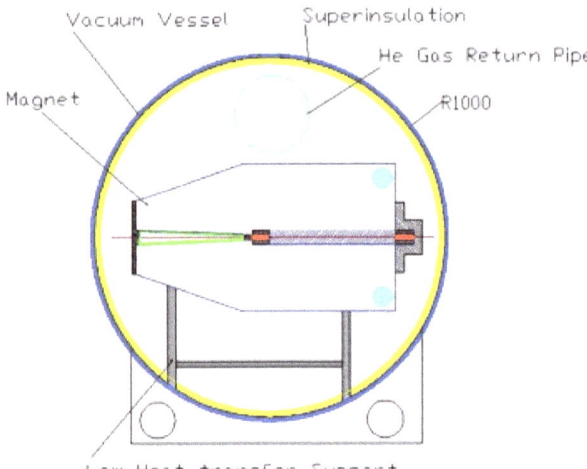

Fig. 18. A magnet design for the second nonlinear NS-FFAG. The aperture of ~65 cm is shown in green.

The additional concentric ring is designed for carbon $^{12}C^{6+}$ ion acceleration from 60 to 400 MeV/u. It is madeup of eight triplet cells with superconducting magnets. The field varies radially, reaching the maximum of 4 T. A magnet design is shown in Fig. 18. The tune stabilization achieved is presented in Fig. 19.

5.3. Linear NS-FFAG with a smaller aperture

The advantages of the linear NS-FFAG are: reduced magnet weight, ease in changing the final energy, high repetition rate, linear dependence of the magnetic field, and very strong focusing with small aperture size. Acceleration of the nonrelativistic particles has to be very fast as the vertical and horizontal tunes vary with energy and cross

integer resonances. Even very advanced ferromagnetic materials cannot fulfill frequency modulation requirements for fast acceleration. An alternative acceleration with a fixed frequency requires adjustment of the RF phase after each turn [26]. The RF cavities and amplifiers are digitally controlled and programmable, so the phase adjustment should not be a problem. The NS-FFAG design for the proton and carbon cancer therapy [27] assumed 3 concentric rings with 48 periodic doublet cells, as shown in Fig. 20. The presented rings could operate simultaneously for either proton or carbon therapy treatment. The two smallest rings are to be used for proton acceleration. The first ring "A" accelerates only protons in the kinetic energy range between 7.95 and 31 MeV. The second ring "B" accelerates either protons from 31 to 250 MeV or carbon $^{12}C^{6+}$ ions from 7.9 to 69 MeV/u. The third ring "C", with radius 8.25 m, accelerates only carbon ions from 69 to 400 MeV/u. The radii of the "A," "B," and "C" rings are $r_A = 5.5$, $r_B = 6.87$, and $r_C = 8.25$ m, respectively. The circumferences of the "A," "B," and "C" rings are $C_A = 34.560$, $C_B = 43.200$, and $C_C = 51.840$ m, respectively. The lengths of the cells in the three rings are $L_A = 0.72$, $L_B = 0.9$, and $L_C = 1.08$ m. Magnets and betatron functions in one cell are shown in Fig. 21. Each cell has two combined function magnets, a focusing unit with reverse bending and a defocusing unit with larger positive bending, and a 0.29-m-long drift for the cavity or the kicker dipole placement. Orbit

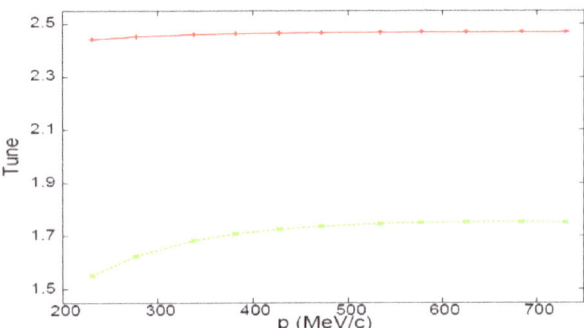

Fig. 19. Tune stabilization: tune dependence on momentum, in the second nonlinear NS-FFAG madeup of superconducting large aperture magnets. The maximum momentum of 729.1 MeV/c corresponds to the kinetic energy of 250 MeV.

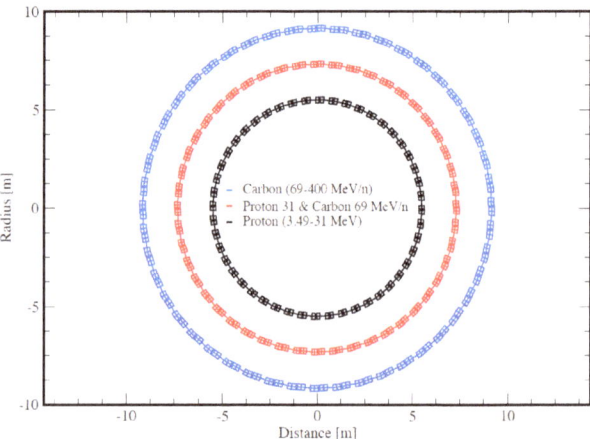

Fig. 20. Layout of the three NS-FFAG rings of the proton–carbon cancer therapy facility. Two smaller rings are used for proton acceleration, while the central and the largest ring are used for carbon ion acceleration.

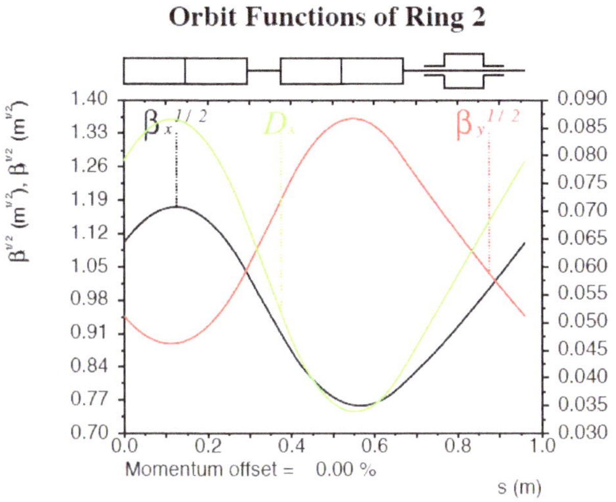

Orbit Functions of Ring 2

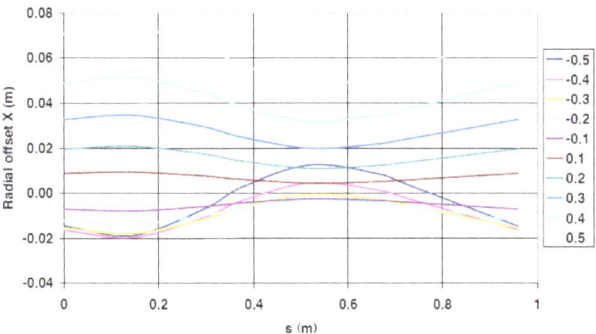

Fig. 21. Betatron functions, magnets, and a cavity in the doublet cell used to accelerate protons from 31 to 250 MeV or carbon ions $^{12}C^{6+}$ from 7.9 to 69 MeV/u, in the NS-FFAG accelerator complex for the proton and carbon therapy.

Fig. 22. Orbit offsets in the 0.9-m-long cell of the second ring with a circumference of $C = 43.2$ m — a part of the NS-FFAG accelerators for the proton and carbon cancer therapy.

offsets during acceleration for the carbon ring are -20 mm $\leq x \leq 70$ mm, with the largest horizontal aperture of 100 mm (see Fig. 22), less than half of that in any other FFAG design.

5.3.1. *Magnet properties and injection–extraction*

The two smaller rings are madeup of combined function warm magnets with the maximum field exceeding 1.69 T, while the largest ring for carbon ion acceleration to 400 MeV/u is madeup of small size superconducting combined function magnets with maximum fields of $+3.35$ T and -2.40 T. The magnet lengths in the three rings are 17, 26, and 35 cm, and for the defocusing 18, 27, and 36 cm, respectively. An example of a possible superconducting magnet [28] made by the Advanced Magnet Lab (AML) is shown in Fig. 23. Tune dependence on momentum is shown in Fig. 24. The systematic half-integer resonances $\nu_x = \nu_y = 0.5$ at low energy and $\nu_x = \nu_y = 0$ at high energy are avoided in the basic cell by the gradient adjustment. The ring has 48 cells. If the integer horizontal tune of the whole accelerator is $\nu_x = 10.00$, for example, then the tune in the cell is $\nu_x = 0.20833$. The variation of the tune per turn ν_τ depends on the RF voltage per turn or the total number of turns required to reach the maximum energy. The allowed error in the magnetic field is defined by $B_n/\bar{B} = 2\nu\sqrt{\nu_\tau}\Delta A/C$ [Eq. (20)]. When the tune is equal to the n_{th} harmonic $\nu = n$, the allowed tolerances for the error could be estimated: for the 1000-turn acceleration the change of the tune per turn is $\nu_\tau = 0.0127$ and allowed tolerances $1/(B_1/\bar{B}/\Delta A) \simeq 0.078$ are easily achievable. If the allowed amplitude change is of the order of $\Delta A/A = 0.10$, and for the beam size of ~ 0.5 mm ($\Delta A \leq 0.05$ mm), the magnetic field variation or misalignment-induced field errors should be less than $B_1/\bar{B} \leq 4 \cdot 10^{-3}$ [29].

The extraction at variable energy, for the largest carbon NS-FFAG ring, requires two kickers as the tune varies with momentum. The optimum tune advance is $\nu_x \approx (2n + 1)/4$, with integer $n \geq 0$

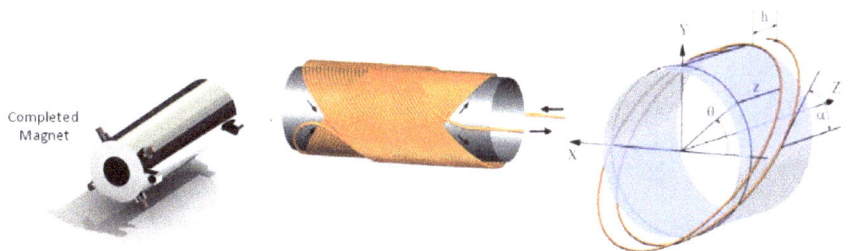

Fig. 23. The Advanced Magnet Lab combined function magnet [28].

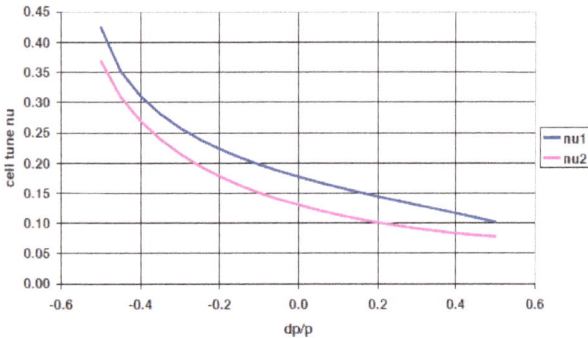

Fig. 24. Tune dependence on momentum in the cell for the 400 MeV/u ^{12}C^{6+} ring.

Table 1. Extraction and injection for the NS-FFAG rings — kicker properties.

Ring	1	2	3
Kick angle (mrad)	11.5	7.6	4.5
Rise time (ns)	120	80	80
Aperture width (mm)	52	107	94
Aperture height (mm)	28	36	19
Kicker length (m)	0.2	0.2	0.2
Kicker field (T)	0.047	0.092	0.14

between kicker and septum [29]. The efficiency of the different number of cells ($k = 1, 2, \ldots, 5$) between the kickers and the septum is shown in Fig. 25. Extraction at any energy is reduced with respect to the extraction at the maximum energy, by using two kicker magnets, to $E_f \geq \sqrt{3}/2$. The extraction is made in two stages: the full aperture kicker deflects the extracted beam by $2\sigma'\sqrt{5}$ and the septum deflects the extracted beam, such that it misses components downstream, and sends it into a transfer line. The kicker angles, rise time in ns, aperture widths and heights, kicker lengths in m, and kicker field in T for all three rings are shown in Table 1.

The injection uses components in reverse order; the septum magnet is two cells from the kicker magnet and the close-to-optimum phase is $3\pi/2$.

5.3.2. Acceleration

Acceleration is performed with the phase jump after each turn. The phase jump during acceleration with a fixed frequency has previously been shown at CERN

[30] and recently re-examined by M. Blaskiewicz [26] from the BNL. The RF frequency needs to be in a high range, ~ 370 MHz, because of the required large number of RF cycles between the passages of bunches, in order to achieve higher values of Q and to limit the frequency swing. The total stored energy in the cavity is related to the amplitude V_{RF} of the RF voltage as $U = V_{\mathrm{RF}}^2/2\omega_r(R/Q)$, where ω_r is the angular resonant frequency, Q the quality factor, and R the resistance. The cavity voltage dependence on the klystron voltage (driven by the low level drive) is shown in Fig. 26. During the phase jump the frequency is almost constant. It is affected by the phase jump, as shown in Fig. 27. The bunch train fills half of the ring. For proton acceleration to 250 MeV, the β value changes from injection 0.251 to 0.614 and there is a beam gap interval of ~ 80 ns during which the correct phase of the voltage wave form may be set. With $Q = 50$ and $f = 374$ MHz the exponential decay time for the field is 43 ns, where $R/Q \sim 30\,\Omega$. If the peak accelerating voltage is 19 kV per turn, the number of proton revolutions required is ~ 1200.

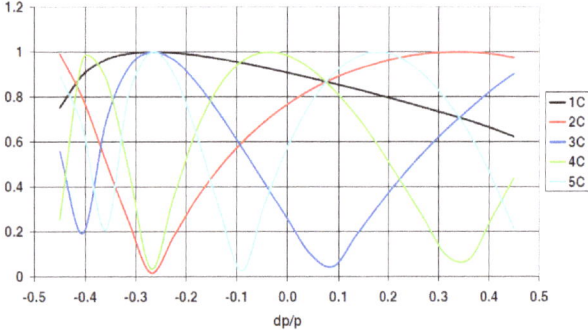

Fig. 25. The efficiency of the extraction system from the carbon NS-FFAG ring with respect to the number of cells between the kickers and the septum for different energies. The efficiency is shown on the vertical axis.

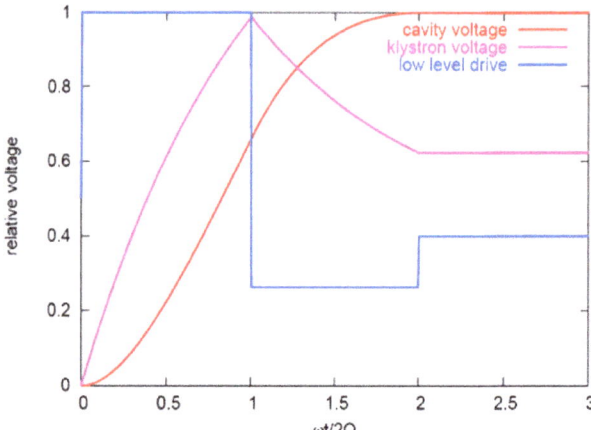

Fig. 26. Cavity voltage during the phase jump acceleration. The klystron voltage is shown in pink, while the low level RF drive is shown in blue.

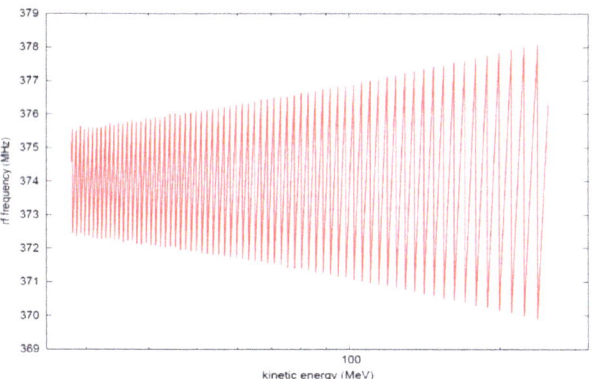

Fig. 27. The RF frequency is kept constant during acceleration. Small variations in frequency during acceleration are shown on a scale with a suppressed zero.

Assuming the use of 12 cavities, the total RF power involved is 1.2 MW and this is a consequence of the rapid acceleration. Faster resonance crossing would involve even more power. The presented solution for proton acceleration, with the phase jump adjustment for each turn, uses very reasonable and easily achievable values for the cavity, and RF amplifiers are available off the shelf.

If the same parameters, like $N_o = 1200$ turns, with the peak accelerating voltage of 19 kV, are to be used for $^{12}C^{6+}$ carbon ion acceleration in the third ring shown in Fig. 20, then a total of 37 cavities need to be distributed within the 48 periods. The total power would be 3.7 MW.

5.4. *Small proton therapy ring*

A smaller, linear NS-FFAG proton ring has also been designed [31], in which the mean radius of the central orbit is reduced from 6.87 to 4.278 m, and the number of cells from 48 to 24. The combined function doublet pattern is retained, as shown in the ring presentation of Fig. 28. The design leads to a doubling in size of the aperture. The magnet lengths are $l_f = 44$ cm and $l_d = 22$ cm for focusing and defocusing magnets, respectively. One of the novelties of this design was that the two combined function magnets bend in the same direction for the central momentum. The drift size for the cavities and extraction and injection kickers is 38 cm, while the cell length is increased from 0.9 to 1.12 m. The tune dependence on momentum, shown in Fig. 29, avoids crossing through $\nu_{x,y\,\text{cell}} < 1/3$ to avoid amplitude growth.

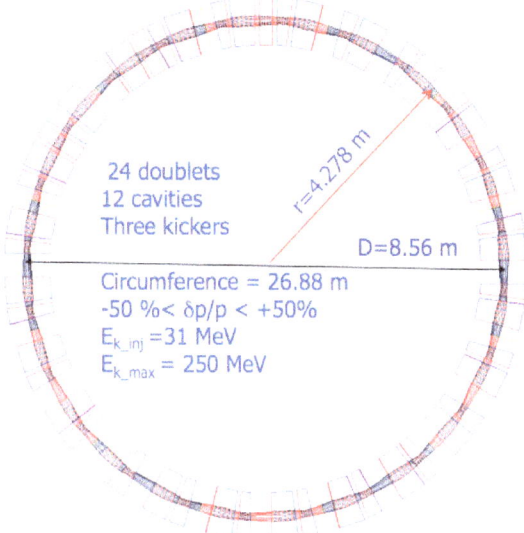

Fig. 28. The NS-FFAG with 24 cells, 12 cavities (thin lens kicks, in green); injection and extraction kickers (red rectangular boxes); and doublet magnets (blue).

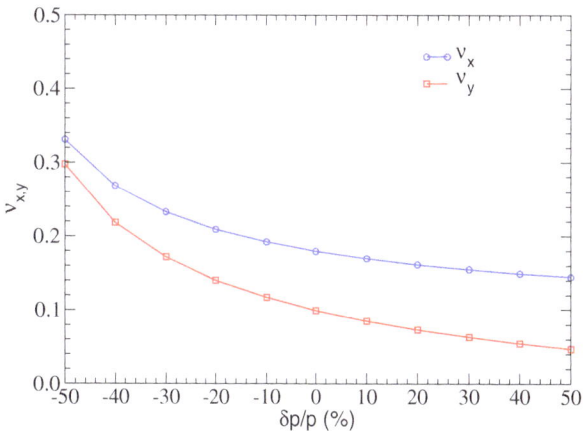

Fig. 29. Tune dependence on momentum in the cell of the 26.8 m proton cancer therapy accelerator.

6. Tracking Simulation Results

6.1. *NS-FFAG C = 27 m*

A comprehensive simulation study of six-dimensional phase space acceleration, using the PTC by E. Forest [9], was reported at the FFAG workshop in Manchester (2008) [31]. The NS-FFAG has linear dependence of the magnetic field along the radial axis — a linear accelerator. The tune shift does not depend on amplitude unless we introduce the magnetic field errors $\Delta B / \bar{B}_n$ or misalignment errors. Most of the NS-FFAG designs include tunes of 1/3 in the range of allowed tune variation (0.4–0.1).

The resonance crossing of the third integer, $m = 3$, when the cell tune includes tunes of $1/3$, can become very difficult if there is a slight systematic sextupole component in the magnetic field [24]. The third integer resonance occurs when $3\nu = n$. The strength of the resonance $b_{n,3}$ becomes large when $n = N_p$, a condition for the intrinsic or structural resonance.

- In the first acceleration simulation study, the random magnetic field errors were absent and no emittance growth was observed.
- The next study introduced small random errors into the magnetic field of $\Delta B/\bar{B}_n = 10^{-4}$ with the cell tunes varying in the range of 0.4–0.1, including $1/3$ tunes. The speed of acceleration was varied:

  (1) The total number of turns was first $N_o \sim 6000$. At the beginning of the acceleration cycle, the horizontal tune in the ring crosses the integer value of $\nu_x = 8.00$ or the tune in the cell is $1/3$. This is a problem as $3\nu_x = 24$ and with $n = 24$. The slower acceleration rate of ~ 6000 turns has shown very large amplitude-emittance growth with a loss of particles (see Fig. 30).

  (2) When the proton acceleration rate was raised to $N_o \sim 1300$ turns, for the kinetic energy range of 31–250 GeV, the crossing of tunes of $1/3$ per cell did not produce a dramatic emittance blowup and beam loss. With the magnetic field error of $\Delta B/\bar{B}_n = 10^{-3}$, the horizontal and vertical emittance blowup ratios were, respectively, $\Delta \epsilon_x/\epsilon_x \sim 1.3$, as shown in Fig. 31, and $\Delta \epsilon_y/\epsilon_y \sim 1.9$ times, as shown in Fig. 32.

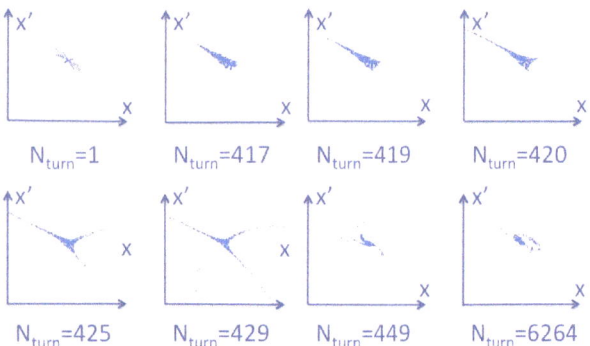

Fig. 30. Slower proton acceleration in the smaller ring with $C = 27$ m, with the total number of turns $N_o \sim 6000$. When the turn number reached a value of $N_{\text{turn}} = 428$, the horizontal tune in the ring was equal to $\nu_x = 8.00$.

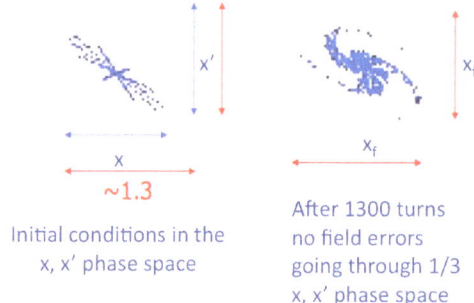

Fig. 31. Six-dimensional acceleration study with $\nu_{x,y}$ crossing $1/3$ in the cell, without magnetic field errors, shown in x, x' phase space.

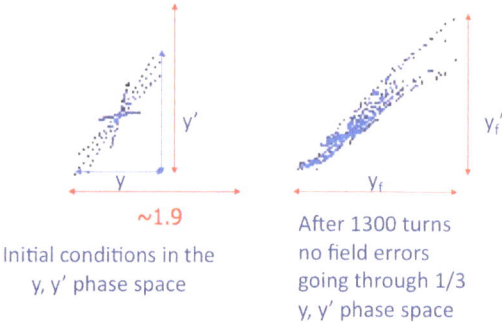

Fig. 32. Six-dimensional acceleration study with $\nu_{x,y}$ crossing $1/3$ in the cell, without magnetic field errors, shown in y, y' phase space.

- In the next study both horizontal and vertical tunes avoided crossing the value of $1/3$ in the cell. The horizontal and vertical tunes $\nu_{x,y}$ in the ring were in the range of 7.82–3.41 and 7.2–1.01, respectively. The horizontal and the vertical emittance blowup due to the crossing of many integer resonances, with the introduction of the random magnetic field error of $\Delta B/\bar{B} = 10^{-3}$, were $\Delta \epsilon_x/\epsilon_x \sim 1.8$, as shown in Fig. 33, and $\Delta \epsilon_y/\epsilon_y \sim 1.4$, as shown in Fig. 34, respectively.

Tracking results were in agreement with the analytical predictions. The $\Delta B_n/\bar{B}$ field errors of $\leq 10^{-3}$ are easily achievable for the short magnets, so the results do not indicate a significant problem.

6.2. NS-FFAG $C = 43.17$ m and $C = 26.7$ m

In this case tracking simulations were made to investigate the effect of misalignment on the resulting orbit distortions. Results have been obtained by S. Sheehy [23] for both linear NS-FFAG rings; those

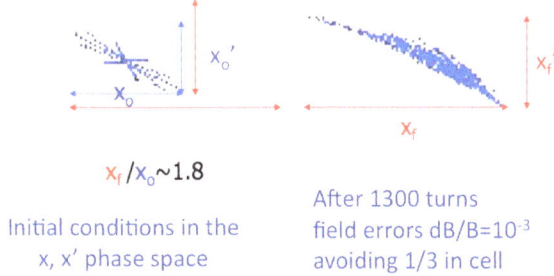

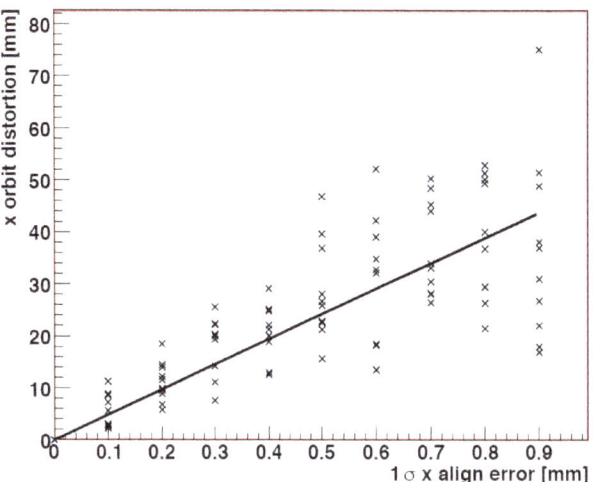

Fig. 33. Six-dimensional acceleration study with ν_x avoiding 1/3 per cell, with magnetic field errors of $\Delta B/\bar{B} = 10^{-3}$, shown in x, x' phase space.

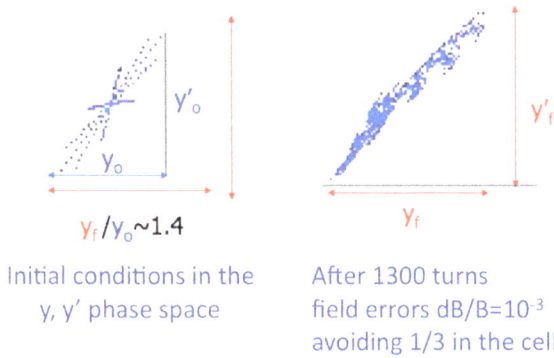

Fig. 34. Six-dimensional acceleration study with ν_y avoiding 1/3 per cell, with magnetic field errors of $\Delta B/\bar{B} = 10^{-3}$, shown in y, y' phase space.

Fig. 36. Studies of orbit distortions during acceleration in the smaller $C_B = 26.8$ m NS-FFAG proton cancer therapy ring. The horizontal axis shows very large misalignment.

in alignment of $\sigma = 0.4$ mm causes a maximum orbit distortion during acceleration of 1.7 mm in the larger ring, and ~ 1 mm in the smaller one. Alignment of the rings to a σ value $< \sim 0.1$ mm is considered achievable. The fitted amplification factor A_x is

$$A_x = \frac{\langle \text{Orbit distortion (mm)} \rangle}{\langle 1\sigma \text{ alignment error (mm)} \rangle} \sim 4.45. \quad (24)$$

6.3. NS-FFAG e-model, EMMA

S. Machida [21] employed the parameters of the electron model EMMA for a simulation study of resonance crossing and dynamic aperture during the acceleration period of a linear NS-FFAG. EMMA uses electrons to simulate conditions needed for fixed frequency, rapid acceleration of large emittance, high energy muon beams. Electron normalized beam emittances are ~ 40 times larger than those proposed for the $C = 43.17$ m and $C = 26.88$ m proton therapy rings. The findings of the EMMA simulations are that even misalignments as low as $\sigma = 0.015$ mm lead to small dynamic apertures and large orbit distortion growth on integer resonance crossing. A previous publication of S. Machida and D. Kelliher [32] reported similar conclusions for simulations of rapid acceleration of large emittance muon beams in a muon NS-FFAG ring. Extra RF cavities were found necessary to increase the speed of the resonance crossings.

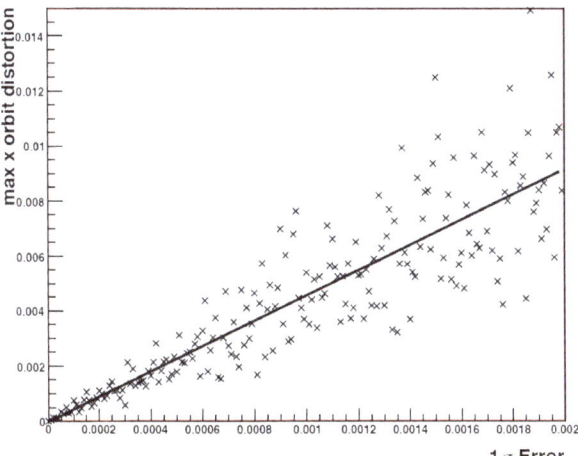

Fig. 35. Orbit distortions during acceleration study versus error in misalignment of the NS-FFAG proton cancer therapy ring with a circumference of $C_A = 43.17$ m, from the design of the carbon/proton complex facility. Both axes are in meters.

for the $C = 43.17$ m ring are presented in Fig. 35 and those for $C = 26.8$ m ring in Fig. 36. Analysis of the results [23] shows that tolerances in alignment have to be better than 0.1 mm. The result of an error

Studies are needed to see if EMMA may be modified to allow tests relevant to the ion cancer therapy rings. An additional EMMA proton acceleration simulation study made by using the nonrelativistic electrons with installation of a different cavity is under consideration.

The average radius of the machine is $\bar{R} = 2.5\,\text{m}$ and the circumference of $C_{\text{EMMA}} \simeq 16.6\,\text{m}$. The "scaled" $B\rho = 0.05\,\text{Tm}$. The maximum orbit offsets during acceleration are within $\Delta x_{\max} \leq \pm 16\,\text{mm}$, which may be compared with the $-20\,\text{mm} \leq \Delta x_{\max} \leq 80\,\text{mm}$ of the $C = 43.17\,\text{m}$ ring and the $-80\,\text{mm} \leq \Delta x_{\max} \leq 180\,\text{mm}$ of the $C = 26.88\,\text{m}$ ring. For simulations, beam amplitudes should scale with the size of the orbit offsets and so the beam sizes and emittances in EMMA should be reduced by factors of ~ 6 and ~ 40, respectively. Thus, the rms, normalized e-beam emittances should be lowered from $20 \cdot 10^{-6}\pi$ rad m to $0.5 \cdot 10^{-6}\pi$ rad m, which is a typical value for a proton therapy ring. The NS-FFAG is a linear machine without tune dependence on amplitude, but the size of the initial emitance is important, as presence of the magnetic field errors will induce the tune dependence on amplitude.

To prevent an increase in transverse space charge levels, the longitudinal extent of the beam should increase by the same factor, which implies that the single EMMA bunch should be replaced by a train of 40 bunches and the need for a much lower frequency acceleration system. The design of such a system and the dynamics of the rotation of the 40 bunches during the acceleration need to be evaluated.

An additional accelerator simulation study using the ZGOUBI code [10] was reported [22]. The results obtained were that the emittance blowup was less than 2 ($\Delta \epsilon / \epsilon \leq 2$) for the input emittances lower than 2π mm mrad if the alignment limits and magnetic field errors were smaller than 0.1 mm and $\Delta B / \bar{B}_n \leq 2 \cdot 10^{-3}$, respectively.

7. NS-FFAG Application for the Carbon and Proton Gantries

The highest cost in the proton or carbon facilities is that of the delivery systems with isocentric gantries. The transport magnets in both proton and carbon cancer therapy facilities are very large warm magnets. The ion dose has to be delivered to the patients from various positions, with precision, high reliability and stability. The preferable treatment is by spot scanning with energy and intensity variation

capabilities. Ion positions are varied transversely by the scanning magnets and longitudinally by varying the energy to define precisely the Bragg peak at the cancerous tumor where most of the energy is deposited.

The main motivation for applying the NS-FFAG concept to the isocentric gantries is to reduce the enormous weight of the transport elements and to make the operation easier. The reduction in size and weight at the same time simplifies construction and reduces the weight of the supporting structure. Two NS-FFAG designs for the isocentric gantry are presented [33] — one with superconducting combined function magnets, to allow carbon and proton ion transport and delivery, and the other with separated function permanent Halbach magnets.

The present world-class facility at Heidelberg for carbon/proton and other ion cancer treatments is already operating with a 630-ton isocentric gantry, where the transport element weight is 135 tons [34]. The NS-FFAG concept provides a reduction of the transport elements for the carbon ions to about 1.5 tons.

To understand better the requirements for the beam size and scanning system of the gantry, it is important to include effects of the beam straggling and multiple Coulomb scattering effects during the propagation of the ions inside the patient. The size of the beam grows along its way to the tumor. The size at the Bragg peak, where most of the energy is deposited, rises with input beam energy. For example, for the 200 MeV proton input energy a position of the Bragg peak is at $\sim 26\,\text{cm}$, the beam size due to multiple Coulomb scattering reaches a value of $\sigma_{\text{MC}} = 6.5\,\text{mm}$, while due to beam straggling the beam size is $\sigma_{\text{ST}} = 7.6\,\text{mm}$, they contribute to the total beam size as $\sigma_T^2 = \sigma_{\text{OPT}}^2 + \sigma_{\text{MC}}^2 + \sigma_{\text{STR}}^2$. The optical beam size $\sigma_{\text{OPT}}^2 \sim \epsilon\beta$ is dependent on the ϵ emittance and the β betatron amplitude function, and it is defined by the last elements in the nozzle. The effective source-axis distance parameter (SAD) is important in the patient treatment. The delivery systems usually have the source point in the beam line designed before the very large bending magnets so as to be able to produce parallel beams to the patient to reduce the skin radiation. The NS-FFAG gantry produces the focal source point at the end of the gantry. This allows placement of the scanning magnets close to it. The difference between "parallel" and "angle" beam scanning is graphically shown in

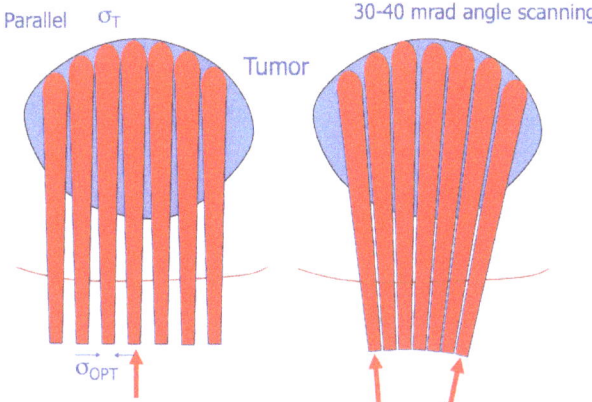

Fig. 37.   Parallel and angle scanning with SAD = 2.5 m.

Fig. 37. The figure shows that there is unavoidable large skin deposition in the "angle" scanning case but it could be reduced by the beam and step size adjustment at the skin.

The NS-FFAG combined function magnet carbon isocentric gantry, presented in Ref. 33 and in Fig. 37, meets the requirements discussed. The carbon treatment at one magnet setting allows transport of ions within the momentum range of $\delta p/p = \pm 30\%$ or in the kinetic energy range of 150–400 MeV/u. Carbon/proton ions of different energies reach the end of the gantry within ± 6 mm and the position at the patient is adjusted with the scanning and triplet focusing magnets for each energy separately. The size and strong focusing of the NS-FFAG make the betatron functions and dispersion small, allowing momentum acceptance for $\pm \delta p/p = 30\%$ with orbit offsets less than 19 mm, as shown in

Fig. 38. The maximum values of the magnetic field for carbon ions are 3.4–3.9 T. Carbon ions tracked at different energies, shown in Fig. 39, arrive with small offsets at the end of the gantry.

For the proton cancer facilities, a permanent Halbach separate function magnet design is shown in Fig. 41. The maximum magnetic field in the center of the Halbach magnet is $B_g = B_r \ln(\text{OD}/\text{ID})$, where B_r is the material permanent magnetic field value, while OD and ID are the outside and inside diameters of the material modules. The range of proton energies under the fixed magnetic field is 68–250 MeV. Magnets are made of neodymium–iron–boron compounds (Nd–Fe–B), assuming the maximum operating temperature of 70°C and with the magnetic field of $B_r = 1.35$ T. The NS-FFAG isocentric gantries provide simple solutions; they are easy to operate as the field is fixed for all treatment energies; they reduce the cost; they are made of light elements — the weight of the carbon transport line is 1.5 tons, while the permanent proton gantry weighs ~ 500 kg; and there is small power consumption in the carbon superconducting design and none at all in the proton permanent magnet design. The scanning and focusing system (shown in Fig. 42) is above the patient, with SAD ~ 3 m.

8. Advantages and Disadvantages of FFAG Accelerators and Gantries

Two S-FFAG accelerators for applications in ion cancer and for boron neutron capture therapy have already been commissioned. Many S-FFAG designs

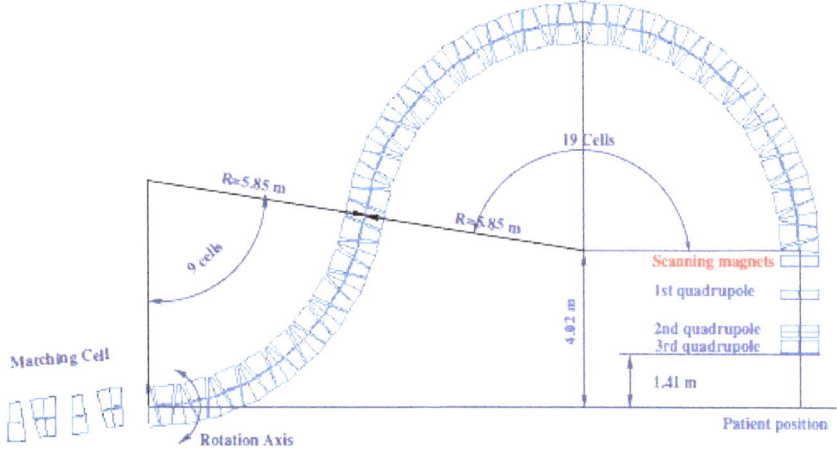

Fig. 38.   The carbon/proton isocentric gantry with superconducting magnets. There is a preliminary design of the magnets.

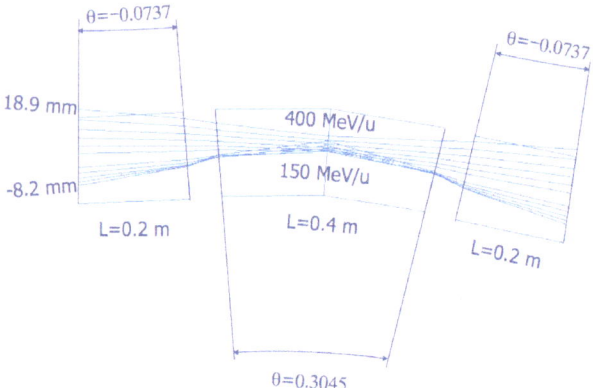

Fig. 39. Magnets and orbit for different energies in the basic cell of the carbon/proton gantry.

for future facilities already exist. The NS-FFAG and designs for acceleration of ions are showing immense progress.

8.1. Advantages of S-FFAGs

The advantages of S-FFAG machines are:

- Fixed tunes during acceleration. Resonance crossing should not be a problem.
- They are easy to operate, which is an advantage with respect to slow-cycling synchrotrons, but as proton therapy requires a chain of three S-FFAGs there are reliability issues to consider, related to the number of kickers and RF systems and the number of beam transfer elements. Cyclotrons

are easier to operate and very reliable. Maintenance of cyclotrons might be a problem due to radioactivity.

- Zero chromaticity.
- Large transverse acceptance.
- Very large energy acceptance.
- The high repetition rate, up ∼ 1 kHz, is larger than in synchrotrons.
- High average current — space charge and collective effects are below threshold. There is a smaller space charge problem than with slow-cycling synchrotrons. The cyclotrons operate CW.

8.2. Disadvantages of S-FFAGs

The disadvantages of S-FFAG machines are:

- Very large aperture and large magnet size. A lot of steel is required, possibly more than for cyclotrons.
- A chain of S-FFAG rings is required for the medical facilities.

8.3. Advantages of NS-FFAGs

The advantages of NS-FFAG machines are:

- Variable energy is obtained by the single turn extraction choosing in advance a required number of turns. This allows fast spot scanning due to the very high rate of operation and removes the need for the degraders used in cyclotrons. Fast-cycling

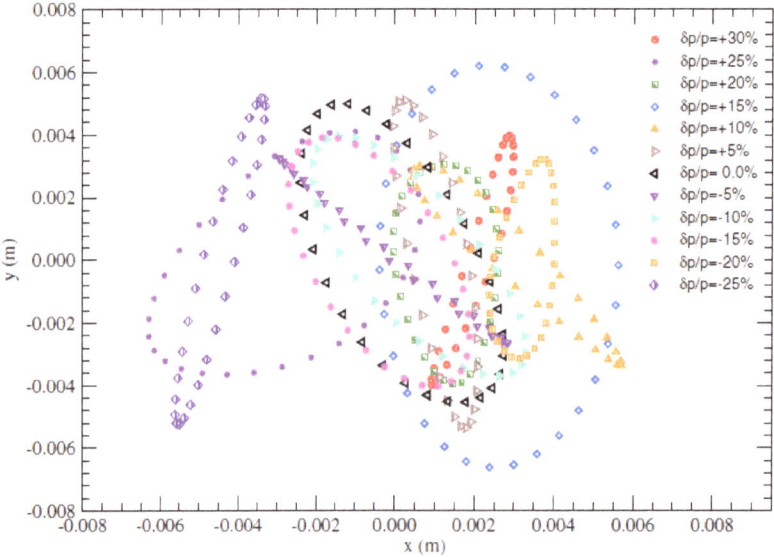

Fig. 40. Tracking particles through the carbon/proton gantry. Particles, with normalized emittance of $\epsilon_n \sim 0.5$ mm mrad, enter the gantry and are tracked through it, and collected and plotted at the end of it. This is the x–y regular space.

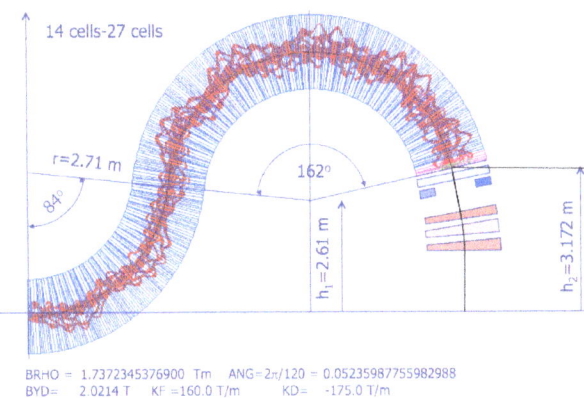

Fig. 41. The proton isocentric gantry designed with permanent Halbach magnets. The magnets have previously been shown [35].

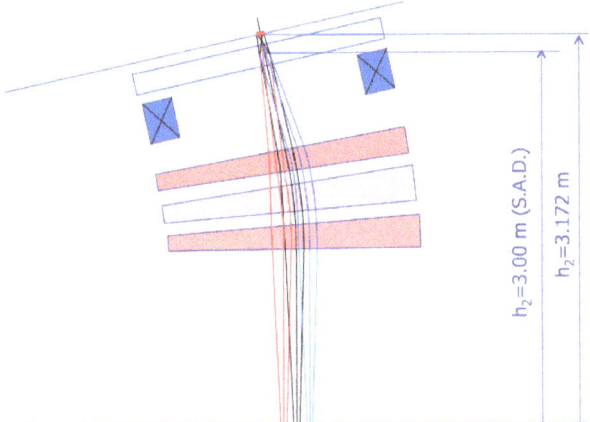

Fig. 42. This is the scanning and focusing system at the end of the NS-FFAG gantry. The scanning system is in two planes. The triplet, madeup of the combined function magnets, provides focusing of the beam to the required size at the patient. This is done separately for each energy.

synchrotrons also have single turn extraction, but slow-cycling rings use slow resonant extraction.

- Small aperture and magnet size.
- Large momentum range.
- High repetition rate, $\sim$1–2 kHz, for patient treatment.
- Simpler magnets than in the S-FFAG but more complex than those for synchrotrons.

8.4. *Disadvantages of NS-FFAGs*

The disadvantages of NS-FFAG machines are:

- Tunes vary with momentum; integer resonances are crossed.

- High cost of the RF power amplifiers required for the fast acceleration.
- Chromaticities vary with momentum.
- Tight alignment tolerances.
- The need to operate an accelerator chain; the need to change the kicker fields and to correct the divergence of the extracted beam in ms time intervals when changing the extraction energy.

A simpler permanent magnet solution for proton therapy is presently being developed at BNL to reduce the cost. A nonlinear NS-FFAG of nearly constant tunes has the advantage of slower acceleration and hence cheaper RF systems than the linear NS-FFAG ring.

8.5. *NS-FFAG beam delivery systems*

The advantages include the following:

- Small aperture and magnet size.
- Large energy range for protons and light ions.
- Fixed fields in beam line and gantry magnets for energy ranges required for the treatment.
- Substantial reduction in weight for the gantry magnets.
- Reduced weight and cost for the gantry structures.
- Reduced operating power consumption and cost.
- Reduced size and cost of the support structures and buildings.

The possible disadvantages include:

- Small space in the compact lattice for all diagnostics needed.
- Required matching in dispersion and betatron functions from the ring to the beam line and from the beam line to the rotating gantry.
- Inability to change the gantry lattice parameters in the case of permanent magnets.

The time required for the scanning of large tumors depends on the repetition rate used for the scanning. If restricted to a 1 kHz rate of the ring, the time will be similar to that of the existing facilities. To do better, it is necessary to scan during the duration of a single-turn extracted beam, and the accuracies for doing this need to be considered.

9. R&D

- Modification of the NS-FFAG electron model EMMA to allow testing of concepts relevant to the ion cancer therapy ring.
- Acceleration: both the S-FFAG and the NS-FFAG require RF development to improve the efficiency, reduce the cost, raise the Q factor and accelerating voltage, and so make FFAGs more competitive with other ion therapy accelerators. In addition, improved solution with phase jumping would allow even faster acceleration.
- Detailed design and prototyping of permanent Halbach magnets is necessary the proton therapy, FFAG accelerators and gantries.
- A suitable insertion design, with the betatron and dispersion function matched to the arcs of an FFAG, for each energy, is an important development, as this would simplify their use and reduce the aperture for the RF and kickers.

Acknowledgments

I would like to add a special note of appreciation to Andrew Sessler for his continuing scientific support, advice, and friendship. I would also like to express my gratitude to Michael Craddock, Eberhard Keil, Francois Meot, Etienne Forest, Michael Blaskiewicz, Adam Rusek, Yoshiharu Mori, Carol Johnstone, Takeichiro Yokoi, Shinji Machida, Scott Berg, Rick Bartmaan, Suzane Sheehy, Steve Peggs, Jay Flanz, Chris Prior, Alessandro Ruggiero, and Grahame Rees, who directly contributed to this article and granted permission to incorporate their results.

Supported by the US Department of Energy under contract No. DE-AC02-98CH10886.

References

[1] S. Peggs, *IEEE Trans. Nucl. Sci.* **51**(3), 677 (2004).
[2] M. K. Craddock and K. R. Symon, Cyclotrons and fixed-field alternating-gradient accelerators, in *Reviews of Accelerator Science and Technology*, eds. A. Chao and W. Chou, Vol. 1 (World Scientific, 2008), pp. 65–97.
[3] T. Ohkawa, University of Tokyo, Japan, FFAG structure suggested earlier at a symposium on nuclear physics of the Physical Society of Japan in 1953. Private communication.
[4] A. Kolomensky *et al.*, *Zh. Eksp. Teor. Fiz.* **33**, 298 (1957).
[5] K. R. Symon, *Phys. Rev.* **100**, 1247 (1955).
[6] E. D. Courant, A. A. Garren, S. J. Berg, R. Talman and D. Trbojevic, A comparison of several lattice tools for computation of orbit functions of an accelerator, in *Proc. 2003 PAC* (Portland, Oregon, USA; May 12–16, 2003), pp. 3485–3487.
[7] D. W. Kerst, K. M. Trwilliger, K. R. Symon and L. W. Jones, *Phys. Rev.* **98**, 1153(A) (1955).
[8] D. Trbojevic, Linear spiral non-scaling FFAG, FFAG workshop in Grenoble, France, Apr. 12–17.
[9] E. Forest, E. Mcintosh and F. Schmidt. KEK Report 2002–3, CERN-SL-2002-044 (AP) **44**, 3 (2002).
[10] F. Meot and S. Valero, *ZGOUBI User's Guide*, CEA DSM DAPNIA-02-395 (2002), CEA Saclay, DSM, DAPNIA/SEA, July 11, 2003.
[11] F. Schmidt, MAD-X PTC integration, in *Proc. PAC 2005* (Knoxville, Tennessee, USA), pp. 1271–1274. http://mad.web.cern.ch/mad/Introduction/doc.html
[12] M. Tanigaki, M. Inoue, Y. Mori *et al.*, *Proc. EPAC'06*, Vol. 950 (Edinburgh, UK, 2006), pp. 2367–2369. JACoW, http://jacow.org/
[13] F. Meot, RACCAM research project. Invited talk at *Workshop on Hadron Beam Therapy of Cancer* (Erice, Sicily, Italy; Apr. 24–May 1, 2009). http://erice2009.na.infn.it/programme.htm
[14] T. Misu, Y. Iwata, A. Sugiura, S. Hojo, N. Miyahara, M. Kanazawa, T. Murakami and S. Yamada, *Phys. Rev. S.T. Accel. Beams* **7**, 094701 (2004).
[15] M. Ankenbrandt *et al.*, *Phys. Rev. S.T. Accel. Beams* **2**, 081001 (1990).
[16] L. C. Teng, *Rev. Sci. Instrum.* **27**, 1051 (1956).
[17] C. Johnstone, W. Wan and A. Garren, *Proc. 1999 Particle Accelerator Conference* (New York, USA, 1999), p. 3068.
[18] D. Trbojevic, E. D. Courant and M. Blaskiewicz, *Phys. Rev. S.T. Accel. Beams* **8**, 050101 (2005).
[19] J. S. Berg, *Nucl. Instrum. Methods Phys. Res. A* **596**, 276 (2008).
[20] C. R. Prior, International committee for future accelerators, *Beam Dynam. Newslett.* **43**, 19 (2007).
[21] S. Machida, *Phys. Rev. S.T. Accel. Beams* **11**, 094003 (2008).
[22] T. Yokoi, J. Cobb, K. Peach and S. Sheehy, Beam acceleration studies of proton NS-FFAG, in *Proc. EPAC'08* (Genoa, Italy, 2008), THPP011.
[23] S. Sheehy, Dynamics of the machida lattice, Presented at FFAG Workshop (University of Manchester, UK, Sept. 1–5, 2008). http://www.cockcroft.ac.uk/events/FFAG08
[24] R. Baartman, Resonance crossing topics. Presented at FFAG Workshop (Vancouver, Canada, 2004). http://legacyweb.triumf.ca/ffag2004
[25] C. Johnstone, S. Koscielniak, M. Berz, K. Makino, P. Snopok and F. Mills, Non-scaling FFAG variants for HEP and medical applications, in *Proc. PAC'09* (Vancouver, Canada, TU6PFP080, May 4–8, 2009).
[26] M. Blaskiewicz, private communication.
[27] E. Keil, A. M. Sessler and D. Trbojevic, *Phys. Rev. S.T. Accel. Beams* **10**, 054701 (2007).

[28] C. Goodzeit, R. Meinke and M. Ball, Combined function magnet using double-helix coils, in *Proc. PAC'07* (Albuquerque, New Mexico, USA, MOPAS055, 2007), pp. 560–562.

[29] E. Keil, invited talk at *Cyclotrons 2007: 18th Int. Conf. Cyclotrons and Their Applications* (Giardini-Naxos, Sicily, Italy, Oct. 1–5, 2007).

[30] D. Boussard, RF for $p\bar{p}$ (PARTIII). CERN/SPS/B4-2 ARF (Geneva, Jan. 1984), pp. 1–31.

[31] D. Trbojevic, Small proton therapy accelerator by non-scaling FFAG. Presented at FFAG Workshop (University of Manchester, UK; Sept.

1–5, 2008). http://www.cockcroft.ac.uk/events/FFAG08/programme.htm

[32] S. Machida and D. J. Kelliher, *Phys. Rev. S.T. Accel. Beams* **10**, 114001 (2007).

[33] D. Trbojevic, B. Parker, E. Keil and A. M. Sessler, *Phys. Rev. S.T. Accel. Beams* **10**, 053503 (2007).

[34] U. Weinrich, *Proc. EPAC'06* (Edinburgh, UK, 2006), 964.

[35] D. Trbojevic, Innovative gantry design with non-scaling FFAG. Workshop on Hadron Beam Therapy of Cancer (Erice, Sicily, Italy; Apr. 24–May 1, 2009). http://erice2009.na.infn.it/programme.htm

---

Dejan Trbojevic is a tenured physicist at Brookhaven National Laboratory, BNL, Upton, New York. He received his Ph.D. in Physics at Georgetown University 1984. He was an accelerator physicist at Fermi National Laboratory from 1984–1992 where he designed and built the vertical overpass over the D0 detector in the Main Ring. He presented in 1990 at EPAC new synchrotron lattice without transition. As a head of the Main ring group at Fermilab he moved 1992 to the Brookhaven National Laboratory to work on designing, building and commissioning of Relativistic Heavy Ion Collider (RHIC). From his involvements with the muon collider and neutrino factory came 1999 a non-scaling Fixed Field Alternating Gradient lattice. He is an APS fellow.

Reviews of Accelerator Science and Technology
Vol. 2 (2009) 253–263
© World Scientific Publishing Company

The Dielectric Wall Accelerator*

George J. Caporaso[†], Yu-Jiuan Chen[‡] and Stephen E. Sampayan[§]

Lawrence Livermore National Laboratory 1-410,
PO Box 808, Livermore, CA 94551, USA
[†]*gjcaporaso@aol.com*
[‡]*chen6@llnl.gov*
[§]*sampayan1@llnl.gov*

Dielectric wall accelerators, a class of induction accelerators, employ a novel insulating beam tube to impress a longitudinal electric field on a bunch of charged particles. The surface flashover characteristics of this tube may permit the attainment of accelerating gradients on the order of 100 MV/m for accelerating pulses on the order of a nanosecond in duration. A virtual traveling wave of excitation along the tube is produced at any desired speed by controlling the timing of pulse-generating modules that supply a tangential electric field to the tube wall. Because of the ability to control the speed of this virtual wave, the accelerator is capable of handling any charge-to-mass-ratio particle; hence it can be used for electrons, protons and any ion. The accelerator architectures, key technologies and development challenges will be described.

Keywords: Dielectric wall accelerators; hadron therapy; proton therapy; high gradient insulators; photoconductive switches.

1. Introduction

1.1. *A new type of induction accelerator*

Most particle accelerators employ metallic structures to produce the electric fields necessary for energy gain. These structures include cyclotrons, synchrotrons, RF linear accelerators and induction accelerators. They could be resonant cavities, accelerating cells, or traveling or standing wave structures. Advanced techniques under active development use plasmas as the accelerating medium in which a large "wakefield" is excited by the passage of an intense laser pulse or a charged particle bunch and accelerates a subsequent, less intense "witness bunch." A version of this technique employs a dielectrically lined conducting tube and is known as a dielectric wall wakefield accelerator (not to be confused with the dielectric wall accelerator, the subject of this article).

This article will describe accelerator concepts in which the beam tube is largely an insulator (the dielectric wall) that is energized by a pulsed power system. In order to prevent the accelerating voltages from appearing on the outside of the structure, the accelerator is of necessity an induction accelerator [1]. While this type of accelerator is most suited for applications involving large beam currents, versions of it may be able to reach very high gradients that would make it suitable for low current applications like hadron therapy for cancer [2, 3]. Although some of these ideas are not new, emerging technologies will permit their realization in the near term. And new ideas have been generated to capitalize on these emerging technologies.

1.2. *Induction accelerator*

The dielectric wall accelerator (DWA) grew out of attempts to increase the gradient of induction accelerators, which were used primarily to handle large beam currents. Induction accelerators comprise electrically independent modules whose exterior surfaces are at ground potential. The acceleration is produced nonresonantly and individual pulses may persist for up to microseconds. This is accomplished by

---
*This work was performed under the auspices of the US Department of Energy by Lawrence Livermore National Laboratory under contract DE-AC52-07NA273444. Patents pending.

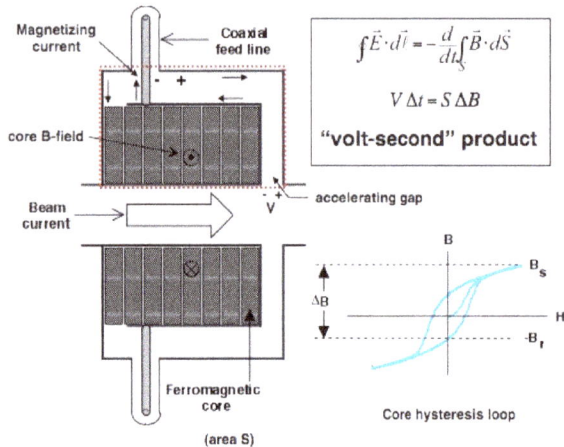

Fig. 1. A typical induction cell. The integration contour (dotted line) can be used to apply Faraday's law. All contributions to the loop integral of the tangential electric field vanish except along the gap, since all surfaces are conductors and the field reverses polarity in the cable. The core material has a B-H loop and is usually "reset" to a remnant value of magnetization in a direction opposite to that in which the accelerating pulse would drive it to increase the available flux swing (ΔB). Integrating Faraday's law with respect to time yields the "volt-second" product that is the time integral of the voltage, which can be sustained across the gap. This product is equal to the cross section area of the core multiplied by the flux swing.

employing inductive isolation provided by ferro- or ferrimagnetic cores inside the modules [4].

Figure 1 is a schematic of a typical induction module that is the basic building block of the accelerator. A voltage pulse enters the cell from a number of coaxial cables spaced azimuthally around the structure. There is a direct path for the voltage pulse down to the accelerating gap. However, there is also a DC short circuit from the center conductor of the cable around the inner surface of the cell back out to the shield of the cable. If the region encircled by this path contains magnetically permeable material, substantial impedance to the flow of current around this path will exist as long as the material is not magnetically saturated.

While achievable electric field stresses across the (vacuum) accelerating gap may be in the range of 10–20 MV/m, the overall, average gradient of machines made from these cells is usually less than 1 MV/m because of the relatively large amount of space occupied by the core. Since it is the cross section area of the core that is important for the volt-second product, it is tempting to use a core that is short axially and larger radially to make up the area. However, all core material has substantial energy loss

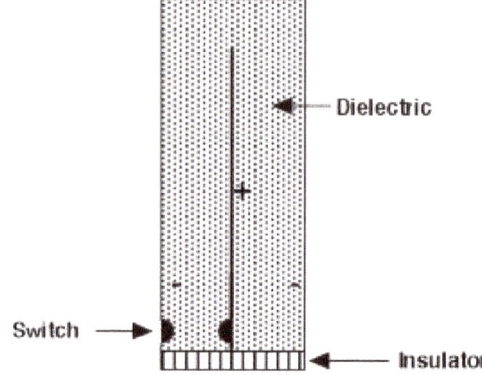

Fig. 2. A coreless induction cell. The structure shown is azimuthally symmetric and is filled with a dielectric. The center conductor is charged with respect to the outer can. A closing switch is placed in close proximity to an insulator that separates the dielectric from the vacuum of the beam tube. There is no net voltage across the insulator until the switch closes. This voltage persists until the wave launched in the radial line by the switch closure has time to propagate up to the top and down the other side to the insulator.

that is proportional to the core volume. Using cores that are axially long and thin in the radial direction will minimize the energy loss. Induction accelerators are low impedance devices that are well matched to beam currents on the order of kiloamperes.

1.3. Coreless induction accelerator

Scientists in the former Soviet Union developed a coreless induction accelerator concept in 1970 [5] that had a more favorable aspect ratio than the previously described magnetic core version. In the schematic shown in Fig. 2 the isolation is provided by the transit time of an electromagnetic wave from one line of the cell to the other.

The original cells use deionized water as the dielectric. The gradient achieved was approximately 1 MV/m. The radial lines had a relatively large radius, to achieve a 25 ns pulse width. The performance of this line was limited by breakdown in the dielectric and vacuum insulator. Improvements in vacuum insulators, solid dielectrics and switches promise to increase the performance of such coreless cells by a large factor.

2. Concepts and Technologies to Increase the Gradient of Coreless Induction Accelerators

This section will discuss the innovations that have occurred in vacuum insulators, solid dielectrics and switches that will permit operation at elevated gradients.

2.1. *High gradient vacuum insulators*

A key electrical weak point in the coreless induction cell of Fig. 2 is the vacuum insulator. Vacuum surface flashover is not definitively understood. However, general trends can be discerned by examining published data. Failure of the insulator, i.e. surface flashover, occurs at a value of field stress that has an inverse power dependence on the pulse width [6]. This key observation is the principal motivation for the accelerator mode to be discussed in Sec. 3.

A novel insulator configuration, called a high gradient insulator (HGI), has the same inverse dependence of flashover field strength on pulse width but has superior performance [6, 7]. The HGI comprises alternating layers of conductor and insulator with periods on the order of a mm or less.

It is generally believed that field emission of electrons that repeatedly bombard the insulator surface and desorb contaminant gas molecules is an essential precursor to surface flashover [8]. Studies of the trajectories of electrons near the surface of HGIs reveal a type of periodic focusing that tends to deflect these electrons away from the insulator surface [9] if the ratio of conductor-to-insulator thickness is in a certain range. This situation is depicted in Fig. 4.

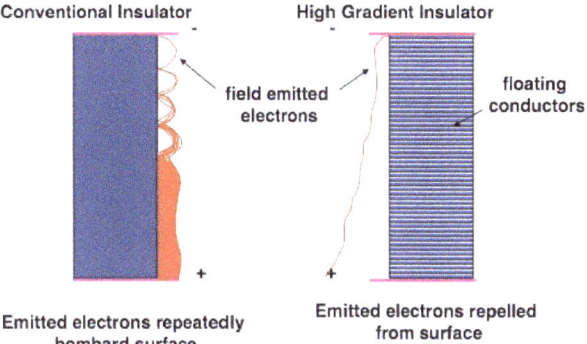

Fig. 4. The flashover mechanism on a monolithic insulator is shown on the left hand side of the figure. A high gradient insulator is shown on the right hand side. Studies by Leopold and coworkers [9] showed that the periodic nature of the microstructure of the fields near the surface tends to deflect electrons away from the insulator surface. This observation may be important in explaining the improved performance of HGIs as compared to monolithic insulators.

2.2. *Solid dielectrics*

It is very desirable to replace the liquid dielectric used in the original coreless induction cells with a solid dielectric. Development of a castable dielectric with high bulk breakdown strength and adjustable permittivity has been proceeding for some time [10]. This particular material is composed of nanoparticles of $BaSrTiO_2$ blended into various bases, such as epoxy or silicone. The material is cast under vacuum and is virtually void-free. Relative dielectric constants can vary from about 3 up to 45 by varying the density of nanoparticles. The material can be cast into large structures (over 1 m in length) and arbitrary shapes.

The bulk breakdown strength of the material is dependent upon the sample thickness and has demonstrated 400 MV/m for submillimeter thicknesses for low dielectric constant epoxy-based materials. A sample transmission line measuring 4 cm × 56 cm with embedded electrodes and an electrode separation of 0.8 mm is shown in Fig. 5. This transmission line was charged repeatedly with 40-ns-wide

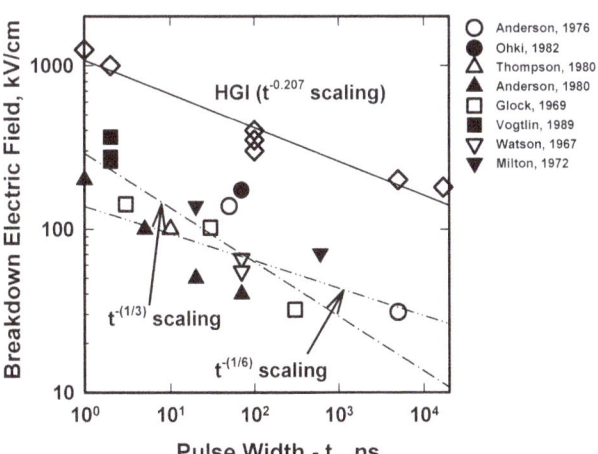

Fig. 3. Dependence of vacuum surface flashover threshold vs. pulse width for conventional insulators (lower two curves) and for HGIs (upper curve). The general trend is an inverse dependence of the flashover threshold on a power of the pulse width. The HGIs flashover threshold is several times that of monolithic insulators.

Fig. 5. A cast dielectric transmission line with embedded electrodes. The electrodes are separated by 0.8 mm and are 1-cm-wide by 56-cm-long. The dielectric is epoxy-based, with a relative dielectric constant of approximately 3. The sample failed at an average stress of 170 MV/m.

pulses at increasing voltages and finally failed at 141 kV or an average field stress of 170 MV/m.

2.3. Wide band gap photoconductive switches

The coreless induction cell in Fig. 2 has a closing switch whose function is to initiate the output voltage pulse. In order to achieve a high gradient in the cell, the switch must be able to operate at a high electrical field stress.

There are other approaches employing opening switches that can be used for the DWA; they will be discussed in Sec. 4.

Photoconductive switching of wide band gap materials such as SiC or GaN is compatible with very high voltage gradients [11]. The bulk breakdown strength of good quality SiC and GaN is in the range of 200 MV/m or higher.

The use of laser illumination having photon energies below the band gap permits one to take direct advantage of the high bulk breakdown strength of these materials. Light with energy below the band gap can propagate on the order of 1–2 cm in these materials, permitting electrodes to be placed on opposite sides of a thin, large area wafer. By doping the material with appropriate concentrations of the right impurities, photoconductivity can be generated by laser illumination.

These switches can conduct over relatively wide areas (1–4 cm^2) and have a nearly ideal configuration for placement in pulse generating lines [12]. A switch schematic is depicted in Fig. 6.

An important feature of these devices is that the carriers recombine very rapidly (< 1 ns) after removal of the laser illumination and recover even against full voltage. Thus, they act not only as closing switches but as opening switches also. More precisely, they are light-controlled resistors and can be operated in a linear regime at lower illumination levels [13].

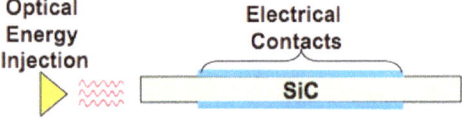

Fig. 6. A typical wide band gap photoconductive switch (in this case SiC). Placing electrodes on opposite sides of the wafer enables one to take direct advantage of the high bulk breakdown strength of these materials. Doping the wafer with the appropriate impurities leads to absorption of the laser light and generation of photoconductivity.

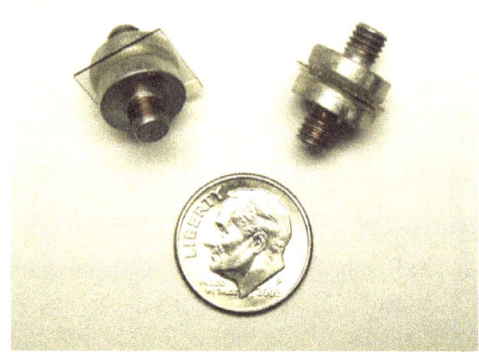

Fig. 7. SiC wafers configured as switches for testing. They measure 12 mm × 12 mm × 1 mm and are cut from the A-plane 6H-SiC. Similar devices have been fabricated from 4H-SiC and GaN.

A typical switch configured for testing in isolation (not in a pulse-generating line) is shown in Fig. 7.

The ultimate design goal for a switch of the type and size shown in Fig. 7 is about 100 kV, corresponding to an average field stress in the material of 100 MV/m. Careful management of enhancements of the electric field at the electrode edges will be necessary for achieving this goal.

3. Sequential Pulse, Virtual Traveling Wave Accelerator

Because of the general inverse dependence of the flashover threshold with pulse width shown in Fig. 3, it would seem desirable to design pulse-generating lines with the minimum possible output pulse width. However, in order to make the acceleration process efficient the particles should always be embedded in an accelerating field. In order that this be true the region of the dielectric wall exposed to a high electric field must move along with the accelerating particles.

This can be accomplished if the coreless modules are relatively thin and are activated in sequence to produce a region of excitation along the wall that maintains synchronism with the charge bunch.

Maxwell's equations can then be used to impose a restriction on how short the pulse width may be in order to avoid dilution of the accelerating field on the axis of the tube [14].

3.1. Pulse width constraint

We start by finding the axial electric field given its value along the tube wall. Inside the tube E_z satisfies the wave equation

$$\left(\nabla^2 - \frac{1}{c^2}\frac{\partial^2}{\partial t^2}\right)E_z = 0. \tag{1}$$

Here E_z is the axial component of the electric field. We seek a solution that is a function only of the "retarded time" variable $\tau = t - z/u$, where u is the speed of the excitation that is produced along the wall by sequential activation of the coreless induction modules. We consider the case where there is no azimuthal dependence. Then, inside the tube, $E_z = E_z(r,\tau)$. Substituting this form into Eq. (1) and Fourier-transforming in the variable τ to ω gives

$$\tilde{E}_z(r,\omega) = \tilde{E}_0(\omega)\frac{I_0\left(\dfrac{\omega r}{\gamma u}\right)}{I_0\left(\dfrac{\omega b}{\gamma u}\right)}. \tag{2}$$

Here b is the radius of the tube, $E_0(\tau)$ is the z component of electric field along the wall of the tube and γ is the usual Lorentz factor, $(1 - u^2/c^2)^{-1/2}$. The quantity I_0 is the modified Bessel function of order 0.

Taking the inverse transform of Eq. (2) for a variety of trial functions for $E_0(\tau)$, such as a Gaussian, a super-Gaussian and a hyperbolic secant, we obtain a nearly universal curve that expresses the ratio of E_z on the axis to that on the wall of the tube as a function of the parameter $\theta = b/(\gamma u \tau_0)$. The parameter τ_0 is the temporal full width at half maximum of the field at the wall at any given location. The spatial full width at half maximum of the field value at the wall is just $u\tau_0$.

From Fig. 8 it can be seen that the on-axis field is comparable to that at the wall provided that $\theta < 0.3$.

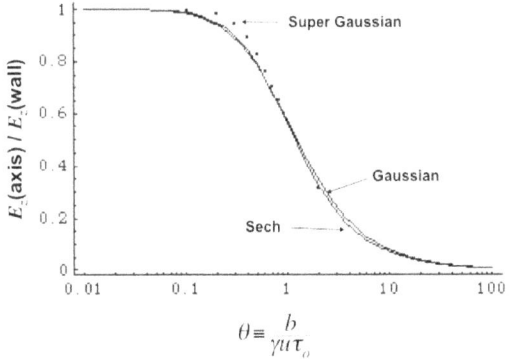

Fig. 8.   A nearly universal curve illustrating the dependence of the ratio of the on-axis value of the accelerating field to its value at the tube wall on the parameter θ. For values of $\theta < 0.3$ the on-axis field is comparable to its value at the wall.

For highly relativistic particles this criterion can be satisfied for very short pulses. However, for nonrelativistic particles where γ is effectively unity, the criterion implies that the spatial full width at half maximum of the wall excitation $> 3b$. A typical value for b is $\approx 2\,\mathrm{cm}$.

4. Architectures for the DWA

There are several possible architectures for the DWA which involve different geometrical arrangements of pulse-generating lines and which can employ closing or opening switches.

4.1. *Geometrical choice*

There are a very wide variety of pulse-generating lines employing closing switches which we generically refer to as "Blumleins." These lines are made up of two or more transmission lines [15].

Fundamental to the discussion of geometrical configurations is the characteristic impedance of a parallel plate transmission line, which depends on the dimensions of the line and the dielectric constant (we will assume that the relative permeability of the dielectric material is unity).

4.1.1. *Classical radial line*

For an azimuthally symmetrical structure the characteristic impedance is a function of the cylindrical radius and is given by

$$Z(r) = \sqrt{\frac{\mu_0}{\varepsilon}}\frac{d}{2\pi r} = \frac{60}{\sqrt{\varepsilon_r}}\frac{d}{r}\,\mathrm{ohms}. \tag{3}$$

Here Z in the characteristic impedance, d the axial separation between the plates, r the cylindrical radius, ε the actual dielectric constant and ε_r the relative dielectric constant. Equation (3) reveals that the characteristic impedance of the radial line varies with r. However, by tapering either d or ε or both with the radius in the appropriate manner, nonvarying characteristic impedance can be achieved [14]. This equation implies that massive currents must flow when such a transmission line operates at a high gradient. For example, if the gradient associated with a wave propagating in the line is $100\,\mathrm{MV/m}$, and the relative dielectric constant is 3, a tube radius of $2\,\mathrm{cm}$ will produce an associated current flow of $58\,\mathrm{kA}$.

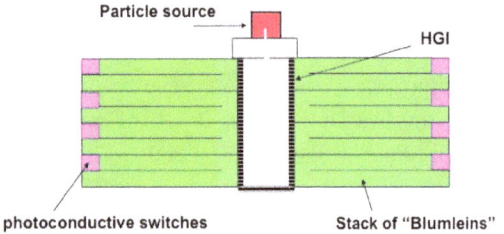

Fig. 9. An example of a DWA using two stacks of Blumleins placed on opposite sides of the beam tube.

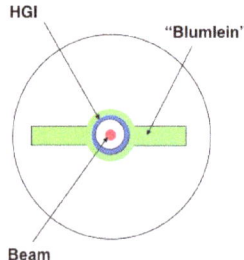

Fig. 10. End view of the example in Fig. 9, showing two stacks of strip Blumleins placed on opposites sides of the beam tube.

4.1.2. *Strip transmission line configuration*

For a parallel plate transmission line the characteristic impedance is given by

$$Z = \sqrt{\frac{\mu_0}{\varepsilon}} \frac{d}{w} = \frac{120}{\sqrt{\varepsilon_r}} \frac{d}{w} \text{ ohms.} \qquad (4)$$

Here w is the transverse width of the plates and d is their separation.

An accelerator configured from strip Blumleins is shown in Fig. 9. The end view is shown in Fig. 10.

The strip Blumlein configuration can have considerably higher impedance than the radial line, and

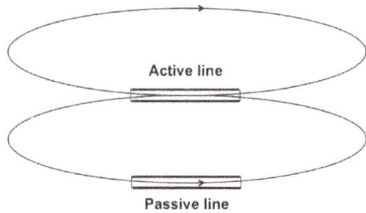

Fig. 11. End view of two planar transmission lines. The active line is carrying a current produced by a switch closure or injection of a signal. The magnetic field lines produced by the current flow close out of the plane of the line and intercept nearby lines (the passive line). The intercepted magnetic field induces a current in the passive line. The net effect on a stack of pulse-generating lines is a reduction in the output gradient and pulse distortion.

requires fewer switches but suffers from parasitic coupling between different lines in the stack. The reason for this can be gleaned from examination of Fig. 11.

A transmission line that is propagating a wave (launched by a switch closure, for example) generates a magnetic field attendant to the current flow in the line (active line). Since the magnetic field is divergence-free, the magnetic field lines must close.

In a radial line these field lines close azimuthally, but in a strip transmission line they close out of the plane of the line. In Fig. 11 the line with current flow due to switch closure (the active line) produces magnetic field lines that close out of the plane of the line. If there is a nearby line (the passive line) that intercepts these field lines, a current will be induced in that line. The result of this current will be a reduction in the overall gradient of the stack and distortion of the temporal pulse produced by the lines [16].

The effect of the parasitic coupling can be minimized by increasing the aspect ratio of the lines (i.e. by lowering their impedance). This problem motivated the invention of the architecture to be discussed in Subsec. 4.3.

4.2. *Pulse generation choices*

Up to this point we have considered schemes employing closing switches. Schemes exist that employ opening switches as well [17].

One notable example (SLIM — SLAC Induction Module) uses drift step recovery diodes as opening switches, which are embedded in transmission lines [17]. These diodes conduct current and can then be controlled to open, generating a large voltage pulse that can be applied across an HGI. An example is shown in Fig. 12.

The line shown in Fig. 12 can be composed of separated strip transmission lines or can be made into a radial line [18].

4.3. *Moving virtual gap induction concentrator*

A possible solution to the problem of parasitic coupling and the relatively low impedance of a stacked Blumlein configuration is to use a structure that is placed entirely within conventional induction cells [16].

Consider the structure shown in Fig. 13. It is an inductive voltage adder, a configuration that has

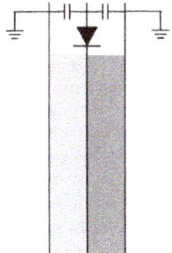

Fig. 12. Illustration of an opening switch architecture using drift step recovery diodes (SLIM). When current flow in the diode is interrupted, a large, pulsed voltage appears across it that sends a wave down both transmission lines. The darker one (on the right hand side) is terminated in a short circuit, while the lighter one (left hand side) is open and exposed to the beam tube. The former provides transit time isolation for the voltage pulse that appears across the bottom of latter.

been used for decades to sum up the voltages of a number of individual induction cells and impress it across a load. In the figure, the voltage of nine cells appears across a single vacuum gap in an electron injector. The voltage of the entire structure is concentrated into the relatively small gap.

Suppose that we could arrange for the gap in Fig. 13 to move in synchronism with an accelerating charge bunch. Then the particles would experience a continuous acceleration.

We could imagine replacing the interior of the inductive voltage adder with a stalk made of a material whose conductivity could be varied on command. To create a virtual gap, the conductivity in the gap region would be relatively low, while everywhere else it could be arranged to be very high so as to approximate a conducting tube with a vacuum gap.

This variable conductive material could be placed on the outer diameter of an HGI beam tube. If we could modulate the conductivity locally and rapidly enough, we could create a moving, virtual gap that would concentrate the voltages of the

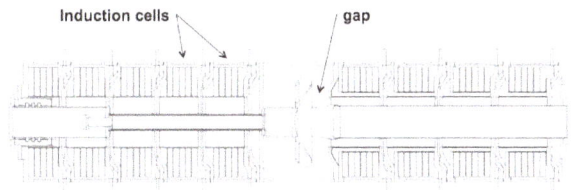

Fig. 13. Diagram of an inductive voltage adder, in this case an electron injector. The voltages of nine induction cells are summed up and appear across a relatively narrow vacuum gap. Thus, the voltage of the entire structure has been concentrated into a small region, resulting in an electric field that is at least an order of magnitude greater than the average gradient of the structure.

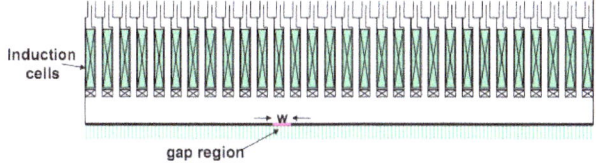

Fig. 14. Inductive voltage adder schematic with a variable conductivity tube. The virtual gap is of width w. The stalk is electrically connected to the induction cells at either end of the structure so that the electromagnetic fields are completely enclosed and no voltage appears on the exterior of the assembly.

induction cells in a localized region. Photoconductive switches, in a variety of possible configurations, could be placed around the tube to provide this capability.

It is clear that this concentration can occur for DC voltages. In order to assess to what degree this mechanism is viable for fast pulses, it is convenient to use a transmission line model of the induction system and conductive stalk.

In the simplest possible model we represent the induction cells by ideal voltage sources and use series inductance per unit length to describe the vacuum inductance of the stalk inside the cells. We also add a shunt capacitance per unit length to describe the distributed capacitance between the stalk and the inner surface of the cells. Finally, we represent the conductivity of the tube by a variable series resistance, R, per unit length.

Using the circuit in Fig. 15, we may write down equations for the voltage and current as

$$\frac{\partial V}{\partial x} = g(t, x) - L\frac{\partial i}{\partial t} - R(t, x)i, \tag{5}$$

$$\frac{\partial i}{\partial x} = -C\frac{\partial V}{\partial t}. \tag{6}$$

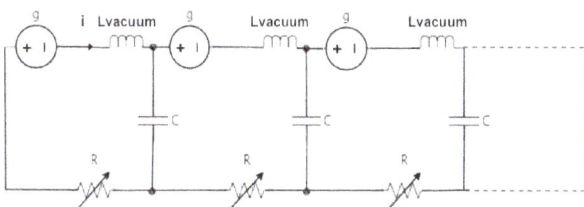

Fig. 15. Simple circuit model of the inductive voltage adder with a variable conductivity stalk. The variable resistors represent the locally variable conductivity of the stalk. The induction cells are represented by ideal voltage sources of g volts per unit length. The coaxial distributed inductance and capacitance are represented by the series L and the shunt C, respectively.

Here x is the distance along the beam axis, g represents the voltage produced by the induction cells, L and C represent the distributed stalk inductance and capacitance respectively, and R represents the variable conductivity tube.

The electric field parallel to the tube E_a, which is just the accelerating field, is given simply by

$$E_a = -R(t, x)i. \qquad (7)$$

It will prove convenient to define dimensionless dependent and independent variables,

$$\eta = \frac{R_0 t}{L}, \quad \xi = \frac{x}{w}, \quad \psi = \frac{R_0 i}{g_0}, \quad \Omega = \frac{V}{g_0 w}, \qquad (8)$$

which are, respectively, dimensionless time, axial distance, current and line voltage. Here g_0 is the maximum source gradient, R_0 the minimum resistance per unit length of the tube, and w the axial extent of the virtual gap. The resistance of the tube and the source voltage per unit length (source function) are given by

$$R(\eta, \xi) = R_0 f(\eta, \xi), g(\eta, \xi) = g_0 \hat{g}(\eta, \xi). \qquad (9)$$

We are seeking a localized pulse of the electric field that occupies a relatively small fraction of the induction system. As an approximation, we seek a traveling wave solution of invariant form for Eqs. (5)–(9). Numerical solutions for a finite length system with the proper boundary conditions verify the validity of this approach. We specify a traveling wave of resistance along the tube as

$$f(\eta, \xi) = f\left(\frac{Lu}{R_0 w}\eta - \xi\right) = f(\sigma), \qquad (10)$$

where u is the speed of the virtual gap and σ is the independent variable. With this specification the system of equations (5)–(10) can be reduced to

$$\frac{1 - LCu^2}{wR_0 Cu}\frac{\partial \psi}{\partial \sigma} - f(\sigma)\psi = 1 \qquad (11)$$

with

$$E_a = -g_0 f(\sigma)\psi(\sigma), \qquad (12)$$

where we have set the source function to 1.

Examination of Eq. (11) reveals that there are two distinct regimes: "subluminal," where $1 - LCu^2 > 0$; and "superluminal," where $1 - LCu^2 < 0$. This terminology arises from the fact that the speed of an electromagnetic wave along the coaxial system formed by the stalk and the inner surface of the induction cells is given by $1/(LC)^{1/2}$. The subluminal (superluminal) regime occurs when the speed of the virtual wave is less than (greater than) the wave propagation speed.

By demanding that the electric field be constant over the interval $0 < \sigma < 1$ (the gap), we can use Eq. (11) to solve for the current. Equation (12) can then be used to determine the functional form for $f(\sigma)$. Analysis of this case reveals that in the strongly subluminal regime the maximum gain, or field concentration, is

$$\frac{E_a}{g_0} = 1 + \frac{1}{wR_0 Cu}, \qquad (13)$$

while for the strongly superluminal regime the maximum gain is given by

$$\frac{E_a}{g_0} = 1 + \frac{Lu}{R_0 w}. \qquad (14)$$

As the particle speed increases, it can be seen from Eq. (13) that the gain decreases. At some point, a transition to the superluminal regime would be appropriate. In order to produce a large potential gain, the stalk could be loaded with magnetically permeable cores to increase the series inductance per unit length, as is depicted in Fig. 16.

Note that the circuit diagram corresponding to Fig. 16 is identical to that of Fig. 15 with an increased value of series inductance.

There is a dual to the circuit of Fig. 15 in which the series resistance and inductance are interchanged. The series inductance can be provided by a helix, while the switches can be placed inside the induction cells. The switches are placed in series with a charged capacitor bank and the induction gaps to power the cells. This arrangement is shown in Fig. 17.

For this configuration we must also consider the loss in the cores inside the cells, which appear in parallel with the switches. It is possible to add a core

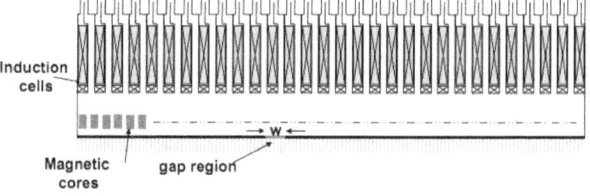

Fig. 16. Magnetically permeable cores have been added around the conducting stalk to increase the series inductance per unit length. This can increase the potential gain in the superluminal regime.

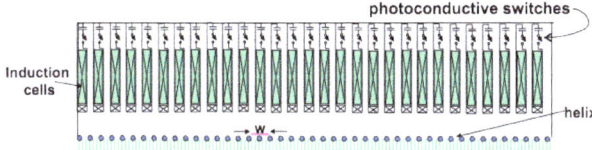

Fig. 17. Dual configuration of the one in Fig. 16 for the superluminal regime. The series inductance per unit length of the system is now provided by a helical conductor around the dielectric beam tube. The switches are now inside the induction cells, where they provide power to the cells by connecting a charged capacitor bank to the induction gaps.

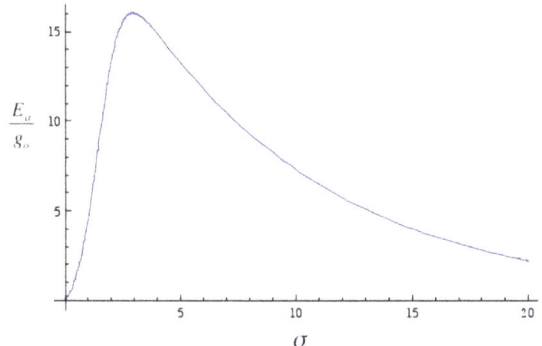

Fig. 18. The traveling wave solution for the configuration of Fig. 17. The gain is plotted as a function of the traveling wave similarity variable for the case of exponential increase in switch resistance after removal of laser light from the photoconductive switches.

resistance in parallel with the switch resistance and obtain another traveling wave solution.

A solution for this case is shown in Fig. 18, corresponding to an exponential increase in switch

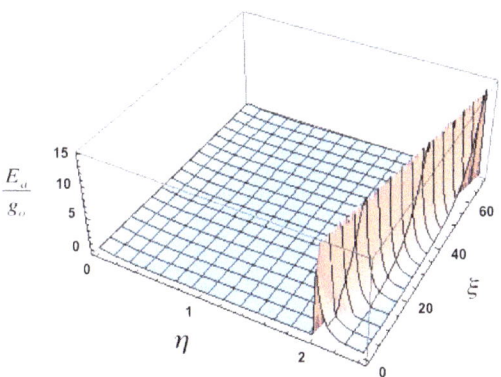

Fig. 19. The full numerical solution for the architecture shown in Fig. 17 is plotted as a function of dimensionless time and distance. The model includes a resistance in parallel with the switches to account for the loss of the magnetic cores in the induction cells. The amplitude and shape of the accelerating field agree very well with the traveling wave solution shown in Fig. 18.

Table 1. Summary of accelerator architecture features for the options discussed.

Architecture	Stripline	Radial line	Induction concentrator
Parasitic coupling	Yes	No	No
Impedance	Low	Very low	Low to moderate
Switch count	High	Very high	Moderate to high
Input power	High	Very high	Moderate to high
Ringing output	Yes*	Yes*	No
Gradient	High	Highest	Core may limit†
Wave form control	Limited	Moderate	Moderate
External focusing	No‡	No‡	Yes

*Unless terminated with a matched resistive or beam load.
†In addition to switch, dielectric material and HGI limitations of the other architecture.
‡Unless the cells are separated to allow focusing between them.

resistance after the removal of laser light. This is an appropriate approximation for describing the resistance change of a switch as the carriers recombine.

The result shown in Fig. 18 may be compared with a numerical solution to a finite length problem with the same parameters shown in Fig. 19 for the case of constant u.

Note that the gain in this scheme is dependent upon rapid increases in the resistivity of the switches upon interruption of laser illumination. In this respect, the scheme is similar to those employing opening switches.

5. Challenges and Prospects

Some features of the various approaches that have been discussed are summarized in Table 1. Both the stripline and the radial line options refer to either the use of "Blumleins" with closing switches or geometries similar to that of Fig. 12 with opening switches. It appears that with either of the first two options external focusing can only be used if the structure is interrupted to insert lenses between the cells.

These lenses could be electric or magnetic quadrupoles or solenoids. Solenoids of several-kilogauss field strength can be used inside the cells of the induction concentrator without difficulty and can focus the beam against its space charge field and emittance.

It should be noted that the induction concentrator is a relatively recent concept and is not as mature in its development as the stacked Blumlein approach.

5.1. *Factors in beam quality*

Certain limits on energy spread and emittance are necessary in order to realize the full treatment potential of a compact accelerator for cancer therapy [19]. The desire for a small (several mm) diameter spot at the tumor imposes a requirement on the order of $10\,\pi$ mm-mrad normalized emittance for reasonable beam sizes exiting the accelerator.

The energy spread is dictated by the actual treatment strategy. For example, a spread-out Bragg peak may be desired which could be furnished on a single pulse by deliberately imposing an energy variation of several percent. On the other hand, the desire to make the Bragg peak as sharp as possible would necessitate limiting the energy spread to being on the order of 0.5%. For shallower tumor depths this figure can be relaxed.

The DWA has several features that make fulfilling these requirements challenging. Chief among these is the non-flat top accelerating wave form that results from the parasitic coupling described in Subsec. 4.1.2. This generates a potential for energy spread and transverse defocusing. The parasitic effects can be reduced at the expense of increasing the stored energy in the Blumleins with attendant increases in the size of the photoconductive switches and the laser system required to trigger them.

A similar tradeoff exists in achieving good azimuthal symmetry of the accelerating field. The HGI has closely spaced (< 1 mm) conducting rings that enforce azimthual symmetry over "long" time scales. Simulations show that adequate symmetry is attained when the wave fronts from each Blumlein stack move halfway around the rings and begin to overlap. Particle in cell simulations show that low emittances can be preserved over meters of transport. The condition that the axial extent of the wall excitation be about three times the tube radius and not much larger in order to limit the potential for surface flashover implies that the temporal pulse width shrinks as the protons accelerate to higher energy. This reduction in pulse width would reduce the stored energy in the Blumleins and the required laser energy but would infringe on the overlap requirement. Balancing these conflicting requirements to achieve a practical system will be necessary.

Electrical strengths of newly developed dielectric and switch materials are consistent with gradients of 100 MV/m, and small sample tests of HGIs have shown vacuum surface flashover thresholds of 100 MV/m for 3 ns pulses. These results must be extended to larger samples so as to be consistent with sizes dictated by the gradient criterion in Subsec. 3.1.

Accelerator architectures have been devised that should be able to capitalize on these technological improvements and lead to the advent of short pulse, high gradient accelerators.

Acknowledgments

It is a pleasure to acknowledge the support of Drs. Cherry Murray and Bill Goldstein, as well as the support from members of the Beam Research Program at Lawrence Livermore National Laboratory, and our coworkers at TomoTherapy, Inc. and CPAC. We would also like to thank Dr. Ralph deVere White, Director of the U.C. Davis Cancer Center, for his outstanding support and encouragement. G. J. C. is indebted to Dr. Anatoly Krasnykh of SLAC for providing him with information on his SLIM concept. Support for this work was provided by Lawrence Livermore National Laboratory, U.C. Davis Cancer Center, TomoTherapy, Inc. and CPAC.

References

[1] C. Kapetankos and P. Sprangle, *Phys. Today* Feb, 58 (1985).

[2] M. Goitein, A. Lomax and E. Pedroni, *Phys. Today* Sep., 45 (2002).

[3] G. Caporaso *et al.*, *Phys. Medica* **24**, 98 (2008).

[4] N. Christofilos, R. Hester, H. Lamb, D. Reagan, H. Sherwood and R. Wright, *Rev. Sci. Instrum.* **35**, 886 (1964).

[5] A. Pavlovski *et al.*, *Sov. J. At. En.* **28**, 549 (1970).

[6] S. Sampayan *et al.*, *IEEE Trans. Diel. Elec. Insul.* **7**(3), 334 (2000).

[7] J. Harris *et al.*, *Appl. Phys. Lett.* **93**, 241502 (2008).

[8] R. Anderson and J. Brainard, *J. Appl. Phys.* **51**(3), 1414 (1980).

[9] J. Leopold *et al.*, *IEEE Trans. Diel. Elec. Insul.* **12**(3), 530 (2005).

[10] D. Sanders, E. Cook, R. Anaya, L. Wang, S. Sampayan, G. Caporaso, K. Selenes, J. Jacquin and R. Fuente, Breakdown performance statistics of a nanoparticle composite system, in *Proc. 2007 IEEE Pulsed Power Conf.* (Albuquerque, New Mexico, USA), p. 1826.

[11] W. C. Nunnally and M. Mazzola, Opportunities for employing silicon carbide in high power

photo-switches, in *Proc. 2003 IEEE Pulsed Power Conf.* (Dallas, Texas, USA), p. 823.

[12] J. Sullivan and J. Stanley, 6H-SiC photoconductive switches triggered below bandgap wavelengths, in *Proc. 27th Int. Modulator Symposium and 2006 High Voltage Workshop* (Washington, D.C., USA, 2006), p. 215.

[13] J. Sullivan and J. Stanley, *IEEE Trans. Diel. Elec. Insul.* **14**(4), 980 (2007).

[14] G. Caporaso *et al.*, High gradient induction accelerator, in *Proc. 2007 Particle Accelerator Conf.* (TUYC02), p. 857.

[15] M. Rhodes, Ferrite-free stacked Blumlein pulse generator for compact induction linacs, in *Proc. Pulsed Power Conf.* (2005), p. 1192.

[16] G. Caporaso *et al.*, Status of the dielectric wall accelerator, in *Proc. 2009 Particle Accelerator Conf.* (TH3GAI02).

[17] F. Arntz, A. Kardo-Sysoev and A. Krasnykh, SLIM, short-pulse technology for high gradient induction accelerators. SLAC-PUB-13477 (2008).

[18] A. Krasnykh, Evaluation of SLIM for hadron therapy. Private communication (2009).

[19] W. T. Chu *et al.*, Performance specifications for proton medical facility. Lawrence Berkeley Laboratory Report LBL-33749, (Mar. 1993).

George J. Caporaso has been a physicist at Lawrence Livermore National Laboratory since 1977 working primarily on high current beam dynamics, accelerators and pulsed power and has led the Beam Research Program since 1997. He is a fellow of the American Physical Society and has taught numerous courses in high current beams at the U.S. Particle Accelerator School. He is the principal investigator for the Dielectric Wall Accelerator.

Yu-Jiuan Chen has been a theoretical physicist at Lawrence Livermore National Laboratory since 1981 working on plasma instabilities and beam transport in induction accelerators. She is a fellow of the American Physical Society and has made pioneering contributions to increasing the beam quality of induction machines, especially for flash x-ray radiography and has taught many courses at the U.S. Particle Accelerator School. She is the theory group leader of the Beam Research Program.

Stephen E. Sampayan has been an electrical engineer at Lawrence Livermore National Laboratory since 1988 working on high average power and ultra compact pulsed power systems. He pioneered the development of advanced technologies for the Dielectric Wall Accelerator such as high gradient insulators, wide band gap photoconductive switches and high electrical strength, cast dielectric materials. He is developing new programs based on Dielectric Wall Accelerator technologies.

Reviews of Accelerator Science and Technology
Vol. 2 (2009) 265–301
© World Scientific Publishing Company

The Supercollider: The Texas Days

A Personal Recollection of Its Short Life and Demise

Stanley Wojcicki

Department of Physics, Stanford University
Stanford, CA 94305, USA
sgwojcicki@gmail.com

This article is the second in a two-part account of the history of the Superconducting Super Collider (SSC). The narrative starts with the move of the SSC activities to its Waxahachie site in Texas and the subsequent organization of its administrative structure. The technical design changes incorporated into the site-specific design are described together with their impact on costs. The principal activities at the SSC — technical progress, conventional construction and planning of the experimental program — are described briefly. The article then discusses the efforts to obtain international collaboration, the growth of the opposition both among the public at large and in Congress, and the final events leading to termination of the SSC. It ends with some subjective views on what went wrong with the SSC and on the prospects for construction of large scientific facilities in the US in the future.

Keywords: SSC; Supercollider; accelerator; Texas; DOE; Congress.

1. Introduction

This is the second of two articles [1] describing the history of the Superconducting Super Collider (SSC), or Supercollider for short. It spans a little over four-and-a-half years, from early 1989, the start of the transfer of the Central Design Group (CDG) activities in Berkeley, California, to Waxahachie,Texas, to 1993, the year of the SSC termination.

The history recounted here has two closely interwoven threads. The first one is the effort to build, on the "green site" at Waxahachie, a frontier high energy laboratory with a first class physics program. The second one is the struggle to convince Congress to fund this multibillion-dollar project in the time of changing administrations, large budget deficits and significant opposition from various quarters. Thus the story is not only about science, technology and organizational matters but even more so about politics and public relations.

Whereas I was intimately involved with the SSC during its CDG phase (in Berkeley), during the Texas days most of my involvement was from outside, in my role as a member of the High Energy Physics Advisory Panel (HEPAP) and later as its chair. I did spend eight months at the SSC Laboratory (SSCL)

on my sabbatical in 1993, just before SSC cancelation, and obtained additional insight from that vantage point.

This account is based very much on my own personal recollections but in writing it I also consulted extensively other sources: articles in popular scientific journals and newspapers, government reports, personal notes, correspondence between major players, minutes of meetings, technical notes and transcripts of Congressional hearings. I also profited from a number of conversations about these events with colleagues who were intimately involved with the SSC. Nevertheless, the article should be viewed as a personal account and no claim is made that it is a true scholarly historical work. While I try to relate the events as accurately as possible, I also occasionally express my own personal opinions, always attempting to identify them as such.

The story and the lessons of the SSC might be especially timely today, since the international high energy physics (HEP) community is proposing to build a new facility, the International Linear Collider (ILC), which would enable deeper probes into the ultimate nature of matter and forces that govern it. There are many parallels with the SSC — in cost, technological challenge, and difficulty of its realization. Study of what went wrong, but also what went right, with the SSC might be instructive as we grapple with how one might best achieve the HEP goals.

2. SSCL Beginnings

The first SSC construction funds, totaling US$126.6 million, were approved in the fall of 1989 [2–4] for FY90. An additional US$87.9 million was appropriated for R&D and auxiliary equipment. Construction of the SSCL could now be formally begun. The initial challenge was to effect a smooth transition from CDG to SSCL, capture all the knowledge and experience acquired during the CDG lifetime, and create a smoothly functioning team with well-defined goals accepted by all.

One of the first tasks was to find a temporary home for the SSCL staff and the initial SSCL activities. The final collider would eventually be located about 35 miles south of downtown Dallas and would encircle the city of Waxahachie, as shown in Fig. 1. For the initial period, before the appropriate land acquisition was finalized, the Laboratory rented a large warehouse on the north side of the future ring and converted that space into the required offices, laboratories, library, cafeteria, etc. Later on, as the staff grew and more technical activities were undertaken in Texas, a second warehouse, closer to Waxahachie, was rented for technical activities. The initial address of the SSCL was 2550 Beckleymeade Avenue, Dallas, TX 75237, USA.

The recruitment of the senior staff did not go as smoothly. The bypassing of CDG members in the process of identifying the Director in August 1988 (as discussed in Ref. 1) created a deeply felt hurt among many CDG members. I do not think that this omission was done maliciously. When I pointed it out to Boyce McDaniel, chair of the Board of Overseers (BOO) at the time, he was obviously

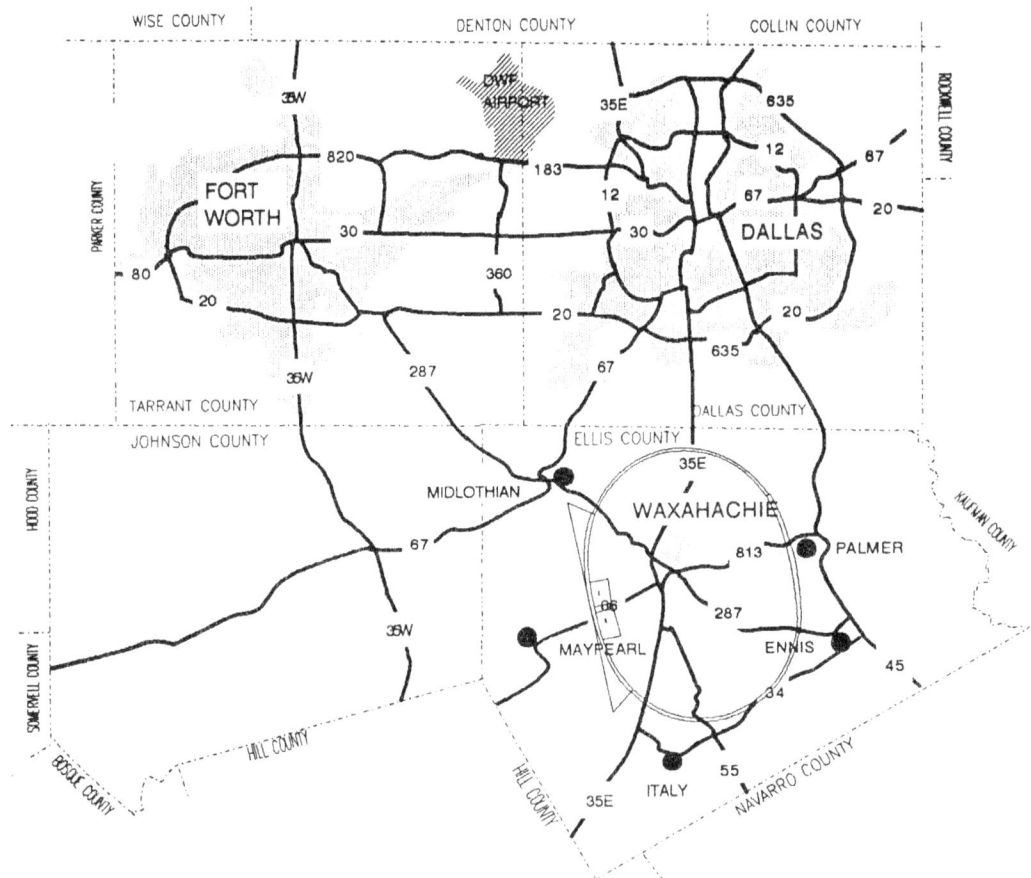

Fig. 1.   Map of the SSCL footprint and of the environs.

embarrassed and I got the impression that the BOO did not even realize this shortcoming of the search process. McDaniel visited the CDG after the fact on September 27 and talked to the senior people there; the comments he received were pretty unanimous in expressing great unhappiness about the process and the outcome. There was also strong concern that what happened would make the transition from the CDG to the SSCL much more difficult [5].

Indeed, the nature of the search process and the subsequent handling of the transition from CDG to SSCL did have a strong negative impact. I think that Schwitters, as well as the BOO and the URA, underestimated the effort required to recruit truly competent and congenial staff. Many key people of the CDG did not move to the SSCL. Individual reasons for these decisions varied but invariably a contributing factor was that there was insufficient effort made to recruit them. This may have been partly due to lack of recognition of their unique abilities and experience but probably also partly because of the desire of the new Directorate to build their own team. Several of those who did join the SSCL went with some amount of disappointment and bitterness.

By far the most important loss to the project was Maury Tigner. Schwitters did offer him the positions of Technical Director/Project Manager and Deputy Director, with responsibility for the construction of the machine [6]. From talking to Tigner shortly afterward, I had the impression that a *modus vivendi* could be worked out. However, after several conversations with Schwitters, Tigner decided that he could not work within the framework that was being proposed (the main issue being whether he would be the ultimate authority in deciding on key personnel in the accelerator area). This decision must have been taken before November 4, 1988 (proposal submission date), since Helen Edwards was listed on the proposal as being in charge of the Accelerator Division. Tigner submitted his resignation to the URA on January 18, 1989, effective as of February 20, 1989. It was a real tragedy that there was not enough effort to recruit Tigner. Given his talents, experience, previous achievements and stature in the community, he would have been a great asset to the project. I feel confident that any disagreements between him and Schwitters regarding relative responsibilities and authority were not beyond resolution given sufficient goodwill on the part of both

individuals. In retrospect, the BOO and the URA can be faulted for not being more forceful in trying to broker an arrangement satisfactory to both individuals.

The CDG became one of the divisions of the SSCL and its activities were going to be overseen by the SSCL. During the transition period lasting until 1990, Chris Quigg was to be in charge of those activities as SSCL Associate Director. These administrative changes and Tigner's resignation were communicated to the CDG staff, numbering 116 people, in January 1989, via a memo [7] from the URA President, Ed Knapp. Quigg would have been another valuable CDG member who could have helped in transferring some of the knowledge gained there to the SSCL but he chose to return to Fermilab, telling me that he was not offered anything of sufficient interest and challenge at the SSCL. Another key person who chose not to go to the SSCL was Peter Limon, who also decided to return to Fermilab, largely for personal reasons. I had made a decision some time earlier to return to Stanford once the site was chosen and the management team selected, and followed up on that plan when the CDG activities were being terminated. Schwitters did invite me later, in July 1989, to join the SSCL as an Associate Director at Large but for a number of different reasons I was unable to accept [8].

Schwitters had more difficulty than anticipated in recruiting top level staff. I think his efforts were made more difficult by what was seen as a rather perfunctory search process for the key people. He was apparently turned down by his top choice for the Research Director and eventually appointed M. K. Gilchriese, a physics professor at Cornell, to that position. Without detracting anything from Gilchriese, he was at that stage of his career where it would make more sense for him to lead one of the new SSC experiments than be mired in administrative work. He resigned after a short time in that position and Fred Gilman from SLAC accepted the job in January 1990. Bruce Chrisman from Fermilab was identified as the Associate Director for Business already in the initial M&O proposal but left the SSCL in July 1989 to go into the private sector.

The industrial partnership with Sverdrup did not work out satisfactorily [9]. Robert Robbins, after a short stay as head of Conventional Construction, was replaced by another Sverdrup employee, Lew Smith, and soon afterward by Jim Sanford on an acting basis. It is a mystery to me why Sverdrup was chosen as a partner in the M&O proposal. They had essentially no experience in underground tunneling. Their involvement had a negative result of shunting off experienced CDG members like Jim Sanford, Tim Toohig and Chris Laughton to secondary roles. The partnership with Sverdrup was terminated by the URA within a year.

Schwitters went outside the HEP community to appoint the Deputy Director and the Head of the Magnet Division. Tom Kirk, a former member of the CDG and then at Argonne National Laboratory, was considered for the latter position but Schwitters eventually decided to go with Tom Bush, a retired Navy captain who had extensive experience with missile design, development and production. Richard Briggs, originally Associate Director at Livermore National Laboratory in charge of fusion activities, became the Deputy Director of the SSCL.

Politically most sensitive and most important was the position of Project Manager. Initially it was filled by Doug Pewitt for a short time and then by Briggs, followed by Ted Kozman from LBNL on an acting basis. Such a position has been traditionally occupied by a member of the high energy community with significant accelerator project management experience. However, the new Secretary of the DOE in the Bush Administration, retired Navy admiral James D. Watkins, had strong and somewhat different views on this subject which were influenced by his experience with military contractors. The key ingredients was the belief that "academics" do not have competence and experience to manage a large scale construction project and that one needs strong management by federal officials. Schwitters and the URA proposed Paul Reardon as Project Manager. Reardon was an accelerator physicist with extensive management experience, both at BNL and at Fermilab, and was then Vice-President of Scientific Applications International Corporation (SAIC). Watkins, however, wanted the position given to Edward Siskin, a former vice-president of a large construction company, Stone and Webster, and someone he knew from his Navy days.

This impasse was finally resolved in October 1990, by creating the new position of General Manager for Siskin. Reardon was made Project Manager

reporting to Siskin and responsible for construction of the accelerator. George Robertson, a retired major general in the US Army Corps of Engineers, became Reardon's Deputy. Siskin's position created another bureaucratic layer between the Director and Project Manager and Division Heads (except for the head of Physics Research, which officially reported directly to the Director and was connected to Siskin only by a dashed line). Other key personnel changes occurred in the same time frame: John Ives, a retired admiral, became head of Conventional Construction and Ted Kozman head of Accelerator Systems [10].

This new management scheme at the SSCL created more complex reporting lines between the accelerator designers — certainly one of the prime activities of the Laboratory at that time — and the Directorate. There was a position of Technical Director reporting to the Project Manager, which was held initially by Helen Edwards from Fermilab, who resigned in 1991, reportedly because of conflicts with the management [11]. Also under Reardon was an Accelerator Design and Operations Division, headed by Don Edwards, also from Fermilab. One of the four departments in that division was Accelerator Physics, responsible for theoretical calculations. Thus the input from that very important aspect of the accelerator design was separated from the Director — at least bureaucratically — by three or four layers. This full SSCL organization chart is shown in Fig. 2.

The SSCL tried to merge two cultures at the top management levels — industrial/military and academic/laboratory, with very different styles and different ways of doing things. This two-culture situation persisted and intensified with time and seriously handicapped the functioning of the Laboratory [12]. The free exchange of ideas at low levels, typical of HEP laboratories, was stifled and in some areas expressly forbidden. This distinction was probably most pronounced between the magnet and accelerator designers, and this was the area where the conflicts were most severe. The appointment of Siskin accentuated this situation. He had personal access to Watkins, which resulted in his having an extraordinary amount of authority and autonomy.

The overall situation was not helped by Schwitters' required heavy involvement in various SSC-related interactions outside the Laboratory. Most people I talked to felt that this seriously limited his

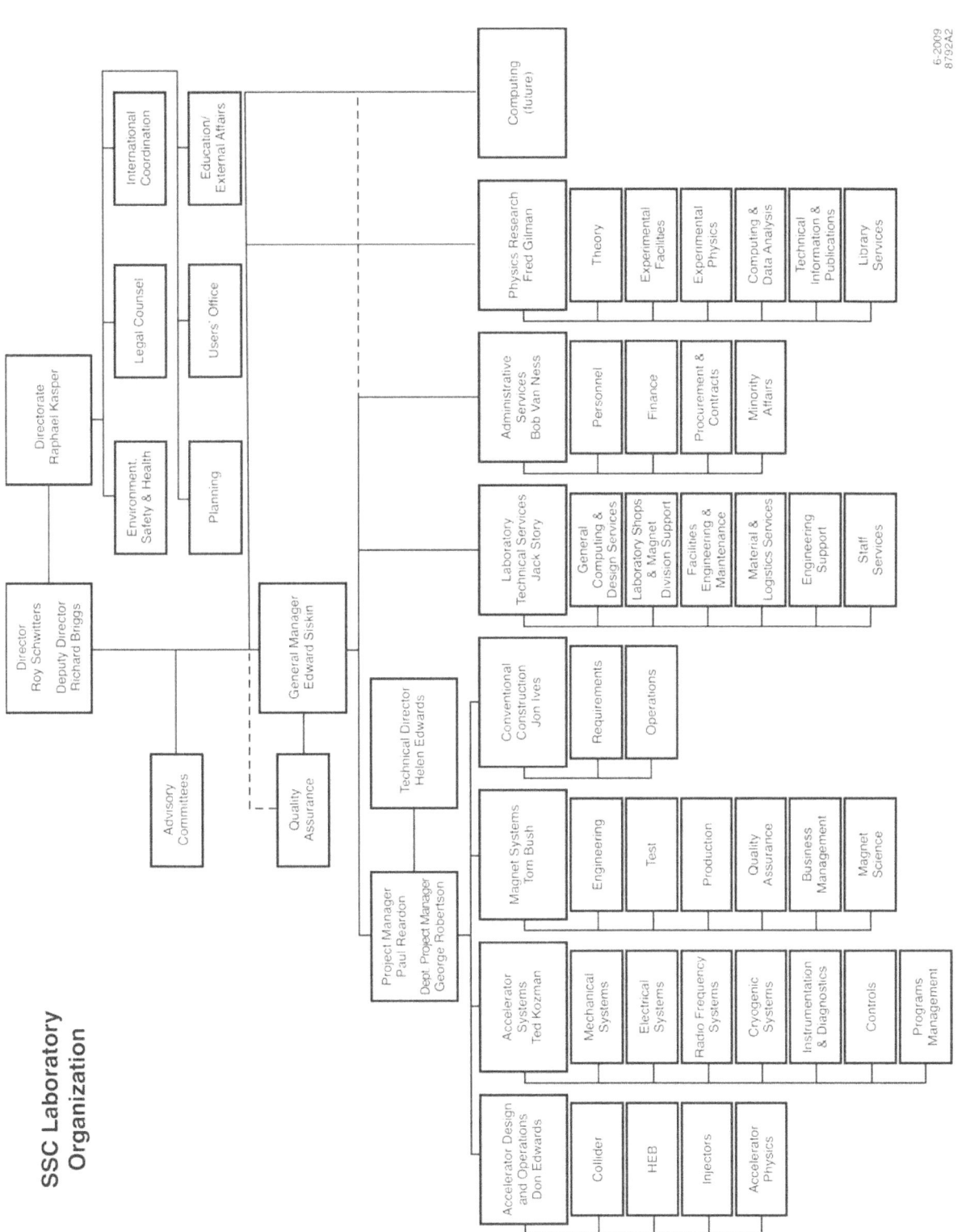

Fig. 2. Initial organization chart of the SSC Laboratory.

ability to be on top of the most important SSCL issues and carry out his intention of being a Director with an "open door policy." Furthermore, as time went on, the future physics program at the SSCL, an area that was of paramount interest to Schwitters, was attracting more and more of his attention at great cost to his primary responsibility, namely creating a well-functioning laboratory dedicated to building the collider as its principal goal.

3. Site-Specific CDR and Its Validation Process

An early SSCL activity was re-examination of the status of the magnet R&D program and basic machine parameters, taking into account site-specific information [13]. This was done mainly in light of what had been learned since the CDG CDR but there was also a tendency sometimes to look at some issues from scratch. Thus, for example, the possibility of bore tube corrections, which were being developed at the CDG and which appeared to indicate that one could live with 4 cm magnet apertures, was not pursued at the SSCL because of its perceived added complexity. The most relevant new information came from the operational experience of the Tevatron and HERA [14]. After ramping down from high energy to much lower energy for injection, there were residual circulating currents which distorted the magnetic field and were not reproducible from cycle to cycle. New programs were developed to investigate these issues, and the availability of more computing power enabled one to follow the proton trajectories over a longer period of time [15], providing new information regarding the required magnetic field quality.

Schwitters appointed an Aperture Task Force to study these effects. They reported their findings to a standing Machine Advisory Committee (MAC), which recommended several changes to the machine design [16–18]. It was argued that without these changes machine commissioning would be much more difficult and operation of the machine could be compromised. There was also the argument that future upgrades, for example to higher luminosities, might not be possible without these changes because a 4 cm aperture would preclude insertion of a liner that might be needed to intercept synchrotron radiation. A rough estimate suggested that these modifications and recalculation of labor, material and R&D

costs, necessary for assuring reliable machine operation, would generate a 20–30% cost increase [19]. The three major recommended changes were:

- Increase the diameter of the dipole aperture from 4 to 5 cm. That would give a larger volume with a uniform field.
- Increase the energy of the HEB from 1 to 2 TeV and at the same time make a corresponding change in MEB (from 100 to 200 GeV) and in LEB (from 8 to 12 GeV).
- Provide stronger focusing in the main ring, necessitating an increase in the circumference of the main ring from 52 to 54 miles.

Strangely, the diameter of the quadrupole aperture remained in this iteration at 4 cm. Several people with whom I discussed this issue claimed that it was apparently due to a miscommunication between Schwitters and the MAC. Whatever the reason, it did become subsequently a source of significant debate and controversy [20]. It was still unresolved in January 1991, when at the HEPAP meeting Don Edwards gave a presentation enumerating potential options and their pros and cons in this area [21].

These proposed design changes were discussed within the Laboratory as well as with the SSC Scientific Policy Committee, the Users Executive Committee and an ad hoc Committee on SSC Physics [22]. They were also presented to the SSC BOO and URA Board of Trustees. The BOO Executive Committee was leaning initially [23] toward trying to keep the original price and finding a "solution focused on 17 TeV energy" rather than the original 20 TeV. The final BOO consensus, however, was that the changes proposed by the Laboratory should be adopted. The other groups also supported the Laboratory decision. The ultimate decision rested formally with the Director, Roy Schwitters, but he was clearly constrained by the opinions of both his advisory groups and his oversight bodies. The final decision was controversial then and remains so today. Even some members of the MAC had doubts about the wisdom of that recommendation. Clearly, the suggested changes made the design more conservative, facilitating magnet fabrication and making the machine easier to commission and easier to operate. But many are still convinced that the original design would have worked, especially if one allows for advances in instrumentation, computing and, most importantly, human ingenuity. On the whole, I found from talking to the people involved, that this is an issue on which there is a significant divergence of opinion.

The next question was whether one should descope the machine so as to recover the original cost figure of US$5.9 billion (see later section for cost discussion). The easiest way to achieve any significant savings was to reduce the energy by about 25%. To obtain guidance on this issue, the DOE asked HEPAP to appoint a blue ribbon Subpanel to address the question "How much physics potential would be lost by reducing the energy below 20 TeV?" The Subpanel was chaired by Sid Drell from SLAC and included 15 members, 5 of them Nobel laureates in physics. The Subpanel concluded that some of the physics objectives might not be attained if energy were reduced to 15 TeV per beam. Furthermore, they did not see any other reasonable alternative which would give a significant-enough cost saving [24, 25].

The report was presented to HEPAP by T. D. Lee (since Drell could not be there) at its January 12, 1990, meeting. The meeting was notable for the presence of Sen. Phil Gramm (R-TX), probably the only time that a sitting Senator participated in a HEPAP discussion. Gramm was supportive of the changes if they were really necessary. He cautioned the community against "bells and whistles" in the machine and hoped that the design was "a sturdy Chevrolet and not a Cadillac." In retrospect, I wonder if we tried to build a Buick. After some discussion, HEPAP unanimously endorsed the report [26].

As a result of these proposed changes [27], the DOE called for another HEPAP Subpanel to provide an independent assessment of the SSC cost estimate (US$7.8 billion) and the realism of the proposed schedules and funding profiles. The Subpanel had six members and was chaired by John Townsend, Jr., Director of the Goddard Space Flight Center. In report, delivered to HEPAP on July 19, 1990, the Subpanel stated its main conclusions [28]:

- The cost-estimating methodology was appropriate and was being properly applied.
- The estimates were complete and credible to the extent that they were determinable. But the low contingency used made the overall cost estimate not credible.
- The proposed schedule and assumed funding profile were not realistic. The schedule was found to

have no float for events not under project control and the projected huge increase in Congressional appropriation from FY91 to FY92 seemed highly unlikely. The Subpanel recommended adding 6–12 months to the early part of the schedule.
- The DOE and the SSCL had failed to date to staff all key managerial and technical positions.
- The estimate of cost to completion should be raised from US$7.8 billion to US$8.6 billion. An additional US$300 million for detectors was recommended.

Within errors, these conclusions paralleled those reached by the 82-member DOE review panel [29] which met at the SSCL on June 25–30, 1990. Their estimate was US$8.37 billion. A third cost estimate was obtained by DOE independent cost estimators (ICEs), who arrived at US$9.33 billion. They also suggested, however, that US$2.492 billion in additional cost elements be considered for inclusion in the total project cost. Almost half of that (US$1.18 billion) was for an increased detector scope. Based on all this input the DOE decided to present to Congress a baseline TPC estimate of US$8.25 billion with a completion date at the end of FY1999 [30, 31].

This significant cost increase was a serious blow to the SSC project and gave powerful ammunition to the SSC opponents. Undoubtedly, it also had a strong impact on the future DOE–SSC relationship, discussed in the next section. It is tempting to examine, with the benefit of hindsight, whether the situation could have been handled better. One big shortcoming in the Drell Subpanel process was the narrowness of the charge, i.e. limiting it to pure physics considerations. Certainly, their recommendation made sense on pure physics grounds, especially considering that the SSC would be the main focus of US HEP for at least a quarter of a century. However, a very important question — what would be the risk of the SSC not happening at all if such a design change were to be implemented — was not part of the charge and, to my knowledge, was never considered in any significant depth by the SSCL Directorate, by the Subpanel, by HEPAP or by the DOE. A good yardstick by which one could have appraised the potential future difficulties was the fact that the required annual expenditures for the SSC construction would have to be significantly higher than the total annual HEP budget at that time. In an ideal world, that risk should have been weighed against the risk of longer commissioning time and more difficult operation. An alternative charge, which might have given a recommendation that would lead to a more successful course of future events, would have been to ask how much physics one could do within the constraints of the initial cost estimate.

The Drell Subpanel report argued that all possible cost-saving options were looked at and they were all unsatisfactory. But some of the Subpanel's assumptions, such as the statement "... present detector technology does not allow use of higher luminosities for a broad range of experiments," ignored the possibility of progress. Developments over the last decade showed that the capability to deal with order-of-magnitude higher luminosities can be achieved (the LHC design luminosity is above $10^{34}\,\mathrm{cm^{-2}\,s^{-1}}$). To reduce initial costs, experimental areas and detectors could have been staged or the level of their sophistication could be coupled to foreign contributions. If necessary, energy could have been raised later by lowering the temperature and raising the field. Similar staged strategy was adopted by the CERN Directorate in 1993, with the missing magnet scheme, in getting the Council to approve the LHC within a certain budgetary cap. Small savings could probably have been achieved by cutting back or postponing some of the campus buildings or trying to get some of the buildings built with private funds.

One of the arguments against reducing the circumference, such as for a 17 TeV machine, was that "a smaller ring would require a new footprint and environmental study, causing delays, which implies increased cost." Apparently, however, no comparison, was made of the delay caused by this versus the delay caused by the cost increase, and no allowance was made for the time gained due to the resulting shorter construction period. The process of acquiring land was still in very early stages at that time, so there would not be much wasted effort in that area. Furthermore, the Townsend report argued for the necessity of a 6–12-month delay in the first half of the schedule. And obviously the footprint already had to be modified somewhat for an increased circumference.

One of the functions of the oversight and management groups like the BOO and the URA was to provide advice on issues broader than Laboratory governance, like the potential repercussions of such a design change. I discussed this issue with Panofsky while the BOO deliberations were taking place and he told me that he was the only member of the BOO arguing until the end for energy reduction to 17 TeV [32, 33]. In hindsight, a "cocktail" approach of reducing the energy slightly, scaling back on detectors and experimental areas, economizing on the campus area, modifying staffing levels and not going all the way on MAC recommendations which aimed for a very conservative design, might have made it possible to limit the cost increase without prejudicing the physics goals. For example, it was argued that the energy ratio between successive accelerators could not be greater than 10 for adequate operation; HERA used bore tube correctors and managed to run efficiently with a 40 GeV injector and an 820 GeV proton ring energy — a factor of over 20. They were able to handle the persistent current problem by connecting two identical reference magnets into the main magnet circuit [34]. It is true, however, that HERA injection time and hence coasting time at low energy were much shorter and thus the stability problem was less severe. On the whole, I feel that there was not sufficient attention given to potential cost-saving solutions; the oversight system appears to have failed in this instance.

4. SSCL–DOE Relationship

Besides the internal SSCL organizational problems, there were major issues connected with the DOE–SSCL relationship and the DOE management structure for the SSCL. These issues were very closely interrelated. The initial DOE–SSC organization, adopted still in the CDG days, had the Office of SSC, headed by Robert Diebold, reporting to the OER Associate Director for the Office of Nuclear and High Energy Physics (part of the OER), a position held at that time by Wilmot Hess. This arrangement was modified by Robert Hunter, the OER Director appointed late in Reagan's second term to fill a position that was held for over a year after Al Trivelpiece's resignation on an acting basis by James Decker. Hunter wanted to strengthen the direct control by the DOE over the SSCL and for that purpose he adopted a system where the SSC Office at Headquarters (OSSC) reported directly to the OER Director. The OSSC, in turn, would direct the SSC site office at the SSCL. This system would bypass the Office of High Energy and Nuclear Physics and effectively exclude those in the DOE who were most familiar with the SSC project and the field of high energy physics.

DOE headquarters wanted the OSSC to be able to exert strong technical direction and management of the project rather than the more traditional oversight. This was strongly resisted by the SSCL and the URA as being contrary both to the signed contract and to the practice in HEP in the past [35]. The OER would have liked to manage the SSCL as a procurement where the overall integration responsibility for the project would rest with DOE headquarters. There would be two contractors — a scientific one, i.e. the URA, and a general industrial contractor (GIC), responsible for the construction of the Laboratory. This plan was presented to a group of senior reviewers on May 4–5, 1989, and their negative comments led to rejection of the scheme. On September 7, 1989, Watkins directed Hunter to establish the OSSC at the DOE headquarters mainly for oversight [36]. This was viewed as a victory by the BOO but turned out to be only the end of the first chapter in the DOE/SSCL management struggles. The issue as to what was to be the exact role of the OSSC was a contentious one throughout the rest of the SSCL's life.

Hunter managed to antagonize a number of key people in Congress, especially members of the Texas and New Jersey delegations: the former over his heavy-handed approach to SSC management and the latter over proposed reductions in the Princeton Plasma Lab funding. As a result, late in 1989 Hunter was forced to resign and Decker again assumed the position of an acting director of the OER. This new situation made it easier for Watkins to mold the formal organizational structure in such a way as to give him much closer personal control over the SSCL. Watkins had the DOE SSC (on site) Project Director, Joseph Cipriano (another Watkins colleague from his Navy days), report directly to him, short-circuiting the normal lines of authority in the DOE [37]. This allowed him to exercise much more direct control but marginalized even further the influence of the DOE people with the HEP experience. The OSSC [38], which was to be headed by an Associate Director of the OER, was directed for a long time on an acting basis by Garry Gibbs, a long

term DOE official, until Cipriano assumed that role in June 1991, continuing also to serve as the on-site DOE Project Manager. The anticipated staffing level of SSCL offices at the headquarters and on site was 110 DOE personnel. All of these changes and innovations and the manner in which they were instituted caused a great deal of friction between the URA, the SSCL Directorate and the DOE, friction which persisted and grew throughout the life of the project.

Watkins was undoubtedly strongly influenced in making these personnel actions by the significant increase in the SSC projected cost, discussed in the previous section. The cost estimate was released by the SSCL in June 1990, but was known by Watkins already in the fall of 1989 [39]. Also influential was probably the fact that the management of the project by the DOE had been the subject of criticism at various levels from the very beginning. These criticisms were clearly expressed both in an early DOE IG report [40] and in a GAO report [41] commissioned by the Senate Budget Committee in March 1990. They were quite critical of the DOE for its slowness in building up an oversight organization that could respond promptly and effectively to SSCL-proposed actions requiring DOE approval. From his response [42] to the GAO report, it is clear that Watkins' intention was to reshape the management system so as to allow him to keep tight control over the SSCL. In his May 8, 1991 letter to Jamie Whitten, chair of the House Committee on Appropriations, Watkins wrote: "Deputy Secretary Henson Moore is involved in the program on a daily basis.... I personally look to Mr. Joseph Cipriano and Mr. Edward Siskin to manage this project on a daily basis and report to me...."

It was in that time frame that I was appointed by Watkins as the new HEPAP chair, to replace Francis Low from MIT, as of September 1, 1990. It was already clear then that the management issues at the SSCL were of paramount importance and potentially a real trouble spot. On October 4, 1990, I wrote a letter [43] to the DOE Deputy Secretary, Henson Moore, citing my charge as the HEPAP Chair to "provide the best possible advice on the High Energy program" and expressing my concerns about the formal SSCL–DOE relationship. I quote a few sentences from it, because they pretty much summarize the concerns at that time of many in the HEP community: "The SSC situation exemplifies an ever-increasing trend toward shifting more and more of the authority for key decisions away from the scientific and technical performers in the field and towards Washington. Such a course must lead to a disaster... adversarial relationship that follows naturally from this course, results in time, energy, and resources being spent fighting unnecessary battles rather than being directed towards the project itself."

Such concerns were expressed earlier by the SSC BOO. In a letter [44] to the Chairman of the URA Board of Trustees, the BOO stated: "...the approach of the DOE to the oversight of the URA contract for the establishment and management of the SSC Laboratory appears to be developing in the direction of detailed DOE management of the project rather than oversight of the performance and accountability of the contractor."

5. Planning the Physics Program

Planning of the SSC physics program was initiated with a letter [45] sent by the Director to the US HEP community early in 1989 outlining his views on the initiation of the program. The key components were the establishment of the Scientific Policy Committee (SPC), composed of 17 senior members of the international HEP community, to advise him on general policy issues and of the Program Advisory Committee (PAC) to provide advice on more specific details, including review of detector initiatives. The PAC was composed of 18 senior scientists from all over the world. Jerry Friedman from MIT was appointed as chair of the SPC and Jack Sandweiss from Yale as chair of the PAC.

The PAC met for the first time on February 9–10, 1990. Guided by their advice, the SSCL issued a call and guidance for expressions of interest (EOIs), to be due on May 25, 1990. By that time 15 EOIs were received, authored by 1946 collaborators from 303 institutions, demonstrating broad, worldwide interest. There was a strong emphasis in the EOIs on both high-P_T and B physics, but there were also ideas for smaller experiments. The EOIs were considered at the next two meetings of the PAC, in June and July 1990. Based on advice from the SPC and the PAC, it was decided that the initial program should have two large high-P_T detectors (plus smaller experiments) that would be complementary

and overlapping [46]. One of them should be a general purpose detector with tracking capability in the magnetic field; the other should emphasize calorimetry and muon identification. The funding allocation (within the new US$8.25 billion total SSCL cost) for the initial SSC experimental program was US$842 million; based on that, the US part of the cost of each large detector was capped at US$250 million, to allow funds for other initiatives. With these guidelines, letters of intent (LOIs) for the large detectors were requested with a deadline of November 30, 1990.

A few explanatory words may be appropriate here, explaining the relationship between the large detector collaborations and the host laboratory. These collaborations become semiautonomous international organizations, with their own leaders and governance, and they are the ones that are responsible for the design and construction of the detectors and the subsequent analysis and publication of the data taken. A large fraction of the funds for the construction comes from the funding agencies supporting the collaborating institutions, whether in the US or abroad. The Laboratory's principal involvement, once a given collaboration is approved for their physics program, is in the financial area. The Laboratory has the responsibility for making sure that the approved design is consistent with the funds available and that the detector will be completed on time and on budget. This will often create some friction when the physics ambitions of the Collaboration exceed the funds available for their detector.

Three LOIs for such large high-P_T detectors were received [47] and considered by the PAC at their meeting on December 13–15, 1990. Based on discussions at that meeting and PAC recommendations, the following decisions were taken by the Director [48]:

- The Solenoidal Detector Collaboration (SDC), with George Trilling, professor at UC Berkeley, as spokesperson, was approved for support in its development of a formal proposal/design report.
- The L* (Lone Star) Collaboration, with Sam Ting, professor at MIT, as spokesperson, was not approved even though it was seen as potentially competitive and complementary to the SDC. Concerns were raised about the reliability of its cost estimate, the adequacy of US participation, and the composition and mode of governance of the

Collaboration. A definite recommendation would await the resolution of these issues.

- The EMPACT/TEXAS proposal, with Mike Marx, professor at New York State University at Stony Brook, as spokesperson, was not recommended to proceed to a technical proposal.
- The deadline for formal proposals for the two large detectors was set for April 1992.

The L* proposal never materialized [49]. There were a number of crucial issues where significant differences existed between the SSCL and the L* management teams and between Schwitters and Ting. The principal ones had to do with the Laboratory's insistence on greater transparency of what funds (and from where) were available for the construction of the detector, as well as the SSCL pressure for greater involvement of US groups in the L* management [50]. This latter requirement did not sit well with some of the non-US groups. This unhappiness was voiced in letters to Schwitters from V. G. Shevchenko, Vice-Director of ITEP in Moscow (February 18, 1991), and K. Lubelsmeyer, Professor and Institute Director at Aachen, Germany (March 4, 1991). In response to the requirement for strengthened US management, Barry Barish, professor at Caltech, was asked to join the experiment as a coleader with Ting. A tentative management plan was presented to Ting in a letter from Barish on March 5. Ting would be the spokesman and Barish the chair of the Management Board and US Group Leader. Six out of seven seats on the Management Board would be held by the Europeans. This scheme was accepted by the L* Collaboration at its meeting the following day. At its March 10–12 meeting, the SSCL PAC still expressed some reservations about L* but recommended [51] that "L* be supported to proceed toward the development of a technical stage report, subject to the prompt development of scientific and technical leadership that is acceptable to the SSCL Director." Before any serious attempt could be made to resolve the outstanding issues, both Hans Hofer, professor at ETH in Zurich, Switzerland (on March 13), and Lubelsmeyer (on March 28) announced the withdrawal of their institutes from the Collaboration. After his fruitless trip to CERN for a meeting with the L* principals to try to resolve the impasse, Schwitters announced [52] on May 3, 1991 that L* would not be supported.

Later that May the Laboratory invited the HEP community to propose a second detector to do high-P_T physics and complement the SDC [53]. Shortly afterward, Barish joined with Bill Willis, professor at Columbia, to form a new collaboration [54] with the same goals as L*, called Gammas, Electrons, Muons (GEM). Initially the Collaboration was composed mainly of the former members of L* and EMPACT/TEXAS. Both the SDC and GEM proceeded and eventually obtained a more formal go-ahead from the SSCL: the SDC Technical Design Report (TDR) was approved [55] by the Director on July 31, 1992; GEM, somewhat further behind, submitted its TDR [56] on April 30, 1993, which was to be reviewed by the PAC later that summer (a meeting that was later canceled in light of SSCL funding uncertainties). Based on these proposals, the Laboratory initiated work on the design of interaction regions, both the underground halls and the surface facilities. Some decisions on smaller experiments were anticipated for the 1993/94 time frame.

Both large detectors relied in their financial planning on significant contributions from abroad. Indeed, there were a large number of physicists from abroad (over a hundred from Japan in the SDC) who were collaborators on these experiments but there was no corresponding commitment of funds. Because it was more mature, the SDC's plans were more developed: of its US$589 million estimated cost, there was the expectation that US$231 million would come from abroad [57]. Thus the scope of the eventual SSC physics program was very much dependent on success in negotiations for foreign contributions [58].

The very successful generic detector R&D program started at the CDG was discontinued after FY90 (US$17 million were allocated to it in FY90). In its place, the SSC Detector Subsystem R&D Program was initiated, designed to address technical issues relevant to the approved and potential experimental program [59]. This was also proposal-driven; 30 proposals were considered by the Laboratory's R&D committee in October 1990. Twenty-four of them were recommended for support, with a sum total of US$18.4 million [60]. As part of the detector R&D effort, the SSCL organized [61] the Symposium on Detector Research and Development for the SSC at Fort Worth, Texas, held on October 15–18, 1990. About 500 scientists participated.

The SSCL also started a physics research program, in addition to work on building the collider. Initially the work was centered on detector R&D for the SSCL detectors, both the SDC and GEM. As part of the GEM R&D program, that collaboration had constructed at the Laboratory an instrumented cosmic ray muon spectrometer [62] (Texas Test Rig — TTR) in which one could test and compare various muon detector prototypes. There was also some involvement in the CDF experiment at Fermilab. Longer range plans involved submitting a proposal for support of a broader research program. In its explanatory document of August 1992, the DOE explicitly stated that the DOE HEP program would support efforts of SSCL physicists in their physics research activities on a competitive basis with other proposals submitted to the US HEP program [63].

The SSCL hosted the XXVI International Conference on High Energy Physics (ICHEP) in Dallas in July 1992. On the positive side, this conference helped to stimulate international interest in the SSC and many visitors from abroad were able to see for themselves the progress being made in building the Laboratory. There were 1286 delegates, from 47 countries. On the other hand, that time frame overlapped with the period of Congressional deliberations on the SSC's fate (see below) and thus undoubtedly reinforced the doubts in some people's minds whether the SSC would come to fruition.

I should mention one other important activity in the physics area. Some US$100 million of the US$1 billion pledged to the SSC project by the state of Texas via the Texas National Research Laboratory Commission (TNLRC) was devoted to supporting SSC-related research activities by young people, at the SSCL and at other institutions. The awards were given on a competitive basis — based on proposals submitted to a committee formed for that purpose by the TNRLC. It was an excellent program which filled a real need in the time of funding shortfalls and helped to nurture a pro-SSC climate in the whole country. The first 20 of these awards were made in 1990/91, with a total value of US$11 million; several of the awardees subsequently went on to distinguished careers in particle physics, The detectors at the SSCL would confront many technical challenges due to the high rates, high energies and high radiation environment. The TNLRC program provided significant help in funding the R&D required to address these challenges.

6. Technical Progress

Something that is frequently ignored and forgotten when the SSC's history is discussed is the rather impressive progress in building the collider and the Laboratory. Regardless of any possible management shortcomings and rough SSCL–DOE interactions, significant progress was being made [64]. This great progress in the early stages of the SSCL life was made possible by many collaborative efforts from other national laboratories (both in the US and abroad), which assumed major or total responsibility for many subsystems and R&D activities. Limited space does not allow me to do justice to the many accomplishments at the Laboratory during its short lifetime, but a brief enumeration of them is needed to at least give a flavor of that aspect of the SSC's history. My choice of items to mention here is undoubtedly quite subjective and for a much more complete story the reader is referred to a very comprehensive description of the SSCL achievements in the document "A Retrospective Summary," prepared within the framework of the SSCL closeout activities [65]. This 400-plus-page document, written by the technical leaders of the SSCL construction project, gives a very detailed account of the technical accomplishments during the SSCL's existence in Texas.

There was significant progress in the conventional construction area. Modifications were made in the details of the collider tunnel placement and the location of the experimental halls, to better conform to the site. Tradeoff studies of the number, size and configuration of the collider shafts resulted in an improved design and significant cost savings. There were claims by the DOE IG office that the SSC architectural design, engineering and construction management contractor, the Parsons Brincker-hoff/Morrison Knudsen joint venture (PB/MK), had significant cost overruns [66, 67] on early surface work. These claims were disputed by the SSCL, on the grounds that they were based on misinterpreted data. The underground work on the collider tunnel was proceeding very well and most of the contracts were bid at or below the estimated price. At the time of the termination, four tunnel boring machines (TBMs) were in operation and 14.7 miles (27%) of the total tunnel was excavated. Several new tunneling records were established: 420 feet in a day, 1294 feet in a week and 4269 feet in a month [68].

The 110,000-square-foot Magnet Development Laboratory, to be used for production of special purpose magnets, was completed in the fall of 1991. The somewhat smaller Magnet Test Laboratory, for testing industrially produced magnets, was completed a year later. At the time of termination, it was nearly ready for operation. The contract for the linac tunnel and linac buildings was awarded in May 1992. The tunnel work was completed and turned over to the Laboratory on June 21, 1993, and there was good progress on the buildings. Most of the components had also been ordered; the first three, H-ion source, low energy beam transport (LEBT) and the radio frequency quadrupole (RFQ), which accelerated the beam to 2.5 MeV, were installed and operating by April 1993; the first full energy 600 MeV linac beam was scheduled for April 1995.

An important achievement of the SSCL was significant progress on superconducting magnets, both dipoles and quadrupoles. Initial work was mainly at the national labs where the expertise resided. As facilities at the SSCL became available, more of these activities were moved there. In parallel, there were industrialization efforts, starting with the participation of industry people in the work at BNL, LBNL, Fermilab and SSCL. The magnet performance was good, both of the older 4-cm-diameter design and of the new 5-cm-diameter design; the first six 5-cm-diameter design short magnets achieved 110–118% of the required operating current; in tests at 1.8 K, dipoles reached 10,000 A, indicating excellent mechanical robustness. A milestone was reached on August 14, 1992, six weeks ahead of schedule [69], with the completion of the Accelerator Systems String Test (ASST). It involved operating a collider half-cell consisting of five dipoles, one quadrupole and two spool pieces at the design current of 6500 A. The magnets in the test were assembled at Fermilab by General Dynamics personnel. Subsequently, a full cell, with eight General Dynamics–produced dipoles and two Westinghouse dipoles, was assembled and cooled down. It was ready for electric and magnetic tests when the SSC was terminated. It was planned to have the ASST serve in the future as a test bed for further string tests, accelerator systems,

development of operations procedures, and personnel training.

Based on the responses to the RFP [70], General Dynamics and Westinghouse were selected as the leader and follower contractors for the dipole magnets [71]. Contracts were signed with both General Dynamics and Westinghouse for US$166 million and US$101 million, respectively, for the initial complement of some 500 prototype, preproduction and low rate production magnets [72]. The adopted leader/follower arrangement was a departure from the standard HEP procedure with multiple vendors and was somewhat controversial. The leader contractor had the responsibility for the design of the magnets and the associated tooling. Both companies would produce a number of magnets, at which point a decision would be made about allocation of the rest of the order. At the time of termination, the first dipoles assembled at the new GD plant at Hammond, LA, were on the verge of completion. Babcock and Wilcox won the bid for the US$62 million contract for the final design and initial production of the quadrupole magnets for the collider [73].

There was also work on the magnets for the HEB, contracted to Westinghouse, and on magnets for the warm machines. Some of the contracts for their production were given to the labs or industries abroad, frequently resulting in significant savings. Thus, for example, the Budker Institute for Nuclear Physics (BINP) in Novosibirsk assumed responsibility for the design, engineering and fabrication of the LEB magnets. The Moscow Radiotechnical Institute was given the contract for quadrupoles in MEB [74].

An important issue was the quality of the collider ring vacuum, specifically the impact of the desorption induced by the synchrotron radiation photons striking the beam tube. This potential problem was known during the CDG days and several experiments were performed to get a better understanding of the phenomenon. The issue, however, turned out to be more complex than initially anticipated. Because this was a critical path item affecting magnet design a two-prong approach was taken. A number of experiments were initiated to understand better this effect and its possible remediation, and simultaneously an effort was started to design an optimum liner for the magnet bore tube.

The analytical and numerical accelerator physics studies continued with a focus on optimizing the collider lattice. The SSCL management decided eventually to increase the quadrupole aperture from a 4 cm to a 5 cm coil winding diameter. This simplified the lattice, and several beam components included in the 1989 design could be eliminated. A very important development came as a result of more extensive numerical tracking calculations. It was learned that the dynamic aperture available did not decrease as quickly with the lower injection energy as one thought in 1989 and thus there were plans to propose reducing the HEB energy to 1.5 TeV from the nominal 2 TeV [75].

7. International Collaboration Efforts

In the 1989–1993 time frame there continued to be major efforts to enlist other countries in SSC participation. But it was not the coherent effort that it needed to be, with all the major players — Department of Energy, State Department, both Houses of Congress and the US HEP community — following a well-defined strategy. From the beginning it was clear that it was not going to be an easy task, especially with Western Europe. Carlo Rubbia became CERN DG in 1989 and he harbored strong hopes of an early and quick LHC start even though LEP construction was not yet finished. In a letter [76] to Francis Low, current HEPAP chair, Rubbia wrote in May 1989: "...it seems a logical step to consider the construction of the LHC, *on the fastest possible timescale*, as a very effective precursor of the SSC." Even for some time afterward CERN management continued to project highly optimistic (and unrealistic) cost and time scale projections for the LHC without any specific design to back them up [77]. As late as 1993 the stated plan was still to locate the new superconducting magnets above the LEP magnets and to preserve the electron–proton (e–p) collider option.

There was a meeting on September 7, 1990, between the senior Laboratory staff, a few members of the SPC (including three from abroad), and DOE and State Department representatives, to lay out some general guidelines and discuss possibilities. I attended that meeting as HEPAP chair. It was recognized that because of the LHC, commitments from Europe were unlikely at that time, and that foreign participation in construction most likely would take the form of "in kind" contributions. It was agreed

that ideally contribution to construction should not affect ability to participate in the scientific program. Mutual visits of principals were strongly encouraged. The issue of coupling participation to contributions was not trivial. The stated policy that there should be no connection between possible exploitation participation and construction contributions was the norm in the worldwide HEP, but costs were becoming so large that people were beginning to question it. In a second (July 31, 1989) letter [78] to Low, Rubbia wrote: "I doubt that the governments would be willing to pay large sums for the accelerator if they knew that the proposals from their scientists will be accepted or rejected with no correlation to whether or not they have contributed to the funding of the machine."

The initial efforts in the international area involved laboratory-to-laboratory agreements on collaboration to construct SSC accelerator components, and negotiations within the detector collaborations on how the construction costs of the detector were to be apportioned. Already early in his tenure as Director, Schwitters made several trips abroad to discuss possible collaborations [79]. Several protocols were signed, but no SSC funds (except for US$50 million pledged by India) materialized at that time as a result. Thus, for example, there was an agreement signed by Schwitters with Rubbia regarding interactions between CERN and the SSCL [80] and a similar agreement with Erich Vogt [81], Director of TRIUMF, and with several Russian institutes. Any serious discussion with CERN about collaboration was made difficult by Rubbia's very optimistic view of the LHC time scale. In the second letter to Low he mentioned 1996/7 as the tentative completion date for the LHC and 1999 as the date for the e–p option.

There were also initial efforts to conclude more formal government-to-government agreements. In June 1990 Deputy Secretary Hensen Moore visited Japan and Korea, where he delivered letters from President Bush to Prime Minister Kaifu (Japan) and President Roh (Korea), inviting them to participate in the SSC. There was positive response from the Koreans, and the Japanese said that it would be at least a year before they could decide. A relatively slow response to any new initiative was consistent with standard practice in Japan, where an effort was always made to first reach a consensus. In September 1991, Secretary Watkins and his Korean counterpart agreed to form a joint working group to discuss possible collaboration on the SSC. A similar working group was established with the Russians in July 1991, during Moore's visit to Moscow. It was followed on January 6, 1993, with an agreement (signed by Watkins) between the DOE and Russia's Ministry of Atomic Energy for a program of collaboration on the SSC.

Japan had always been viewed as the most likely significant contributor to the SSC, and thus the principal effort in enlisting foreign contributions was directed toward the Japanese. The Japanese high energy physicists expended significant effort toward participating in the SDC, making up about 20% of the collaboration. There was a segment of that community interested in building a domestic linear collider but most people realized that the time scale for it was significantly later than for the SSC [82]. Subsequent to Moore's trip already mentioned, there was a visit by H. Sugawara, DG of KEK, to the SSC late in 1990. He was not optimistic about Japanese contribution and emphasized that negotiations must be done at a very high level and involve the Finance Ministry [83].

A major effort toward Japan was made in 1991 with a series of trips, with higher level contact on each succeeding visit. It was generally accepted that for a large contribution at the level of US$1 billion, the request has to be made by the President himself to the Prime Minister. Thus the goal of the initial trips was to lay the groundwork for an eventual meeting between President Bush and the new Japanese Prime Minister, Miyazawa. The first visit took place in early October 1991, and was led by Will Happer, Director of Energy Research in the DOE at that time. I participated in that trip, together with Roy Schwitters, Steve Weinberg and Jerry Friedman. We met with middle/high level officials and also had a number of interactions with the Japanese HEP community, the general public and the press. We were given a good reception but the main point that was made to us was that the top priority of the Japanese was to refurbish the infrastructure in their universities. Thus, just like in the US, any SSC support had to be "on top" of funding for their basic research and education needs and therefore probably had to be approved at the highest levels [84]. At one of those meetings Will Happer urged the Japanese to press for such an ideal course of events, namely strengthening

the university support and also becoming a partner in the SSC.

This first visit was followed up with a trip by D. Allan Bromley, President Bush's science advisor and Director of OSTP, on October 14–18 and by Watkins on December 3–5. Bromley offered the Japanese "true partnership" in the SSC and invited them to participate in the planning, construction and management of the SSC in return for investment of cash and provision of Japanese-built components. In his memoirs [85] he claims that he received enthusiastic support from the Science Council of Japan for his suggestion of a significant increase in the Japanese science budget, part of it to be allocated for upgrading the Japanese universities and part for participation in the SSC at the level of about US$1.5 billion. Subsequently, on his visit, Watkins met with Prime Minister Miyazawa and also made a strong pitch for Japanese participation in the SSC.

President Bush made a trip to Japan on January 7–10, 1992, and it was widely expected that a formal agreement would be reached with the new PM regarding the Japanese contribution to the SSC. However, the topic of the Japanese support for the SSC apparently never came up in the direct Bush–Miyazawa talks. Bromley claims that it was the White House Chief of Staff, Samuel Skinner, who was responsible for this agenda omission by changing its emphasis to the trade issues. The fact that Bush became sick and threw up during the formal state dinner was undoubtedly a damper on the negotiations. The only substantive outcome of the meetings that was relevant to the SSC was the establishment of a US–Japan working group on the SSCL. According to Japanese press reports, one of the goals of the working group would be to decide by the end of 1992 whether and how Japan should participate in the SSC [86, 87]. This group was led on the US side by Will Happer and Jim Decker, Deputy Director of the DOE OER, and it met several times during the Bush term, starting with a meeting on April 9–10, 1992. The initial discussions focused on resolving significant differences in estimated construction and component costs and establishing common costing methodology; some progress in that area was apparently achieved.

These studies were put on the back burner while awaiting news regarding the Democratic presidential nominee Bill Clinton's position on the SSC.

The negative House Congressional vote of 1992 (discussed below) undoubtedly reinforced this "go slow" attitude. After winning the November 1982 election Clinton supported the SSC with a US$640 million budgetary request for FY94 but also planned a three-year delay in its completion. He met with Miyazawa in Washington on April 16, 1993, but the SSC was not discussed at that meeting. The fact that the US was engaged at that time in rather difficult trade talks with the Japanese, which were high on Clinton's priority list, probably had a role in eliminating the SSC from the agenda. Clinton certainly had no comparable interest in the SSC that Bush did and undoubtedly did not want to use up his credit with the Japanese on a controversial project that was not among his priorities. The delays in reaching an agreement worked against the possibility of eventual success of the project; as time went on, more and more contracts were being awarded and hence there was less and less of an incentive for the Japanese to get involved in the construction.

The situation was made more complicated by the fact that there were influential voices against seeking foreign contributions in kind, which were the only ones that might materialize [88]. The Senate, led by Bennett Johnston from Louisiana, argued for a purely national effort. Initially there was also a sentiment against such contributions in the House [89], but with time, cost-cutting sentiments prevailed there. The GAO pointed out that "any such subcontracts with foreign suppliers will further reduce the multiplier effect of the investment on the US economy." A number of industry leaders expressed concern about the fact that foreign companies were allowed to compete with the US ones, resulting in sending many jobs overseas. Thus there was ambivalence about signing production contracts with foreign laboratories with significantly lower price, which could be viewed as an in-kind contribution to the SSC construction [90].

Had the SSCL survived, some international contributions would have most likely been forthcoming even though nowhere near the one-third of the total cost initially claimed by the DOE. Going over press reports on these negotiations during 1990–1993, I am struck by the naiveté on the American part about the workings of the Japanese decision-making process and the lack of understanding of how to read the Japanese "signals." Negotiations were

much more complex, difficult, and time-consuming than anticipated. They required strong coherent high level effort from the Administration, Congress, the SSCL and the HEP community, an effort which at best was sporadic. I have no doubt that the Congressional vacillations regarding the SSC funding played an important role in discouraging foreign collaborations. Barry Barish, GEM co-leader, said that a lot of their foreign collaborators together with their contributions "evaporated" after the House action of 1992 (discussed below). For example, Taiwan withdrew their offer of a US$50 million contribution to the GEM detector. It was a vicious circle, a true catch-22 situation, well summarized by Rep. Brown [91]: "Major foreign participation has remained elusive because of uncertainty about the US commitment to the project, yet our own commitment has wavered in large part because of the absence of foreign funding."

8. Change in the Political and Public Opinion Landscape

By 1990 there was a significant deterioration of the public sentiment toward the SSC. There were many factors that contributed to this change, and they tended to reinforce one other. Certainly, the decision to site the SSCL in Texas cooled the ardor of other site-candidate states. But the most important factor was probably the growing budget deficit. The concerns about the deficit were beginning to be present already when Reagan announced his support for the SSC. On the political front they reflected themselves in the Gramm–Hollings–Rudman Act, officially the Balanced Budget and Emergency Deficit Control Act, which was enacted on December 12, 1985. The law was modified in 1987 so as to comply with the court ruling which found some of its provisions unconstitutional. The act provided for automatic spending cuts if the proposed budgets failed to reach established targets for reducing the deficit [92].

This act failed to curb the deficits and as a result was replaced by a Budget Enforcement Act in 1990. The first provision of the act was the establishment of a limit on the level of discretionary spending, divided into defense, international and domestic sectors. The second one established the pay-as-you-go procedure requiring that increases in direct spending or reductions in revenues due to adopted legislation in a given sector be offset by other legislative actions so that there is no net increase in the deficit. As far as the SSC was concerned, the implication of the law was that any increase in spending would have to be either offset by cuts in other discretionary spending or funded by new revenue. This clearly provided additional fuel to many who opposed the SSC on the grounds that it would eat into financial support for other sciences.

The first significant impact of this new law occurred in 1991, when the resulting decrease in the DOE OER budget forced its Director, Will Happer, to convene a special Panel chaired by Charles Townes, Nobel Prize winner from UC Berkeley, to establish priorities across all the fields in the OER's portfolio [93]. Projected budget levels were such that drastic cuts were necessary; in HEP, the recommendations of the Panel were to delay the start of both the Main Injector at Fermilab and the B-Factory at SLAC. The SSC was placed outside the Panel purview as a presidential initiative, which caused understandable resentment among many physicists. This decision was probably a strategic mistake; for the SSC to be viable in this climate it had to be judged on its scientific merits *vis-a-vis* other potential science projects and come out significantly better [94]. One of the outcomes of the Panel's recommendations was the convening of another HEPAP Subpanel, chaired by Michael Witherell, a professor at the University of California in Santa Barbara. In view of the very pessimistic budgetary guidelines given to the Subpanel, it was not surprising that the resulting recommendations advocated rather severe curtailment of the HEP program, even though it did urge proceeding with the SSC on a fast track schedule [95, 96].

The concern about funding was especially prevalent among the non-HEP scientists, even those who supported the SSC on intellectual grounds. It was clearly stated already in Congressional testimony in April 1987 by many prominent scientists: Sheila Widnall, president of the American Association for Advancement of Science, wrote in her testimony [97]: "We believe that the nation should undertake the construction and operations of SSC, but that it must not be done at the expense of investment in other areas of basic research." Dan Kleppner, an atomic physicist from MIT and generally quite sympathetic to HEP goals, testified [98] at the same hearing: "I believe that the nation can and should build the

SSC. However, if we are to have a future in science, it is essential to provide our universities with the resources to maintain their roles as world leaders in education and research." Senator Bennett Johnston, one of the strongest supporters of the SSC in the Senate, was quoted as saying: "I don't want to cannibalize everything in science for this one project." The Administration tried to alleviate some of these concerns by proposing rather robust budgets for civilian science R&D [99]. For example, the proposed increase for the NSF basic research funds for FY93 was 21%.

Given this emphasis on fiscal austerity and the concerns about deficits, aggravated by the first indications of a recession, the escalating cost of the SSC became a major issue. Another contributing factor was the lack of any significant commitment from abroad to contribute to the SSCL construction and thus reduce US expenditures. Together, these two factors drove the required federal allocations to such high levels that they were deemed by many to be unacceptable.

The situation was aggravated by a previously mentioned, rather critical 1991 GAO report on the SSCL and the DOE, prepared at the request of Sen. Jim Sasser (D-Tenn), chair of the Senate Budget Committee [100]. The report noted that the DOE Office of the SSC "lacked stable leadership since it was created in January 1989" and identified a number of technical areas where they felt that there was a considerable performance and cost risk. The DOE tried to vigorously rebut the report point by point in a letter [101] by Secretary Watkins to Jamie Whitten, Chairman of the House Committee on Appropriations, but undoubtedly the report did influence the opinion of many in Congress and hence subsequent deliberations there.

The conclusion of the site competition significantly decreased the ranks of SSC proponents — one notable example was Rep. Boehlert from New York State. A way to describe this shift in sentiment could be: "What might have been a potential US$4.4 billion exciting science project in our backyard turned out to be a US$8.3 billion pork barrel boondoggle in Texas." The SSC contractor, the URA, was ill-suited to lead a strong campaign for the SSC, being composed of some 80 universities, none of which had a strong stake in the SSC. Many of them had strong local opposition against it within its own non-HEP faculty.

The SSC was also hurt by being lumped in public discussion with the International Space Station [102]. Articles and editorials were written arguing that we cannot afford two huge science projects, even though there never was a good case made that the Space Station was motivated on scientific grounds. The Space Station also had numerous opponents in Congress and one year its appropriation survived in the House by only one vote. However, in a direct competition, the SSC had to lose out for a number of reasons. It was much easier for a layman to understand the Space Station, the stars being visible, unlike the inner structure of a nucleus. Paradoxically, its much higher cost had some advantages, as it provided opportunity for large and widely dispersed industrial involvement in it and hence a larger army of potential lobbyists arguing its case. NASA had a strong tradition of involving industrial concerns in the development and construction of its projects and learned how to use this connection. It certainly was much more experienced and more successful in lobbying Congress than the DOE Office of Research. Finally, not being associated with any single state, it was harder to characterize the Space Station as a local pork barrel project, as opposed to the SSC, for which most of the new jobs were going to be in either Texas or Louisiana.

False information about the SSC played a major role in turning the opinion of many against it. I have already mentioned the unrealistic hopes about the high temperature superconductors (in the first article [1]) and the lure of the LHC at CERN as a cheap early alternative. A new potential option that was gaining more prominence was a possible e^+e^- linear collider [103] in the 1 TeV energy range. There was vigorous worldwide R&D in this area, especially in the US and Japan, and SLAC was beginning to observe collisions with its prototype SLAC Linear Collider (SLC). However, it was apparent to everyone involved in this work that construction of a high luminosity TeV collider was still more than a decade away [104]. Nevertheless, these three potential alternatives were frequently cited as arguments for delaying or canceling the SSC.

The general change in the mood of the country is well illustrated by an editorial [105] in *The New York Times* in 1990 titled "The Behemoth and the Boson." This presents a remarkable contrast to the one published less than seven years earlier [106]. A few quotes illustrate this new, somewhat naive and misinformed anti-SSC point of view: "By buying into the Europeans' Large Hadron Collider, which would cost only US$1 billion to build, American physicists would be assured of a ringside seat... its magnets might use the newly discovered materials that become superconducting at high temperatures... why not invest in linear colliders instead of a machine at the dead end of evolution?... nothing of scientific interest may lie in the energy range the supercollider opens up.... These risks would be worth running in better times, but not in the straitened circumstances that now prevail."

Even the once-unequivocal support from the HEP community for the SSC began to weaken. The 1983 panel, after extensive discussion, decided that an intermediate time scale machine was not essential for the health of the US HEP. That decision was based on an estimated time scale for the SSCL of about 12 years from then. In 1990, the projected SSC completion time was September 1999, 4 years later than the original estimate, and even that projection was based on unrealistically high funding levels during the peak years. Thus many in the community began turning their attention to other areas. There were great pressures not to sacrifice any elements of the existing program (which might be required by the newly enacted budgetary caps) and even to expand it by construction of the Main Injector at Fermilab and the B-Factory at SLAC [107]. Such attitudes reinforced among other scientists the image of greedy HEP physicists. In addition, the uncertainty in the completion date and/or doubt whether it would ever be completed made recruitment of staff to the SSCL quite difficult.

Important changes also occurred in the national political landscape shortly after the establishment of the SSCL in Texas. The first one was the resignation of House Speaker James Wright on June 6, 1989, induced by charges of ethics violations. Wright represented the 12th Texas Congressional District, which included Fort Worth, not very far by the Texas standards from the SSCL location. He had served as Speaker since 1987, and previously as majority leader since 1976. He was replaced as Speaker by Tom Foley from the State of Washington (which also submitted a bid for the SSC). Wright was a strong supporter of the SSC and his resignation from the House created a climate that was much more favorable to anti-SSC sentiments. In the early 1990's the activities of the Texas delegation probably had a negative impact on the SSC. While they argued for the SSC funding, a majority of them also pushed for budgetary cuts in social services and supported the constitutional amendment requiring a balanced budget.

The other important change arose from the results of the November 1992 elections. Bush lost his bid for the second term to Bill Clinton, and thus the SSCL lost another powerful advocate. Bush was strongly committed to the SSC and clearly wanted to carry on the Reagan-initiated project; Texas was his home state; he appears to have been genuinely intrigued and interested by the scientific issues the SSC was going to address. Soon after the SSC concept was initiated, during his Vice-Presidency, he was given a briefing on the SSC by the TAC people. Before the crucial SSC votes in Congress in 1992, Bush actively lobbied Senators and House members who were on the fence regarding the SSC. He visited the Laboratory in July 20, 1992. Bush's science advisor, D. Allan Bromley, was a nuclear physicist who actively worked to make the SSC a reality.

In contrast, Clinton was probably initially neutral on the SSC and was very much focused on balancing the budget. Already in the initial FY94 budget proposal the SSC completion date was delayed by three years [108]. Vice-President Gore had voted against the SSC as a Senator in 1991. The new Energy Secretary, Hazel O'Leary, stated that she was not "passionate" about the SSC [109]. The new Director of Office of Management and Budget, Leon Panetta, voted against the SSC when he was a Congressman from California and argued initially for canceling the SSC. Clinton chose Lloyd Bentsen, Senator from Texas, as his new Treasury Secretary, thus removing from the Senate an ardent and highly influential supporter of the SSC. Finally, 114 new members were elected to the House in 1992, many of them on a budget-balancing platform. Thus their natural instinct might be to vote against the SSC. Approval of a major project in Congress requires strong commitment from one or more key people in the Government who are willing to do horse-trading and arm-twisting to get the required votes for it. There was no one in the new Administration or new Congress who would be doing that for the SSC.

The final relevant event here was the end of the Cold War. With the fall of the Berlin Wall in 1989,

there also went away the argument dating back to the Manhattan Project that physicists might be useful in case of a national emergency. A corollary of this change was that now support for science would have to be justified in terms of clear immediate payoffs. In addition, nationalistic arguments, professing that being on the technological and scientific frontier is essential for a superpower, which were so important for getting the SSC initiated in the Reagan era, were no longer persuasive to most people, especially for pure science projects without any obvious immediate benefits.

9. Cost History Analysis

Much has been said about the cost escalation of the SSC and it was the major issue in the Congressional SSC debates. I summarize here the relevant history [110] but caution the reader that the numbers do not tell the whole story and could be misleading. The cost numbers quoted at different times have been heavily influenced by the methodology used for the different estimates, as well as by the assumed length of the construction period and the construction start date.

Very rough cost estimates were made during the Snowmass '82 meeting, but the first serious cost estimate was made at the Cornell workshop in 1983, with a range of US\$2.7–3.05 billion in 1983 dollars for the collider only. The first "official" cost estimate for the collider was US\$3.01 billion in the CDR. At that time it was becoming standard, at the urgings of the GAO [111], to quote the so-called total project cost (TPC), which included, besides accelerator construction, supporting R&D, detectors and computers, and operating costs incurred prior to project completion. Using this criterion, the SSC cost range, as estimated by the CDG, was US\$3935–4247 million in *1986 dollars*. The CDG estimated that the construction time was $6\frac{1}{2}$ years. Based on their own and ICE reviews, the DOE increased the total cost somewhat and extended the schedule by two years to 1996. Using the new schedule and minor cost changes, the DOE developed the first TPC estimate in *then year dollars* as US\$5.32 billion. It was based on the assumption that the construction would start in 1988 and follow the stipulated funding profile. This was the budget submitted to Congress soon after Reagan's endorsement of the SSC. For the sake of comparison with future estimates, that is the number which should be used as the initial official cost estimate.

Congress did not appropriate any construction funds until FY90; this made it impossible to maintain the FY96 completion date. The date was moved by two years to FY98. This delay and stretch-out of the schedule resulted in a cost increase of about 10%; the TPC figure submitted to Congress as part of the FY90 budget request was US$5.89 billion [112].

The next DOE SSCL cost estimate of US$8.25 billion was generated in 1990 after the site specific CDR and the associated cost estimate were prepared by the SSCL and their design and cost estimates extensively reviewed by several committees. It took into account significant design modifications and also a one-year delay in the SSC completion till FY99. Very roughly, of the total cost increase of about 55% over the base design (8.25 vs. 5.32), a little over one-third, could be attributed to inflation and schedule stretch-out, and the rest to changes in design. The state of Texas agreed to contribute US$1 billion to the construction of the SSC [113], so the federal cost would be US$7.25 billion minus any foreign contributions.

In Fig. 3 I reproduce the cost history as given in Elioff's summary [110], where he reduces all the numbers to FY93 dollars. This gives a fairer measure of the actual cost increase due to technical changes and/or extension of the schedule by ignoring the part of the increase due to inflation. The construction cost of the accelerator itself, the so-called total estimated cost (TEC), is indicated as the solid line. The TPC, which in addition takes into account associated R&D, equipment, detectors and preoperations, is shown as the dashed line. At the beginning of the SSC's history it was standard to use just the accelerator construction cost (TEC); at the end the standard was the total cost (TPC). As the figure shows, the only significant cost increase due to technical modifications was in 1990 and was associated with the previously discussed site-specific redesign. The figure does not consider potential subsequent cost increases (discussed below) which might have occurred had the SSC been allowed to continue.

There was no subsequent formal SSCL cost estimate. In the first three years of construction (FY91–93), the appropriated funds were US$233 miillion less than in the projected profile. That did not bode well for completion of the project within the baseline cost.

The new Clinton administration was under great pressure to reduce the expenditures across the board, including the SSC. To reduce annual expenditures, they proposed a stretch-out of the SSCL schedule by three years [114], with completion in FY2002, and requested the Laboratory to provide another baseline estimate taking into account the new funding profile. The DOE estimate [115] of this change was that it would result in a US$2 billion increase in the as-spent project cost, this estimate being accurate to 20%. In parallel, the DOE appointed a 75-member committee to review the status of the SSC (DOE Review Committee on the Baseline Validation of the SSC) and provide a written report in August 1993. Their charge was to compare their findings with the projections in the SSCL monthly report from May 1993 — the Cost Performance Report (CPR). The committee spent over two weeks at the SSCL looking at the SSCL records and interviewing SSCL officials.

In general, the committee found that most of the technical systems were 4–12 months behind schedule and identified a number of critical path items that had to be delayed because of funding shortages. They flagged the collider magnet construction as a major item with potentially significantly higher eventual cost than was then stated. The only possible identified technical unknown was the issue of synchrotron radiation in the beam pipe which was under study at that time. They estimated a potential cost increase of up to a total of US$9.94 billion if changes in management organization and some new procedures were not implemented. In addition, they identified US$1.21 billion in costs that were related to the SSC but not included in the TPC (spares, preoperating costs, foreign contribution to the detectors). They also suggested adding US$219 million to the contingency.

The report of this Review Committee formed the basis of the claim that the SSCL costs went up to over US$11 billion. But it must be noted that the Review Committee did not do an independent cost re-evaluation but merely compared their cost estimates of "to go" items (i.e. parts of the project yet to be completed) with the May 1993 SSC Cost Performance Review (essentially the DOE monthly review of the SSCL). No optimization of schedules, trade-offs, procedures, staffing levels, construction models or design modifications were part of this review. The initial SSCL study of possible ways to mitigate

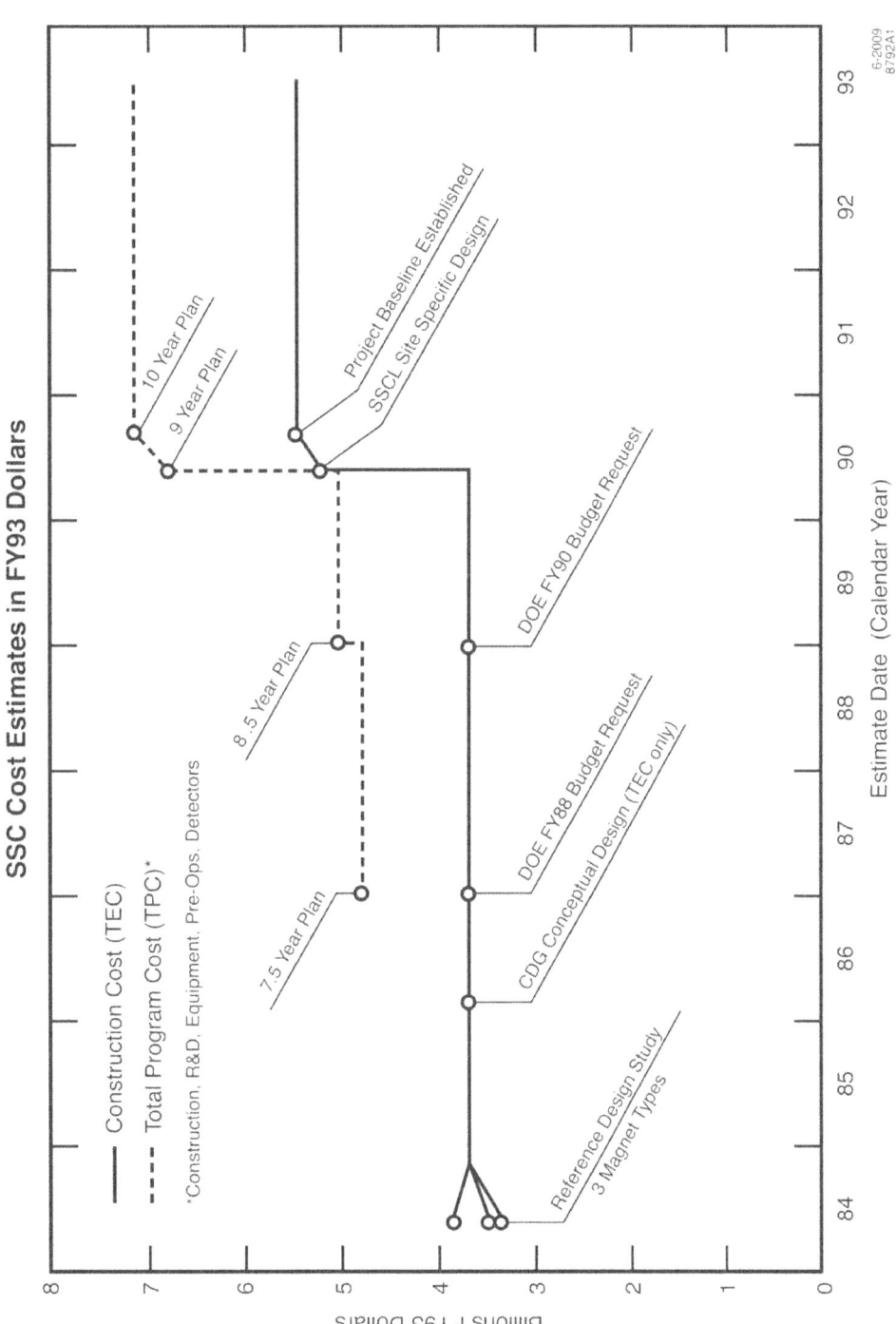

Fig. 3. Estimated SSC costs in FY93 dollars shown as a function of time. Both TEC (accelerator construction cost) and TPC (total project cost) are shown.

potential cost risks indicated that significant reduction of these risks could be achieved [116]. This value engineering exercise indicated that the US$1.354 billion total cost risk estimated by the Review Committee could be reduced by US$852 million. These studies were never completed or reviewed because of the SSC cancelation in October 1993.

What the cost of the SSCL would have been will never be known, since a full-blown rebaseline effort was never completed and the Laboratory was not allowed to continue. Quite likely, the biggest unknown would still have been the funding profile that the Administration and Congress would decide on [117]. SSC opponents postulated even higher costs (Sen. Bumpers put forward US$39 billion), based on including interest on the extra debt incurred by the federal government by deficit funding of the SSC and/or including operations for 25 years, with interest on the cost of those operations. The cost issue was murky and easily amenable to different interpretations and emotional excesses.

Besides the cost of construction, the issue of SSCL operating cost was also frequently distorted. There were a number of cost estimates for SSC operation and they changed very little with time. The cost calculated by CDG [118] was US$224 million (FY85 dollars) without capital equipment, Accelerator Improvement Projects or General Plant Projects. The latest estimates from the SSCL ranged from US$317 million (FY2001) to US$369 million (FY2005) [119]. These estimates were quite consistent with what one might obtain by applying simple scaling from the operating costs of the existing laboratories. However, some of the SSC opponents inflated them significantly without providing any rationale for it. For example, Jim Krumhansl, in his letter to Secretary Herrington [120] wrote: "An additional operating budget of $1–2 billion a year at least will be required."

One final point needs to be emphasized. A project on a green site is much more vulnerable to schedule stretch-out than one in an ongoing laboratory. On a green site, every dollar spent is counted against the TPC. Expenses that would be part of operating costs in an existing laboratory (most of laboratory administration, library, upkeep of infrastructure, general R&D, etc.) are charged to the project. If a project is stretched out, there is some flexibility within an existing lab to shift costs to other efforts and thus reduce the impact of a stretch-out. No such flexibility exists on a green site.

10. Congressional Roller Coaster

The country's new mood was also reflected in the Congressional attitudes. Whereas the Senate was generally favorable to the SSC, strong vocal opposition arose in the House shortly after selection of the Texas site. The four Members most prominent in this opposition were Sherwood Boehlert (R-NY), James Slattery (D-KS), Dennis Eckart (D-OH) and Howard Wolpe (D-MI). Whereas the last three had no scientific background or significant involvement in science issues, Boehlert had been a supporter of science and basic research during his Congressional career. He supported the SSC initially, but turned against it in 1988, after the New York state sites were no longer in competition (the transborder site was disqualified by the DOE, the Newburgh site was withdrawn early, and the Rochester site was withdrawn upon the BQL announcement). He was also persuaded by Krumhansl about the negative impact of the SSC on support of other sciences. None of the other three Members remained in the House after 1994; Slattery and Wolpe left (in 1995 and 1993, respectively) to run for Governor in their respective states (both unsuccessfully) and having fiscally conservative credentials was clearly important for those races. Eckart chose not to run for the 93/94 term and left political life altogether.

Initially there was a strong majority in the House in favor of the SSC. The first substantive vote took place in June 1989 on the proposed FY90 US$190 million appropriation that included the first construction funds for the SSC. Rep. Eckart's amendment to remove the US$100 million in construction funds failed by a vote of 330 to 93 [121]. A year later, a similar motion to stop SSC construction was also defeated [122] by a very lopsided vote of 309 to 109 even though the proposed SSC allocation rose to US$340 million.

A clear sign of changing mood occurred in 1991; Eckart again proposed an amendment to strike the SSC funds (US$483 million) from the budget, which was again defeated, but by a significantly narrower margin of 251 to 163. A similar amendment in the Senate introduced by Sen. Bumpers from Arkansas lost by a 62–37 margin [123]. These were the first votes after the design change and announcement of an US$8.25 billion price tag for the SSC.

The SSC opposition grew substantially in the House during the following year. 1992 was an election year and in times of deficits showing fiscal austerity was a definite plus. Furthermore, the negative publicity about the SSC (technical alternatives, cost escalation, mismanagement accusations, lack of foreign contributions) began to have an impact.

The President's budget request for FY93 was US$650 million. The House Appropriations Subcommittee on Energy and Water, chaired by Tom Bevill (D-AL), voted to allocate US$484 million to the SSC but the Eckart amendment [124] to strike all but US$33.7 million from that proposal passed the House by a surprisingly large majority of 232 to 181. The House vote occurred just a few days after the vote on a constitutional amendment to require balanced federal budgets. Many Representatives who voted against that amendment wanted to restore their fiscal credentials by voting against the SSC.

This House action galvanized the SSC proponents, many of them abroad, into sending letters to key individuals in the government urging them to try to reverse this vote [125]. The result was positive and the Senate voted to allocate US$550 million to the SSC, helped to a large extent by the personal intervention of both Bush [126] and Bromley. As is the standard procedure, a conference committee was appointed to arrive at a compromise. The House and Senate leaders pick the committee, and only SSC proponents were chosen as House conferees. Thus it was not surprising that the recommendation was to approve the SSC funding, with a figure of US$514.5 million (about US$135 million below the President's request). The House Appropriations Committee, in allocating this lower sum, recognized in the associated language [127] that such action "will adversely impact both final cost and schedule." Subsequently, both Houses approved the bill coming out of the conference.

The SSC was saved for another year, but it was severely wounded, and short of drastic action it would be even more vulnerable the following year. For some time now there had been simmering resentment in Congress against the Appropriators, who tended to use their positions to obtain funding for "pork barrel" projects in their districts. Going completely against the will of the House in the conference

undoubtedly amplified this feeling. Generally members of Congress are reluctant to vote against energy and water bills because of the "goodies" that they bring to their districts or their states; this undoubtedly explains the positive House vote on the final bill. But one could expect that the SSC opponents would try to come up with a more effective strategy for the following year.

11. The Death Blow

As the time approached for the FY94 Congressional appropriations activities, it was clear that the SSC question — to fund or to kill — would come up again [128]. Furthermore, several other important events happened during the intervening year — essentially all of them boding poorly for the SSC's fate. Maybe the most important one was the re-examination by the Clinton Administration, as part of its budget planning process, of the SSC schedule. Its outcome, stretch-out of the construction through 2002 and the resultant rise of the projected total SSC cost to about US$11 billion, fueled the arguments of the SSC opponents that its costs were out of control.

There was intensified criticism of the DOE oversight and SSCL management. A report by the GAO published in February 1993 was highly critical of the DOE, URA and SSCL in those areas. It argued that the SSC was over budget and behind schedule. A major complaint had to do with the absence of standard management tools used to track progress and expenditures, e.g. the Cost and Schedule Control System. The report accused the SSCL staff of impeding the GAO investigation by being less than forthright in providing requested information. In the same time frame the office of the Inspector General (IG) of the DOE performed their own investigation of the SSCL and came up with similar conclusions.

On February 25, *ABC Prime Time Live* devoted a segment to the SSC, which focused entirely on the negative news in these reports. The other investigation ongoing at the Laboratory early in 1993 was by the staff of the House Subcommittee on Oversight and Investigation, chaired by Rep. John Dingell (D-MI), in preparation for subsequent hearings. Looking back on these investigations, it is clear that the focus was on bureaucratic issues rather than on the level of performance. The criticism was that the proper procedures were not followed, that formal control systems were not established, that various actions were not documented adequately. There was very little attention paid to how well the goals were being accomplished.

These investigations impacted the morale in the Laboratory and were a significant distraction for everyday work; things were also aggravated by the ever-increasing antagonism between the DOE SSC officials and the SSCL managers and also among the SSC staff. There were serious efforts by several DOE officials to remove some of the high level URA SSCL managers but such action was vetoed by Watkins. In a leaked memo from his computer, Cipriano was reported to argue to higher level DOE officials that Schwitters had to be replaced to save the SSC [129]. As a result of all of these events a number of SSCL employees started to leave. There was a net decrease in the headcount at the SSC in September of that year. The level of frustration is well illustrated in the following quote from Schwitters [130]: "Every one of us here spends most of his or her time answering questions from Washington or trying to comply with the latest twists in accounting rules dreamed up by people who want to kill the SSC. We should be devoting ourselves to completing this machine as rapidly and cheaply as possible and getting on with real science. Instead our time and energy are being sapped by bureaucrats and politicians. The SSC is becoming a victim of the revenge of the C students."

The IG report was dated May 19, 1993, and its content leaked shortly before the House vote on the appropriation bill for energy and water. That vote was 280 to 150 to terminate the SSC; the 114 House freshmen voted almost 3 to 1 in favor of termination [131].

The Dingell Subcommittee hearing was held on June 30 and it highlighted the issues that both the GAO and the IG found amiss [132]. The tenor of Dingell's introductory remarks was very scathing of the DOE and SSCL management: "SSC ranks among the worst projects that we have seen in terms of plain contract mismanagement and failed government oversight... The inability of the Department and the inability of the prime contractor to take control of this program is a problem that has persisted for far too long, and, like cancer, it has grown." There were certainly some management issues that were a fair target for criticism. But many other complaints

and accusations showed either naivete, pettiness or deliberate intention to cast the SSC in the worst possible light. It was a perfect example of "guilty until proven innocent." Happer's paraphrase [133] of an ancient Greek proverb, "He whom the gods would destroy, they first make to appear foolish," seems especially on the mark here.

The Laboratory was criticized for improper subcontracting without bids for contracts valued at US$216 million; US$156 million of that sum was for work by BNL and Fermilab on prototype magnets, for which no industrial concern had the required expertise. The discussion of budgetary issues ignored the contingency held by the DOE, of which only 3.5% has been spent by the summer of 1993. The Laboratory was accused [134] of "lavish spending of taxpayers' money on luxuries and entertainment"; one of the highlighted items was the US$21,000 spent for annual maintenance of plants used to brighten up the warehouses modified for use as office space. The use of warehouses, rather than conventional office space, was said by the SSCL to save the lab US$500,000 annually [135]. Incidentally, the idea that one would spend some money for esthetic enhancement of buildings is not radical. Several states require that a certain fraction of construction funds for state-funded buildings be spent on artworks. According to the DOE IG, John C. Layton, in his testimony before the Dingell Subcommittee, the DOE requested an audit "to determine the allowability of the $250 million in incurred costs. The audit identified about $233,000 in questioned costs." This represented 93 cents out of each $1000 spent.

Clinton's Secretary of Energy, Hazel O'Leary, testified at that hearing. She did not defend the SSC very strongly in her testimony and agreed with a lot of the criticism, especially on management issues [136]. The general tenor of her remarks was that most of the current difficulties at the SSCL were due to the lax oversight by the previous administration and rather unusual reporting procedures set up by them. She promised to study all the problems and said that she was prepared to revamp the management drastically.

On August 3, 1993, Sen. Pryor from Arkansas urged another investigation of the SSC [137]. In a letter to the DOE IG, he expressed concern that some individuals at the SSCL were engaged in improper and/or illegal lobbying activities, via various letters, meetings or phone calls. The SSC was terminated before this investigation could be initiated.

SSC was confronted with additional problems on September 1, when the DOE released a report from its review committee [138] (DOE Review Committee on the Baseline Validation of the SSC) which outlined the potential cost risks discussed previously. But the report also said many positive things. To quote O'Leary [139]: "The committee confirmed that the project is 20 percent complete, which is where we should be on the project... also found that the 73 major subcontracts awarded to date, in aggregate, have come in at approximately seven percent under budget." The report mentioned the difficulty of obtaining solid cost estimates and reliable schedules due to unsatisfactory records and bookkeeping even though they stated that "SSCL was most responsive to the committee." On the whole the report was rather objective but quite detailed and lengthy; as such it lent itself very well to 30-second sound bites by the SSCL opponents.

The URA and the SSCL made rather detailed written and oral rebuttals [140] about the charges made in the IG and GAO reports and during the Dingell Subcommittee hearing. Rep. Bryant from Texas, a member of the Dingell committee, also tried to refute some of the charges during the hearing. But the charges were given much more attention by the press than the rather complex explanations.

Secretary O'Leary tried to save the SSC by initiating several steps designed to contain the SSC costs. Five planned actions were enumerated in the DOE news release [141] of September 1, 1993, one of them being reduction of the cost through downscoping. The plan was to appoint a task force, to be chaired by the DOE's Wilmot Hess, which would identify by November 30, 1993, items in the current design that could be eliminated. As HEPAP chair, I was asked [142] on September 20, 1993, to appoint a HEPAP Subpanel which would "rank them in order of potential impact on the scientific objectives of the SSC. These project changes will be implemented, in the order established by HEPAP, to the extent that they are necessary to avoid cost overruns of the project." The subsequent events, however, made these plans irrelevant.

The Secretary tried to deflect the management criticisms by a drastic action: replacing the URA in most of its duties by an industrial contractor experienced in construction of large projects. The URA would retain the responsibility for design and operation, but the new contractor would oversee conventional construction, magnet production and installation [143, 144].

During that spring and summer a significant effort was made by the advocates of the SSC to influence Congress to vote in favor of continuation of the SSC. There were numerous letters and visits by physicists and other SSC supporters to their Congressional representatives. The TNLRC made an organized effort to shift the public opinion in favor of the SSC. This was done by sponsoring trips by teams of 3–5 members (typically a TNLRC member, a member of the SSCL, a physicist from the area who was involved in the SSC, and a member of a local industrial company with an SSCL contract) for meetings with editorial boards of local newspapers. There was certainly a need for some activity along those lines; it seemed that a favorite sport of many editorial boards at that time was to write negative opinions about the SSC with cute titles. Some examples from California (where I participated in one of those trips) were: "Science in a Pork Barrel" and later "Atomize this Boondoggle" (*Santa Ana Orange County Register*), "Superconducting Superporc" (*Sacramento Bee*) and "Quashing the Atom-Smasher" (*La Habra Daily Star Progress*). But these activities were "too little and too late." By the time they were initiated the opposition was well organized [145] and general opinions were pretty much set; these efforts had very little noticeable influence.

There was essentially no significant lobbying for the SSC by the upper levels of the Administration. Clinton's support was limited to a letter written to the House leadership in which he expressed his continuing support for the project [146]. The overall White House support was lukewarm at best. Speaking of the SSC and the Space Station, the press secretary Dee Dee Myers stated [147] that they remain in the President's budget "at this point" but added that the two projects are not among essential "principles."

Thus by the summer of 1993 the situation looked very bleak for the SSC but Sen. Bennett Johnston (D-LA), chief strategist on the Senate side and the SSC's staunchest supporter, thought that the situation could still be saved. He wanted to have a strong pro-SSC vote in the Senate with the hope that a stacked conference committee would again be willing to go along with the Senate recommendation [148, 149]. Johnston organized a joint hearing on the SSC of the Senate Committee on Energy and Natural Resources and of the Senate Appropriations Subcommittee on Energy and Water Development, with the idea that distinguished scientists would testify to the benefit of the SSC and sway the Senate to approve the full funding.

The initial steps went as hoped for. A number of prominent scientists, among them Jerry Friedman, Leon Lederman and Steve Weinberg, testified in favor of the SSC. Roy Schwitters, in his initial remarks, tried to rebut the allegations made against the SSC during the previous months [150]. There was some opposition led by Sen. Bumpers, who was supported in his downplaying of possible SSC spinoffs by N. Bloembergen, Nobel Prize–winning physics professor from Harvard. How influential the hearing was is not clear; what was important was that the Senate voted 57 to 42 for full US$640 million funding for the SSC. The next step was also successful. The conference committee went along with the Senate and reported out US$640 million appropriation for the SSC.

But after this, the whole plan unraveled. When the conference committee report was brought back to the full House on October 19 for the vote, Slattery introduced an amendment to send the report back to the Conference with a specific instruction to eliminate SSC funding. There was a heated debate on the floor, with both opponents and proponents repeating the arguments that had been made previously. The escalating costs, the need to balance the budget, the poor management, the lack of foreign support, and the esoteric nature of the science were all put forward in the arguments against the SSC. I do not think that the debate influenced many people to change their vote. The perceived need to show fiscal austerity overwhelmed all other considerations. The SSC was doomed to lose this one even before the debate began. The final vote was 242 to 183 in favor of Slattery's amendment; 81 of 113 freshmen voted with the majority [151].

When the conferees met on October 21 it was clear that the opposition had won. This was acknowledged by the leadership when a rather unusual step was taken of inviting both Slattery and Boehlert (and also SSC proponents from Texas) to the conference committee meeting. The opposition to the SSC in the House was so strong that there was no chance that termination could be avoided. The debate was reduced to a discussion on what fraction of the US$640 million should be used for the closeout and how the SSC employees should be treated. There was also a discussion on how to make the best use of what had been learned and developed at the SSCL. Both Johnston and the Texas delegation argued passionately for fair treatment of all the staff, especially in light of the suddenness of the termination. In the end, the full $640M was allocated for termination costs. The conference report was formally approved by both Houses and signed into law by President Clinton a few days later. The SSC was dead [152].

I mention only briefly some of the subsequent SSC-related events [153]. Roy Schwitters resigned on November 5, 1993, as the Director and took a faculty position at the University of Texas in Austin. He was replaced by John Peoples, Director of Fermilab, who was given the painful job of orderly termination of the SSC. Of the 1943 employees at the end, many were able to return to their previous places of employment, laboratories or universities. A number found new positions in their line of work. But a number of others suffered serious professional and financial setbacks [154]. There was also a pretty universal feeling in the US HEP community that the field had suffered a devastating blow [155]. The closeout activities lasted two years, coming to an end only near the end of 1995 [156]. The total cost of the SSC termination was US$736 million.

The first HEPAP meeting after the SSC termination took place on November 8–9. Secretary O'Leary wrote to me as HEPAP chair: "... we would like recommendations on how to make the best use of the investment that has been made in the SSC," and also "DOE would like HEPAP to turn its attention immediately to the task of defining a long term program to pursue the most important high energy physics goals... we request that a dedicated subpanel be constituted to address these issues." At that HEPAP meeting, the CERN Director of Research, Lorenzo Foa, extended an invitation to the US HEP community to participate in exploitation of the LHC. An effort was made to dispose of SSCL tangible assets but they brought only a fraction of the original investment.

The most significant subsequent event was the convening of the Drell Subpanel in the spring of 1994, which recommended US participation and contribution to the LHC [157]. That led to initiation of lengthy but eventually successful negotiations about US involvement. The US contribution enabled immediate construction of the full LHC rather than the staged approach initially approved by the CERN Council [158]. The US financial contributions to the detectors enabled the currently planned LHC experimental program. Many ideas developed within the framework of the SSC detector R&D program influenced significantly the design of both LHC detectors: large scale silicon strip trackers and silicon pixel detectors deserve special mention. Nevertheless, even with the large scale US involvement, the LHC construction was completed only in the summer of 2008, some 24 years after Schopper's discussion of it at the AAAS meeting [159]. The first collisions are currently (summer 2009) planned for late fall of 2009. Clearly, the long time scales for construction of major science projects are a fact of life, a fact which needs to be kept in mind when planning such future initiatives.

12. Why Did It Fail?

My primary goal for this article was to describe the facts to the best of my knowledge in as objective a way as possible. I end by giving some subjective conclusions. The SSC demise generated a number of postmortem papers with explanations as to what went wrong with the SSC and what caused its demise [160–162]. I will give my views on this question but first I will address an equally important one, one that is seldom asked but should be: What went right with the SSC? We sometimes forget that the SSC almost succeeded; considering the challenge it presented and the long odds it faced at its beginnings in 1983, many things must have obviously been done right. Al Trivelpiece told me some time around 1986: "There are hundreds of different ways that we can fail with the SSC but probably only one way to make it happen. We need to find that way."

Several things allowed the SSC to progress as far as it did. First, in the CDG there was a group of highly dedicated and highly competent people, under inspired centralized leadership, who devoted several years of their professional careers to making the SSC

happen. Their ability to work together toward a common goal only increased with time. The milestones were met in spite of limited budgets. Secondly, in the CDG days there was still a relatively collegial relationship with the DOE; there were innumerable reports, reviews and conflicts, but the performers on the whole were allowed to perform without daily micromanagement. There was no strong feeling of "us vs. them," but rather of two groups with different obligations working toward a common goal. Similarly, in these early times there was no adversarial relationship with Congress; the key people in Congress who took interest in the SSC saw it as an important scientific tool and wanted it to succeed. Undoubtedly many others were driven by more parochial interests, i.e. hopes of landing the SSC in their state; but there were also others, like Vic Fazio and Ron Packard from California (with whom I had many interactions), who remained steady SSC supporters until the end, well after California was eliminated from the site competition. Finally, and this is somewhat controversial, the SSC got as far and as fast as it did because it was initially conceived as a national enterprise and "sold" to the public and the government as such. Key decisions required for initiation of the project could be taken without having to engage in international negotiations. But there was a price to pay for this fast start later on: future negotiations for international participation and material support would be much more difficult.

So why did it eventually fail? There was a spectrum of reasons. Almost every player in the SSC "game" bears a share of the blame and I will address that shortly. I feel that there were three key branch points at which decisions went the wrong way; alternative actions might well have saved the SSC. This is a very personal, hindsight view; a number of my colleagues, whose views I deeply respect, see these events differently.

The first one involved the choice of the site. In retrospect, choosing a green site in Texas was a mistake. There were many hidden costs and additional complexities generated by not using an existing national laboratory site. One of the most important ones was the need to build up scientific staff from scratch. One of the shortcomings of the site selection process was that the ability to recruit first class technical staff was not given sufficient weight. Whether the Tevatron energy was or was not high enough was almost irrelevant; one could have always replaced the old magnets with new, higher field ones and have a 1.5 TeV injector. Funding shortfalls would have been easier to absorb. Even though it might have been harder to get political support for the SSC initially, in the long term the political battles in Congress would have been minimized. The SSC at Fermilab would have been much more difficult to criticize as a pork barrel project. And, very importantly, one would have avoided a very serious and potentially divisive future problem for the field: What does one do with Fermilab when the SSCL is fully operating? Finally, a cancelation of the SSC at Fermilab by one Congress could conceivably have been reversed by another a few years later.

The second key branch point involved the decisions regarding the SSCL management team and the transition from CDG to SSCL. There were a number of missteps, some of which I alluded to in the previous article. First of all, there was the question of partnership with Sverdrup and EG&G. This changed the culture of the management and of the Laboratory from what was traditional in HEP and it was not clear that it contributed much in either technical expertise or political support; it also was a major factor in generating a large "standing army" from the very beginning, a deadly situation when the project had to be stretched out. Secondly, as described in the first article, there was a somewhat behind-the-scenes selection process of the Directorate which created a lot of antagonism in the community. In addition, the process was too hurried and there never was enough thought given to the qualifications that the Director must have to be able to adequately face the challenges associated with the job. Thirdly, there was a lack of appreciation of the collective wealth of technical information residing in the CDG. An effort should have been made to build the new organization with the CDG people as its basis. Such a smooth transition would have made the site-specific planning more efficient. Even if there were difficulties in creating a working arrangement between Schwitters and Tigner that was satisfactory to both of them, the BOO and the URA should have worked much harder to find a satisfactory solution.

The final ill-fated decision point had to do with the design modifications. As mentioned earlier, the possibility that a resulting cost increase could doom the project was apparently never considered

seriously. There were many negative repercussions of this cost escalation. This escalation stretched out the time scale of the project, resulting in additional cost increase due to the "standing army" effect. At termination, SSCL staff numbered 1943 people (not including contractors' people on site), projected to rise to about 2500, requiring an annual expenditure for salaries and benefits probably over US$150 million. The needed annual appropriation was now so close to the maximum feasible that significant stretch-outs (and hence additional cost increases) were unavoidable. Equally important was that it fueled the argument that the "egghead academics" could not manage a big project without cost overruns. Stretch-out of the project influenced the research plans of the HEP community, resulting in increased demand for other facilities and in growing difficulties in hiring people for the SSCL. It also decreased the chances of getting foreign support.

How could the different parties have done better? The URA and the BOO could have done a better job in deciding on industrial partners, in choosing the Directorate and in negotiating details of the M&O contract. The URA should have used the influence of the universities to keep the record straight regarding the SSCL and its science and provide stronger advocacy for the SSC at all levels. The SSCL management should have been more aggressive in keeping a lid on the costs; specifically, it can be faulted for not anticipating better the potential repercussions due to the cost escalation caused by the design changes. It could have done a much better public relations job, especially in the area of refuting the ill-informed anti-SSC attacks. It can also be faulted for staffing too quickly before the funding pattern was known. Greater reliance on subcontracts with the other national laboratories would have obviated the necessity of an immediately large staff and would have increased political support. The HEP community collectively should have been more steadfast in their support for the SSC, more involved in their advocacy and more prepared to sacrifice short term physics opportunities. They were also too careless in discussing the HEP and SSC benefits to the society as a whole, allowing others to blow them out of proportion. SSC opponents then seized on these exaggerated claims to discredit the arguments for the SSC.

The DOE should have maintained the tradition of partnership and mutual respect between overseers and performers, rather than introducing a micromanagement system that short-circuited traditional oversight mechanisms. Congress can be faulted for not making a greater effort to understand the cost history and repercussions of annual funding shortfalls (due to Congressional actions), the status and goals of the SSC, and for not taking a long term view of what is best for the country rather than focusing on short term political expediency. The non-HEP science community as a whole were rather shortsighted in their view as to how the SSC would affect long term science support in the US. They should have been more responsible and professional in their discussion of the pros and cons of the SSC. The press failed in their responsibility to objectively report on controversial issues and do its own investigation of the allegations, rather than succumbing to the temptation of "sensational" journalism.

I turn next to the issue of international collaboration. Lack thereof is cited by almost all the postmortem papers as a principal cause of the SSC failure [163]; based on that, the conclusion is then drawn that all future large scale science projects must be international. These are two distinct arguments and not necessarily related. Here I deal with the first one, which refers to the past history, i.e. the SSC. I will address the second one, the future situation, in the next section. After all, things do evolve and the future situations will not necessarily be identical to the SSC situation.

The SSC was able to get off the ground quickly because it was a national project, seen as being of benefit to the US through stimulating science and technology as well as the interest in these fields among the young, generating high-tech jobs and increasing the technical capabilities of US industry. As a US project, the SSC could be constructed relatively expeditiously because no complex international negotiations were required, either for site selection, for defining management and oversight organization, for creating laboratory directorate structure, or for apportioning the budgetary responsibilities. Resolving all of these would have taken years and the process might never have converged. A lot of the SSC support was due to the fact that it was a US national project. Already there were complaints from several US industrial concerns that foreign companies received a significant fraction of high-tech contracts. A HERA type model could have

worked and it almost did with the Japanese providing the principal contribution from abroad. But the cost offset, especially after allowing for increased management complexities, would have been minimal — 10–15% at most — and thus would not be significant enough to help counter the anti-SSC arguments being made on fiscal austerity grounds. In my opinion, lack of international participation was an excuse, not a reason, for the votes against the SSC.

When we talk about the reasons for the SSC's demise, we must not ignore the bad political luck which befell it. I have already discussed how the most important events in that arena worked against the SSC. The SSC might well have materialized if a few of them had turned out the other way. Where would the SSC be today had Bush I won the 1992 election?

I will enumerate some other important factors. The renewed emphasis on fiscal austerity and budget deficits forced many Members of Congress to give at least an appearance of being concerned about this issue. Even though the proposed FY94 SSC appropriation was about 0.3% of the projected budget deficit for that year, being against it made for a good sound bite. To quote Sen. Bumpers: "It would be nice to know the origin of matter. It would be even nicer to have a balanced budget." The SSC was a natural scapegoat — big enough to be visible but not big enough (like the Space Station) to have attracted strong industrial involvement and hence strong lobbying support. There clearly was cost escalation and it was a big factor in turning opinion against the SSC. But a lot of this escalation was due to shortfalls in the funding profile and the cost increase did not even come close to that of the Space Station, which also had more serious management problems [164–166]. The 1993 House SSC vote came a day after the House voted to continue the Space Station by a 1-vote margin — 216 to 215. Many Representatives did not want to be seen as having voted for two multibillion-dollar "boondoggles."

Finally, but very importantly, we have an inadequacy of our funding system which played a major role in the SSC cost overruns and hence its termination. I make this point by quoting from one of Watkins' last letters [167] as Secretary of Energy (written on January 14, 1993), to Rep. George Brown, Chair of the House Committee on Science, Space, and Technology: "A new way of doing business should be pursued that will support the long term planning necessary to execute a large construction program. Annual Congressional debate and uncertainty over the proper funding level can be inordinately expensive. ... Experience in military has application here. Once a shipbuilding project is mature enough to move out of R&D, it is *fully funded* for construction of the entire ship at one time. The project director is given the flexibility needed to make real time trades to meet objectives. I strongly believe that we should pursue in FY 1994 a similar 'full funding' strategy for the SSC."

13. Lessons for the Future?

What can we learn from the SSC's history? What are the chances of succeeding with another big science project in the future? More specifically, what does the SSC's history tell us about the prospects of the ILC? I will end by giving my views on these issues.

Most importantly, several general conditions must be satisfied before a big new basic research project, in which the US plays a major part, can have a reasonable chance of becoming a reality:

- The US as a whole has to recognize the value of basic research. This appreciation has to be deep and not dependent on how many jobs are created directly by the project or by how many new or improved widgets the project will provide for us tomorrow.
- The deficit has to be brought under control. The SSC's history illustrates that a large deficit is like a suspended guillotine, ready to fall if properly exploited by the enemies of the project.
- A funding strategy has to be found which gives reasonable certainty that a long term project, once started, can be completed close to the original schedule. As Admiral Watkins argued, a big science project cannot be subjected to the vagaries of the annual congressional appropriations.

None of these conditions was satisfied at the time of the SSC, and this failure played a major part in its eventual demise.

I turn now to specifics. Starting a lab on a green site generated more difficulties than originally anticipated. Thus working within the framework of an existing laboratory is highly desirable. Excessive optimism, either about the total cost, schedule, technical readiness or benefits to society can be very damaging. One should have a conservative cost estimate at the beginning, with costing ground rules clearly spelled out and complying with the current procedures. One must expect attacks by the project's opponents, many of them far from factual, and be prepared to respond to them as soon as possible.

It is accepted today that an international collaboration is a must in planning and executing such a project. I agree that any future large HEP project will be so large and expensive that it will need to be planned in the context of a growing internationalization of all big science. But the execution in such a framework will lead to serious complications and introduce a host of new challenges. It is imperative to be aware of them in the planning process. There are several possible models which might work if the three prerequisites cited above are satisfied. None of these options will be easy to execute and all of them will take a significant amount of time to come to fruition:

- A modified HERA model with the US providing the site and playing a leading role is one possibility. The project should be planned internationally, but the US would make a decision to go ahead with only an informal commitment from other countries to contribute. This is basically a modification of the SSC situation with more upfront collaborative efforts and negotiations. The disadvantage is that the US would have to put up most of the money, probably exceeding the current threshold of what is possible; many of the present stringent immigration, visa, employment, etc. rules would have to be modified. But getting the project started would be much simpler and hence it might proceed faster. Internationalization of the operations once the construction is complete would be a desirable option.
- Starting a new international laboratory, either on a green site or within the framework of an existing one, either in the US or elsewhere [168]. All the arrangements for its construction and operation would have to be settled upfront; this probably would take at least a decade. This is the model followed by ITER. I have some skepticism that one could pull this off easily, in the present international climate. Due to the current distrust of the US in many countries which might be potential partners, especially in regard to the US' long term

reliability, I foresee a great reluctance to join in an enterprise sited in the US [169]. Similarly, the US Congress today is unlikely to approve major expenditures for a project on foreign soil.

- The US joining an international laboratory abroad as a full-fledged (or associate if one is willing to play a secondary role) member, with that laboratory committing itself to such broader membership and to the construction of the project. Today the only possible such candidate laboratory is CERN. This could be the mode that the US participation in the LHC would evolve to eventually but it is certainly not there today, the US having essentially no role in the management of the laboratory. This scheme gives up on having a frontier energy facility in this country and would probably result in a decline in the HEP activity in the US. It might be difficult to get Congressional and Administration approvals for such a scheme in the present climate. Approval might require a complementary action, involving construction in the US of a facility in some other field, with the other countries participating in its funding and management. Such a scheme is an attractive and sensible option, but rather difficult to carry out because of the need for a great deal of coordination [170].

The current strategy for building the ILC adopts essentially the second model. This is a model that probably requires the most negotiations, and thus I have always been skeptical that it could be accomplished on a time scale even close to what its proponents are projecting (if at all). I also see very little chance that the three conditions I listed above as prerequisites for success will be satisfied in the near future. Certainly, the budgetary action in 2008 — zeroing out the ITER contribution, stopping the ILC R&D program, and stopping the investments in the future of the US HEP — has indicated that the value given to the basic research in this country is rather low, that the current budgetary deficits put a big damper on any new significant initiatives, and that the US funding system has difficulty supporting any long term project. Thus none of my three conditions are satisfied today and they do not appear likely to be in the near future.

I end with a personal anecdote illustrating some of the difficulties that the US will encounter in the future when trying to build a large scale project in

HEP. At the 1992 ICHEP conference in Dallas, during a tour of the SSCL, I sat on a bus next to Chris Llewellyn-Smith, a British particle theorist, who was to become CERN DG in 1994. CERN was trying hard at that time to get its Council's approval for the LHC construction. I asked Chris what was the justification for CERN building the LHC in light of the SSC. His answer was that there were two strong arguments for it: it was needed for CERN self-preservation and there was a possibility that the US would never complete the SSC. This answer clearly points to two of the issues that have to be confronted and resolved if we are to have any chance of future success.

Acknowledgments

In writing this history I tried to check my own recollections by reference to contemporary sources and to other people's recollections. I would like to express my gratitude to a number of individuals who have helped me in this area.

I am very grateful to Adrienne Kolb, Fermilab archivist, for facilitating my access to Fermilab Archives, the current repository for a lot of the original SSC materials. The SLAC Library personnel, especially Abraham Wheeler, have been extremely helpful in retrieving articles and government documents not readily available on the shelves. I thank Maury Tigner and David Corson for facilitating access to the materials in the Cornell Library Archives. Ezra Heitowit and URA staff helped by providing copies of several relevant URA documents from the SSC era.

I have profited greatly from a number of interviews, in person or by phone, with many of my colleagues involved in the SSC during its Texas days. I would like to acknowledge the very informative conversations with Barry Barish, Alex Chao, Roger Coombes, Helen Edwards, "Gil" Gilchriese, Tom Kirk, Chris Laughton, Vera Luth, Robert Matyas, Chris Quigg, John Rees, Jim Sanford, Jim Siegrist, Roy Schwitters, Jenny Thomas and George Trilling. I thank them all for their generosity with their time and for their willingness to share with me their recollections of those times.

Finally, I would like to thank Gil Gilchriese, J. David Jackson, Adrienne Kolb, Jim Sanford, Jim Siegrist and George Trilling for reading an early draft of this article and giving me their critical comments.

I am very grateful to Michael Riordan for reading a more recent draft and offering many useful comments and suggestions. The responsibility for any factual errors or erroneous conclusions is, however, entirely mine.

This work was partially supported by the National Science Foundation grant PHY-0354945.

List of Abbreviations

AAAS — American Association for the Advancement of Science

ASST — Accelerator Systems String Test

BINP — Budker Institute for Nuclear Physics

BNL — Brookhaven National Laboratory

BOO — Board of Overseers (for the SSC)

BQL — Best-Qualified List

CDG — Central Design Group

CDR — Conceptual Design Report

CERN — European Organization for Nuclear Research (this acronym is for a former name in French: Conseil Européen pour la Recherche Nucléaire)

CPR — Cost Performance Review

DESY — Deutches Electronen-Synchrotron (high energy physics laboratory in Hamburg, Germany)

DG — Director General

DOE — Department of Energy

EOI — expression of interest

FY — fiscal year

GEM — Gammas, Electrons, Muons (proposed SSC detector)

GAO — General Accounting Office

GIC — general industrial contractor

HEB — high energy booster

HEP — high energy physics

HEPAP — High Energy Physics Advisory Panel

HERA — Hadron Electron Ring Accelerator

ICE — independent cost estimator

ICHEP — International Conference on High Energy Physics

IG — Inspector General

ILC — International Linear Collider

ITEP — Institute of Theoretical and Experimental Physics (in Moscow)

ITER — International Thermonuclear Experimental Reactor

KAON — Kaon–Antiproton–Other-hadron–Neutrino Factory

KEK — National Laboratory for High Energy Physics, in Tsukuba, Japan (Japanese acronym)

LBL — Lawrence Berkeley Laboratory

LBNL — Lawrence Berkeley National Laboratory (new name for the former LBL)

LEB — low energy booster

LEBT — low energy beam transport (element in LEB)

LEP — Large Electron–Positron collider

LHC — Large Hadron Collider

LOI — letter of interest

MAC — Machine Advisory Committee

MEB — medium energy booster

MIT — Massachusetts Institute of Technology

M&O — Management and Operations

NSF — National Science Foundation

OER — Office of Energy Research (in the DOE)

OSSC — Office of SSC (at DOE headquarters)

OSTP — Office of Science and Technology Policy

PAC — Program Advisory Committee

PB/MK — Parsons Brinckerhoff/Morrison Knudsen joint venture

R&D — research and development

RFP — request for proposals

RFQ — radio frequency quadrupole

SAIC — Scientific Applications International Corporation

SDC — Solenoidal Detector Collaboration

SLAC — Stanford Linear Accelerator Center

SLC — SLAC Linear Collider

SPC — Scientific Policy Committee

SSC — Superconducting Super Collider

SSCL — Superconducting Super Collider Laboratory

TAC — Texas Accelerator Center

TBM — tunnel-boring machine

TDR — Technical Design Report

TNLRC — Texas National Research
  Laboratory Commission
TPC — total project cost
TRIUMF — Triuniversity Meson Facility
  (Canada's National Laboratory
  for Particle and Nuclear Physics)
URA — Universities Research
  Association

References and Notes

[1] The pre-Waxachachie SSC-related events are described in: S. Wojcicki, "The Supercollider — the pre-Texas days, *Reviews of Accelerator Science and Technology* (World Scientific, 2008), Vol. 1, pp. 259–302.

[2] M. Crawford, *Science* **245**, 25 (1989).

[3] I. Goodwin, *Phys. Today* **42**, 51 (1989).

[4] I. Goodwin, *Phys. Today* **43**, 45 (1990).

[5] McDaniel's handwritten notes from the September 27,1988, interviews at the CDG (from the Cornell archives) and comments made to the author by some of the interviewees.

[6] Letter from R. Schwitters to M. Tigner, dated September 19, 1988.

[7] Memorandum from Edward A. Knapp, URA President, to the staff of the URA/SSC Central Design Group, dated January 25, 1989.

[8] Letter from R. Schwitters to S. Wojcicki dated August 23, 1989, and Wojcicki's response of August 31.

[9] Everyone I talked to who was at the SSC during its early Texas days agreed that the Sverdrup partnership was a disaster. A generous way to summarize their opinions about Sverdrup staff at the SSC would be to say that they were not very competent and rather uninterested.

[10] It was this organization that was presented to HEPAP at HEPAP's first meeting on the SSC site, on January 4 and 5, 1991.

[11] *Science* **251**, 1551 (1991).

[12] For a detailed discussion on the two-culture syndrome at the SSCL, see M. Riordan, *Historical Studies in the Physical and Biological Sciences* **32**, Part 1, 125 (2001).

[13] Report of the SSC Collider Dipole Review Panel, eds. G. Voss and T. Kirk (June 1989, SSC-SR-1040).

[14] S. Dickman and G. C. Anderson, *Nature* **343**, 343 (1990).

[15] T. Garavaglia, S. K. Kaufmann, R. Stiening and D. M. Ritson, SSCL-268 (Apr. 1990).

[16] SSC Laboratory report SSC-SR-1040.

[17] M. Crawford, *Science* **245**, 809 (1989).

[18] B. Schwarzschild, *Phys. Today* 47 (1990).

[19] M. Crawford, *Science* **247**, 153 (1990).

[20] D. H. Hamilton, *Science* **249**, 731 (1990).

[21] Accelerator design status. Presentation by Don Edwards at the HEPAP Meeting, Jan. 4, 1991.

[22] Report of the Ad Hoc Committee on SSC Physics, SSC-250/Rev., Dec. 1989 (revised: Apr. 1990). This report did not make a very strong case for the need to go to 40 TeV. It stated that "total energy anywhere in the range from 30 to 40 TeV would provide an enormous increase in physics capability." The loss in the rate for production of heavy particles was estimated at 25% for each 5 TeV decrease.

[23] Minutes of the BOO's Executive Committee meeting on November 21, 1989.

[24] Report of the 1990 HEPAP Subpanel on SSC Physics (DOE-ER-0434, Jan. 1990).

[25] R. J. Smith, Full-scale super collider urged despite rise in cost, *The Washington Post* Jan. 11 (1990).

[26] I. Goodwin, *Phys. Today* **43**, 67 (1990).

[27] SSCL Site-Specific Conceptual Design (SSCL-SR-1056, July 1990).

[28] Report of the HEPAP Subpanel on SSC cost estimate oversight (DOE/ER-0464P, July 1990).

[29] Report of the DOE Office of Energy Research Review Committee on the Site-Specific Conceptual Design of the Superconducting Super Collider (DOE/ER-0463P, Sept. 1990).

[30] DOE Report on the SSC cost and schedule baseline (DOE/ER-0468P, Jan. 1991).

[31] D. P. Hamilton, *Science* **251**, 741 (1991).

[32] Personal conversation with Panofsky and minutes of the Joint Informational Meeting of the URA Board of Trustees and the SSC Board of Overseers (Arlington, Virginia, USA; Dec. 19, 1989).

[33] W. K. H. Panofsky, *Panofsky on Physics, Politics and Peace* (Springer, 2007), p. 144.

[34] *CERN Courier* (Jan./Feb. 1992), p. 12.

[35] This account of these issues and events is based on minutes of the BOO's meetings in 1989, Panofsky's memos to the BOO file, and Panofsky–McDaniel correspondence (in the Cornell archives).

[36] Minutes of the Meeting of the SSC Board of Overseers (Oct. 4–5, 1989).

[37] Interview with Joe Cipriano, *Super Collider News* (May 1991), pp. 3–7.

[38] High energy physics research and the Superconducting Super Collider: DOE policies and practices. DOE document, Aug. 1992.

[39] I. Goodwin, *Phys. Today* **43**, 45 (1990).

[40] Office of Inspector General, Special Report on the Department of Energy's Superconducting Super Collider Program (DOE-IG-0291, Nov. 16, 1990).

[41] Status of DOE's Superconducting Super Collider, (GAO Report GAO/RCED-91-116, Apr. 1991).

[42] Letter from Secretary Watkins to Jamie L. Whitten, Chairman, House Committee on Appropriations, dated May 8, 1991.

[43] Letter from S. Wojcicki, HEPAP chair, to Hensen Moore, Deputy Secretary of DOE, dated Oct. 4, 1990.

[44] Minutes of the Meeting of the SSC Board of Overseers, BNL (July 28, 1989).

[45] Letter from R. Schwitters to the US HEP community, dated Apr. 10, 1989.

[46] Summary of the Meeting and Recommendations of the Superconducting Super Collider Program Advisory Committee (Snowmass, Colorado, USA; July 1990). The need to have two new general purpose detectors at the start was not held universally. The SPC recommended at least one general purpose detector. Leon Lederman argued for "recycling" one of the Fermilab Tevatron detectors.

[47] R. Crease, *Science* **250**, 1648 (1990).

[48] R. Schwitters, Decision Memorandum — on aspects of the initial scientific program for the SSC (Jan. 4, 1991).

[49] D. P. Hamilton, *Science* **252**, 908 (1991).

[50] The history of the L* Collaboration is detailed in "Report to the L* Collaboration," by Hans Hofer, Deputy Spokesman, and Samuel C. C. Ting, Spokesman, dated May 15, 1991. The perspective is one from the L* point of view but a number of relevant documents, discussed here, are included there.

[51] Report of the Mar. 10–12, 1991, meeting of the SSCL PAC.

[52] R. F. Schwitters, Report to the PAC (May 3, 1991).

[53] Letter from R. F. Schwitters to the US HEP community, dated May 10, 1991.

[54] D. P. Hamilton, *Science* **252**, 1610 (1991).

[55] Solenoidal Detector Collaboration Technical Design Report (SSCL-SR-1215, Apr. 1, 1992).

[56] GEM Technical Design Report (SSCL-SR-1219, Apr. 30, 1993).

[57] J. Mervis, D. Dickson and D. Swinbanks, *Nature* **362**, 385 (1993).

[58] House vote cooled international ardor for GEM, *Super Collider News* (Oct. 1992), p. 3.

[59] F. Gilman, presentation to HEPAP on the SSC Experimental Program on June 4, 1991.

[60] Presentation by M. Gilchriese at the January 4–5, 1991, HEPAP meeting at the SSCL.

[61] *Proc. Symp. Detector Research and Development for the Superconducting Super Collider* (Oct. 15–18, 1990; Fort Worth, Texas, USA) (World Scientific), eds. T. Dombeck, V. Kelly and G. Yost.

[62] Texas Test Rig Project, Presentation by G. Mitselmakher at the Sept. 11, 1992, HEPAP meeting.

[63] High energy physics research and the Superconducting Super Collider: DOE policies and practices (DOE/ER, Aug. 1992), distributed at the Sept. 11–12, 1992, HEPAP meeting.

[64] For a summary of the SSC's technical status in Aug. 1992, see: R. F. Schwitters, *Proc. XXVI Int. Conf. High Energy Physics* (Aug. 6–12, 1992; Dallas, Texas, USA), ed. J. Sanford, pp. 306–320.

[65] *Superconducting Super Collider: A Retrospective Summary 1989–1993* (Apr. 1994), eds. G. F. Dugan and J. R. Sanford.

[66] 'Tail wagging dog' overrun picture drawn by IG, *Super Collider News* (Aug. 1992), p. 3.

[67] URA, PB/MK contractors severely criticized, *Super Collider News* (May 1992), p. 3.

[68] Tunneling records broken, *SSC News* (June 1993), p. 4.

[69] I. Goodwin, *Phys. Today* **45**, 54 (1992).

[70] SSC magnet (RFP SSC-90-A-01701).

[71] Magnets: GD, Westinghouse beat out Grumman, *Super Collider News* (Nov. 1990), p. 1.

[72] SSCL news release (July 19, 1991).

[73] SSCL news release (July 31, 1991).

[74] Russians to supply LEB accelerator magnets, *Super Collider News* (Mar. 1992), p. 1.

[75] R. Meinke *et al.*, Chap. 8. of Ref. 65.

[76] Letter from C. Rubbia to F. Low, dated May 18, 1989.

[77] F. Flam, *Science* **256**, 466 (1992).

[78] Letter from C. Rubbia to F. Low, dated July 31, 1989.

[79] For example: R. F. Schwitters, Foreign travel report: Soviet Union and Germany (Nov. 1990).

[80] C. Rubbia and R. Schwitters, Memorandum of Understanding between CERN as Scientific Institution and the SSC Laboratory, signed April 8, 1991.

[81] Memorandum of Understanding between TRIUMF and the SSC Laboratory, signed by Erich Vogt and Roy Schwitters, August 22, 1991.

[82] D. Swinbank, *Nature* **344**, 8 (1990).

[83] R. Schwitters, Conclusions from visit of H. Sugawara to SSCL. Memo to file, Dec. 12, 1990.

[84] D. Hamilton, *Science* **255**, 279 (1992).

[85] D. A. Bromley, *The President's Scientists: Reminiscences of a White House Adviser* (Yale University Press, 1994), pp. 211–214; J. Mervis, *Science* **265**, 1357 (1994). I am grateful to M. Riordan for bringing Bromley's memoirs to my attention.

[86] US/Japan SSC working group being formed, *Super Collider News* (Feb. 1992), p. 1.

[87] Japan's interest is strong but path is not clear, *Super Collider News* (May 1992), p. 1.

[88] I. Goodwin, *Phys. Today* **44**, 52 (1991).

[89] M. Crawford, *Science* **246**, 557 (1989).

[90] M. Crawford, *Science* **252**, 25 (1991).

[91] G. Taubes, *Science* **259**, 756 (1993).

[92] I. Goodwin, *Phys. Today* **42**, 49 (1989).

[93] I. Goodwin, *Phys. Today* **44**, 53 (1991).

[94] Potentially dire consequences of this action for the SSC were emphasized by Rep. George Brown, chair of the House Committee on Space, Science and Technology: *Phys. Today* **45**, 45 (1992).

[95] Report of the 1992 HEPAP Subpanel on the US Program of High Energy Physics Research (Apr. 1992, DOE/ER - 0542P).

[96] I. Goodwin, *Phys. Today* **45**, 54 (1992).

[97] Prepared statement of Sheila Widnall for the April 7, 1987, hearing of the House Committee on Science, Space and Technology.

[98] Testimony of Daniel Kleppner at the April 7, 1987, hearing of the House Committee on Science, Space and Technology.

[99] C. Norman, *Science* **255**, 672 (1992).

[100] Status of DOE's Superconducting Super Collider (GAO Report, Apr. 1991, GAO/RCED-91-116).

[101] Letter from Secretary Watkins to Jamie Whitten, dated May 8, 1991.

[102] Yes, big science. But which projects? *The New York Times* (May 20, 1988).

[103] F. Dyson, *Phys. Today* **41**, 77 (1988).

[104] A. M. Sessler, *Phys. Today* **41**, 26 (1988).

[105] *The New York Times* editorial (Mar. 21, 1990).

[106] *The New York Times* editorial titled "Europe 3, U.S. not even zero" (June 8, 1983).

[107] It should be emphasized that the leadership of the HEP, in its formal statements and reports, did maintain a united front regarding the proposed new facilities. See for example the joint statement by the Directors John Peoples (Fermilab) and Roy Schwitters (SSCL) dated May 29, 1990.

[108] I. Goodwin, *Phys. Today* **46**, 43 (1993).

[109] A. Reifenberg, Clinton may trim '94 collider funding, *The Dallas Morning News* (Feb. 4, 1993).

[110] The DOE summary of the early cost history is discussed in "Report on the Superconducting Super Collider cost and schedule baseline" (DOE/ER-0468P); for a detailed breakdown of the costs at different stages of the project see: T. Elioff, A Chronicle of costs (SSCL-SR-1242, Apr. 1994).

[111] Nuclear science: information on DOE accelerators should be better disclosed in the budget (GAO/RCED-86-79, Apr. 9, 1986).

[112] BOO claimed that the appropriate figure should have been US$6.243 billion, the difference being due to the DOE's failure to apply proper escalation to the R&D component of the project and to take into account the extra cost in management expenses due to the stretch-out (minutes of the BOO Executive Committee meeting, Nov. 21, 1989).

[113] Memorandum of Understanding between the DOE and the TNRLC, signed on Nov. 9, 1990, by Henson Moore and Morton Meyerson, Chairman of the TNRLC.

[114] The out year guidance given to the DOE for Federal Budget Authority for FY94–FY98 was US$640 million, US$551 million, US$570 million, US$591 million and US$812 million (Robert Diebold's presentation at the Apr. 7, 1993, HEPAP meeting). It

is not clear whether the project could converge at that rate (see quote in Ref. 117).

[115] H. R. O'Leary, United States Department of Energy budget highlights (DOE/CR-0014, Apr. 1993).

[116] T. Elioff, in Ref. 110.

[117] In his testimony at the Dingell Committee hearing on June 30,1993, Victor Rezendez of the GAO stated: "The project could not be completed at a US$550 million federal funding level because overhead costs and reductions in buying power would consume most of the available funds after fiscal year 2000."

[118] Report of the Task Force on SSC Commissioning and Operations (SSC-SR-1005, Apr. 1985).

[119] Reference 94 and SSCL-SR-1216; Garry Gibbs, in his testimony on May 9, 1991, to the Subcommittee on Investigations and Oversight of the House Committee on Science, Space, and Technology, quoted an estimate of the annual operating cost of US$380 million in FY92 dollars.

[120] J. Krumhansl, letter to Secretary John S. Harrington, dated February 19, 1987; quoted in I. Goodwin, *Phys. Today* **40**, 50 (1987).

[121] M. Crawford, *Science* **245**, 25 (1989).

[122] D. H. Hamilton, *Science* **249**, 731 (1990).

[123] I. Goodwin, *Phys. Today* **44**, 52 (1991).

[124] D. Hamilton, *Science* **256**, 1752 (1992).

[125] For example, 40 prominent physicists signed strong pro-SSC letters sent on July 15 to both President Bush and Senator Johnston; subsequently 2032 other scientists added their signatures — I. Goodwin, *Phys. Today* (Aug. 1992) 59.

[126] A. McKenzie, President to rally support for collider with Thursday visit, *Dallas Morning News* (July 28, 1992); during his visit to the Waxahachie site Bush described the SSC in rather glowing terms: "And when you talk basic research, this is the Louvre, the Pyramids, Niagara Falls all rolled into one," quoted in: I. Goodwin, *Phys. Today* (1992), 55.

[127] I. Goodwin, *Phys. Today* **45**, 53 (1992).

[128] C. Anderson, *Science* **260**, 1421 (1993).

[129] S. LaFraniere, Energy dept official urges firing super collider chief, *The Washington Post* (Aug. 2, 1993).

[130] M. Browne, Scientist at work: Roy F. Schwitters; building a behemoth against great odds, *The New York Times* (Mar. 23, 1993, C11).

[131] C. Anderson, *Science* **288**, 288 (1993).

[132] Hearing before the House Committee on Energy and Commerce, Subcommittee on Oversight and Investigations, held on June 30, 1993.

[133] T. Beardsley and R. Ruthen, *Scientific American* (Sept. 1993), pp. 20–24.

[134] C. Frampton, "Audit questions supercollider costs as project faces crucial House vote, *The Wall Street Journal* (23 June, 1993).

[135] C. MacIlwain, *Nature* **364**, 92 (1993).

[136] V. Kiernan, *New Scientist* (July 10, 1993), p. 5.

[137] Letter from Sen. David Pryor, member of the Senate Committee on Government Affairs, to John Layton, Inspector General of the US Department of Energy, dated August 3, 1993.

[138] Report of the DOE Review Committee on the baseline validation of the Superconducting Super Collider (Sept. 1, 1993).

[139] DOE press release: Secretary O'Leary announces plan for containing costs of Superconducting Super Collider (Sep. 1, 1993).

[140] For example Roy F. Schwitters [150]; memo from W. K. H. Panofsky to SSC staff dated Juky 27, 1993 and R. Schwitters' All hands talk on July 21, 1993; "Some comments on the SSC costs and the GAO report," SSC publication (Mar. 11, 1993); "Comments on the draft Inspector General report on SSC Laboratory subcontractor expenditures (SSC press release, June 23, 1993).

[141] Reference 139.

[142] Letter from James Decker, Acting Director of Office of Energy Research, to S. Wojcicki, HEPAP chair, dated Sept. 20, 1993.

[143] DOE press release: DOE to strengthen super collider management (Aug. 4, 1993).

[144] I. Goodwin, *Phys. Today* **46**, 52 (Sept. 1993).

[145] One such anti-SSC coalition was Organizations Opposing the Super Collider (OOPS), aiming to stop SSC construction. Some of the participating organizations were the Council for Citizens Against Government Waste, Friends of the Earth, National Taxpayers Union, and Citizens for a Sound Economy.

[146] Letter from President Clinton to William Natcher, Chairman of the House Committee on Appropriations (June 16, 1993).

[147] E. Littlejohn, Foes try to kill supercollider funds, *Portland Oregonian* (June 9, 1993).

[148] J. Mervis and K. Fox, *Science* **261**, 288 (1993).

[149] I. Goodwin, *Phys. Today* **46**, 43 (1993).

[150] R. Schwitters' testimony before the Senate Committee on Energy and Natural Resources and the Subcommittee on Energy and Water Development of the Senate Committee on Apropriations, (Aug. 4, 1993).

[151] C. Krauss, Knocked out by the freshmen, *The New York Times* (Oct. 26, 1993).

[152] I. Goodwin, *Phys. Today* **46**, 77 (1993).

[153] I. Goodwin, *Phys. Today* **47**, 87 (1994).

[154] M. Browne, The Supercollider's demise disrupts many lives and rattles a profession, *The New York Times* (Nov. 14, 1993).

[155] F. Flam, *Science* **262**, 644 (1993).

[156] Report on the closure of the Superconducting Super Collider Laboratory (URA document, Sept. 30, 1995).

[157] Report of the High Energy Physics Advisory Panel's Subpanel on Vision for the Future of High Energy Physics (May 1994, DOE-ER-0614P).

[158] The CERN Council approved the LHC in a "1/3 missing magnets" configuration on Dec. 16, 1994 (*CERN Courier*, Jan./Feb. 1995, p. 1). The US participation was approved by Congress in 1995.

[159] D. Dickson, *Science* **224**, 1216 (1984).

[160] D. Ritson, *Nature* **366**, 607 (1993).

[161] W. K. H. Panofsky, *Phys. Today* **47**, 13 (1994).

[162] M. Riordan, *Phys. Perspective* **2**, 411 (2000).

[163] See for example *Nature* **365**, 771 (1993).

[164] M. Waldrop, *Science* **235**, 965 (1987).

[165] E. Marshall, *Science* **246**, 1110 (1989).

[166] M. Waldrop, *Science* **250**, 364 (1990).

[167] J. D. Watkins, letter to George E. Brown dated January 14, 1993.

[168] Sid Drell argued for full internationalization of the SSC, both for construction funding and for management, just before the SSC was killed. This would be in the spirit of my model (2): S. Drell, *Phys. Today* **46**, 73 (1993).

[169] This difficulty was pointed out by the President's Council of Advisors on Science and Technology Panel on Megaprojects, chaired by Harold Shapiro and John McTague (Dec. 1992). One of its findings was: "The United States has a history of failing to meet certain commitments to international partners for particular megaprojects that have long time horizons." A broad range of negative repercussions from a potential SSC cancelation was eloquently described by Will Happer at the time of his departure from OER — quoted in: I. Goodwin, *Phys. Today* **46**, 90 (1993).

[170] Such a "basket" approach, advocated by Burt Richter, was discussed at the 5th EPS International Conference on Large Facilities in Physics (Lausanne, Switzerland, Sept. 12–14, 1994), eds. M. Jacob and H. Schopper (World Scientific), pp. 457–461.

Stanley Wojcicki is a professor of physics at Stanford University. During the early SSC days he took a 4-year leave of absence to work at Berkeley in the Central Design Group. His current professional interests focus on study of neutrino oscillations. He served as chairman of the DOE High Energy Physics Advisory Panel in 1990–1996. His hobbies include hiking and travel to exotic places.

Reviews of Accelerator Science and Technology
Vol. 2 (2009) 303–312
© World Scientific Publishing Company

A Man for All Seasons: Robert R. Wilson*

Edwin L. Goldwasser

Department of Physics, University of Illinois,
1110 West Green Street, Urbana, IL 61801, USA
egoldwas@uiuc.edu

Robert R. Wilson was the brilliant designer, builder and founding director of the Fermi National Accelerator Laboratory with its series of high-energy physics particle accelerators providing collision energies of 200, 400 and 2,000 GeV, the most powerful facilities in their class over a period of 40 years. He undertook the "impossible" and succeeded. With untrammeled courage he challenged the establishment as he bypassed many conventional practices in accelerator design, construction and cost control. With his remarkable talents he addressed a wide range of important aspects of the relationships of art and science, elegance and efficiency and physics and society. In doing so he always found ways for his pursuit of science to support his strong advocacy for human rights, international collaboration and democracy.

Keywords: Particle physics; accelerator; humanist; frontier; aesthetics; atomic energy; cold war; human rights; prairie; international; Tevatron; URA; Robert Wilson; Fermilab.

1. Introduction

When I was asked for a title for this article, for reasons not completely understood *A Man for All Seasons*, the name of Robert Bolt's 1960 play, immediately came to mind. The play is about Sir Thomas More, a close friend and adviser of Henry VIII, whom Henry ordered to be executed when Sir Thomas opposed Henry's placing himself above the Pope in church affairs. Sir Thomas More is described as a statesman, lawyer, humanist, saint, poet and author. Bob Wilson had many of those talents, but he was certainly neither lawyer nor saint. Instead, he was a physicist, statesman, humanist, sculptor, architect, poet and author; so the title still seems fitting.

First and foremost, Bob was a physicist in the subfield of elementary particles, but he became one of the most important figures in the history of particle accelerator research, development and construction (Fig. 1). I won't list all of his publications or his numerous memberships, awards and other honors. Suffice it to say that he was awarded the Fermi Prize and the President's Medal for Science. He also served as President of the American Physical Society.

Bob was born in Frontier, Wyoming on March 4, 1914. He always treasured his Wyoming roots and often described himself as a cowboy (Fig. 2). He was,

Fig. 1.   Robert R. Wilson at Cornell, 1967.

in fact, a master horseman and could deftly lasso any of his three sons when they were young. As a reminder of his roots, both to himself and to others, he brought a herd of buffalo to the Fermilab site and gave me the task of justifying the cost of that action

---

*Based on a memorial talk at the American Physical Society Meeting in Long Beach, California on May 1, 2000.

Fig. 2.   Young Robert R. Wilson, Wyoming cowboy.

Fig. 4.   Lawrence's group at Berkeley. Robert R. Wilson is in the top row, third from right.

to the AEC. I did so in terms of the expense of grass cutting that would be saved. In fact, though, those buffalo (bison) were and are one of many touches that provide a unique character to Fermilab (Fig. 3).

In spite of Bob's view of himself as a cowboy, he entered the University of California as an undergraduate in 1931 and then studied physics as a graduate student under Ernest Lawrence. He received his PhD in 1940 (Fig. 4). He was immediately recruited as an instructor at Princeton and two years later was promoted to Assistant Professor. In 1943 Bob, together with his Princeton group, joined the Manhattan Project in Los Alamos, where his leadership abilities were soon recognized. He was made leader of a large research group and a year later was appointed Head of the Physics Research Division (Fig. 5). As World War II drew to a close he became very active in the successful effort to establish civilian control of atomic energy and was a founding member of the Federation of American Scientists, of which he soon was elected President.

Fig. 3.   The art, architecture and buffalo of Fermilab.

Fig. 5.   Enrico Fermi and Robert R. Wilson, in 1943 Los Alamos: front row, third and fourth from left.

Following the war Bob joined the faculty at Harvard, where he designed a 150 MeV synchrocyclotron. It was designed so well that for many decades it continued to serve as a dedicated cancer therapy facility. In that connection he was the first to publish a paper recognizing the possible use of a proton beam for cancer therapy [1]. Bob's stay at Harvard was brief. After only one year, Dale Corson, who later became President of Cornell University, recruited him to join that faculty, an accomplishment which Dr. Corson described as one of the most important of his administrative career.

From 1947 to 1967 Bob and his Cornell colleagues built four successively more powerful electron synchrotrons, all the while maintaining the tradition and the spirit within which physics facilities were designed and built by the scientists who wanted to use them. His success in maintaining that spirit was of key importance in establishing Cornell as one of the leading centers for elementary particle physics research — the last of the university-based accelerator laboratories and for many years competitive with the large national laboratories at which builders and users have tended to become separate communities. Bob never permitted that to happen at Fermilab. In fact, he rejected one proposal for a potential experiment when the spokesman said they would not have the time to contribute to the development and construction of the required facilities in the manner that Bob had established as laboratory policy. That experiment went elsewhere and won a Nobel Prize.

The foregoing imposing list of accomplishments would constitute a remarkable and complete career for even the most enterprising and energetic physicist, but Bob was about to embark on the venture that became the crowning accomplishment of his professional career — the creation, in the midst of the Illinois cornfields, of the miracle that has become Fermilab, one of the world's leading laboratories for physics research [2].

My first exposure to Bob's genius was during the period when the then-named National Accelerator Laboratory (NAL) was just taking its first steps. Ideas for magnet design were being born, ideas for a sequential series of accelerators were being discussed, and ideas about the size and structure of a tunnel were being debated. I began to wonder what was Bob's key talent that was making all those things happen? It was not his unerring intuition about magnet design, not his accelerator and engineering expertise, and not his business acumen or his architectural taste. Rather, it was his capacity for leadership that was the key. He repeatedly crafted agreements among highly talented and strong-minded scientists and engineers, initially holding widely different views about very complicated problems. I truly believe that Fermilab would never have come into being had it not been for the courage with which Bob repeatedly took on the impossible, convinced others that it was possible, and then made it happen.

One particularly critical example was his role in the congressional hearings on the possible authorization for construction of the world's most powerful and most expensive particle accelerator. A group at the Radiation Laboratory at Berkeley had completed a design for a 200 GeV accelerator and the accompanying experimental facilities. The cost was estimated to be $350 *million*. At that time the annual national deficit was large and the House Appropriation Committee responded with an unequivocal: "No. The cost cannot exceed $250 *million*." So the Berkeley group prepared a "reduced scope" proposal for a machine with somewhat less energy and with a substantially reduced set of experimental facilities. That proposal triggered a different problem.

Following the advent of nuclear energy the US Congress had created a Joint Committee on Atomic Energy (JCAE) with a formidable set of members representing both the Senate and the House of Representatives. The members of that committee were both remarkably knowledgeable about science and extremely powerful in Congress. When informed about the reduced scope proposal they responded with an unequivocal: "No, the reduced scope laboratory would not be capable of doing research interesting enough to make it worthwhile." So there was the ultimate catch-22 — no more than $250 million but no less than 200 GeV with a full set of experimental facilities. There was only one possible way out, but no one was foolhardy enough to embrace that solution — no one except Bob, who insisted that he could build a 200 GeV accelerator along with a full set of experimental facilities for $250 million. In fact, as long as he was going that far out on a limb, he added that he would also build in a capability for the energy later to be raised to 400 GeV. The making of

such a claim was an unbelievable act of courage —
some might say of irresponsible foolhardiness (and
many did) — but in Congress it worked. (And, in
fact, in typical Wilson fashion, the energy was briefly
pushed to 500 GeV.)

During the final authorization hearings by the
JCAE in 1969, Senator Pastore asked an AEC witness to describe what the proposed laboratory might
contribute to the defense of the country. When the
witness on the stand was dumbfounded and speech-
less, Bob Wilson, sitting next to Norman Ramsey
(then President of the Universities Research Association, Inc., the contractor for the project), asked
Norman if he might volunteer to answer. Norman,
although he was unaware of any precedent for audi-
ence participation in a congressional hearing, told
Bob to go ahead and try to be recognized. Bob
was recognized and his now-famous answer was, in
part, "It has nothing to do directly with defending
our country except to help make it worth defending [3]." It was a statesmanlike performance and one
with a little poetry to boot. In that short sentence,
and in many other eloquent appeals, Bob beauti-
fully expressed the spirit that underlies all of basic
research. Our project was authorized.

Bob saw to it that there was never a dull moment
during the development of Fermilab. The first task
was to produce a design report detailing the design
and the cost of the accelerator and associated facilities. That was done by October of 1967, but the
report needed a cover. Bob consulted with Angela
Gonzales (who was his adviser on all matters of
aesthetics) and came up with the idea of using a
familiar sketch by Leonardo da Vinci. The original
shows a nude man inscribed within a circular frame.
For the cover of our design report the circle was to
be modified to portray the accelerator's Main Ring
(Fig. 6). Today the adoption of that cover would
not cause even the raising of an eyebrow — and
the furor it caused in 1967 is now unimaginable! A
typical reaction was: "Of course I, myself, have no
problem, but you don't know Senator X or Representative Y." Protests of that kind came both from
Congress and from the AEC at all levels, including
the Commissioners themselves. We were told that it
might well scuttle the entire project if we used "that
obscene cover." But Bob remained adamant. That
was going to be the cover of our report no matter
who might object. Finally, some wise man suggested

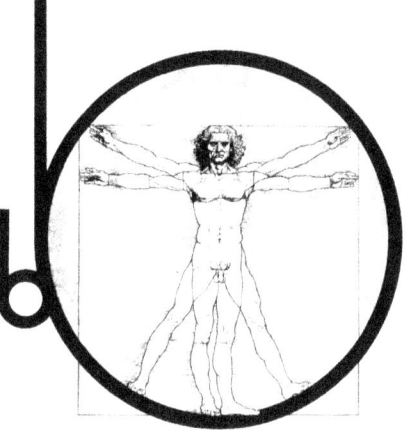

design report

national
accelerator
laboratory

Fig. 6.   Cover of the NAL Design Report, 1968.

that a first set of copies could be sent to Congress
without any cover whatsoever. They would be told
that the cover was still being prepared. Then all subsequent copies would have the Leonardo cover. Bob
uncharacteristically agreed, and that's the way it was
done. Bob remained intransigent and, fortunately,
he won.

Bob knew that he'd have to take risks in order to
complete the construction at two-thirds of the cost
that had been originally estimated, but the taking of
risks and the courage required to be willing to make
mistakes lay at the heart of Bob's approach to completing a project at minimum cost. He realized that
he couldn't include comfortable contingency cushions
in every component. That was not his style. Rather,
he believed that every component would have to be
designed to the very edge of what might not work. It
should then be expected that some elements would
perform successfully, and at great savings. Others
would undoubtedly fail and would have to be fixed at
some cost. He was correct that, on balance, savings
would be realized.

One example was his decision not to drill deep
holes and sink pylons into bedrock to insure that the
tunnel and the accelerator would remain accurately
in place. To save money he decided to take a chance
and "float" the tunnel on the firmly compressed
glacial till that lies under the topsoil in northern

Illinois. The international community of accelerator builders rose up in indignation, saying that if the tunnel began to settle there would be nothing to do but tear everything apart in order to install the traditional pylons, and the cost of such an operation would be prohibitive. Bob said: "Nonsense." If the tunnel were not stable he would simply install motorized jacks on every magnet and actuate the motors with signals from detectors that would be "servoed" to the proton beam. That solution would have been difficult to implement, at best, but its concept was good enough to quiet the uprising. In fact, glacial till proved to be a completely stable base, and the Rube Goldberg system of jacks was never needed.

Then there was a great controversy about the *size* of Bob's tunnel. He had designed much smaller magnets than had been foreseen in the original proposal. (That was made possible by his decision to go to a "separated function," rather than a "combined function" magnet system. Independent sets of bending and focusing magnets could be much more compact.) Bob felt that his magnets could be comfortably housed in a tunnel that was only 10 feet wide, leaving ample room for access, repair, removal and replacement. The experts (who had originally designed large combined function magnets) stated that 14 to 16 feet was the minimum acceptable width. So Bob asked our model builder to build a full scale length of tunnel, 12 feet wide, in our Oak Brook office quarters, and also build some model full-scale separated function magnets along with a model magnet transporter and to place them inside the tunnel (Fig. 7). He then invited the experts to come for a showing. They walked through the model, turned to Bob and honestly confessed that it was larger than necessary. The size was reduced to Bob's original plan and turned out to be completely adequate.

Tradition dictated that laboratories should manage major construction projects by establishing their own construction engineering divisions. Then all designs, drawings and blueprints done by the prime contractor would be reviewed, modified as desired, and then reissued as laboratory documents that would go to the subcontractors.

Bob, much to the AEC's consternation, felt that was a waste of money and refused to create a large, in-house, construction engineering group. Instead, he said he would look very carefully at the people the prime contractor appointed to key positions

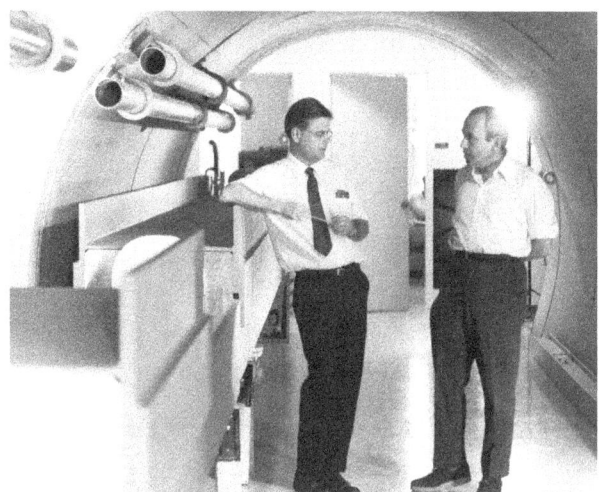

Fig. 7. Robert R. Wilson and the author in the mockup of the Main Ring, Oak Brook, Illinois, 1968.

and would demand the right to veto any proposed appointment or to request the termination of an unsatisfactory person. Once that was agreed, the prime contractor's architect engineering staff would become the laboratory's.

In fact, in practice, Bob did demand that the first project manager be replaced, and, to pacify the AEC, the replacement, a wonderful man named Parke Rohrer, eventually was given a nominal appointment in the laboratory at no pay. Enormous savings were realized as a result of that approach, and there was never any reason to regret the trust placed in Parke Rohrer and his staff (Fig. 8).

As suggested above, some risky design decisions were expected *not* to work. In fact, if none failed, the accelerator would have been overdesigned in Bob's view. When hundreds of magnets did fail in 1971, Bob's directorship was questioned and the laboratory was in serious jeopardy. Yet, with unflagging support from Norman Ramsey, President of URA, and from K. C. Brooks, the on-site manager for the AEC, we survived that crisis, and Bob demonstrated that his absolutely essential philosophy of living "on the brink" could work (Figs. 8 and 9).

Speed is of the essence in reducing the cost of a construction project. When construction of the Main Ring tunnel was lagging and only a short section of concrete slabs had been poured in a muddy trench and covered with still fewer precast concrete arches, Bob complained to Parke Rohrer that magnets were coming off the production line and we had no place

Fig. 8.   Parke Rohrer (left) and Norman Ramsey at the NAL Groundbreaking Ceremony, on December 1, 1968.

Fig. 9.   Robert R. Wilson and Kennedy C. Brooks at the NAL Village, in September 1968.

to put them. They were supposed to go directly down access ramps into the completed tunnel. To drive home his complaint about the subcontractor, Bob had Parke arrange to have a crane brought over to

Fig. 10.   Installing a magnet in the Main Ring tunnel, 1971.

the construction site, and the first, so-called, "magnet," one that could never possibly have worked, was lifted off a truck and rigged into place in the open tunnel (Fig. 10). That exercise, in terms of real progress, was sheer nonsense, but a clear message was delivered to the "sub" and to everyone else as well.

In business matters, Bob was equally creative. When some element had to be mass-produced he decided that one-third of the total number of units should be contracted to each of *two* different companies under an arrangement where whichever contractor performed better in quality, speed and price would get the award of the third third (and that's where the big profit is to be made).

When the project was complete we (reluctantly on my part) returned to the US Treasury almost $7 million of the $250 million that had been appropriated for construction. In that case, though, as well as in most others, I eventually decided that I had been wrong. No project, ever before, had returned appropriated funds to the government following completion. Bob's highly publicized flourish earned us much more than $7 million in the years to come.

The project also gave Bob an opportunity to exercise his architectural and other aesthetic talents. He felt that progress in science was a triumph of the human spirit and that the visual impact of a scientific laboratory should, itself, convey that philosophy. The high-rise building with its atrium reaching for the sky is testimony to Bob's aesthetic taste.

Bob saw to it that the importance of honoring that taste was not lost on the contractors. When the first 50-ton concrete unit of the basement of the high-rise was to be poured, Bob gave detailed instructions as to how the concrete surface was to look, reflecting the details of a randomly assembled wooden form (Fig. 11). After that first huge hunk of concrete was poured and had set, Parke invited Bob and me to come down to inspect the result. Bob took one look and expressed his dissatisfaction in no uncertain terms, ordering that the whole thing be redone. That made an unforgettable impression on all contractors. They had to toe the mark on aesthetic details as well as on the more standard items.

But Bob was the architect of far more than the mere physical structures that are only one facet of Fermilab's unique character. He was always an eloquent spokesman for a broad view of the mission of science and of its connections with the world of art and beauty and its relationship with the entire human endeavor. He was the architect of the reconstituted prairie that fills the center of the Main

Fig. 12. Robert and Jane Wilson restoring the NAL prairie with Prof. Robert Betz of Northeastern Illinois University, 1976.

Ring. He felt that if we were to take 6800 acres of prime farmland out of its productive use and into the future, we should also work to see that some portion of the land be returned to its past (Fig. 12). Most importantly, Bob was the architect of the human spirit that characterized the lab from its very start.

At the Fermilab memorial service for Bob, Bob Mau, one of the first accelerator operators, movingly told how his 30 years at Bob's Fermilab had affected his life. He mentioned how much it meant to him that nobody at the lab, including the director, had a reserved parking space. He recalled how important it was to him that Bob, as director, took notice of *his* work as well as that of all employees and made them aware of his appreciation of what they were doing. He remembered how he appreciated it when Bob issued an edict that, in order to save money, there would be no partitions within the laboratory building. The edict applied to the directorate as well as to everyone else.

Less than a year after an embryonic staff met in our new offices at Oak Brook, in the summer of 1968, a time of social upheaval and rioting in our cities and before affirmative action or even equal opportunity had become national policy, Bob and I issued our "Policy Statement on Human Rights." It started thus: "It will be the policy of the National Accelerator Laboratory to seek the achievement of its scientific goals within a framework of equal employment opportunity and of a deep dedication to the

Fig. 11. Desired impression in Wilson Hall concrete.

fundamental tenets of human rights and dignity." And it finished with these words: "Our support of the rights of minority groups in our laboratory is inextricably intertwined with our goal of creating a new center of technical and scientific excellence. The latter cannot be achieved unless we are successful in the former." Ten years later, when Bob retired, the Fermilab staff of more than 1500 employees included close to 20% African-Americans, many of whom had been recruited from gangs in Chicago, had been trained, and had become excellent technicians.

Before Bob retired, 20 flags of different countries flew in front of Fermilab (Fig. 13). Each one represented a country whose scientists had been active in this laboratory. Bob believed that in this day and age problems of physics as well as many other problems of the world were intrinsically so hard that they should not be made harder by working on them in isolation. He felt that both the work and the world would benefit if all people would join together to learn more about the natural sciences. So, from the very start he saw Fermilab as an international facility which should be available to qualified scientists regardless of the place from which they came.

Initially there was a good deal of opposition to this concept in Washington, where many congressmen felt that since the laboratory was built

Fig. 13.   Flags in front of Wilson Hall, 1978.

with American money, its use should be reserved for American scientists. However, on the international scene, Bob successfully pressed for a uniform policy under which all laboratories, worldwide, agreed to evaluate proposals only on the basis of their scientific merit and technical feasibility [4]. With that international agreement firmly in place, Bob was able to get approval for the internationalization of Fermilab's program. In fact, our government and that of the USSR, recognizing that elementary particle physics presented no promise of immediate military application, established a Joint Committee on Cooperation with a mission to organize free and open collaboration and exchanges in high energy physics. Under that pact, dozens of scientists from the USSR enjoyed the hospitality of Fermilab and of Bob and Jane Wilson at the height of the Cold War. Furthermore, Bob led the effort to bring Chinese scientists to this country to give them experience with modern experimental techniques so that they could begin to focus their untapped talent on the current problems of elementary particle physics.

One day toward the end of our time together at Fermilab, Bob and I were returning from a Washington visit at which he had, for the first time, presented his thoughts about an "energy doubler," a ring of superconducting magnets in the existing Main Ring tunnel that would support a beam energy of 1 TeV. He lamented to me that all his life he had been dreaming up crazy unrealistic schemes, and, luckily, no one had taken him seriously. Back then he noted that, with Fermilab well along the road to success, no sooner did he open his mouth with a *really* insane idea than everyone accepted it as potentially real. In any case, he proposed the energy doubler and he would have to make it come true (Fig. 14). He did, and the Tevatron collider, a direct descendant of the doubler, has embarked on an exploration of territory which, for the past 20 years, has been uniquely accessible at Fermilab.

This article would not be complete, for me, if I did not conclude with some personal remarks. It was in March of 1967 that Bob phoned and asked me to consider joining him, in some undefined capacity, to help build, in the middle of the Illinois cornfields, what has since become Fermilab. In that same telephone call I accepted whatever the position would turn out to be. I went to Ithaca the next week to talk with Bob and to meet his wife,

Fig. 14. Fermilab's piggyback rings of conventional magnets (above) and superconducting magnets (below) that make up the Tevatron, 1979.

Jane. My meeting with Bob was remarkable, in that with all the problems of the start of a huge new laboratory to be discussed, we spent an appreciable fraction of our time discussing the opportunity we would have to take affirmative action to bring motivated young African-Americans into solid employment at the growing laboratory. That Ithaca visit marked the start of the most exciting, satisfying and rewarding experience of my professional life and also the beginning of my deep and unforgettable friendship with Bob, shared by our wives, Lizie and Jane. (And here I must digress for one minute to note that talented though Bob was, he never could have done what he did without the single-minded support and hard-headed encouragement and advice with which Jane influenced him always to do the right thing, to do it better, to do it humanely and to keep his feet firmly on the ground and his head no bigger than normal.)

March became a landmark month in the relationship between Bob and me. Not only did I sign up as his deputy director in March, but, also, both our birthdays are in March, just 5 days apart. During the next 11 years at Fermilab as well as the succeeding 20 years, we celebrated most of our birthdays together — usually at what I thought of as the Wilsons' "island paradise" on Valhalla Key in Florida. Jane tended to think of it as a slum, but she and Lizie enjoyed touring nearby nurseries, buying plants, all the while grumbling about the general

Florida scene. Bob and I, in addition to fixing whatever was falling apart, inside and outside the house, would go out in his boat to the reef, sometimes snorkeling, sometimes fishing, but almost always getting seasick as soon as we left land more than a hundred yards behind.

The last time we were together in Florida was in March of 1997, just two weeks before Bob's stroke. We didn't boat or swim, but we did take long walks on the abandoned portion of the 7-mile bridge. We bemoaned the demise of the SSC, reminisced about the old Fermilab days and marveled that the lab was about to enter what could be another golden age.

Later that month it all came to an end, along with the Bob with whom I had worked as one — through many hair-raising crises but through many more and greater triumphs. Symbolically, the Wilsons' Florida home, where Bob and Jane enjoyed many Christmas holidays with their children and more recently their grandchildren, was literally blown away by a hurricane shortly thereafter. Since then I've missed those joint birthday celebrations and I'll continue to miss Bob in many other ways in the years ahead. So will the world of physics.

Acknowledgments

My thanks go to Adrienne Kolb, Fermilab's archivist since 1983. Adrienne prepared this article for publication. She assembled the accompanying photographs from the Fermilab History and Archives Project Collections. Without her help this article would never have appeared.

Poem by Jane Wilson

When it's late in the day
What can I say?
Except that I'm glad that you came my way?
And we trod together a path rather straight.
It's a comfortable thing to have a mate
Whom one trusts through the thrusts of
What's called "man's fate."
No fantastic fling on this earth's trip.
We sought (modern parlance) relationship.
And — oh my dear!
When the hour comes to part.
What an ache! What a break of
This steadfast heart.

References

[1] R. R. Wilson, *Radiology* **47**, 487 (1946).

[2] Hoddeson, Lillian, Adrienne W. Kolb and C. West-fall, *Fermilab: Physics, the Frontier and Megascience* (University of Chicago Press, 2008).

[3] *Atomic Energy Commission Authorizing Legislation,* Fiscal Year 1970, Part I, Hearings before the Joint Committee on Atomic Energy, Congress of the United States, Ninety-First Congress, Apr. 17–18, 1969, pp. 112–118, quote on p. 113.

[4] T. H. Groves, Physics Advisory Committee summer meeting, *Fermilab Report,* July 1979, 8; L. M. Lederman, International Committee on Future Accelerators, *Fermilab Report,* July 1980, 6.

Edwin L. Goldwasser graduated from Harvard in 1940. He served with the US Navy and then received his PhD from the University of California-Berkeley doing research on cosmic rays. He joined the University of Illinois in 1951, worked on experiments with the newly invented betatron, and collaborated with an MIT group to produce PSSC Physics. Goldwasser was one of the founders of Fermilab. In 1967 he became the Deputy Director of the new laboratory and was in that position during the period of construction and initial operation. He oversaw the development of the research program and contributed in a major way to the equal opportunity programs. In 1978 he returned to the University of Illinois as Vice-Chancellor.